DYNAMICS OF PLATES

Dynamics of Plates

J.S. Rao

Marcel Dekker, Inc.
New York ● Basel ● Hong Kong

Narosa Publishing House
New Delhi ● Madras ● Bombay ● Calcutta

J.S. Rao
Professor, Department of Mechanical Engineering
Indian Institute of Technology, Delhi
New Delhi 110 016, INDIA

Copyright © 1999 Narosa Publishing House

All rights reserved. No part of this publication may be reproduced, stored in a retrieval system, or transmitted in any form or by any means, electronic, mechanical, photocopying, recording or otherwise, without the prior written permission of the publishers.

All export rights for this book vest exclusively with Narosa Publishing House. Unauthorised export is a violation of Copyright Law and is subject to legal action.

Exclusive distribution in Europe and North America by Marcel Dekker, Inc., 270 Madison Avenue, New York, New York 10016

ISBN 0-8247-1977-8 Marcel Dekker, Inc., USA
ISBN 81-7319-250-2 Narosa Publishing House, India

Printed in India.

To
My Beloved Wife Indira

and in the memory of

Professor Henry L. Langhaar
who introduced Variational Calculus and Energy Methods to me

and

Professor William D. Carnegie
with whom I applied the Energy Methods for Rotating Beam Problems

Preface

Plates and shells form a fascinating subject in Engineering Mechanics. It is very customary to simplify many a structures to a one dimensional problem. These one dimensional structures are bars, rods and beams depending on the way the deformation is assumed to take place, axial, torsional and lateral/flexural/bending, respectively. Such an assumption leads to simple and elegant solutions for many practical problems, e.g., shafts, blades amongst others. Considerable work is reported in these areas through several text books, including those of the present author.

Having devoted a considerable part of my life time work to Turbine Blades, I discovered pretty soon that the one dimensional theories of structures are highly inadequate in predicting the dynamic behavior of compressor blades, wide chord turbine blades, particularly in gas turbines. It is essential to consider these structures as two or three dimensional members. The displacements, strains and stresses are functions of two dimensions in the case of thin plates and are functions of three dimensional coordinates in the case of shells. Strangely, there are very few books on this subject. The classical text on *The Theory of Plates and Shells* by S.P. Timoshenko is still the best book available on statics of shells, even though it is written nearly more than four decades ago. To anyone working on Dynamics of Plates and Shells, the monograph *Vibration of Plates* written by Professor A.W. Leissa and published by NASA comes handy. However, this monograph has very little fundamental background material on plate theories, though it gives an extensive coverage of the results available in the literature till the year 1969. Quite a large number of texts on Finite Element Methods have appeared in the last two decades, however, they also emphasize little on the theory of plates itself. They often contain several topics, e.g., structures, fluids, heat transfer etc. with an emphasis on finite element method rather than the basic fundamentals. As a result whenever I gave a course on Vibration of Plates and Shells, it has been difficult to find a suitable text to recommend to the students. While doing this exercise at the National Chung Cheng University, I decided to put my lecture material in the form of a text, which resulted in this book.

Despite the availability of high speed personal computers and good scientific word processors, writing a text book remains a tough task. One requires uninterrupted time, particularly from the administration work and meetings one often ends up in the universities. To my fortune, I had a wonderful academic time during the sabbatical I have taken during 1994-96 at Chia Yi and Sydney. I am very thankful to the National Science Council of Taiwan to have provided a generous research chair at the National Chung Cheng University, particularly to Professor Bernard Shiau and to the University of New South Wales, and to Professor Eric J. Hahn, for a visiting professorship. I am thankful to several of my students, particularly to Professor Bishan Kishore, Professor K. Gupta, Dr. Dhruv Gupta, Dr. Y.D. Yu and Dr. K.K. Rao for their contributions in this work.

My two children have settled well in their life leaving us free from any worries, not only that, our two grand children have been a source best possible relaxation in vacation times. This atmosphere provided an excellent environment to the thinking process which is so essential in writing a text book. It goes without saying, that one needs a significant support from his wife without which, it is practically impossible to undertake such a task. Throughout my academic career, my wife Indira has put up with a lot of inconveniences at home during the periods of text book writing and other academic works. I am very thankful for her understanding and support.

J. S. Rao

June 24, 1998
New Delhi

Contents

Preface   vii

1. **Introduction**   1

2. **Variational Principles**   5
   2.1 First variation of a double integral   *5*
   2.2 Use of delta operator   *15*
   2.3 Integral with two space coordinates and time as independent parameters   *24*
   2.4 Lagrange method   *25*
   2.5 Ritz method   *28*
   2.6 Galerkin method   *31*
   2.7 Levy's method   *34*
   2.8 Kantorovich method   *39*

3. **Rectangular Plates**   42
   3.1 Bending of long rectangular plates   *42*
   3.2 Bending of rectangular plates: Sophie Germain theory   *47*
   3.3 Navier's solution for simply supported rectangular plates   *66*
   3.4 Free vibration of simply supported rectangular plates   *74*
   3.5 Forced vibration of simply supported rectangular plates   *77*
   3.6 Nadai solution for simply supported plates   *80*
   3.7 Clamped plates   *84*
   3.8 Mindlin's plate theory   *91*
   3.9 Plate on viscoelastic foundation   *103*

4. **Circular plates**   110
   4.1 Equations of bending of plates in polar coordinates   *110*
   4.2 Strain energy expression in polar coordinates   *114*
   4.3 Clamped circular plates   *115*
   4.4 Simply supported circular plates   *127*
   4.5 Natural frequencies of a circular plate with free boundary   *131*
   4.6 Mindlin's theory for axisymmetric plates   *136*
   4.7 Circular plates on viscoelastic support   *142*

5. Finite element methods for thin plates — 152
   5.1 Tocher's nonconforming triangular element *153*
   5.2 ACM nonconforming rectangular element *167*

6. Plates with combined Lateral and in-plane forces — 179
   6.1 Von Karman's theory *179*
   6.2 Rectangular membranes *192*
   6.3 Circular membranes *195*
   6.4 Initially stressed rectangular plates *199*
   6.5 Clamped circular plates *206*
   6.6 Berger's approximation for rectangular plates *211*
   6.7 Berger's approximation for circular plates *221*
   6.8 Von Karman rectangular plate on a viscoelastic foundation *227*
   6.9 Von Karman circular plate on a viscoelastic foundation *234*

7. Finite element methods for initially stressed thin plates — 239
   7.1 Constant strain triangular element *239*
   7.2 Rectangular membrane element *265*
   7.3 Triangular plate element with in-plane stresses *275*
   7.4 Isoparametric formulation of quadrilateral element *280*
   7.5 Eight noded isoparametric element *289*
   7.6 Eight noded isoparametric Mindlin plate element *295*
   7.7 Example of a rotating plate *305*

8. Anisotropic plates — 312
   8.1 Generalized Hooke's law *312*
   8.2 Transformation laws *313*
   8.3 Material symmetries *317*
   8.4 Engineering constants (orthotropic materials) *320*
   8.5 Stiffness transformations and reduced stiffness matrix *322*
   8.6 Specially orthotropic plate equations *326*
   8.7 Simply supported orthotropic rectangular plates *330*
   8.8 Plates supported at $y = 0, b$ and clamped at $x = \pm a/2$ or $x = 0, a$ *338*
   8.9 Clamped plates *342*
   8.10 Fiber reinforced laminae *346*
   8.11 Laminated plates *358*
   8.12 Simply supported cross ply laminates *378*
   8.13 Simply supported angle ply laminates *384*
   8.14 Berger's approximation for orthotropic plates *390*
   8.15 Von Karman orthotropic plates *398*
   8.16 Eight noded isoparametric orthotropic plate element *406*
   8.17 Rotating orthotropic laminated plate *411*

9. Pre-twisted Plates — 415
    9.1 Theory of a surface *415*
    9.2 Strain-displacement relations *432*
    9.3 Forces and moments *455*
    9.4 Strain energy *465*
    9.5 Pre-twisted rectangular plate *473*
    9.6 Rotating pre-twisted plate *481*

10. Finite Element Method for Shells — 490
    10.1 Ahmad's super-parametric element *490*
    10.2 Rotating pre-twisted plate *505*
    10.3 Pre-twisted shell finite element *511*
    10.4 Rotating pre-twisted plate finite element *524*

References — 533
Index — 557

1

Introduction

The study of plate vibrations has its origin with the experimental work of Chladni in the early nineteenth century who demonstrated the nodal patterns of a square plate in lateral bending. A theoretical explanation of this vibratory behavior was sought by the French Academy in 1811 and this challenge was accepted by Sophie Germain. The theory thus produced was corrected by the celebrated Lagrange in 1816 and French Academy awarded Sophie Germain for this effort. Thus, the theoretical background of the theory of plates was laid by Sophie Germain and Lagrange. However, there was considerable discussion on the boundary conditions produced in this work and it was left to Kirchoff and Kelvin around 1850 to settle this matter.

Navier's solution for simply supported rectangular plates is perhaps the first complete solution followed later by Levy for plates with two opposite edges simply supported and the other two with arbitrary conditions. Several mathematical techniques, e.g., Ritz, Galerkin and other variational methods have been widely used to determine the static and dynamic behavior of plates of different shapes with different boundary conditions. The classical plate theory was subsequently improved by von Karman (1910), Reissner, (1944), Mindlin (1951) amongst others.

With the advent of high speed computers and the introduction of finite element method, plate finite elements are developed, some of the earliest ones being Melosh (1963), Zienkiewicz and Cheung (1964). With the help of finite element methods, it is now practically feasible to handle non uniform plates and plates with arbitrary shapes.

Plates are the common structural elements in many engineering applications, civil engineering roof structures, ship structures, airplane fuselage and wings, pressure vessels, missiles, liquid containers etc. The application of plate and shell theories has recently become very important with high speed turbomachines. Turbomachines employ rotating plates (discs) mounted on the shaft, gas turbine and compressor blades are typically rotating cantilever plates and they are most severely stressed elements subjected to highly fluctuating forces. The design of these elements is of critical importance for safe operation of these units. Hence, the importance of the study of theory of plates to practical engineers.

An attempt is made in this book to introduce basic theories concerning with plates. The solution procedures used include classical energy methods as well as finite element methods. The book assumes knowledge of at least two books: Introductory Course on Theory and Practice of Mechanical Vibrations by J.S. Rao and K. Gupta and Advanced Theory of Vibration, One Dimensional Structures by J.S. Rao. In addition, the reader should have basic knowledge of Theory of Elasticity and a good background of Engineering Mathematics.

In chapter 2, variational principles concerned with two independent space coordinates and time parameter are introduced. This is an extension of the principles given in *Advanced Theory of Vibration*. The delta operator principle that facilitates the derivation of differential equations and boundary conditions is explained. Euler-Lagrange equations are derived and Lagrange method is outlined. Ritz and Galerkin energy methods for systems with two independent space coordinates and time are explained in detail and examples are given illustrating these principles. Two special principles, viz., Levy's method and Kantorovich method applicable for plate problems are given at the end of this chapter.

Rectangular plates is the subject of chapter 3. Bending of long rectangular plates is considered first to extend the one dimensional theories of beams for the case of two dimensional problems. Bending of rectangular plates is then considered to derive the energy expressions, differential equations and boundary conditions through the principles outlined in chapter 2. Classical Navier's solution is then presented. Basic principles of free and forced vibration of simply supported plates are then explained using modal analysis. Nadai solution of rectangular plates is also presented. Clamped plate solutions are then given. The classical theory is then extended to Mindlin's plate theory. Before closing this chapter, plates on viscoelastic foundation are introduced.

Chapter 4 is concerned with thin circular plates. First the equations of motion are derived in polar coordinates. The strain energy expression is then given in polar coordinates. Clamped and simply supported circular plates and circular plates with free boundary are considered. Mindlin's theory is derived for axisymmetric plates and finally circular plates on viscoelastic support are considered. In chapter 5, the finite element method is introduced. Tocher's triangular element and Adini, Clough and Melosh rectangular elements are derived in detail and examples are given.

Von Karman's theory is derived in chapter 6. First initially stressed membranes are considered. Both rectangular as well as circular membranes examples are provided. The membrane analysis is then extended to initially stressed rectangular plates with small deformations without any coupling between extensional and lateral deformations. Stability of these plates is illustrated. Clamped circular plates are next considered in a similar manner. Then, an approximation due to Berger is introduced for rectangular as well as circular plates. Using this approximation, the nonlinear problem solutions for plates on viscoelastic foundation are given. Finally, large deformations are considered and nonlinear solutions of Von Karman rectangular and circular plates on a viscoelastic foundation are given.

Chapter 7 gives the finite element formulation for initially stressed thin plates. First membrane elements are considered, two elements in particular are introduced, 1. constant strain triangular element and 2. rectangular membrane element. The plate and membrane element properties are then combined to give the triangular plate element with in-plane stresses. Then, the isoparametric formulation is introduced for a quadrilateral element followed by the eight noded isoparametric element. The classical plate elements are then extended to the case of eight noded isoparametric Mindlin element. Finally, the case of a rotating plate is considered with an example.

Anisotropic plates are considered in chapter 8. The generalized Hooke's law and the transformation laws for stress and strain are presented. Different material symmetries for special anisotropic materials are then introduced. Then, different engineering constants for orthotropic materials are given. For the case of orthotropic plates, the stiffness transformations and reduced stiffness matrix are then derived. The case of specially orthotropic plates is considered and the equations of motion are derived. Rectangular plates with different boundary conditions are considered and solutions presented. Fiber reinforced laminae are then introduced and solutions for laminated plates with different arrangements are presented. Both cross ply and angle ply laminates are introduced. Berger's approximation for orthotropic plates is then presented. Next, Von Karman orthotropic plates are dealt with. Finally, the eight noded isoparametric orthotropic plate element is presented and the effect of rotation is considered through examples.

Chapter 9 begins with a treatment on the theory of surfaces, which is necessary to understand the shell theories. The strain-displacement relations and equilibrium conditions of a three dimensional thin shell are next given. The strain energy expression for a shell is next derived. Though, the theory presented here is general and applicable to any shell, pre-twisted plates are only considered in detail. The case of rectangular cantilever plates and rotating plates are next considered. Ritz method is applied in solving these problems.

Finite element methods for shell type structures is next considered in chapter 10. To begin with Ahmad's super-parametric element is derived. The case of rotating pre-twisted plates by this finite element is next presented. A shell finite element based on the theory given in chapter 9 is next derived and applied to the case of pre-twisted plates. Finally, a finite element for rotating pre-twisted plates using the thin shell theory is derived and some examples given.

2

Variational Principles

In *Advanced Theory of Vibration* book, basic principles of variational calculus have been given. There, only one dimensional structures were considered, i.e., only one independent space parameter was accounted in deriving the Euler-Lagrange equations and related energy methods. Here, the variational calculus of a functional with two independent space parameters, as in the case of a plate, is first outlined.

2.1 FIRST VARIATION OF A DOUBLE INTEGRAL

In the case of bars, rods or beams, we defined the structure as a one dimensional structure with independent parameter x at which the linear deflection u or angular displacement θ or lateral deformation y is taken a dependent parameter. In general, the potential energy is a double integral as in the case of a plate or a triple integral for a general structure like a shell. Hence, let us consider here a functional of the form

$$F(x, y, w, w_{,x}, w_{,xx}, w_{,y}, w_{,yy}, w_{,xy})$$

(2.1.1)

where w is the dependent variable with x and y as independent variables. Here, it is convenient to represent partial differentiation by a comma followed by the corresponding independent variable, i.e., $w_{,x} = \frac{\partial w}{\partial x}$, $w_{,xx} = \frac{\partial^2 w}{\partial x^2}$, $w_{,xy} = \frac{\partial^2 w}{\partial x \partial y}$ etc. Let us consider a region R bounded with a closed curve $\Gamma(x, y)$ shown in Fig. 2.1.1 and find the conditions of extremum property of an integral I of the functional F over this region R

$$I = \iint_R F(x, y, w, w_{,x}, w_{,xx}, w_{,y}, w_{,yy}, w_{,xy}) \, dx \, dy$$

(2.1.2)

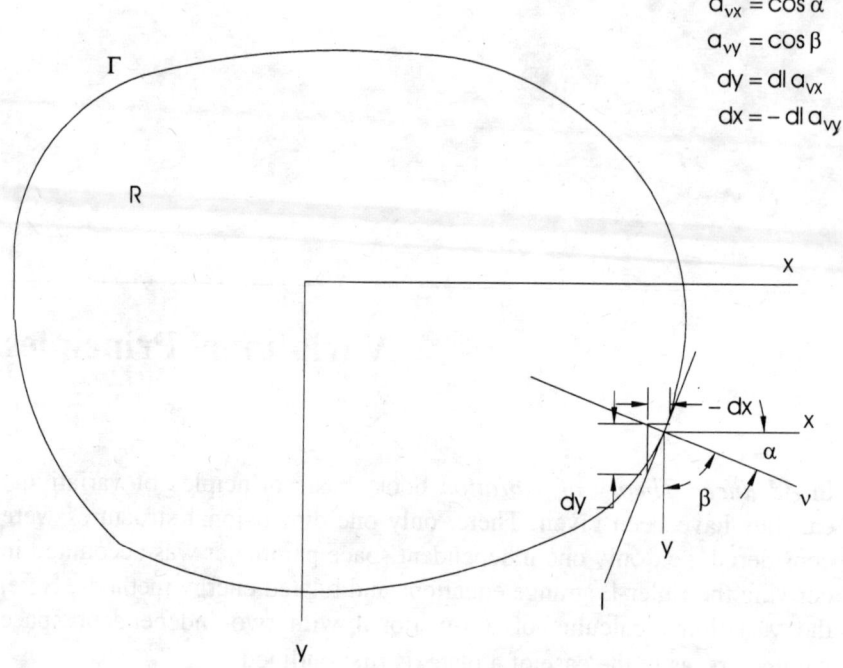

Fig. 2.1.1 A Region R bounded by Γ

For the above integral to be an extremum, consider the varied path $\tilde{w}$ in the neighborhood of the admissible path $w(x,y)$ given by

$$\tilde{w}(x,y) = w(x,y) + \epsilon\, \eta(x,y) \tag{2.1.3}$$

with its derivatives

$$\tilde{w}_{,x}(x,y) = w_{,x}(x,y) + \epsilon\, \eta_{,x}(x,y)$$
$$\tilde{w}_{,y}(x,y) = w_{,y}(x,y) + \epsilon\, \eta_{,y}(x,y)$$
$$\cdots \tag{2.1.4}$$

where ϵ is a small parameter and η are a class of differentiable functions with respect to x and y at least two times with the condition that they satisfy linear forced boundary conditions of the type

$$aw + b\frac{\partial w}{\partial v} = c \tag{2.1.5}$$

in which a, b and c are given point functions on the boundary Γ and $\frac{\partial w}{\partial v}$ denotes the normal derivative of w at the boundary. The integral taken over the varied path is

$$\tilde{I} = \iint_R F(x, y, \tilde{w}, \tilde{w}_{,x}, \tilde{w}_{,xx}, \tilde{w}_{,y}, \tilde{w}_{,yy}, \tilde{w}_{,xy}) \, dx \, dy \tag{2.1.6}$$

which can be written as

$$\tilde{I} = \iint_R F(x, y, w + \epsilon\, \eta, w_{,x} + \epsilon\, \eta_{,x}, w_{,xx} + \epsilon\, \eta_{,xx}, w_{,y} + \epsilon\, \eta_{,y}, w_{,yy} + \epsilon\, \eta_{,yy}, w_{,xy} + \epsilon\, \eta_{,xy}) \, dx \, dy \tag{2.1.7}$$

Expanding the functional by Taylor's series to first powers of ϵ, we get

$$\tilde{I} = \{\tilde{I}\}_{\epsilon=0} + \left\{\frac{d\tilde{I}}{d\epsilon}\right\}_{\epsilon=0} \epsilon + \cdots$$

i.e.,

$$\tilde{I} - \{\tilde{I}\}_{\epsilon=0} = \left\{\frac{d\tilde{I}}{d\epsilon}\right\}_{\epsilon=0} \epsilon + \cdots \tag{2.1.8}$$

For $I = \tilde{I}(\epsilon = 0)$ to be an extremum, we readily recognize that

$$\left\{\frac{d\tilde{I}}{d\epsilon}\right\}_{\epsilon=0} = 0 \tag{2.1.9}$$

From equation (2.1.7), noting that $\left\{\frac{\partial F}{\partial \tilde{w}}\right\}_{\epsilon=0} = \frac{\partial F}{\partial w}$ etc., equation (2.1.9) above becomes

$$\iint_R \left\{\frac{\partial F}{\partial w}\eta + \frac{\partial F}{\partial w_{,x}}\eta_{,x} + \frac{\partial F}{\partial w_{,y}}\eta_{,y} + \frac{\partial F}{\partial w_{,xx}}\eta_{,xx} + \frac{\partial F}{\partial w_{,yy}}\eta_{,yy} + \frac{\partial F}{\partial w_{,xy}}\eta_{,xy}\right\} dx \, dy$$

$$= 0 \tag{2.1.10}$$

8 Dynamics of Plates

To integrate the above by parts, consider the Gauss' theorem given in Appendix of the book *Advanced Theory of Vibration* for a first order tensor V_j in the body

$$\iiint_V V_{j,j} dV = \iint_S V_j v_j dA$$

i.e.,

$$\iiint_V \nabla \cdot \overline{V} dV = \iint_S \overline{V} \cdot d\overline{A}$$

and in expanded form

$$\iiint_V \left(\frac{\partial V_x}{\partial x} + \frac{\partial V_y}{\partial y} + \frac{\partial V_z}{\partial z} \right) dV = \iint_S (V_x a_{vx} + V_y a_{vy} + V_z a_{vz}) dA \tag{a}$$

The above equation is *Divergence theorem*. Consider first the case when $V_y = V_z = 0$ and let V_x be product of two functions u and w. Then equation (a) becomes

$$\iiint_V \frac{\partial (uw)}{\partial x} dV = \iiint_V u \frac{\partial w}{\partial x} dV + \iiint_V w \frac{\partial u}{\partial x} dV = \iint_S (uw) a_{vx} dA$$

i.e.,

$$\iiint_V u \frac{\partial w}{\partial x} dV = \iint_S (uw) a_{vx} dA - \iiint_V w \frac{\partial u}{\partial x} dV \tag{b}$$

Equation (b) in general can be written as

$$\iiint_V u \frac{\partial w}{\partial x_i} dV = \iint_S (uw) a_{vx_i} dA - \iiint_V w \frac{\partial u}{\partial x_i} dV \tag{c}$$

For a two dimensional case as in our present context, the above is

$$\iint_R u \frac{\partial w}{\partial x_i} dA = \oint_\Gamma (uw) a_{vx_i} dl - \iint_R w \frac{\partial u}{\partial x_i} dA \tag{2.1.11}$$

Consider the second and third terms in equation (2.1.10) and apply the above rule to integrate them by parts with respect to x and y respectively to obtain (note from Fig. 2.1.1 that $a_{vx} dl = dy$ and $a_{vy} dl = -dx$)

$$\iint_R \frac{\partial F}{\partial w_{,x}} \eta_{,x} dx\, dy = \oint_\Gamma \left(\eta \frac{\partial F}{\partial w_{,x}}\right) dy - \iint_R \eta \frac{\partial}{\partial x}\left(\frac{\partial F}{\partial w_{,x}}\right) dx\, dy$$

$$\iint_R \frac{\partial F}{\partial w_{,y}} \eta_{,y} dx\, dy = -\oint_\Gamma \left(\eta \frac{\partial F}{\partial w_{,y}}\right) dx - \iint_R \eta \frac{\partial}{\partial y}\left(\frac{\partial F}{\partial w_{,y}}\right) dx\, dy$$

(2.1.12)

The fourth term upon integration by parts with respect to x gives

$$\iint_R \frac{\partial F}{\partial w_{,xx}} \eta_{,xx} dx\, dy = \oint_\Gamma \left(\eta_{,x} \frac{\partial F}{\partial w_{,xx}}\right) dy - \iint_R \eta_{,x} \frac{\partial}{\partial x}\left(\frac{\partial F}{\partial w_{,xx}}\right) dx\, dy$$

$$= \oint_\Gamma \left(\eta_{,x} \frac{\partial F}{\partial w_{,xx}}\right) dy - \oint_\Gamma \eta \frac{\partial}{\partial x}\left(\frac{\partial F}{\partial w_{,xx}}\right) dy$$

$$+ \iint_R \eta \frac{\partial^2}{\partial x^2}\left(\frac{\partial F}{\partial w_{,xx}}\right) dx\, dy$$

(2.1.13)

Similarly, the fifth term upon integration with respect to y gives

$$\iint_R \frac{\partial F}{\partial w_{,yy}} \eta_{,yy} dx\, dy = -\oint_\Gamma \left(\eta_{,y} \frac{\partial F}{\partial w_{,yy}}\right) dx - \iint_R \eta_{,y} \frac{\partial}{\partial y}\left(\frac{\partial F}{\partial w_{,yy}}\right) dx\, dy$$

$$= -\oint_\Gamma \left(\eta_{,y} \frac{\partial F}{\partial w_{,yy}}\right) dx + \oint_\Gamma \eta \frac{\partial}{\partial y}\left(\frac{\partial F}{\partial w_{,yy}}\right) dx \quad (2.1.14)$$

$$+ \iint_R \eta \frac{\partial^2}{\partial y^2}\left(\frac{\partial F}{\partial w_{,yy}}\right) dx\, dy$$

Finally the last term in (2.1.10) is integrated by parts, we can do this in either of the two ways 1. integrate first with respect to x and then with respect to y, or 2. integrate first with respect to y and then with respect to x to give

$$\iint_R \frac{\partial F}{\partial w_{,xy}} \eta_{,xy} dx\, dy = \oint_\Gamma \left(\eta_{,y} \frac{\partial F}{\partial w_{,xy}}\right) dy - \iint_R \eta_{,y} \frac{\partial}{\partial x}\left(\frac{\partial F}{\partial w_{,xy}}\right) dx\, dy$$

$$= \oint_\Gamma \left(\eta_{,y} \frac{\partial F}{\partial w_{,xy}}\right) dy + \oint_\Gamma \eta \frac{\partial}{\partial x}\left(\frac{\partial F}{\partial w_{,xy}}\right) dx$$

$$+ \iint_R \eta \frac{\partial^2}{\partial x \partial y} \left(\frac{\partial F}{\partial w_{,xy}}\right) dx\, dy$$

(2.1.15)

10 Dynamics of Plates

or

$$\iint_R \frac{\partial F}{\partial w_{,xy}} \eta_{,xy} \, dx \, dy = \oint_\Gamma \left(\eta_{,x} \frac{\partial F}{\partial w_{,xy}} \right) dx - \iint_R \eta_{,x} \frac{\partial}{\partial y} \left(\frac{\partial F}{\partial w_{,xy}} \right) dx \, dy$$

$$= -\oint_\Gamma \left(\eta_{,x} \frac{\partial F}{\partial w_{,xy}} \right) dx - \oint_\Gamma \eta \frac{\partial}{\partial y} \left(\frac{\partial F}{\partial w_{,xy}} \right) dy$$

$$+ \iint_R \eta \frac{\partial^2}{\partial x \partial y} \left(\frac{\partial F}{\partial w_{,xy}} \right) dx \, dy$$

(2.1.16)

Note that in equations (2.1.15) and (2.1.16), the result is same for the double integral term, whereas the line integrals both are different. In arriving at the boundary conditions, we can choose a convenient form of either (2.1.15) or (2.1.16). With the help of equations (2.1.12) to (2.1.16), we can write equation (2.1.10) as

$$\iint_R \left\{ \frac{\partial F}{\partial w} \eta + \frac{\partial F}{\partial w_{,x}} \eta_{,x} + \frac{\partial F}{\partial w_{,y}} \eta_{,y} + \frac{\partial F}{\partial w_{,xx}} \eta_{,xx} + \frac{\partial F}{\partial w_{,yy}} \eta_{,yy} + \frac{\partial F}{\partial w_{,xy}} \eta_{,xy} \right\} dx \, dy$$

$$= \oint_\Gamma \left(\eta \frac{\partial F}{\partial w_{,x}} \right) dy - \oint_\Gamma \left(\eta \frac{\partial F}{\partial w_{,y}} \right) dx + \oint_\Gamma \left(\eta_{,x} \frac{\partial F}{\partial w_{,xx}} \right) dy$$

$$- \oint_\Gamma \eta \frac{\partial}{\partial x} \left(\frac{\partial F}{\partial w_{,xx}} \right) dy - \oint_\Gamma \left(\eta_{,y} \frac{\partial F}{\partial w_{,yy}} \right) dx + \oint_\Gamma \eta \frac{\partial}{\partial y} \left(\frac{\partial F}{\partial w_{,yy}} \right) dx$$

$$+ \left. \begin{cases} \oint_\Gamma \left(\eta_{,y} \frac{\partial F}{\partial w_{,xy}} \right) dy + \oint_\Gamma \eta \frac{\partial}{\partial x} \left(\frac{\partial F}{\partial w_{,xy}} \right) dx \\ -\oint_\Gamma \left(\eta_{,x} \frac{\partial F}{\partial w_{,xy}} \right) dx - \oint_\Gamma \eta \frac{\partial}{\partial y} \left(\frac{\partial F}{\partial w_{,xy}} \right) dy \end{cases} \right\} \text{either of the rows}$$

$$+ \iint_R \eta \frac{\partial F}{\partial w} \, dx \, dy - \iint_R \eta \frac{\partial}{\partial x} \left(\frac{\partial F}{\partial w_{,x}} \right) dx \, dy$$

$$- \iint_R \eta \frac{\partial}{\partial y} \left(\frac{\partial F}{\partial w_{,y}} \right) dx \, dy + \iint_R \eta \frac{\partial^2}{\partial x^2} \left(\frac{\partial F}{\partial w_{,xx}} \right) dx \, dy$$

$$+ \iint_R \eta \frac{\partial^2}{\partial y^2} \left(\frac{\partial F}{\partial w_{,yy}} \right) dx \, dy + \iint_R \eta \frac{\partial^2}{\partial x \partial y} \left(\frac{\partial F}{\partial w_{,xy}} \right) dx \, dy$$

$$= 0$$

(2.1.17)

Regrouping the terms, we have

$$\iint_R \left[\begin{array}{c} \dfrac{\partial F}{\partial w} - \dfrac{\partial}{\partial x}\left(\dfrac{\partial F}{\partial w_{,x}}\right) - \dfrac{\partial}{\partial y}\left(\dfrac{\partial F}{\partial w_{,y}}\right) + \dfrac{\partial^2}{\partial x^2}\left(\dfrac{\partial F}{\partial w_{,xx}}\right) \\ + \dfrac{\partial^2}{\partial y^2}\left(\dfrac{\partial F}{\partial w_{,yy}}\right) + \dfrac{\partial^2}{\partial x \partial y}\left(\dfrac{\partial F}{\partial w_{,xy}}\right) \end{array} \right] \eta\, dx\, dy$$

$$+ \oint_\Gamma \left[\dfrac{\partial F}{\partial w_{,x}} - \dfrac{\partial}{\partial x}\left(\dfrac{\partial F}{\partial w_{,xx}}\right) \right] \eta\, dy - \oint_\Gamma \left[\dfrac{\partial F}{\partial w_{,y}} - \dfrac{\partial}{\partial y}\left(\dfrac{\partial F}{\partial w_{,yy}}\right) \right] \eta\, dx$$

$$+ \oint_\Gamma \left(\eta_{,x} \dfrac{\partial F}{\partial w_{,xx}} \right) dy - \oint_\Gamma \left(\eta_{,y} \dfrac{\partial F}{\partial w_{,yy}} \right) dx$$

$$+ \left\{ \begin{array}{c} \oint_\Gamma \left(\eta_{,y} \dfrac{\partial F}{\partial w_{,xy}} \right) dy + \oint_\Gamma \eta \dfrac{\partial}{\partial x}\left(\dfrac{\partial F}{\partial w_{,xy}}\right) dx \\ - \oint_\Gamma \left(\eta_{,x} \dfrac{\partial F}{\partial w_{,xy}} \right) dx - \oint_\Gamma \eta \dfrac{\partial}{\partial y}\left(\dfrac{\partial F}{\partial w_{,xy}}\right) dy \end{array} \right\} = 0$$

either of the rows

(2.1.17a)

The Euler-Lagrange equation for this problem can now be written as

$$\dfrac{\partial F}{\partial w} - \dfrac{\partial}{\partial x}\left(\dfrac{\partial F}{\partial w_{,x}}\right) - \dfrac{\partial}{\partial y}\left(\dfrac{\partial F}{\partial w_{,y}}\right) + \dfrac{\partial^2}{\partial x^2}\left(\dfrac{\partial F}{\partial w_{,xx}}\right)$$

$$+ \dfrac{\partial^2}{\partial y^2}\left(\dfrac{\partial F}{\partial w_{,yy}}\right) + \dfrac{\partial^2}{\partial x \partial y}\left(\dfrac{\partial F}{\partial w_{,xy}}\right)$$

$$= 0$$

(2.1.18)

Consider for the present discussion that the region R is made of edges parallel to either x or y axes; then on the part of the boundary represented by x = constant (parallel to y axis) $dx = 0$ and the line integrals containing dx vanish. In this case, it is convenient to use equation (2.1.15) for integration by parts of the last term in equation (2.1.10). Similarly, on the part of the boundary represented by y = constant (parallel to x axis) $dy = 0$ and the line integrals containing dy vanish and here we use equation (2.1.16) for integration by parts of the last term in equation (2.1.10). The boundary conditions for the respective edges are

12 *Dynamics of Plates*

$$\left[\frac{\partial F}{\partial w_{,x}} - \frac{\partial}{\partial x}\left(\frac{\partial F}{\partial w_{,xx}}\right) - \frac{\partial}{\partial y}\left(\frac{\partial F}{\partial w_{,xy}}\right)\right]\eta = 0 \text{ on the edge } x = \text{constant}$$

$$\eta_{,x}\frac{\partial F}{\partial w_{,xx}} = 0 \text{ on the edge } x = \text{constant}$$

$$\left[\frac{\partial F}{\partial w_{,y}} - \frac{\partial}{\partial x}\left(\frac{\partial F}{\partial w_{,xy}}\right) - \frac{\partial}{\partial y}\left(\frac{\partial F}{\partial w_{,yy}}\right)\right]\eta = 0 \text{ on the edge } y = \text{constant}$$

$$\eta_{,y}\frac{\partial F}{\partial w_{,yy}} = 0 \text{ on the edge } y = \text{constant}$$

(2.1.19)

We can separate the above into kinematic and natural boundary conditions:

Kinematic boundary conditions

$$\text{Specify } w \text{ or } \frac{\partial w}{\partial x} \text{ on the edge } x = \text{constant}$$

$$\text{Specify } w \text{ or } \frac{\partial w}{\partial y} \text{ on the edge } y = \text{constant}$$

(2.1.20a)

Natural boundary conditions

$$\frac{\partial F}{\partial w_{,x}} - \frac{\partial}{\partial x}\left(\frac{\partial F}{\partial w_{,xx}}\right) - \frac{\partial}{\partial y}\left(\frac{\partial F}{\partial w_{,xy}}\right) = 0 \text{ on the edge } x = \text{constant}$$

$$\frac{\partial F}{\partial w_{,xx}} = 0 \text{ on the edge } x = \text{constant}$$

$$\frac{\partial F}{\partial w_{,y}} - \frac{\partial}{\partial x}\left(\frac{\partial F}{\partial w_{,xy}}\right) - \frac{\partial}{\partial y}\left(\frac{\partial F}{\partial w_{,yy}}\right) = 0 \text{ on the edge } y = \text{constant}$$

$$\frac{\partial F}{\partial w_{,yy}} = 0 \text{ on the edge } y = \text{constant}$$

(2.1.20b)

As an example of application of the above Euler-Lagrange equations, consider the following functional

$$F = \frac{D}{2}\left[(\nabla^2 w)^2 + 2(1-v)\{w_{,xy}^2 - w_{,xx}w_{,yy}\}\right] - qw$$

(2.1.21)

in a rectangular region R enclosed by the edges $x = 0$, $x = a$; $y = 0$, $y = b$, where D and q are constants. The terms in equation (2.1.18) for the above functional are evaluated below.

$$\frac{\partial F}{\partial w} = -q \tag{2.1.22a}$$

$$-\frac{\partial}{\partial x}\left(\frac{\partial F}{\partial w_{,x}}\right) = 0 \tag{2.1.22b}$$

$$-\frac{\partial}{\partial y}\left(\frac{\partial F}{\partial w_{,y}}\right) = 0 \tag{2.1.22c}$$

$$\frac{\partial F}{\partial w_{,xx}} = D\,\nabla^2 w - D(1-v)\,w_{,yy}$$

$$\frac{\partial^2}{\partial x^2}\left(\frac{\partial F}{\partial w_{,xx}}\right) = D\,\frac{\partial^2}{\partial x^2}(\nabla^2 w) \tag{2.1.22d}$$

$$\frac{\partial F}{\partial w_{,yy}} = D\,\nabla^2 w - D(1-v)\,w_{,xx}$$

$$\frac{\partial^2}{\partial y^2}\left(\frac{\partial F}{\partial w_{,yy}}\right) = D\,\frac{\partial^2}{\partial y^2}(\nabla^2 w) \tag{2.1.22e}$$

$$\frac{\partial F}{\partial w_{,xy}} = 2D(1-v)w_{,xy}$$

$$\frac{\partial^2}{\partial x \partial y}\left(\frac{\partial F}{\partial w_{,xy}}\right) = 0 \tag{2.1.22f}$$

Summing up the above quantities in equations (2.1.22a) to (2.1.22f), we get the differential equation for the system governed by the functional in equation (2.1.21) as

$$D\,\nabla^4 w = q \tag{2.1.23}$$

14 Dynamics of Plates

The boundary conditions are obtained from equations (2.1.20a) and (2.1.20b). The following possible boundary conditions exist.

Fixed-Fixed Conditions

$$w = 0; \quad \frac{\partial w}{\partial x} = 0 \text{ on the edge } x = 0 \text{ and } x = a$$

$$w = 0; \quad \frac{\partial w}{\partial y} = 0 \text{ on the edge } y = 0 \text{ and } y = b$$

(2.1.24)

Simply Supported Conditions

$$\frac{\partial F}{\partial w_{,xx}} = D(w_{,xx} + v w_{,yy}) = 0 \text{ on the edge } x = 0 \text{ and } x = a$$

$$w = 0 \text{ on the edge } x = 0 \text{ and } x = a$$

and

$$\frac{\partial F}{\partial w_{,yy}} = D(w_{,yy} + v w_{,xx}) = 0 \text{ on the edge } y = 0 \text{ and } y = b$$

$$w = 0 \text{ on the edge } y = 0 \text{ and } y = b$$

(2.1.25)

Free-Free Conditions

$$-\frac{\partial}{\partial x}\left(\frac{\partial F}{\partial w_{,xx}}\right) = -D(w_{,xxx} + v w_{,yyx})$$

$$\frac{\partial F}{\partial w_{,xy}} = 2D(1-v)w_{,xy}$$

$$-\frac{\partial}{\partial y}\left(\frac{\partial F}{\partial w_{,xy}}\right) = -2D(1-v)w_{,xyy}$$

$$\frac{\partial F}{\partial w_{,x}} - \frac{\partial}{\partial x}\left(\frac{\partial F}{\partial w_{,xx}}\right) - \frac{\partial}{\partial y}\left(\frac{\partial F}{\partial w_{,xy}}\right) = -D\left\{\frac{\partial}{\partial x}(\nabla^2 w) + (1-v)w_{,xyy}\right\} = 0$$

$$\frac{\partial F}{\partial w_{,xx}} = D(w_{,xx} + v w_{,yy}) = 0 \text{ on the edge } x = 0 \text{ and } x = a$$

and

$$-\frac{\partial}{\partial x}\left(\frac{\partial F}{\partial w_{,xy}}\right) = -2D(1-v)w_{,xxy}$$

$$\frac{\partial F}{\partial w_{,y}} - \frac{\partial}{\partial y}\left(\frac{\partial F}{\partial w_{,yy}}\right) - \frac{\partial}{\partial x}\left(\frac{\partial F}{\partial w_{,xy}}\right) = -D\left\{\frac{\partial}{\partial y}(\nabla^2 w) + (1-v)w_{,xxy}\right\} = 0$$

$$\frac{\partial F}{\partial w_{,yy}} = D(w_{,yy} + v w_{,xx}) = 0 \text{ on the edge } y = 0 \text{ and } y = b$$

(2.1.26)

A combination of the fixed or supported or free conditions of the edges is possible, e.g., a plate fixed on two edges and free on two edges, supported on two edges and free on two edges, fixed on one edge and free on other three edges etc.

2.2 USE OF DELTA OPERATOR

Here, we follow the application of δ operator similar to that of section 5.3 of *Advanced Theory of Vibration*. Let us recall that δw is the variation of the admissible path w and the variation of any other function such as $w_{,x}$ can be written as

$$\delta w_{,x} = \tilde{w}_{,x} - w_{,x} = (\delta w)_{,x}$$

(2.2.1)

meaning thereby that the δ operator is commutative. Now, consider the functional in equation (2.1.1) over the varied path

$$\tilde{F} = F(x, y, w + \delta w, w_{,x} + \delta w_{,x}, w_{,xx} + \delta w_{,xx}, w_{,y} + \delta w_{,y}, w_{,yy}$$
$$+ \delta w_{,yy}, w_{,xy} + \delta w_{,xy})$$

(2.2.2)

Using Taylor's series, the above functional can be written as

$$\tilde{F} = F(x, y, w, w_{,x}, w_{,xx}, w_{,y}, w_{,yy}, w_{,xy})$$
$$+ \frac{\partial F}{\partial w}\delta w + \frac{\partial F}{\partial w_{,x}}\delta w_{,x} + \frac{\partial F}{\partial w_{,xx}}\delta w_{,xx} + \frac{\partial F}{\partial w_{,y}}\delta w_{,y}$$
$$+ \frac{\partial F}{\partial w_{,yy}}\delta w_{,yy} + \frac{\partial F}{\partial w_{,xy}}\delta w_{,xy} + O(\delta^2)$$

(2.2.3)

Therefore, the total variation of the functional is

$$\delta^T F = F(x, y, w+\delta w, w_{,x}+\delta w_{,x}, w_{,xx}+\delta w_{,xx}, w_{,y}+\delta w_{,y}, w_{,yy}$$
$$+\delta w_{,yy}, w_{,xy}+\delta w_{,xy}) - F(x, y, w, w_{,x}, w_{,xx}, w_{,y}, w_{,yy}, w_{,xy})$$
$$= \left\{ \begin{array}{l} \dfrac{\partial F}{\partial w}\delta w + \dfrac{\partial F}{\partial w_{,x}}\delta w_{,x} + \dfrac{\partial F}{\partial w_{,xx}}\delta w_{,xx} + \dfrac{\partial F}{\partial w_{,y}}\delta w_{,y} \\ +\dfrac{\partial F}{\partial w_{,yy}}\delta w_{,yy} + \dfrac{\partial F}{\partial w_{,xy}}\delta w_{,xy} \end{array} \right\} + O(\delta^2)$$

(2.2.4)

We can immediately recognize the first variation

$$\delta^1 F = \left\{ \begin{array}{l} \dfrac{\partial F}{\partial w}\delta w + \dfrac{\partial F}{\partial w_{,x}}\delta w_{,x} + \dfrac{\partial F}{\partial w_{,xx}}\delta w_{,xx} + \dfrac{\partial F}{\partial w_{,y}}\delta w_{,y} \\ +\dfrac{\partial F}{\partial w_{,yy}}\delta w_{,yy} + \dfrac{\partial F}{\partial w_{,xy}}\delta w_{,xy} \end{array} \right\}$$

(2.2.5)

The total variation of the integral in equation (2.1.2) is

$$\delta^T I = \tilde{I} - I$$
$$= \iint_R F(x, y, w+\delta w, w_{,x}+\delta w_{,x}, w_{,xx}+\delta w_{,xx}, w_{,y}+\delta w_{,y}, w_{,yy}$$
$$+\delta w_{,yy}, w_{,xy}+\delta w_{,xy})dx\,dy - \iint_R F(x, y, w, w_{,x}, w_{,xx}, w_{,y}, w_{,yy}, w_{,xy})dx\,dy$$
$$= \iint_R \left\{ \begin{array}{l} \dfrac{\partial F}{\partial w}\delta w + \dfrac{\partial F}{\partial w_{,x}}\delta w_{,x} + \dfrac{\partial F}{\partial w_{,xx}}\delta w_{,xx} + \dfrac{\partial F}{\partial w_{,y}}\delta w_{,y} \\ +\dfrac{\partial F}{\partial w_{,yy}}\delta w_{,yy} + \dfrac{\partial F}{\partial w_{,xy}}\delta w_{,xy} \end{array} \right\} dx\,dy + O(\delta^2)$$

(2.2.6)

We can now recognize the first variation of the integral

$$\delta^1 I = \iint_R \left\{ \begin{array}{l} \dfrac{\partial F}{\partial w}\delta w + \dfrac{\partial F}{\partial w_{,x}}\delta w_{,x} + \dfrac{\partial F}{\partial w_{,xx}}\delta w_{,xx} + \dfrac{\partial F}{\partial w_{,y}}\delta w_{,y} \\ +\dfrac{\partial F}{\partial w_{,yy}}\delta w_{,yy} + \dfrac{\partial F}{\partial w_{,xy}}\delta w_{,xy} \end{array} \right\} dx\,dy$$

(2.2.7)

Variational Principles 17

With the help of equation (2.1.11), we can integrate by parts the second to sixth terms in the equation (2.2.7) above. The second and fourth terms give

$$\iint_R \frac{\partial F}{\partial w_{,x}} \delta w_{,x} dx\, dy = \oint_\Gamma \left(\delta w \frac{\partial F}{\partial w_{,x}}\right) dy - \iint_R \delta w \frac{\partial}{\partial x}\left(\frac{\partial F}{\partial w_{,x}}\right) dx\, dy$$

$$\iint_R \frac{\partial F}{\partial w_{,y}} \delta w_{,y} dx\, dy = -\oint_\Gamma \left(\delta w \frac{\partial F}{\partial w_{,y}}\right) dx - \iint_R \delta w \frac{\partial}{\partial y}\left(\frac{\partial F}{\partial w_{,y}}\right) dx\, dy$$

(2.2.8)

The third term upon integration by parts twice with respect to x gives

$$\iint_R \frac{\partial F}{\partial w_{,xx}} \delta w_{,xx} dx\, dy = \oint_\Gamma \left(\delta w_{,x} \frac{\partial F}{\partial w_{,xx}}\right) dy - \iint_R \delta w_{,x} \frac{\partial}{\partial x}\left(\frac{\partial F}{\partial w_{,xx}}\right) dx\, dy$$

$$= \oint_\Gamma \left(\delta w_{,x} \frac{\partial F}{\partial w_{,xx}}\right) dy - \oint_\Gamma \delta w \frac{\partial}{\partial x}\left(\frac{\partial F}{\partial w_{,xx}}\right) dy$$

$$+ \iint_R \delta w \frac{\partial^2}{\partial x^2}\left(\frac{\partial F}{\partial w_{,xx}}\right) dx\, dy$$

(2.2.9)

Similarly, the fifth term upon integration twice with respect to y gives

$$\iint_R \frac{\partial F}{\partial w_{,yy}} \delta w_{,yy} dx\, dy = -\oint_\Gamma \left(\delta w_{,y} \frac{\partial F}{\partial w_{,yy}}\right) dx - \iint_R \delta w_{,y} \frac{\partial}{\partial y}\left(\frac{\partial F}{\partial w_{,yy}}\right) dx\, dy$$

$$= -\oint_\Gamma \left(\delta w_{,y} \frac{\partial F}{\partial w_{,yy}}\right) dx + \oint_\Gamma \delta w \frac{\partial}{\partial y}\left(\frac{\partial F}{\partial w_{,yy}}\right) dx$$

$$+ \iint_R \delta w \frac{\partial^2}{\partial y^2}\left(\frac{\partial F}{\partial w_{,yy}}\right) dx\, dy$$

(2.2.10)

Finally the last term in (2.2.7) is integrated by parts, we can do this in either of the two ways as in section 2.1, viz., 1. integrate first with respect to x and then with respect to y, or 2. integrate first with respect to y and then with respect to x to give

$$\iint_R \frac{\partial F}{\partial w_{,xy}}\delta w_{,xy}\,dx\,dy = \oint_\Gamma \left(\delta w_{,y}\frac{\partial F}{\partial w_{,xy}}\right)dy - \iint_R \delta w_{,y}\frac{\partial}{\partial x}\left(\frac{\partial F}{\partial w_{,xy}}\right)dx\,dy$$

$$= \oint_\Gamma \left(\delta w_{,y}\frac{\partial F}{\partial w_{,xy}}\right)dy + \oint_\Gamma \delta w\frac{\partial}{\partial x}\left(\frac{\partial F}{\partial w_{,xy}}\right)dx$$

$$+ \iint_R \delta w\frac{\partial^2}{\partial x\partial y}\left(\frac{\partial F}{\partial w_{,xy}}\right)dx\,dy$$

(2.2.11)

or

$$\iint_R \frac{\partial F}{\partial w_{,xy}}\delta w_{,xy}\,dx\,dy = \oint_\Gamma \left(\delta w_{,x}\frac{\partial F}{\partial w_{,xy}}\right)dx - \iint_R \delta w_{,x}\frac{\partial}{\partial y}\left(\frac{\partial F}{\partial w_{,xy}}\right)dx\,dy$$

$$= -\oint_\Gamma \left(\delta w_{,x}\frac{\partial F}{\partial w_{,xy}}\right)dx - \oint_\Gamma \delta w\frac{\partial}{\partial y}\left(\frac{\partial F}{\partial w_{,xy}}\right)dy$$

$$+ \iint_R \delta w\frac{\partial^2}{\partial x\partial y}\left(\frac{\partial F}{\partial w_{,xy}}\right)dx\,dy$$

(2.2.12)

Following the steps given in section 2.1, it is clear that we get Euler-Lagrange equation and the boundary conditions same as in (2.1.18) and (2.1.19) with $\eta = \delta w$.

As an example of application of the delta operator method to derive the differential equation and boundary conditions, consider the extremum of the following integral

$$I = \iint_R \left\{\frac{D}{2}\left[(\nabla^2 w)^2 + 2(1-v)\{w_{,xy}^2 - w_{,xx}w_{,yy}\}\right] - qw\right\}dx\,dy$$

(2.2.13)

First we write the above integral in a convenient form as

$$I = \iint_R \left\{\frac{D}{2}\left[(\nabla^2 w)^2 + (1-v)\{w_{,xy}^2 + w_{,yx}^2 - 2w_{,xx}w_{,yy}\}\right] - qw\right\}dx\,dy$$

(2.2.13a)

then the extremum of the above integral is given by the condition

$$\delta I = \iint_R \left[\begin{array}{c} D(\nabla^2 w)\{\delta w_{,xx} + \delta w_{,yy}\} \\ +D(1-v)\left\{\begin{array}{c} w_{,xy}\delta w_{,xy} + w_{,yx}\delta w_{,yx} \\ -w_{,xx}\delta w_{,yy} - w_{,yy}\delta w_{,xx}\end{array}\right\} - q\delta w\end{array}\right]dx\,dy = 0$$

(2.2.14)

The following integrals in the above are evaluated by using Green's theorem in (2.1.11) and Fig. 2.1.1:

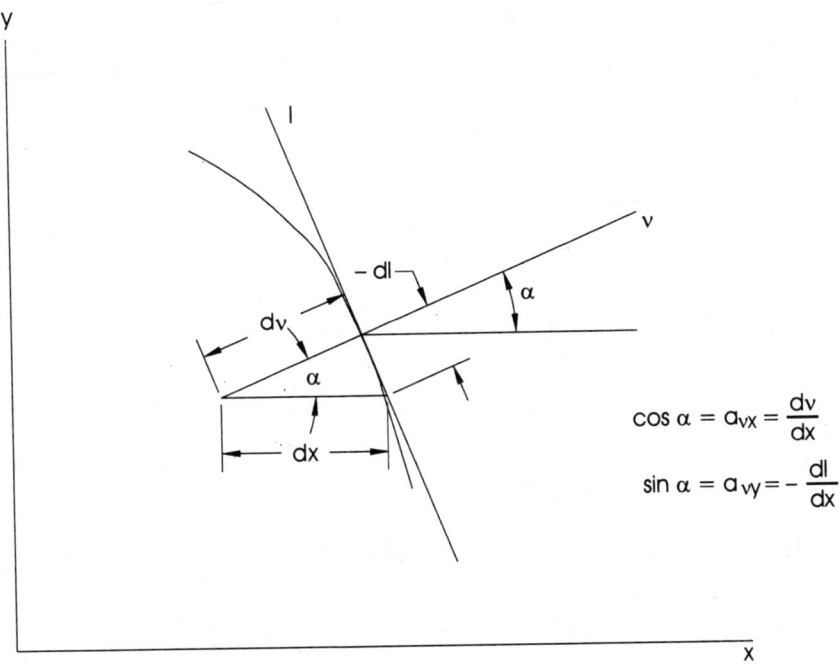

Fig. 2.2.1 Curved Boundary Relations

$$\iint_R (\nabla^2 w)\delta w_{,xx} dxdy = \oint_\Gamma \delta w_{,x}(\nabla^2 w)dy - \iint_R \delta w_{,x}\frac{\partial}{\partial x}(\nabla^2 w)dx\,dy$$

$$= \oint_\Gamma \delta w_{,x}(\nabla^2 w)dy - \oint_\Gamma \delta w \frac{\partial}{\partial x}(\nabla^2 w)dy$$

$$+ \iint_R \delta w \frac{\partial^2}{\partial x^2}(\nabla^2 w)dx\,dy$$

(2.2.15)

$$\iint_R (\nabla^2 w)\delta w_{,yy} dxdy = -\oint_\Gamma \delta w_{,y}(\nabla^2 w)dx - \iint_R \delta w_{,y}\frac{\partial}{\partial y}(\nabla^2 w)dx\,dy$$

$$= -\oint_\Gamma \delta w_{,y}(\nabla^2 w)dx + \oint_\Gamma \delta w \frac{\partial}{\partial y}(\nabla^2 w)dx$$

$$+ \iint_R \delta w \frac{\partial^2}{\partial y^2}(\nabla^2 w)dx\,dy$$

(2.2.16)

$$\iint_R w_{,xy}\delta w_{,xy}\,dx\,dy = -\oint_\Gamma \delta w_{,x} w_{,xy}\,dx - \iint_R \delta w_{,x} w_{,xyy}\,dx\,dy$$

$$= -\oint_\Gamma \delta w_{,x} w_{,xy}\,dx - \oint_\Gamma \delta w\, w_{,xyy}\,dy + \iint_R \delta w\, w_{,xxyy}\,dx\,dy$$

(2.2.17)

$$\iint_R w_{,yx}\delta w_{,yx}\,dx\,dy = \oint_\Gamma \delta w_{,y} w_{,yx}\,dy - \iint_R \delta w_{,y} w_{,xyx}\,dx\,dy$$

$$= \oint_\Gamma \delta w_{,y} w_{,yx}\,dy + \oint_\Gamma \delta w\, w_{,xxy}\,dx + \iint_R \delta w\, w_{,xxyy}\,dx\,dy$$

(2.2.18)

$$\iint_R w_{,xx}\delta w_{,yy}\,dx\,dy = -\oint_\Gamma \delta w_{,y} w_{,xx}\,dx - \iint_R \delta w_{,y} w_{,xxy}\,dx\,dy$$

$$= -\oint_\Gamma \delta w_{,y} w_{,xx}\,dx + \oint_\Gamma \delta w\, w_{,xxy}\,dx + \iint_R \delta w\, w_{,xxyy}\,dx\,dy$$

(2.2.19)

$$\iint_R w_{,yy}\delta w_{,xx}\,dx\,dy = \oint_\Gamma \delta w_{,x} w_{,yy}\,dy - \iint_R \delta w_{,x} w_{,xyy}\,dx\,dy$$

$$= \oint_\Gamma \delta w_{,x} w_{,yy}\,dy - \oint_\Gamma \delta w\, w_{,xyy}\,dy + \iint_R \delta w\, w_{,xxyy}\,dx\,dy$$

(2.2.20)

With the help of the above equations (2.2.15) to (2.2.20), equation (2.2.14) gives

$$\iint_R (D\nabla^4 w - q)\delta w\,dx\,dy$$

$$+ \oint_\Gamma D(w_{,xx} + vw_{,yy})\delta w_{,x}\,dy - \oint_\Gamma D(w_{,yy} + vw_{,xx})\delta w_{,y}\,dx$$

$$+ \oint_\Gamma D(1-v)w_{,xy}\delta w_{,y}\,dy - \oint_\Gamma D(1-v)w_{,xy}\delta w_{,x}\,dx$$

$$+ \oint_\Gamma D(w_{,yyy} + vw_{,xxy})\delta w\,dx - \oint_\Gamma D(w_{,xxx} + vw_{,xyy})\delta w\,dy$$

$$+ \oint_\Gamma D(1-v)w_{,xxy}\delta w\,dx - \oint_\Gamma D(1-v)w_{,xyy}\delta w\,dy$$

$$= 0$$

(2.2.21)

Variational Principles 21

The above gives the same differential equation (2.1.23) and boundary conditions (2.1.24) to (2.1.26). Here, let us define the following quantities

$$M_x = -D(w_{,xx} + vw_{,yy})$$
$$M_y = -D(w_{,yy} + vw_{,xx})$$
$$M_{xy} = D(1-v)w_{,xy}$$
$$V_x = M_{x,x} - M_{xy,y}$$
$$V_y = M_{y,y} - M_{xy,x}$$

(2.2.22)

then, the line integrals of equation (2.2.21) can be expressed as the following to give the boundary conditions.

$$-\oint_\Gamma M_x \delta w_{,x} dy + \oint_\Gamma M_y \delta w_{,y} dx$$
$$+ \oint_\Gamma M_{xy} \delta w_{,y} dy - \oint_\Gamma M_{xy} \delta w_{,x} dx$$
$$- \oint_\Gamma V_y \delta w dx + \oint_\Gamma V_x \delta w dy$$
$$= 0$$

(2.2.23)

In the previous section, we defined the boundary conditions when the region is described by straight edges parallel to x and y axes. In general, the boundary of the region may be described by the curve $\Gamma(x, y)$. To define the boundary conditions on a general surface, it will be judicious to express increments dx and dy in terms of the differentials along the normal and tangential directions to the surface dv and dl respectively. Towards this end, consider the boundary shown in Fig. 2.2.1 from which we find (note that for a positive dl, dx is negative)

$$\cos \alpha = a_{vx} = \frac{dy}{dl}$$

$$\sin \alpha = a_{vy} = -\frac{dl}{dx}$$

(2.2.24)

22 Dynamics of Plates

Similarly, considering Fig. 2.2.2, we have

$$\cos a = a_{vx} = \frac{dl}{dy}$$

$$\sin a = a_{vy} = \frac{dv}{dy}$$

(2.2.25)

With the help of the above two equations (2.2.24) and (2.2.25), we can obtain

$$\frac{\partial}{\partial x} = \frac{\partial}{\partial v}\frac{dv}{dx} + \frac{\partial}{\partial l}\frac{dl}{dx} = a_{vx}\frac{\partial}{\partial v} - a_{vy}\frac{\partial}{\partial l}$$

$$\frac{\partial}{\partial y} = \frac{\partial}{\partial v}\frac{dv}{dy} + \frac{\partial}{\partial l}\frac{dl}{dy} = a_{vy}\frac{\partial}{\partial v} + a_{vx}\frac{\partial}{\partial l}$$

(2.2.26)

We also note from Fig. 2.1.1 that

$$a_{vx}dl = dy$$

$$a_{vy}dl = -dx$$

(2.2.27)

With the help of equations (2.2.26) and (2.2.27), we can now write (2.2.23) as

$$-\oint_\Gamma M_x(a_{vx}\delta w_{,v} - a_{vy}\delta w_{,l})(a_{vx}dl)$$
$$+\oint_\Gamma M_y(a_{vy}\delta w_{,v} + a_{vx}\delta w_{,l})(-a_{vy}dl)$$
$$-\oint_\Gamma M_{xy}(a_{vy}\delta w_{,v} + a_{vx}\delta w_{,l})(a_{vx}dl)$$
$$+\oint_\Gamma M_{xy}(a_{vx}\delta w_{,v} - a_{vy}\delta w_{,l})(-a_{vy}dl)$$
$$-\oint_\Gamma V_y\delta w(-a_{vy}dl) + \oint_\Gamma V_x\delta w(a_{vx}dl)$$
$$= 0$$

(2.2.28)

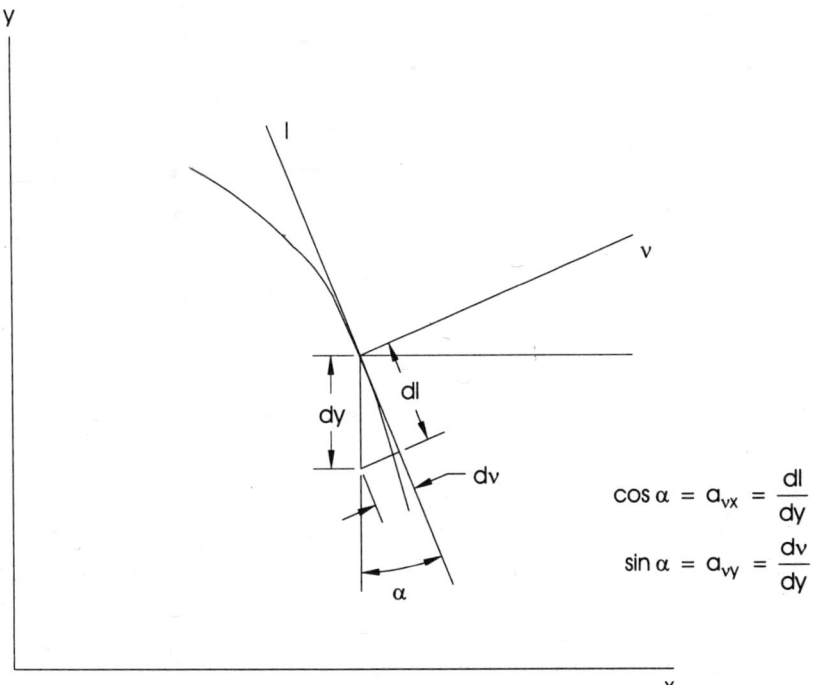

Fig. 2.2.2 Curved Boundary Relations

Regrouping the terms in the above

$$-\oint_\Gamma (M_x a_{vx}^2 + M_y a_{vy}^2 - 2M_{xy} a_{vx} a_{vy}) \delta w_{,v} dl$$
$$+ \oint_\Gamma (M_x a_{vx} a_{vy} - M_y a_{vx} a_{vy} + M_{xy} a_{vx}^2 - M_{xy} a_{vy}^2) \delta w_{,l} dl$$
$$+ \oint_\Gamma (V_y a_{vy} + V_x a_{vx}) \delta w \, dl$$
$$= 0$$

(2.2.29)

Define the following quantities

$$M_v = M_x a_{vx}^2 + M_y a_{vy}^2 - 2M_{xy} a_{vx} a_{vy}$$
$$M_{vl} = (M_y - M_x) a_{vx} a_{vy} - M_{xy}(a_{vx}^2 - a_{vy}^2)$$
$$V_v = V_y a_{vy} + V_x a_{vx}$$

(2.2.30)

so that the boundary conditions in (2.2.29) can be simplified as

$$-\oint_\Gamma M_\nu \delta w_{,\nu} dl - \oint_\Gamma M_{\nu l} \delta w_{,l} dl + \oint_\Gamma V_\nu \delta w\, dl = 0$$
(2.2.31)

The second integral in the above for a smooth boundary can be integrated by parts between any two points, say 1 and 2, to give

$$\int_1^2 M_{\nu l} \delta w_{,l} dl = \{M_{\nu l} \delta w\}_1^2 - \int_1^2 \frac{\partial M_{\nu l}}{\partial l} \delta w\, dl$$
(2.2.32)

For a smooth curve, the curved bracket term in the above equation (2.2.32) vanishes and equation (2.2.31) can be simplified as

$$-\oint_\Gamma M_\nu \delta w_{,\nu} dl + \oint_\Gamma \left(V_\nu + \frac{\partial M_{\nu l}}{\partial l}\right) \delta w\, dl = 0$$
(2.2.33)

Hence the differential equation and boundary conditions for this problem can be written from equations (2.2.21) and (2.2.33) as

$$\nabla^4 w = \frac{q}{D}$$
(2.2.34)

$$M_\nu = 0 \text{ or prescribe } \frac{\partial w}{\partial \nu}$$

$$V_\nu + \frac{\partial M_{\nu l}}{\partial l} = 0 \text{ or prescribe } w$$
(2.2.35)

2.3 INTEGRAL WITH TWO SPACE COORDINATES AND TIME AS INDEPENDENT PARAMETERS

Let the functional in (2.1.1) include the time as another independent parameter

$$F(x, y, t, w, w_{,t}, w_{,x}, w_{,xx}, w_{,y}, w_{,yy}, w_{,xy})$$
(2.3.1)

then, the extremum of the following integral should be considered.

$$I = \int_{t_1}^{t_2} \iint_R F(x, y, t, w, w_{,t}, w_{,x}, w_{,xx}, w_{,y}, w_{,yy}, w_{,xy}) dx\, dy\, dt$$
(2.3.2)

Variational Principles

We can now recognize the first variation of the above integral

$$\delta I = \int_{t_1}^{t_2} \iint_R \left\{ \begin{array}{l} \dfrac{\partial F}{\partial w_{,t}} \delta w_{,t} + \dfrac{\partial F}{\partial w} \delta w + \dfrac{\partial F}{\partial w_{,x}} \delta w_{,x} + \dfrac{\partial F}{\partial w_{,xx}} \delta w_{,xx} \\ + \dfrac{\partial F}{\partial w_{,y}} \delta w_{,y} + \dfrac{\partial F}{\partial w_{,yy}} \delta w_{,yy} + \dfrac{\partial F}{\partial w_{,xy}} \delta w_{,xy} \end{array} \right\} dx\,dy\,dt$$

(2.3.3)

The first term in the above can be integrated by parts with respect to time to give the following:

$$\int_{t_1}^{t_2} \iint_R \left(\dfrac{\partial F}{\partial w_{,t}} \delta w_{,t} \right) dx\,dy\,dt = \iint_R \left\{ \int_{t_1}^{t_2} \left(\dfrac{\partial F}{\partial w_{,t}} \delta w_{,t} \right) dt \right\} dx\,dy$$

$$= \iint_R \left\{ \left[\dfrac{\partial F}{\partial w_{,t}} \delta w \right]_{t_1}^{t_2} - \int_{t_1}^{t_2} \dfrac{d}{dt}\left(\dfrac{\partial F}{\partial w_{,t}} \right) \delta w\, dt \right\} dx\,dy$$

$$= - \int_{t_1}^{t_2} \iint_R \dfrac{d}{dt}\left(\dfrac{\partial F}{\partial w_{,t}} \right) \delta w\, dx\,dy\,dt$$

(2.3.4)

The remaining integrals in (2.3.3) have been considered earlier which led to the Euler-Lagrange equation in (2.1.18) and the boundary conditions (2.1.20). The boundary conditions given in equation (2.1.20) remain same for the extremum of the integral in equation (2.3.2), however, the differential equation should now include the additional term obtained in equation (2.3.4) above. Accordingly, the Euler-Lagrange equation for the present case is

$$\dfrac{\partial F}{\partial w} - \dfrac{\partial}{\partial x}\left(\dfrac{\partial F}{\partial w_{,x}} \right) - \dfrac{\partial}{\partial y}\left(\dfrac{\partial F}{\partial w_{,y}} \right) + \dfrac{\partial^2}{\partial x^2}\left(\dfrac{\partial F}{\partial w_{,xx}} \right)$$

$$+ \dfrac{\partial^2}{\partial y^2}\left(\dfrac{\partial F}{\partial w_{,yy}} \right) + \dfrac{\partial^2}{\partial x \partial y}\left(\dfrac{\partial F}{\partial w_{,xy}} \right) - \dfrac{d}{dt}\left(\dfrac{\partial F}{\partial w_{,t}} \right)$$

$$= 0$$

(2.3.5)

2.4 LAGRANGE METHOD

In chapter 5 of *Advanced Theory of Vibration* book, we have studied the variational principles for one dimensional structures and applied the underlying principles to evolve solution procedures of the statics and dynamics problems. We follow similar ideas here to establish different numerical procedures and obtain solutions for two dimensional structures. First we discuss the Lagrange method in a similar manner to that used in section 5.13 of *Advanced Theory of Vibration* book.

26 Dynamics of Plates

As in the case of one dimensional structures, the approximation is made in assuming shape functions in series form and applying the Euler-Lagrange equations. For this purpose, let us start with the problem governed by the extremum of the integral given in equation (2.2.13)

$$I = \iint \left\{ \frac{D}{2}\left[(\nabla^2 w)^2 + 2(1-v)\{w_{,xy}^2 - w_{,xx}w_{,yy}\}\right] - qw \right\} dx\,dy \tag{2.2.13}$$

subjected to boundary conditions given in equation (2.1.25). Since $w = 0$ on all the edges, we note that $w_{,yy} = 0$ on the edges $x = 0$ and $x = a$ and $w_{,xx} = 0$ on the edges $y = 0$ and $y = b$ and the boundary conditions simplify to

$$w = w_{,xx} = 0 \text{ on the edge } x = 0 \text{ and } x = a$$

$$w = w_{,yy} = 0 \text{ on the edge } y = 0 \text{ and } y = b \tag{2.1.25a}$$

Since Euler-Lagrange equations are derived from the extremum conditions of the integral together with the boundary conditions, it is essential that the shape functions assumed satisfy the above boundary conditions, let us choose

$$w(x,y) = \sum_{m=1}^{\infty}\sum_{n=1}^{\infty} w_{mn} \sin\frac{m\pi x}{a} \sin\frac{n\pi y}{b} \tag{2.4.1}$$

To illustrate the procedure, let us take one term approximation, i.e.,

$$w(x,y) = w_{11}\sin\frac{\pi x}{a}\sin\frac{\pi y}{b} \tag{2.4.2}$$

Taking the derivatives

$$w_{,xx} = w_{11}\frac{\pi^2}{a^2}\sin\frac{\pi x}{a}\sin\frac{\pi y}{b}$$

$$w_{,yy} = w_{11}\frac{\pi^2}{b^2}\sin\frac{\pi x}{a}\sin\frac{\pi y}{b}$$

$$w_{,xy} = w_{11}\frac{\pi^2}{ab}\cos\frac{\pi x}{a}\cos\frac{\pi y}{b}$$

$$\tag{2.4.3}$$

Substituting the above in (2.2.13), we get

$$I = w_{11}^2 \pi^4 \frac{D}{2} \int_0^a \int_0^b \left[\left(\frac{a^2+b^2}{a^2b^2}\right)^2 \sin^2\frac{\pi x}{a} \sin^2\frac{\pi y}{b} + \frac{2(1-v)}{a^2b^2}\left\{\cos^2\frac{\pi x}{a}\cos^2\frac{\pi y}{b} - \sin^2\frac{\pi x}{a}\sin^2\frac{\pi y}{b}\right\}\right] dx\, dy$$

$$- \int_0^a \int_0^b q w_{11} \sin\frac{\pi x}{a} \sin\frac{\pi y}{b} dx\, dy$$

(2.4.4)

The integrals in the above equation (2.4.4) can be evaluated to give

$$\int_0^a \int_0^b \sin\frac{\pi x}{a} \sin\frac{\pi y}{b} dx\, dy = \frac{4ab}{\pi^2}$$

$$\int_0^a \int_0^b \sin^2\frac{\pi x}{a} \sin^2\frac{\pi y}{b} dx\, dy = \frac{ab}{4}$$

$$\int_0^a \int_0^b \cos^2\frac{\pi x}{a} \cos^2\frac{\pi y}{b} dx\, dy = \frac{ab}{4}$$

(2.4.5)

Upon substitution in (2.4.4) and simplifying we get

$$I = \frac{D}{8}\pi^4 \frac{(a^2+b^2)^2}{a^3b^3} w_{11}^2 - \frac{4qab}{\pi^2} w_{11}$$

(2.4.6)

Equation (2.1.18) for the present case simplifies to

$$\frac{\partial I}{\partial w_{11}} = 0$$

which gives

$$w_{11} = \frac{16qa^4b^4}{\pi^6 D(a^2+b^2)^2}$$

giving the one term solution as

$$w = \frac{16qa^4b^4}{\pi^6 D(a^2+b^2)^2} \sin\frac{\pi x}{a} \sin\frac{\pi y}{b}$$

(2.4.7)

The above procedure is akin to Ritz minimization procedure, since the minimum value of I is found with respect to w_{11}. Now, let us consider the functional

$$I = \iint \left\{ \tfrac{1}{2} m \dot{w}^2 - \tfrac{D}{2} \left[(\nabla^2 w)^2 + 2(1-v)\{w_{,xy}^2 - w_{,xx} w_{,yy}\} \right] \right\} dx\, dy$$

(2.4.8)

which represents free vibration problem of a plate with a constant mass distribution m. w is now a function of x, y and t. For the same boundary conditions consider in the problem above, we can use variables separable method and assume the solution with one term approximation to consist of

$$w(x, y, t) = w_{11}(t) \sin \frac{\pi x}{a} \sin \frac{\pi y}{b}$$

(2.4.9)

Following similar steps as in the previous problem, we get

$$I = \frac{mab}{8} \dot{w}_{11}^2 - \frac{D}{8} \pi^4 \frac{(a^2 + b^2)^2}{a^3 b^3} w_{11}^2$$

(2.4.10)

Equation (2.3.5) for this case simplifies to

$$\frac{\partial I}{\partial w_{11}} - \frac{d}{dt}\left(\frac{\partial I}{\partial w_{11,t}}\right) = 0$$

(2.3.5a)

and upon application, we get

$$\frac{mab}{4} \ddot{w}_{11} + \frac{D}{4} \pi^4 \frac{(a^2 + b^2)^2}{a^3 b^3} w_{11} = 0$$

(2.4.11)

Hence, the approximate value of the first mode free vibration frequency of this problem is

$$p = \sqrt{\frac{\pi^4 D (a^2 + b^2)^2}{m a^4 b^4}} \quad \text{rad/s}$$

(2.4.12)

2.5 RITZ METHOD

In this method also, we begin with an assumed shape function of the form

$$w(x,y,t) = \sum_{m=0}^{\infty} \sum_{n=0}^{\infty} w_{mn} f_m(x) g_n(y) q_{mn}(t)$$
(2.5.1)

The time parameter can be eliminated by either averaging the functional over a natural period of time or by employing maximum energy principle (see section 5.14 of *Advanced Theory of Vibration* book). If we employ the averaging energy principle the functional of a system is reduced to

$$\overline{F} = \int_0^{2\pi/p} F dt$$
(2.5.2)

where p is the frequency of motion. If the arbitrary parameters in (2.5.1) are w_{mn}, then using the assumed mode shapes we can integrate the functional over the region R, which reduces to

$$F = F(w_{mn})$$
(2.5.3)

In Ritz method we minimize the above expression with respect to the arbitrary parameters w_{mn} i.e.,

$$\frac{\partial F}{\partial w_{mn}} = 0$$
(2.5.4)

Alternatively we can use maximum energy principle and write (2.5.1) for free vibratory motion as

$$w(x,y,t) = \sum_{m=0}^{\infty} \sum_{n=0}^{\infty} w_{mn} f_m(x) g_n(y) \sin pt$$
(2.5.5)

then

$$w_{,t} = \sum_{m=0}^{\infty} \sum_{n=0}^{\infty} w_{mn} f_m(x) g_n(y) p \cos pt$$

$$\hat{w}_{,t} = \sum_{m=0}^{\infty} \sum_{n=0}^{\infty} w_{mn} f_m(x) g_n(y) p$$

$$\hat{w}_{,x} = \sum_{m=0}^{\infty} \sum_{n=0}^{\infty} w_{mn} \{f_m(x)\}_{,x} g_n(y)$$

...

(2.5.6)

Substituting the above in a given functional and upon integration over the region R, we get the functional of the form (2.5.3). This procedure for the free vibration problem whose functional is given in (2.4.8) of previous section

$$I = \iint \left\{ \frac{1}{2} m \dot{w}^2 - \frac{D}{2}\left[(\nabla^2 w)^2 + 2(1-v)\{w_{,xy}^2 - w_{,xx}w_{,yy}\}\right] \right\} dx\, dy \tag{2.4.8}$$

is illustrated here for a one term approximation.

$$w(x,y,t) = w_{11} \sin \frac{\pi x}{a} \sin \frac{\pi y}{b} \sin pt \tag{2.5.7}$$

The required derivatives with maximum values are

$$\hat{w} = w_{11} \sin \frac{\pi x}{a} \sin \frac{\pi y}{b}$$

$$\hat{\dot{w}} = w_{11} p \sin \frac{\pi x}{a} \sin \frac{\pi y}{b}$$

$$\hat{w}_{,xx} = w_{11} \frac{\pi^2}{a^2} \sin \frac{\pi x}{a} \sin \frac{\pi y}{b}$$

$$\hat{w}_{,yy} = w_{11} \frac{\pi^2}{b^2} \sin \frac{\pi x}{a} \sin \frac{\pi y}{b}$$

$$\hat{w}_{,xy} = w_{11} \frac{\pi^2}{ab} \cos \frac{\pi x}{a} \cos \frac{\pi y}{b}$$

$$\tag{2.5.8}$$

With the help of integrals in (2.4.5), the functional in (2.4.8) with maximum values can be evaluated as

$$I = \frac{D}{8} \pi^4 \frac{(a^2+b^2)^2}{a^3 b^3} w_{11}^2 - \frac{1}{2} m \frac{ab}{4} p^2 w_{11}^2 \tag{2.5.9}$$

Minimizing with respect to w_{11}, i.e.,

$$\frac{\partial I}{\partial w_{11}} = 0 \tag{2.5.10}$$

we get

$$p = \sqrt{\frac{\pi^4 D(a^2+b^2)^2}{ma^4 b^4}} \text{ rad/s}$$

2.6 GALERKIN METHOD

We began our discussion in this chapter with the integral given in (2.1.2)

$$I = \iint_R F(x, y, w, w_{,x}, w_{,xx}, w_{,y}, w_{,yy}, w_{,xy}) dx\, dy \tag{2.1.2}$$

This integral has an extremum when the following is satisfied as given in equation (2.1.17a).

$$\iint_R \left[\begin{array}{c} \dfrac{\partial F}{\partial w} - \dfrac{\partial}{\partial x}\left(\dfrac{\partial F}{\partial w_{,x}}\right) - \dfrac{\partial}{\partial y}\left(\dfrac{\partial F}{\partial w_{,y}}\right) + \dfrac{\partial^2}{\partial x^2}\left(\dfrac{\partial F}{\partial w_{,xx}}\right) \\ + \dfrac{\partial^2}{\partial y^2}\left(\dfrac{\partial F}{\partial w_{,yy}}\right) + \dfrac{\partial^2}{\partial x \partial y}\left(\dfrac{\partial F}{\partial w_{,xy}}\right) \end{array} \right] \eta dx\, dy$$

$$+ \oint_\Gamma \left[\dfrac{\partial F}{\partial w_{,x}} - \dfrac{\partial}{\partial x}\left(\dfrac{\partial F}{\partial w_{,xx}}\right)\right]\eta\, dy - \oint_\Gamma \left[\dfrac{\partial F}{\partial w_{,y}} - \dfrac{\partial}{\partial y}\left(\dfrac{\partial F}{\partial w_{,yy}}\right)\right]\eta dx$$

$$+ \oint_\Gamma \left(\eta_{,x} \dfrac{\partial F}{\partial w_{,xx}}\right) dy - \oint_\Gamma \left(\eta_{,y} \dfrac{\partial F}{\partial w_{,yy}}\right) dx$$

$$+ \left\{ \begin{array}{c} \oint_\Gamma \left(\eta_{,y} \dfrac{\partial F}{\partial w_{,xy}}\right) dy + \oint_\Gamma \eta \dfrac{\partial}{\partial x}\left(\dfrac{\partial F}{\partial w_{,xy}}\right) dx \\ - \oint_\Gamma \left(\eta_{,x} \dfrac{\partial F}{\partial w_{,xy}}\right) dx - \oint_\Gamma \eta \dfrac{\partial}{\partial y}\left(\dfrac{\partial F}{\partial w_{,xy}}\right) dy \end{array} \right\}_{\text{either of the rows}} = 0$$

(2.1.17a)

We can represent the above as

$$\iint_R [Lw(x,y)]\,\eta(x,y) dx\, dy$$

+ A set of line integrals representing boundary conditions

(2.6.1)

where L is a differential operator according to equation (2.1.17a). We can identify

$$Lw(x,y) = 0 \tag{2.6.2}$$

as the governing differential equation of the system subjected to the boundary conditions given by various line integrals. The usual way of obtaining the solution of the system for an extremum condition is to solve the differential equation with the corresponding boundary conditions. Such a solution is not possible except in very simple cases, e.g., rectangular plates with constant thickness supported on knife edges (simply supported). Where an exact solution for the differential equation is not possible, we resort to approximate methods. Galerkin's approximation is directly obtained from equation (2.1.17a). Here, we assume as approximate solution $\tilde{w}(x,y)$ that satisfies all the boundary conditions (making all the line integrals zero) and obtain the error $\in$ in the differential equation

$$\in = L\tilde{w}(x,y) \tag{2.6.3}$$

In order to satisfy equation (2.1.17a), we require

$$\iint_R [L\tilde{w}(x,y)]\,\eta(x,y)\,dx\,dy = 0 \tag{2.6.4}$$

In Galerkin method, the function $\eta(x,y)$ is chosen as the assumed shape function it self, i.e.,

$$\iint_R [L\tilde{w}(x,y)]\,\tilde{w}(x,y)\,dx\,dy = 0 \tag{2.6.5}$$

The above method is illustrated in the following example.

Example
Consider the differential equation

$$D\nabla^4 w = q \tag{2.1.23}$$

with the boundary conditions

$$w = 0; \quad \frac{\partial w}{\partial x} = 0 \text{ on the edge } x = -\frac{1}{2}a \text{ and } x = \frac{1}{2}a$$

$$w = 0; \quad \frac{\partial w}{\partial y} = 0 \text{ on the edge } y = -\frac{1}{2}a \text{ and } y = \frac{1}{2}a$$

(2.1.24)

Let us take a one term approximation that satisfies the above boundary conditions

$$\tilde{w}(x,y) = C(X^4 - 2X^2 + 1)(Y^4 - 2Y^2 + 1)$$

(2.6.6)

where $X = \frac{2x}{a}$ and $Y = \frac{2y}{a}$. The following derivatives are first obtained.

$$\tilde{w}_{,XXXX} = 24C(Y^4 - 2Y^2 + 1)$$
$$\tilde{w}_{,YYYY} = 24C(X^4 - 2X^2 + 1)$$
$$\tilde{w}_{,XXYY} = 16C(3X^2 - 1)(3Y^2 - 1)$$

(2.6.7)

Substituting the above in the differential equation (2.1.23) and noting that $dx = \frac{1}{2}a \, dX$ and $dy = \frac{1}{2}a \, dY$, we get the error

$$\epsilon = \frac{128CD}{a^4} \left\{ \begin{array}{c} 3(Y^4 - 2Y^2 + 1) + 4(3X^2 - 1)(3Y^2 - 1) \\ +3(X^4 - 2X^2 + 1) \end{array} \right\} - q$$

(2.6.8)

According to Galerkin method

$$\int_{-1}^{1} \int_{-1}^{1} \left[\frac{128CD}{a^4} \left\{ \begin{array}{c} 3(Y^4 - 2Y^2 + 1) + 4(3X^2 - 1)(3Y^2 - 1) \\ +3(X^4 - 2X^2 + 1) \end{array} \right\} - q \right]$$
$$\times (X^4 - 2X^2 + 1)(Y^4 - 2Y^2 + 1) dX \, dY$$
$$= 0$$

(2.6.9)

Upon integration, we get

$$6.6873 C = \frac{qa^4}{112.5D} \qquad (2.6.10)$$

Therefore,

$$\tilde{w}(x,y) = \frac{qa^4}{752.32D}(X^4 - 2X^2 + 1)(Y^4 - 2Y^2 + 1) \qquad (2.6.11)$$

The maximum deflection at the center of the plate $X=0$ and $Y=0$ is given by

$$w_{max} = 0.001329 \frac{qa^4}{D} \qquad (2.6.12)$$

2.7 LEVY'S METHOD

Levy's method is an improvement over Lagrange method of section 2.5. In Lagrange, Ritz and Galerkin methods, we assumed the approximate solution in series form of the type

$$w(x,y) = \sum_{m=0}^{\infty} \sum_{n=0}^{\infty} w_{mn} f_m(x) g_n(y) \qquad (2.7.1)$$

Both the functions $f_m(x)$ and $g_n(y)$ are assumed in a suitable form, e.g., trigonometric or polynomial functions and an approximate numerical procedure is obtained consistent with the extremum property of the functional of the system. In Levy's method, one of these functions is assumed in a suitable form and the other function is retained as a general function that will be determined in a closed form. The procedure is best illustrated by an example.

Consider the functional given in equation (2.2.13)

$$I = \iint \left\{ \frac{D}{2} \left[(\nabla^2 w)^2 + 2(1-v)\{w_{,xy}^2 - w_{,xx} w_{,yy}\} \right] - qw \right\} dx\,dy \qquad (2.2.13)$$

subjected to the following boundary conditions

Variational Principles 35

$$w = w_{,xx} = 0 \text{ on the edge } x = 0 \text{ and } x = a$$
$$w = w_{,y} = 0 \text{ on the edge } y = 0 \text{ and } y = b$$

(2.7.1)

For the edge conditions on $x = 0$ and $x = a$, we can write down a function $\sin\frac{m\pi x}{a}$ and keep the function open for the edge conditions on $y = 0$ and $y = b$ as $Y_m(y)$, such that the solution can be assumed as

$$w(x, y) = \sum_{m=1}^{\infty} Y_m(y) \sin\frac{m\pi x}{a}$$

(2.7.2)

The functional can now be written as

$$I = \frac{D}{2} \int_0^b \int_0^a \left[\begin{array}{c} \left[\sum_{m=1}^{\infty}\left\{-Y_m\left(\frac{m\pi}{a}\right)^2 \sin\frac{m\pi x}{a} + Y_m'' \sin\frac{m\pi x}{a}\right\}\right]^2 \\ +2(1-v)\left[\sum_{m=1}^{\infty} Y_m'\left(\frac{m\pi}{a}\right)\cos\frac{m\pi x}{a}\right]^2 - 2(1-v) \\ \times \left[\sum_{m=1}^{\infty} -Y_m\left(\frac{m\pi}{a}\right)^2 \sin\frac{m\pi x}{a}\right]\left[\sum_{m'=1}^{\infty} Y_{m'}'' \sin\frac{m'\pi x}{a}\right] \end{array} \right] dx\, dy$$

$$- \int_0^b \int_0^a q\left(\sum_{m=1}^{\infty} Y_m \sin\frac{m\pi x}{a}\right) dx\, dy$$

(2.7.3)

Noting that

$$\int_0^b \sin\frac{n\pi y}{b} \sin\frac{k\pi y}{b} dy = \delta_{nk}\frac{b}{2}$$

$$\int_0^a \sin\frac{n\pi x}{a} \sin\frac{k\pi x}{a} dx = \delta_{nk}\frac{a}{2}$$

(2.7.4)

equation (2.7.3) after integration with respect to x becomes

$$I = \frac{Da}{4} \int_0^b \left[\begin{array}{c} \left[\sum_{m=1}^{\infty}\left\{-Y_m\left(\frac{m\pi}{a}\right)^2 + Y_m''\right\}\right]^2 \\ +2(1-v)\left\{\left[\sum_{m=1}^{\infty} Y_m'^2\left(\frac{m\pi}{a}\right)^2\right] - \left[\sum_{m=1}^{\infty} -Y_m Y_m''\left(\frac{m\pi}{a}\right)^2\right]\right\} \\ - \sum_{m=1}^{\infty} 2q_m Y_m \end{array} \right] dy$$

(2.7.5)

where

$$q_m = \frac{2}{aD}\int_0^a q \sin \frac{m\pi x}{a} dx$$

$$= \frac{2q}{m\pi D}(1-\cos m\pi)$$

$$= \frac{4q}{m\pi D} \quad m \text{ odd}$$

(2.7.6)

The integral in equation (2.7.5) above is now a functional $F(Y_m, Y'_m, Y''_m, y)$ and for it to be an extremum, we can use the appropriate Euler-Lagrange equations given by

$$\frac{\partial F}{\partial Y_m} - \frac{\partial}{\partial y}\left(\frac{\partial F}{\partial Y'_m}\right) + \frac{\partial^2}{\partial y^2}\left(\frac{\partial F}{\partial Y''_m}\right) = 0 \quad m = 1, 2, \ldots$$

(2.1.18a)

Applying the above to (2.7.5), we get

$$Y_m'''' - 2\left(\frac{m\pi}{a}\right)^2 Y_m'' + \left(\frac{m\pi}{a}\right)^4 Y_m = \frac{4q}{m\pi D} \quad m = 1, 3, 5 \ldots$$

(2.7.7)

The second set of boundary conditions in (2.7.1) are written as

$$Y_m = Y'_m = 0 \text{ at } y = 0 \text{ and } y = b$$

(2.7.8)

We should take into account both the complementary function and particular integral of equation (2.7.7) in finding the solution Y_m. The complementary function can be taken in the form of e^{sy}, which gives the characteristic equation

$$s^4 - 2\left(\frac{m\pi}{a}\right)^2 s^2 + \left(\frac{m\pi}{a}\right)^4 = 0$$

$$\left[s^2 - \left(\frac{m\pi}{a}\right)^2\right]^2 = 0$$

(2.7.9)

The double roots of the above characteristic equation are

$$s = \pm \frac{m\pi}{a}$$

(2.7.10)

Variational Principles 37

and the complementary function is

$$(Y_m)_C = C_1 e^{\frac{m\pi}{a}y} + C_2 e^{-\frac{m\pi}{a}y} + C_3 y e^{\frac{m\pi}{a}y} + C_4 y e^{-\frac{m\pi}{a}y}$$

(2.7.11)

The particular integral is

$$(Y_m)_P = \frac{4qa^4}{m^5\pi^5 D}$$

(2.7.12)

The complete solution of equation (2.7.7) is therefore

$$Y_m = C_1 e^{\frac{m\pi}{a}y} + C_2 e^{-\frac{m\pi}{a}y} + C_3 y e^{\frac{m\pi}{a}y} + C_4 y e^{-\frac{m\pi}{a}y} + \frac{4qa^4}{m^5\pi^5 D}$$

(2.7.13)

and

$$Y'_m = C_1 \frac{m\pi}{a} e^{\frac{m\pi}{a}y} - C_2 \frac{m\pi}{a} e^{-\frac{m\pi}{a}y} + C_3 e^{\frac{m\pi}{a}y}\left(1 + \frac{m\pi}{a}y\right)$$
$$+ C_4 e^{-\frac{m\pi}{a}y}\left(1 - \frac{m\pi}{a}y\right)$$

(2.7.14)

We now use the four boundary conditions in (2.7.8) to determine the constants of integration in the above equation.

At $y=0$

$$C_1 + C_2 = -\frac{4qa^4}{m^5\pi^5 D}$$

$$C_1 \frac{m\pi}{a} - C_2 \frac{m\pi}{a} + C_3 + C_4 = 0$$

(2.7.15)

At $y=b$

$$C_1 e^{\frac{m\pi}{a}b} + C_2 e^{-\frac{m\pi}{a}b} + C_3 b e^{\frac{m\pi}{a}b} + C_4 b e^{-\frac{m\pi}{a}b}$$
$$= -\frac{4qa^4}{m^5\pi^5 D}$$

$$C_1 \frac{m\pi}{a} e^{\frac{m\pi}{a}b} - C_2 \frac{m\pi}{a} e^{-\frac{m\pi}{a}b} + C_3 e^{\frac{m\pi}{a}b}\left(1 + \frac{m\pi}{a}b\right)$$
$$+ C_4 e^{-\frac{m\pi}{a}b}\left(1 - \frac{m\pi}{a}b\right) \tag{2.7.16}$$
$$= 0$$

Let

$$a_m = -\frac{4qa^4}{m^5\pi^5 D} \quad \text{and} \quad \beta_m = m\pi\frac{b}{a} \tag{2.7.17}$$

then equations (2.7.15) and (2.7.16) can be expressed as

$$\begin{bmatrix} 1 & 1 & 0 & 0 \\ \frac{\beta_m}{b} & -\frac{\beta_m}{b} & 1 & 1 \\ e^{\beta_m} & e^{-\beta_m} & be^{\beta_m} & be^{-\beta_m} \\ \frac{\beta_m}{b}e^{\beta_m} & -\frac{\beta_m}{b}e^{-\beta_m} & e^{\beta_m}(1+\beta_m) & e^{-\beta_m}(1-\beta_m) \end{bmatrix} \begin{Bmatrix} C_1 \\ C_2 \\ C_3 \\ C_4 \end{Bmatrix} = \begin{Bmatrix} a_m \\ 0 \\ a_m \\ 0 \end{Bmatrix}$$

$$\tag{2.7.18}$$

Let

$$\Delta = 2 + 4\beta_m^2 - e^{2\beta_m} - e^{-2\beta_m}$$

then

$$C_1 = a_m[1 + 2\beta_m + 2\beta_m^2 - e^{\beta_m}(1+\beta_m) + e^{-\beta_m}(1-\beta_m) - e^{-2\beta_m}] \div \Delta$$

$$C_2 = a_m[1 - 2\beta_m + 2\beta_m^2 + e^{\beta_m}(1+\beta_m) - e^{-\beta_m}(1-\beta_m) - e^{2\beta_m}] \div \Delta$$

$$C_3 = \frac{a_m \beta_m}{b}[-1 - 2\beta_m + e^{\beta_m} + e^{-\beta_m}(2\beta_m - 1) + e^{-2\beta_m}] \div \Delta$$

$$C_4 = \frac{a_m \beta_m}{b}[1 - 2\beta_m + e^{\beta_m}(2\beta_m + 1) - e^{-\beta_m} - e^{2\beta_m}] \div \Delta$$

$$\tag{2.7.19}$$

The solution for w can now be written down from (2.7.2), (2.7.13) and (2.7.17)

$$w(x,y) = \sum_{m=1,3,\ldots}^{\infty} \left[C_1 e^{\frac{m\pi}{a}y} + C_2 e^{-\frac{m\pi}{a}y} + C_3 y e^{\frac{m\pi}{a}y} + C_4 y e^{-\frac{m\pi}{a}y} - a_m \right] \sin \frac{m\pi x}{a}$$

$$\tag{2.7.20}$$

2.8 KANTOROVICH METHOD

The method proposed by Kantorovich is similar to that of Levy, only difference is Galerkin postulation is used as the basis instead of Lagrange method. To illustrate this procedure, consider the differential equation

$$D \nabla^4 w = q \tag{2.1.23}$$

with the boundary conditions

$$w = 0; \quad \frac{\partial w}{\partial x} = 0 \text{ on the edge } x = -\frac{1}{2}a \text{ and } x = \frac{1}{2}a$$

$$w = 0; \quad \frac{\partial w}{\partial y} = 0 \text{ on the edge } y = -\frac{1}{2}a \text{ and } y = \frac{1}{2}a \tag{2.1.24}$$

The solution is assumed in the form

$$w(x,y) = \sum_{m=1}^{\infty} Y_m(y) X_m(x) \tag{2.8.1}$$

in which $X_m(x)$ is taken in the form of a suitable function satisfying the boundary conditions and $Y_m(y)$ is retained in a general form whose solution is sought in a closed form. Consider a one term approximation for the above problem and write (2.8.1) as

$$\tilde{w} = Y(y)(a^4 - 8a^2x^2 + 16x^4) \tag{2.8.2}$$

The Galerkin integral is now written as

$$\int_{-a/2}^{a/2} \left\{ \int_{-a/2}^{a/2} \left[\nabla^4 \tilde{w} - \frac{q}{D} \right] (a^4 - 8a^2x^2 + 16x^4) dx \right\} Y(y) dy = 0 \tag{2.8.3}$$

The above relation is satisfied, if we choose

$$\int_{-a/2}^{a/2} \left[\nabla^4 \tilde{w} - \frac{q}{D} \right] (a^4 - 8a^2x^2 + 16x^4) dx = 0 \tag{2.8.4}$$

40 Dynamics of Plates

Noting

$$\tilde{w}_{,xx} = Y(y)(-16a^2 + 192x^2)$$
$$\tilde{w}_{,xxyy} = Y''(y)(-16a^2 + 192x^2)$$
$$\tilde{w}_{,xxxx} = 384Y(y)$$
$$\tilde{w}_{,yyyy} = Y''''(y)(a^4 - 8a^2x^2 + 16x^4)$$

(2.8.5)

equation (2.8.3) becomes

$$\int_{-a/2}^{a/2} \left[384Y + 32Y''(-a^2 + 12x^2) + Y''''(a^4 - 8a^2x^2 + 16x^4) - \frac{q}{D} \right]$$
$$\times (a^4 - 8a^2x^2 + 16x^4)dx$$
$$= 0$$

(2.8.6)

Upon integration with respect to x above, we get the following differential equation for Y

$$384 \times \frac{8a^5}{15} Y + 32 \times \left(-\frac{32a^7}{105}\right) Y'' + \frac{128a^9}{315} Y'''' - \frac{8a^5}{15} \frac{q}{D} = 0$$

(2.8.7)

i.e.

$$\frac{a^4}{504} Y'''' - \frac{a^2}{21} Y'' + Y = \frac{q}{384D}$$

For the complementary function, we can write the solution as $e^{\lambda \frac{y}{a}}$ and find the roots of the characteristic equation as in the previous section. This gives

$$Y_C = C_1 e^{(4.1503+i2.2858)\frac{y}{a}} + C_2 e^{(4.1503-i2.2858)\frac{y}{a}} + C_3 e^{(-4.1503+i2.2858)\frac{y}{a}}$$
$$+ C_4 e^{(-4.1503-i2.2858)\frac{y}{a}}$$

(2.8.8)

The particular integral is

$$Y_P = \frac{q}{384D}$$

(2.8.9)

In view of the symmetry of the deflection surface, the total solution is given by

$$Y = \frac{q}{384D}\left(\begin{array}{l} 1 + C_1 \cosh 4.1503\frac{y}{a} \cos 2.2858\frac{y}{a} \\ + C_2 \sinh 4.1503\frac{y}{a} \sin 2.2858\frac{y}{a} \end{array} \right)$$

(2.8.10)

From the boundary conditions $Y = Y' = 0$ at $y = \pm\frac{1}{2}a$, we find

$$C_1 = -0.5023$$
$$C_2 = -0.04396$$

(2.8.11)

and

$$Y = \frac{q}{384D}\left(\begin{array}{l} 1 - 0.5023 \cosh 4.1503\frac{y}{a} \cos 2.2858\frac{y}{a} \\ -0.04396 \sinh 4.1503\frac{y}{a} \sin 2.2858\frac{y}{a} \end{array} \right)$$

(2.8.10)

Therefore, the solution for the deflected surface is

$$w = \frac{q}{384D}\left(\begin{array}{l} 1 - 0.5023 \cosh 4.1503\frac{y}{a} \cos 2.2858\frac{y}{a} \\ -0.04396 \sinh 4.1503\frac{y}{a} \sin 2.2858\frac{y}{a} \end{array} \right)(a^4 - 8a^2x^2 + 16x^4)$$

(2.8.11)

The maximum deflection occurs at the center of the plate

$$w_{max} = \frac{0.4977qa^4}{384D}$$
$$= 0.001296\frac{qa^4}{D}$$

(2.8.12)

3

Rectangular Plates

In *Advanced Theory of Vibration* book, one dimensional structures were considered, i.e., structures which can be defined by one independent space parameter alone. Lateral bending of beams is an example of one dimension structures. In such an idealization, the governing differential equation recognizes only two geometrical properties, viz., the area and the second moment of area of the beam at any location along the neutral axis. We accept this idealization for structures whose cross-sectional dimensions are very small compared to the length coordinate. However, when one of the cross-sectional dimensions is comparable to the length of the structure, we need two independent space coordinates to describe its static and dynamic behavior. Such a structure is called a *plate*. The thickness of a plate is very small compared to that of the length and width dimensions.

In the previous chapter, we considered potential functionals with two independent space coordinates (for static deformations) or three independent parameters, two space coordinates and time parameter (for dynamical problems). The procedures outlined in chapter 2 are applied here to understand the behavior of rectangular shaped plates. We begin with a simple extension of classical beam theory to plates which are very long compared to width dimension and then extend to general case of thin plates.

3.1 BENDING OF LONG RECTANGULAR PLATES

Consider a long rectangular plate in y direction with width L in x direction and thickness h. The middle surface of this plate before deformation is shown in Fig. 3.1.1. Since the plate is long in y direction, we can assume that the plate away from the edges parallel to x axis deforms to a cylindrical surface whose axis is parallel to y direction. Then, we can consider a strip of unit width in y direction far away from the edges parallel to *x* axis to represent the plate deformation. The deformation of this strip can be considered to be similar to that of bending of a beam.

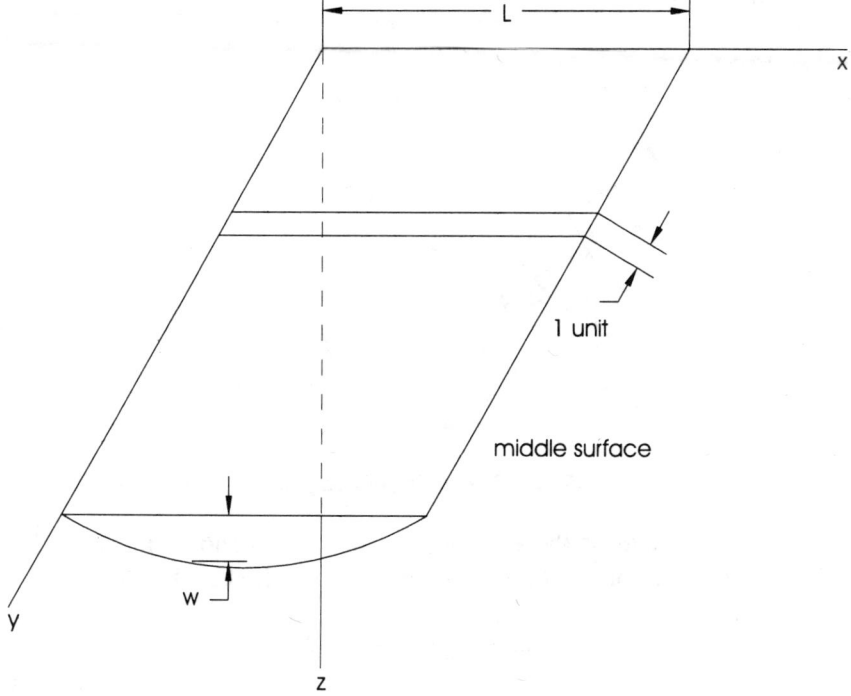

Fig. 3.1.1 A Long Rectangular Plate

The deformed surface of the strip is shown in Fig. 3.1.2, wherein, we assume simple Euler-Bernoulli type of bending, viz., the cross-section of the strip remains plane during bending and undergoes a rotation about the neutral axis. Then, the displacement field can be expressed as

$$u_x = -zw_{,x}$$
$$u_y = 0$$
$$u_z = w$$

(3.1.1)

The strain field is therefore

$$\varepsilon_{xx} = -zw_{,xx}$$
$$\varepsilon_{yy} = 0$$
$$\varepsilon_{zz} = 0$$

(3.1.2)

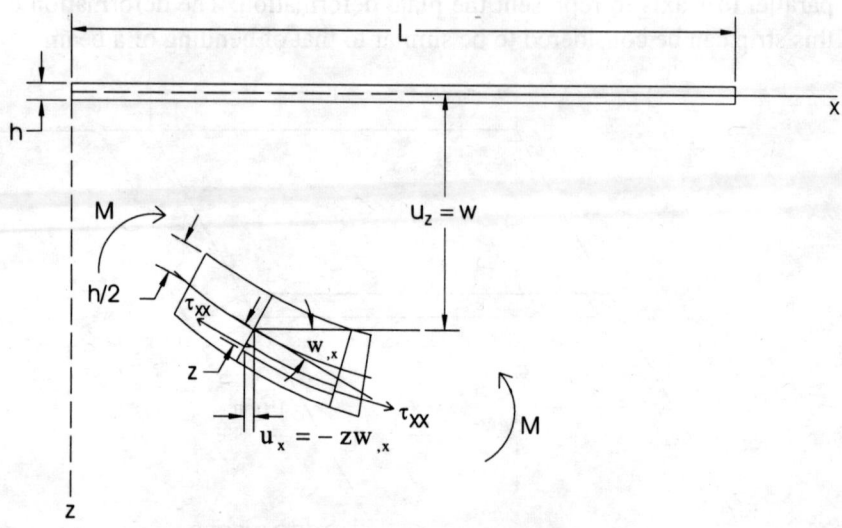

Fig. 3.1.2 Bending of Long Plate

The stress field of the element at a depth z is shown in Fig. 3.1.2. Under this condition of plane stress, we can use Hooke's law to write

$$\varepsilon_{yy} = \frac{\tau_{yy}}{E} - \nu \frac{\tau_{xx}}{E} = 0 \tag{3.1.3}$$

i.e.,

$$\tau_{yy} = \nu \tau_{xx} \tag{3.1.4}$$

where ν is Poisson's ratio. Therefore

$$\begin{aligned}\varepsilon_{xx} &= \frac{\tau_{xx}}{E} - \nu \frac{\tau_{yy}}{E} \\ &= \frac{\tau_{xx}}{E}(1 - \nu^2)\end{aligned} \tag{3.1.5}$$

Hence, the normal stress τ_{xx} is given by

$$\begin{aligned}\tau_{xx} &= \frac{E}{(1-\nu^2)} \varepsilon_{xx} \\ &= -\frac{Ez}{(1-\nu^2)} w_{,xx}\end{aligned} \tag{3.1.6}$$

Rectangular Plates

As in the case of Euler-Bernoulli bending of beams, we note that the normal stress component τ_{xx} gives rise to a *bending moment M* given by

$$M = \int_{-h/2}^{h/2} \tau_{xx} z \, dz$$

$$= -\int_{-h/2}^{h/2} \frac{Ez^2}{(1-v^2)} w_{,xx} \, dz$$

$$= -\frac{Eh^3}{12(1-v^2)} w_{,xx}$$

(3.1.7)

The above relation can be written down in a similar manner to that of a beam,

$$M = -Dw_{,xx}$$

(3.1.8)

where

$$D = \frac{Eh^3}{12(1-v^2)}$$

(3.1.9)

is called *flexural rigidity of the plate* corresponding to EI of a beam.

Referring to Fig. 3.1.3, we can write the radius ρ of the cylindrical middle surface after bending as

$$\varrho(-w_{,xx} dx) = dx$$

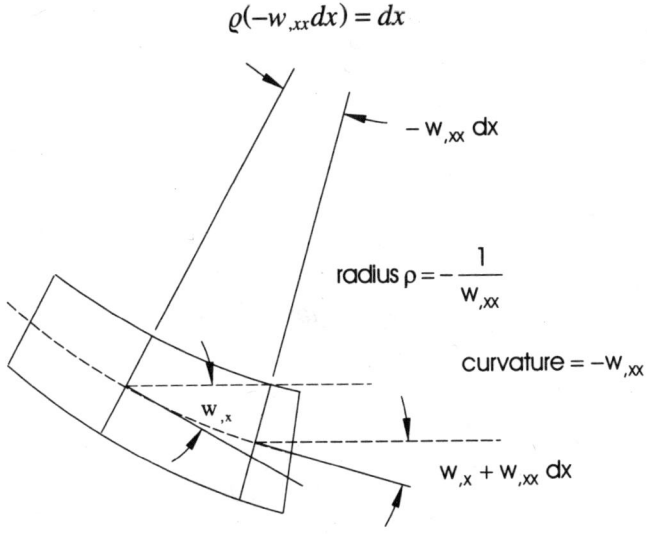

Fig. 3.1.3 **Curvature Relations**

or

$$\varrho = -\frac{1}{w_{,xx}}$$

(3.1.10)

The *curvature* of the surface is defined as in the case of beam by

$$\frac{1}{\varrho} = -w_{,xx}$$

(3.1.11)

The curvature is considered positive if it is convex downward as in Fig. 3.1.3. The minus sign appears in the above equation because the second derivative for the deflection convex downward is negative. The strain energy in the strip is

$$U = \int_0^L \int_{-h/2}^{h/2} \frac{1}{2} \tau_{xx} \varepsilon_{xx} \, dz \, dx$$
$$= \int_0^L \int_{-h/2}^{h/2} \frac{1}{2} \frac{Ez^2}{(1-v^2)} w_{,xx}^2 \, dz \, dx$$
$$= \frac{D}{2} \int_0^L w_{,xx}^2 \, dx$$

(3.1.12)

Let q be the uniformly distributed force acting in the lateral direction of the plate, then

$$V = -\int_0^L w \times q \times 1 \times dx$$

(3.1.13)

The potential functional of the plate is then given by

$$\pi = U + V$$
$$= \frac{D}{2} \int_0^L w_{,xx}^2 \, dx - \int_0^L wq \, dx$$

(3.1.14)

Extremizing the above

$$\delta\pi = \int_0^L D w_{,xx} \delta w_{,xx} \, dx - \int_0^L q \delta w \, dx = 0$$

(3.1.15)

Integrating the first term twice by parts with respect to x we get

$$[Dw_{,xx}\delta w_{,x}]_0^L - [(Dw_{,xx})_{,x}\delta w]_0^L + \int_0^L \{(Dw_{,xx})_{,xx} - q\}\delta w\, dx = 0$$
(3.1.16)

The above leads to the following differential equation and boundary conditions.

$$(Dw_{,xx})_{,xx} = q$$
(3.1.17)

$$Dw_{,xx} = 0 \text{ or prescribe } w_{,x}$$

$$(Dw_{,xx})_{,x} = 0 \text{ or prescribe } w$$

on the edges $x = 0$ and $x = L$
(3.1.18)

3.2 BENDING OF RECTANGULAR PLATES: SOPHIE GERMAIN THEORY

Consider now a general rectangular plate of thickness h with its middle plane before bending lying in xy plane. The plate undergoes small deformation w perpendicular to the xy plane. A section of the plate normal to y axis as shown in Fig. 3.2.1.

Slopes and Curvatures of the Middle Surface

The slope of the middle surface in this plane xy is $w_{,x}$ as in Fig. 3.1.2. In the same manner, if we consider a section of the plate normal to x axis, the slope is given by $w_{,y}$. Taking any direction v making an angle a with x axis, the slope can be obtained from

$$dw = w_{,x}dx + w_{,y}dy$$

$$\frac{\partial}{\partial v} = \frac{\partial}{\partial x}\frac{dx}{dv} + \frac{\partial}{\partial y}\frac{dy}{dv}$$

$$w_{,v} = w_{,x}\frac{dx}{dv} + w_{,y}\frac{dy}{dv}$$

$$= w_{,x}\cos a + w_{,y}\sin a$$
(3.2.1)

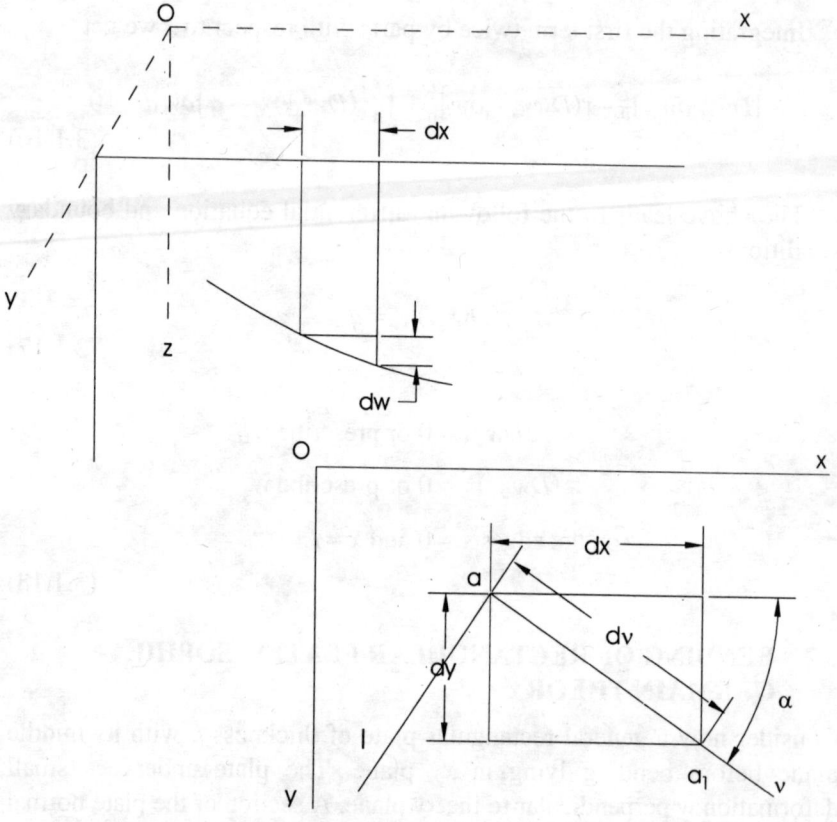

Fig. 3.2.1 Small Deformations of a Rectangular Plate

The direction a_1 in which the slope is maximum can be obtained by taking derivative of the above equation with respect to a and equating it to zero, i.e.,

$$-w_{,x} \sin a_1 + w_{,y} \cos a_1 = 0$$

$$a_1 = \tan^{-1}\left(\frac{w_{,y}}{w_{,x}}\right)$$

(3.2.2)

Substituting $\sin a_1 = w_{,y}/\sqrt{w_{,x}^2 + w_{,y}^2}$ and $\cos a_1 = w_{,x}/\sqrt{w_{,x}^2 + w_{,y}^2}$ in (3.2.1), the maximum slope is given by

$$(w_{,v})_{max} = \sqrt{w_{,x}^2 + w_{,y}^2}$$

(3.2.3)

Let a_2 be the direction in which the slope of the middle surface is zero, then, from (3.2.1) we get

$$w_{,x}\cos a_2 + w_{,y}\sin a_2 = 0$$

$$a_2 = \tan^{-1}\left(-\frac{w_{,x}}{w_{,y}}\right)$$

(3.2.4)

From equations (3.2.3) and (3.2.4), we can infer that the directions of maximum slope and zero slope are mutually perpendicular to each other. The slope of the middle surface of the plate in any direction can be taken as the angle made by the tangent to the surface in that direction with xy plane and the square of the slope is neglected compared to unity. In particular, the slope of the middle surface in a plane parallel to xz plane is $w_{,x}$ as in Fig. 3.1.2 and in a plane parallel to yz plane is $w_{,y}$. The curvature of the middle surface (positive for convex downward deflection) in a plane parallel to xz plane is then given by

$$\frac{1}{\varrho_x} = -w_{,xx}$$

(3.1.11)

Similarly, the curvature of the middle surface in a plane parallel to yz plane is given by

$$\frac{1}{\varrho_y} = -w_{,yy}$$

(3.2.5)

The curvature of the middle surface in any other direction v can be written as

$$\frac{1}{\varrho_v} = -w_{,vv}$$

(3.2.6)

Noting from (3.2.1) $\frac{\partial}{\partial v} = \frac{\partial}{\partial x}\frac{dx}{dv} + \frac{\partial}{\partial y}\frac{dy}{dv} = \frac{\partial}{\partial x}\cos a + \frac{\partial}{\partial y}\sin a$, we get

$$\frac{1}{\varrho_v} = -\left(\frac{\partial}{\partial x}\cos a + \frac{\partial}{\partial y}\sin a\right)(w_{,x}\cos a + w_{,y}\sin a)$$

$$= -(w_{,xx}\cos^2 a + 2w_{,xy}\sin a\cos a + w_{,yy}\sin^2 a)$$

(3.2.7)

The twist of the middle surface is defined as

$$\frac{1}{\rho_{xy}} = w_{,xy} \qquad (3.2.8)$$

and the curvature in any direction v in Fig. 3.2.1 given in (3.2.7) can be written as

$$\frac{1}{\rho_v} = \frac{1}{\rho_x}\cos^2 a - \frac{1}{\rho_{xy}}\sin 2a + \frac{1}{\rho_y}\sin^2 a \qquad (3.2.9)$$

The curvature in a direction perpendicular to v, i.e., l in Fig. 3.2.1 can be written down from the above by substituting $a+90°$ for a, therefore

$$\frac{1}{\rho_l} = \frac{1}{\rho_x}\sin^2 a - \frac{1}{\rho_{xy}}\sin 2a + \frac{1}{\rho_y}\cos^2 a \qquad (3.2.10)$$

From (3.2.9) and (3.2.10), we note that

$$\frac{1}{\rho_v} + \frac{1}{\rho_l} = \frac{1}{\rho_x} + \frac{1}{\rho_y} \qquad (3.2.11)$$

Therefore, the sum of the curvatures of the middle surface at any point in two perpendicular directions is independent of the angle a. This sum is called *average curvature* of the middle surface at a point. The *twist of the surface* with respect to v and l directions is given by

$$\frac{1}{\rho_{vl}} = \frac{d}{dl}\left(\frac{dw}{dv}\right) \qquad (3.2.12)$$

We can substitute $a+90°$ for a in (3.2.1) to give

$$\frac{1}{\rho_{vl}} = \left(-\frac{\partial}{\partial x}\sin a + \frac{\partial}{\partial y}\cos a\right)(w_{,x}\cos a + w_{,y}\sin a)$$

$$= \frac{1}{2}\sin 2a(-w_{,xx} + w_{,yy}) + \cos 2a\, w_{,xy}$$

$$= \frac{1}{2}\sin 2a\left(\frac{1}{\rho_x} - \frac{1}{\rho_y}\right) + \cos 2a\frac{1}{\rho_{xy}} \qquad (3.2.13)$$

The directions for the maximum and minimum curvatures of the middle surface at a point are obtained by taking the derivative of equation (3.2.9) with respect to a and equating it to zero,

$$\frac{d}{da}\left(\frac{1}{\rho_x}\cos^2 a - \frac{1}{\rho_{xy}}\sin 2a + \frac{1}{\rho_y}\sin^2 a\right) = 0$$

i.e.,

$$\frac{1}{\rho_x}\sin 2a + \frac{2}{\rho_{xy}}\cos 2a - \frac{1}{\rho_y}\sin 2a = 0 \tag{3.2.13}$$

or

$$\tan 2a = -\frac{\frac{2}{\rho_{xy}}}{\frac{1}{\rho_x} - \frac{1}{\rho_y}} \tag{3.2.14}$$

Therefore, the maximum and minimum curvature directions differ by 90°. Substituting these angles of a from (3.2.14) in (3.2.9), we get the maximum and minimum curvatures of the middle surface at a given point. These curvatures are called *principal curvatures* and the corresponding planes vz and lz the *principal planes of curvature*.

Displacement Field
The displacements at any point can be expressed as

$$u_x = -zw_{,x}$$
$$u_y = -zw_{,y}$$
$$u_z = w$$

$$\tag{3.2.15}$$

Strain Field
The strain field is

$$\varepsilon_{xx} = -zw_{,xx}$$
$$\varepsilon_{yy} = -zw_{,yy}$$
$$\varepsilon_{xy} = -zw_{,xy}$$

$$\tag{3.2.16}$$

Stress Field

The stress field is determined by applying Hooke's law for plane stress condition,

$$\varepsilon_{xx} = \frac{\tau_{xx}}{E} - v\frac{\tau_{yy}}{E}$$

$$\varepsilon_{yy} = \frac{\tau_{yy}}{E} - v\frac{\tau_{xx}}{E}$$

$$\varepsilon_{zz} = -v\frac{\tau_{xx}}{E} - v\frac{\tau_{yy}}{E}$$

$$\varepsilon_{xy} = \frac{\tau_{xy}}{2G} = \frac{\tau_{xy}}{E}(1+v)$$

(3.2.17)

Therefore

$$\tau_{xx} = \frac{E}{(1-v^2)}(\varepsilon_{xx} + v\varepsilon_{yy})$$

$$\tau_{yy} = \frac{E}{(1-v^2)}(\varepsilon_{yy} + v\varepsilon_{xx})$$

$$\tau_{xy} = 2G\varepsilon_{xy} = \frac{E}{(1+v)}\varepsilon_{xy}$$

(3.2.18)

Bending Moments

In pure bending, such as the case shown in Fig. 3.2.2 with moments applied on the edges, we find xz and yz become principal planes, i.e., $w_{,xy} = 0$. The stress field becomes simplified to

$$\tau_{xx} = \frac{E}{(1-v^2)}(\varepsilon_{xx} + v\varepsilon_{yy}) = -\frac{Ez}{(1-v^2)}(w_{,xx} + vw_{,yy})$$

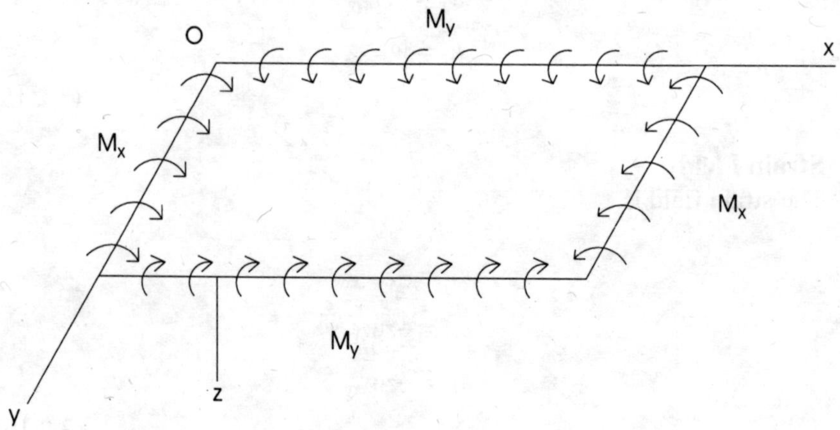

Fig. 3.2.2 Pure Bending of a Plate

$$\tau_{yy} = \frac{E}{(1-v^2)}(\varepsilon_{yy} + v\varepsilon_{xx}) = -\frac{Ez}{(1-v^2)}(w_{,yy} + vw_{,xx})$$
(3.2.18a)

As in the case of Euler-Bernoulli bending of beams, we note that the normal stress components give rise to bending moments (per unit length) M_x and M_y given by

$$M_x = \int_{-h/2}^{h/2} \tau_{xx} z \, dz$$

$$= -\int_{-h/2}^{h/2} \frac{Ez^2}{(1-v^2)}(w_{,xx} + vw_{,yy}) \, dz$$

$$= -D(w_{,xx} + vw_{,yy})$$

$$M_y = \int_{-h/2}^{h/2} \tau_{yy} z \, dz$$

$$= -\int_{-h/2}^{h/2} \frac{Ez^2}{(1-v^2)}(w_{,yy} + vw_{,xx}) \, dz$$

$$= -D(w_{,yy} + vw_{,xx})$$
(3.2.19)

where $D = \frac{Eh^3}{12(1-v^2)}$ is flexural rigidity of the plate.

If $M_x = M_y = M$ in Fig. 3.2.2, then from the above equation (3.2.19)

$$w_{,xx} = w_{,yy} = -\frac{M}{D(1+v)}$$
(3.2.20)

or

$$\frac{1}{\rho_x} = \frac{1}{\rho_y} = \frac{M}{D(1+v)}$$
(3.2.21)

and the plate is bent to a spherical surface.

On the other hand if $M_x = M_1$ and $M_y = M_2$, then equations (3.2.19) give

$$w_{,xx} = -\frac{M_1 - vM_2}{D(1-v^2)}$$

$$w_{,yy} = -\frac{M_2 - vM_1}{D(1-v^2)}$$
(3.2.22)

and the deflected surface is given by

$$w = -\frac{M_1 - \nu M_2}{D(1-\nu^2)}x^2 - \frac{M_2 - \nu M_1}{D(1-\nu^2)}y^2 + a_1 x + a_2 y + a_3$$

Choosing a reference plane tangent to the middle surface at the origin, with the condition $w = w_{,x} = w_{,y} = 0$ at $x = y = 0$, the deflected surface is

$$w = -\frac{M_1 - \nu M_2}{D(1-\nu^2)}x^2 - \frac{M_2 - \nu M_1}{D(1-\nu^2)}y^2 \tag{3.2.23}$$

Reverting back to the case $M_x = M_y = M$, the above equation gives the *deflected surface* to be

$$w = -\frac{M(x^2 + y^2)}{2D(1+\nu)} \tag{3.2.24}$$

which is a paraboloid of revolution rather than a spherical surface as obtained in equation (3.2.21). This discrepancy is because of the use of approximate expressions $w_{,xx}$ and $w_{,yy}$ for the curvatures in (3.2.24) above. However, we find it very convenient to use the second derivatives of deflections, rather than the exact expressions for the curvatures in the theory of plates.

Twisting Moment
Referring to Fig. 3.2.3, we can obtain the stresses on the section whose normal is inclined to x axis by angle a. It is straight forward to obtain

$$\tau_{vv} = \tau_{xx}\cos^2 a + \tau_{yy}\sin^2 a$$
$$\tau_{vl} = \frac{1}{2}(\tau_{yy} - \tau_{xx})\sin 2a \tag{3.2.25}$$

The normal stress τ_{vv} gives rise to the bending moment per unit length acting on this section can be obtained by

$$M_v = \int_{-h/2}^{h/2} \tau_{vv} z \, dz$$
$$= M_x \cos^2 a + M_y \sin^2 a \tag{3.2.26}$$

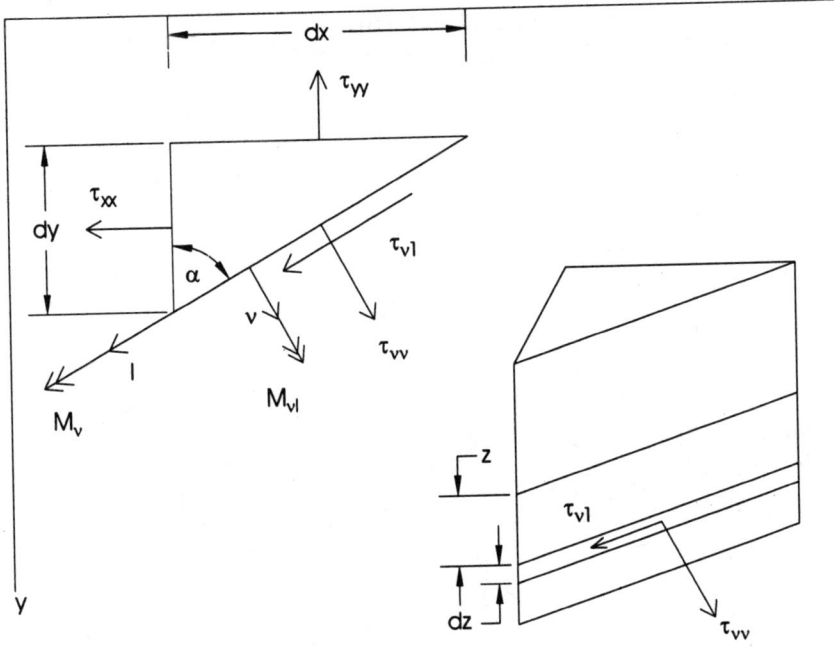

Fig. 3.2.3 Stress Field on an Inclined Section

Using (3.2.19), the above can be written as

$$M_v = -D(w_{,xx}\cos^2\alpha + w_{,yy}\sin^2\alpha) - vD(w_{,xx}\cos^2\alpha + w_{,yy}\sin^2\alpha)$$

With the help of (3.2.7) and (3.2.10) and noting that the twist of the middle surface $\frac{1}{\rho_{xy}}$ is zero, the above equation gives

$$M_v = D\left(\frac{1}{\rho_v} + \frac{1}{\rho_l}\right)$$
$$= -D(w_{,vv} + vw_{,ll})$$

(3.2.26a)

The shear stress τ_{vl} gives rise to the *twisting moment* per unit length acting on this section can be obtained by

56 Dynamics of Plates

$$M_{vl} = -\int_{-h/2}^{h/2} \tau_{vl} z \, dz$$

$$= (M_x - M_y)\sin\alpha\cos\alpha$$

(3.2.27)

Alternatively, we can use equation (3.2.16) for the shear strain $\varepsilon_{vl} = -zw_{,vl}$ and substitute for $\tau_{vl} = -2Gzw_{,vl}$ and obtain

$$M_{vl} = \int_{-h/2}^{h/2} 2Gz^2 w_{,vl} \, dz$$

$$= D(1-\nu)w_{,vl}$$

(3.2.27a)

Strain Energy in Pure Bending

Referring to Figs. 3.1.2 and 3.1.3, the work done by the moment $M_x dy$ through the angle $-w_{,xx}dx$ is

$$-\frac{1}{2}M_x w_{,xx} dx\, dy$$

Similarly, the work done by the moment $M_y dx$ is

$$-\frac{1}{2}M_y w_{,yy} dx\, dy$$

If x and y directions coincide with principal planes of curvature, the strain energy of the element is given by the total work above

$$dU = -\frac{1}{2}(M_x w_{,xx} + M_y w_{,yy})dx\, dy$$

(3.2.28)

With the help of (3.2.19), the above becomes

$$dU = \frac{D}{2}\left(w_{,xx}^2 + w_{,yy}^2 + 2\nu w_{,xx}w_{,yy}\right)dx\, dy$$

(3.2.29)

The total strain energy over the plate of area A is

$$U = \frac{1}{2}DA\left(w_{,xx}^2 + w_{,yy}^2 + 2\nu w_{,xx}w_{,yy}\right)$$

(3.2.30)

The strain energy due to twisting moment $M_{xy}dy$ is obtained by noting the corresponding twist angle $w_{,xy}$ (rate of change of slope $w_{,y}$ as x varies), which is

$$\frac{1}{2}M_{xy}w_{,xy}\,dx\,dy$$

Similarly, the strain energy due to twisting moment $M_{yx}dx$ is

$$\frac{1}{2}M_{yx}w_{,yx}\,dx\,dy$$

and the total strain energy due to twisting moment from (3.2.27a) is

$$M_{xy}w_{,xy}\,dx\,dy = D(1-v)w_{,xy}^2\,dx\,dy \tag{3.2.31}$$

Combining the strain energies due to bending and twisting moments, we get

$$dU = \frac{D}{2}(w_{,xx}^2 + w_{,yy}^2 + 2vw_{,xx}w_{,yy})\,dx\,dy + D(1-v)w_{,xy}^2\,dx\,dy$$

$$= \frac{D}{2}[(w_{,xx} + w_{,yy})^2 - 2(1-v)\{w_{,xx}w_{,yy} - w_{,xy}^2\}]\,dx\,dy \tag{3.2.32}$$

Therefore, the total strain energy for a rectangular plate is

$$U = \iint \frac{D}{2}[(\nabla^2 w)^2 - 2(1-v)\{w_{,xx}w_{,yy} - w_{,xy}^2\}]\,dx\,dy \tag{3.2.33}$$

The strain energy can also be derived from the strain and stress expressions in (3.2.16) and (3.2.18) using

$$U = \iint_R \int_{-h/2}^{h/2} \frac{1}{2}\tau_{ij}\varepsilon_{ij}\,dz\,dx\,dy$$

$$= \frac{1}{2}\frac{E}{(1-v^2)}\iint_R \int_{-h/2}^{h/2}\{\varepsilon_{xx}^2 + 2v\varepsilon_{xx}\varepsilon_{yy} + \varepsilon_{yy}^2 + 2(1-v)\varepsilon_{xy}^2\}\,dz\,dx\,dy$$

$$= \iint_R \frac{D}{2}[(\nabla^2 w)^2 - 2(1-v)\{w_{,xx}w_{,yy} - w_{,xy}^2\}]\,dx\,dy \tag{3.2.33a}$$

Potential Energy for External Force

Let q be the distributed force acting in the lateral direction of the plate, then

$$V = -\iint_R q(x,y) w(x,y)\, dx\, dy \tag{3.2.34}$$

Differential Equation and Boundary Conditions

The potential functional of the plate from (3.2.33a) and (3.2.34) can be written as

$$\pi = U + V$$

$$= \iint_R \frac{D}{2}\left[(\nabla^2 w)^2 - 2(1-\nu)\{w_{,xx}w_{,yy} - w_{,xy}^2\}\right] dx\, dy - \iint_R wq\, dx\, dy \tag{3.2.35}$$

which is same as the integral in equation (2.2.13) containing the functional in equation (2.1.21). The extremization process of this functional is given in chapter 2 to obtain the following differential equation and boundary conditions for a rectangular plate with its edges parallel to x and y axes and dimensions a and b.

$$D\nabla^4 w = q \tag{3.2.36}$$

Fixed-Fixed Conditions

$$w = 0;\quad \frac{\partial w}{\partial x} = 0 \text{ on the edge } x = 0 \text{ and } x = a$$

$$w = 0;\quad \frac{\partial w}{\partial y} = 0 \text{ on the edge } y = 0 \text{ and } y = b$$

$$\tag{3.2.37}$$

Simply Supported Conditions

$$w_{,xx} = 0 \text{ on the edge } x = 0 \text{ and } x = a$$
$$w = 0 \text{ on the edge } x = 0 \text{ and } x = a$$

and

$$w_{,yy} = 0 \text{ on the edge } y = 0 \text{ and } y = b$$
$$w = 0 \text{ on the edge } y = 0 \text{ and } y = b$$

$$\tag{3.2.38}$$

Free-Free Conditions

$$\frac{\partial}{\partial x}(\nabla^2 w) + (1-v)w_{,xyy} = 0$$

$w_{,xx} + vw_{,yy} = 0$ on the edge $x = 0$ and $x = a$

$$\frac{\partial}{\partial y}(\nabla^2 w) + (1-v)w_{,xxy} = 0$$

$w_{,yy} + vw_{,xx} = 0$ on the edge $y = 0$ and $y = b$

(3.2.39)

A combination of the fixed or supported or free conditions of the edges is possible, e.g., a plate fixed on two edges and free on two edges, supported on two edges and free on two edges, fixed on one edge and free on other three edges etc.

Shear Force

The displacement field for the thin rectangular plate is assumed in equation (3.2.15) from which the strain field is derived as given in equation (3.2.16). The plate is assumed to be in a state of plane stress and the stress field is derived in equation (3.2.18). This stress field does not make an allowance for the transverse shear stresses under the lateral force conditions. Despite this inconsistency, we accepted the stress strain field and derived the expression for the strain energy and then the differential equation and boundary conditions for a rectangular plate. (It may be recalled that a similar inconsistency existed while deriving Euler-Bernoulli theory for beams.) However, a simple equilibrium condition tells us that shear stresses in transverse direction exist.

A typical element of the plate is shown in Fig. 3.2.4. In addition to the bending and twisting moments considered earlier, we have the *shear forces* V_x and V_y per unit length due to the shear stresses in the transverse direction τ_{xz} and τ_{yz}.

$$V_x = \int_{-h/2}^{h/2} \tau_{xz} dz$$

$$V_y = \int_{-h/2}^{h/2} \tau_{yz} dz$$

(3.2.40)

The forces acting on the middle surface of the element are shown in Fig. 3.2.5. The equilibrium of vertical forces gives

$$V_{x,x}dxdy + V_{y,y}dxdy + qdxdy = 0$$

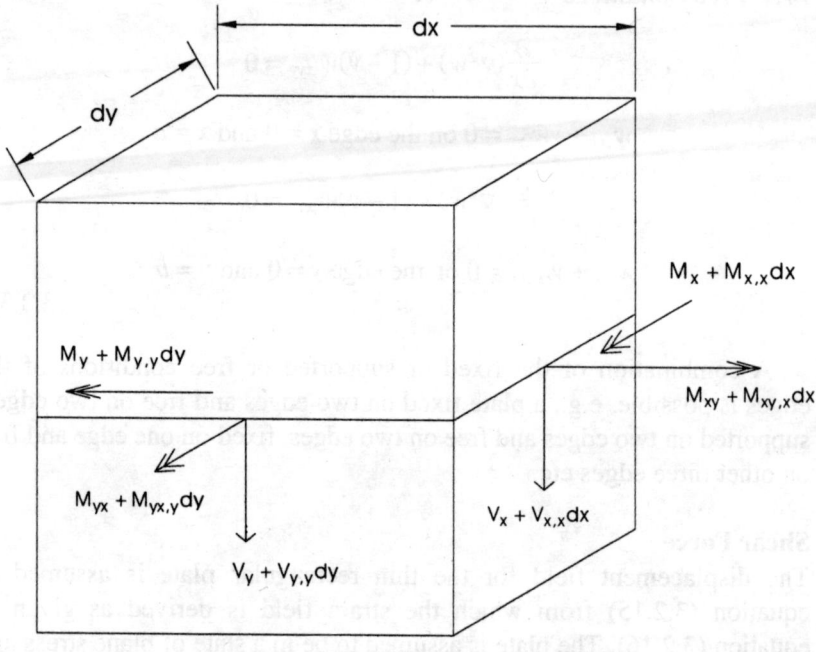

Fig. 3.2.4 Elemental Forces and Moments

i.e.,

$$V_{x,x} + V_{y,y} + q = 0$$

(3.2.41)

Neglecting the moment of the external force q and the higher order terms, the moments about x axis give

$$M_{xy,x}dxdy - M_{y,y}dxdy + V_y dxdy = 0$$

i.e.,

$$M_{xy,x} - M_{y,y} + V_y = 0$$

(3.2.42)

Similarly taking moments about y axis, we get

$$M_{yx,y} + M_{x,x} - V_x = 0$$

(3.2.43)

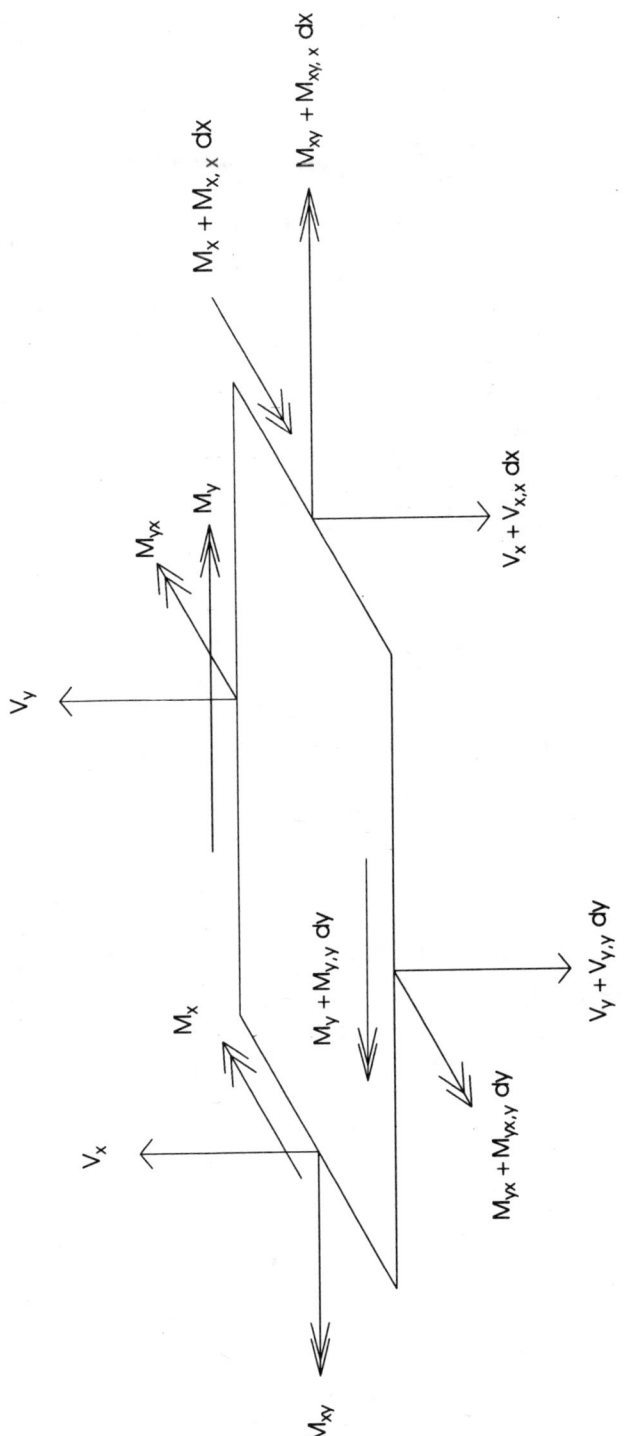

Fig. 3.2.5 Forces on the Middle Plane

Substituting for V_x and V_y from (3.2.43) and (3.2.42) in equation (3.2.41), we get

$$M_{yx,yx} + M_{x,xx} - M_{xy,xy} + M_{y,yy} + q = 0$$

Noting that $M_{xy} = -M_{yx}$, the above simplifies to

$$M_{x,xx} - 2M_{xy,xy} + M_{y,yy} = -q \qquad (3.2.44)$$

With the help of (3.2.26a) and (3.2.27a) and using x and y directions instead of v and l, we can write

$$M_x = -D(w_{,xx} + vw_{,yy})$$

$$M_y = -D(w_{,yy} + vw_{,xx})$$

$$M_{xy} = D(1-v)w_{,xy} \qquad (3.2.45)$$

Equation (3.2.44) can now be simplified to give

$$D\nabla^4 w = q \qquad (3.2.46)$$

which is same as the differential equation for the plate that we derived earlier. We can now define the shear forces from (3.2.43) and (3.2.42) by employing (3.2.45)

$$V_x = M_{yx,y} + M_{x,x}$$
$$= -D\frac{\partial}{\partial x}(\nabla^2 w) \qquad (3.2.47)$$

$$V_y = -M_{xy,x} + M_{y,y}$$
$$= -D\frac{\partial}{\partial y}(\nabla^2 w) \qquad (3.2.48)$$

Before leaving this topic, let us consider the boundary conditions derived earlier for a free edge, say at $x = a$, as given in equation (3.2.39). Poisson assumed that at the free edge,

$$(M_x)_{x=a} = 0, (M_{xy})_{x=a} = 0 \text{ and } (V_x)_{x=a} = 0$$

(3.2.49)

This gives three conditions which is one more than that needed as shown by Kirchoff. The first condition that $(M_x)_{x=a} = 0$ is same as the second condition in equation (3.2.39).

$$w_{,xx} + v w_{,yy} = 0$$

(3.2.39a)

Regarding the second and third conditions in (3.2.49), consider the edge $x = a$ as given in Fig. 3.2.6, wherein the twisting moment $M_{xy} dy$ is replaced by two equal and opposite forces over the distance dy. The twisting moment in the next elemental distance $\left(M_{xy} + \frac{\partial M_{xy}}{\partial y}\right) dy$ is similarly replaced by two equal and opposite forces $M_{xy} + \frac{\partial M_{xy}}{\partial y}$ etc. Therefore at any point, the effect of the twisting moment M_{xy} is to introduce a vertically upward force $\frac{\partial M_{xy}}{\partial y} dy$. Therefore, we can replace the twisting moment by an equivalent force per unit length given by

$$Q_x = -\frac{\partial M_{xy}}{\partial y}$$

(3.2.50)

The effective shear force on the edge is then given by

$$V_{x\text{eff}} = V_x + Q_x$$

$$= V_x - \frac{\partial M_{xy}}{\partial y}$$

(3.2.51)

With the help of equations (3.2.47) and (3.2.45), the above equation reduces to the first condition in (3.2.39)

$$\frac{\partial}{\partial x}(\nabla^2 w) + (1 - v) w_{,xyy} = 0$$

$$\text{or } w_{,xxx} + (2 - v) w_{,xyy} = 0$$

(3.2.39b)

64 Dynamics of Plates

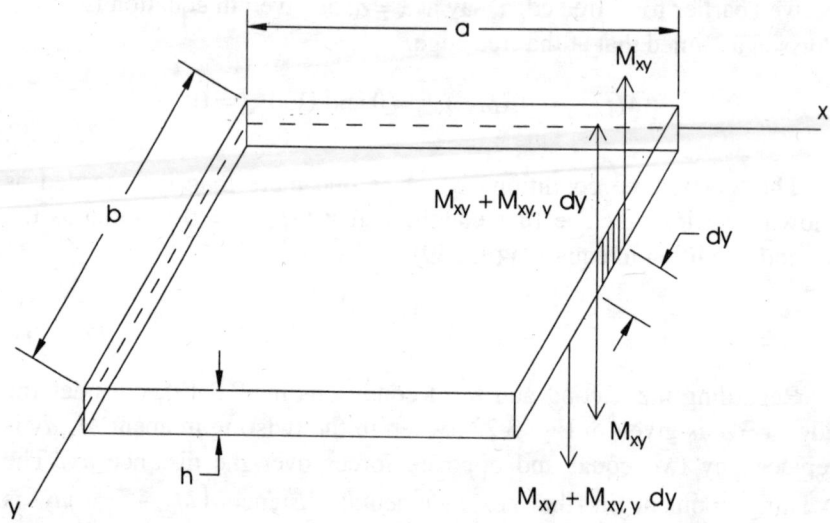

Fig. 3.2.6 Twisting Moment on Edge $x=a$ Replaced by Equal and Opposite Forces

This condition was first derived by Kelvin and Tait. What is significant here is that the Variational Calculus approach predicted exact boundary conditions even with the approximate displacement field, corresponding strain field and the assumption of plane stress condition in the plate without the need for the presence of the shear stress in the lateral direction.

Differential Equation for Bending Vibration of Rectangular Plates

As a last step in this section, we derive the differential equation for lateral vibration of rectangular plates acted on by external force $q(x,y,t)$. All that we need to do is to consider the kinetic energy of the plate in addition to the strain energy and work of external forces and apply the Hamilton's principle to derive the equation of motion. As in the case of simple Euler-Bernoulli theory of beams, we account for kinetic energy due to the velocity $\dot{w}$ only, which is

$$T = \frac{1}{2} \iint_R \rho h \, \dot{w}^2 \, dx\,dy \qquad (3.2.52)$$

Using equation (3.2.35) and the above, we can now set up Hamilton's principle for the plate undergoing small deformations as

$\int_{t_1}^{t_2} (T - U - V) dt$

$$= \int_{t_1}^{t_2} \left[\begin{array}{l} \frac{1}{2} \iint_R \rho h \, \dot{w}^2 \, dx\, dy \\ - \iint_R \frac{D}{2} \left[(\nabla^2 w)^2 - 2(1-v)\{w_{,xx} w_{,yy} - w_{,xy}^2\} \right] dx\, dy \\ + \iint_R wq \, dx\, dy \end{array} \right] dt$$

is stationary

(3.2.53)

The last two integrals have been considered in equation (2.2.13). The first term does not give rise to any line integrals and therefore the boundary conditions remain same as that derived before. Evidently, the differential equation (2.1.13) gets modified to

$$D\nabla^4 w + \rho h \, \ddot{w} = q(x, y, t)$$

(3.2.54)

Corner Reactions

While determining the effective shear forces on the edge of a plate, equation (3.2.51) we have replaced the twisting moment $M_{xy} dy$ by two equal and opposite forces over the distance dy as shown in Fig. 3.2.6. Transforming the twisting moments thus, we obtained the shearing forces Q_x distributed along the edge $x=a$ given by equation (3.2.50). It may be noted here that this process of transformation leaves two concentrated forces $(M_{xy})_{x=a, y=0}$ and $(M_{xy})_{x=a, y=b}$ at the ends of the edge $x=a$ as shown in Fig. 3.2.7.

A similar transformation on the edge $y=b$ creates reaction forces $(M_{yx})_{x=0, y=b}$ and $(M_{yx})_{x=a, y=0}$ as shown in Fig. 3.2.7. Therefore, a rectangular plate supported in some way produces concentrated reactions at the corners.

To determine the directions of the concentrated forces at the corners, consider a simply supported rectangular plate shown in Fig. 3.2.8. The deflection surface for this plate can be obtained, for example, in the case of a uniformly distributed force, the general shape of this deflection is indicated in Fig. 3.2.8, see equations (3.3.10) in the next section. Considering the corner at $x=a$ and $y=b$, we find that the derivative $w_{,x}$ of the deflection surface is negative and decreases numerically with increasing y. Therefore, $w_{,xy}$ is positive at this corner, therefore, M_{xy} is positive here, see equation (3.2.45), and M_{yx} is negative. From Fig. 3.2.7, we can now infer that both the concentrated forces at this corner are downwards and give rise to a net reactive force $R = 2(M_{xy})_{x=a,y=b} = 2D(1-v)(w_{,xy})_{x=a,y=b}$.

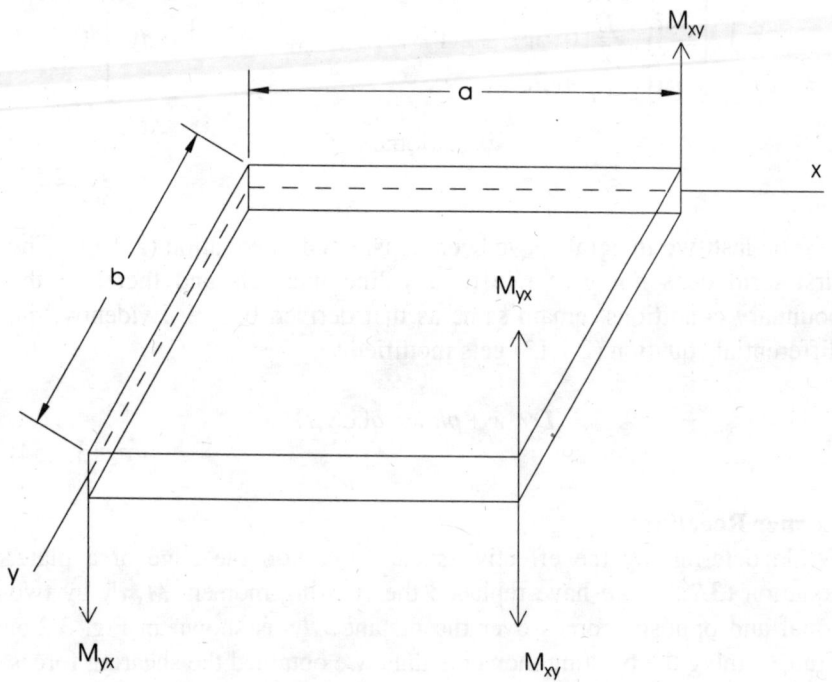

Fig. 3.2.7 Corner Reactions

3.3 NAVIER'S SOLUTION FOR SIMPLY SUPPORTED RECTANGULAR PLATES

Consider a simply supported rectangular plate of width a in x direction and length b in y direction with thickness h acted on by a static force $q(x,y)$. The differential equation for this problem is given by equation (3.2.36) and boundary conditions (3.2.38).

$$D\nabla^4 w = q(x,y) \qquad (3.2.36)$$

$w_{,xx} = 0$ on the edge $x = 0$ and $x = a$

$w = 0$ on the edge $x = 0$ and $x = a$

and

$w_{,yy} = 0$ on the edge $y = 0$ and $y = b$

Rectangular Plates 67

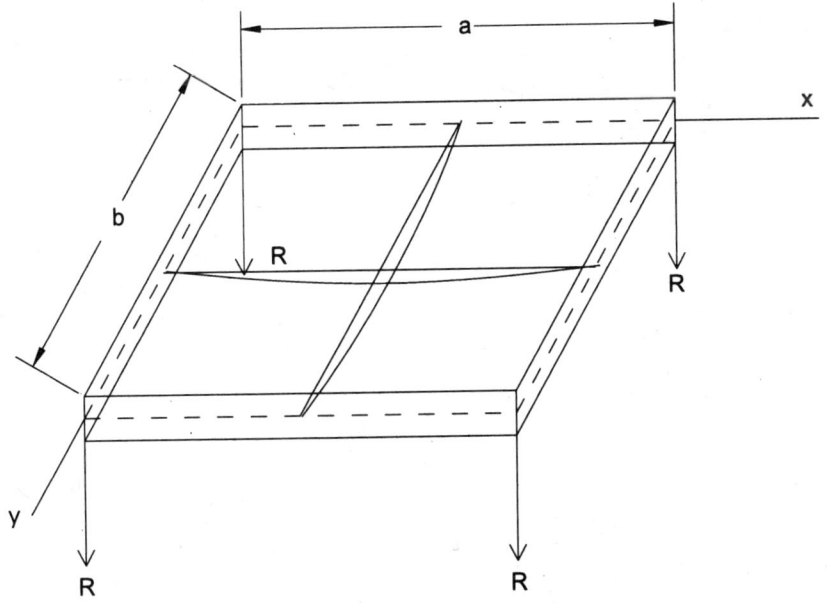

Fig. 3.2.8 Corner Reactions on a Simply Supported Plate

$$w = 0 \text{ on the edge } y = 0 \text{ and } y = b \tag{3.2.38}$$

To solve this problem, Navier represented the external force in the form of a double trigonometric series

$$q(x,y) = \sum_{m=1}^{\infty} \sum_{n=1}^{\infty} q_{mn} \sin \frac{m\pi x}{a} \sin \frac{n\pi y}{b} \tag{3.3.1}$$

The coefficients q_{mn} are determined using the orthogonality property of the sine functions

$$\int_0^b \sin \frac{n\pi y}{b} \sin \frac{k\pi y}{b} dy = 0 \quad \text{for } n \neq k$$

$$\int_0^b \sin \frac{n\pi y}{b} \sin \frac{k\pi y}{b} dy = \frac{b}{2} \quad \text{for } n = k \tag{3.3.2a}$$

$$\int_0^a \sin \frac{n\pi x}{a} \sin \frac{k\pi x}{a} dx = 0 \quad \text{for } n \neq k$$

$$\int_0^a \sin \frac{n\pi x}{a} \sin \frac{k\pi x}{a} dx = \frac{a}{2} \quad \text{for } n = k \tag{3.3.2b}$$

68 Dynamics of Plates

Multiplying both sides of equation (3.3.1) by $\sin \frac{m'\pi x}{a} \sin \frac{n'\pi y}{b}$ and integrating, we get

$$\int_0^a \int_0^b q(x,y) \sin \frac{m'\pi x}{a} \sin \frac{n'\pi y}{b} dx\, dy$$

$$= \int_0^a \int_0^b \left(\sum_{m=1}^{\infty} \sum_{n=1}^{\infty} q_{mn} \sin \frac{m\pi x}{a} \sin \frac{n\pi y}{b} \right) \sin \frac{m'\pi x}{a} \sin \frac{n'\pi y}{b} dx\, dy$$

$$= \frac{ab}{4} q_{mn}$$

(3.3.3)

Therefore

$$q_{mn} = \frac{4}{ab} \int_0^a \int_0^b q(x,y) \sin \frac{m\pi x}{a} \sin \frac{n\pi y}{b} dx\, dy$$

(3.3.4)

Thus the problem is reduced to solving the following differential equation

$$D \nabla^4 w = \sum_{m=1}^{\infty} \sum_{n=1}^{\infty} q_{mn} \sin \frac{m\pi x}{a} \sin \frac{n\pi y}{b}$$

(3.3.5)

where q_{mn} are first obtained from the integral in (3.3.4). Since the problem is linear, we can find the solution for each term in (3.3.4) and sum up the solutions to find the final answer. Therefore, it is sufficient for us to obtain the solution of

$$D(w_{,xxxx} + 2w_{,xxyy} + w_{,yyyy}) = q_{mn} \sin \frac{m\pi x}{a} \sin \frac{n\pi y}{b}$$

(3.3.6)

The solution can be assumed in the form

$$w = C \sin \frac{m\pi x}{a} \sin \frac{n\pi y}{b}$$

(3.3.7)

which satisfies all the boundary conditions in (3.2.38). Substituting the above and its required differentials in (3.3.6), we get

$$C\pi^4 \left(\frac{m^2}{a^2} + \frac{n^2}{b^2} \right)^2 = \frac{q_{mn}}{D}$$

(3.3.8)

Therefore

$$w = \frac{q_{mn}}{D\pi^4 \left(\frac{m^2}{a^2} + \frac{n^2}{b^2}\right)^2} \sin\frac{m\pi x}{a} \sin\frac{n\pi y}{b}$$

(3.3.9)

The solution of equation (3.2.36) for the present case is then given by

$$w = \frac{1}{D\pi^4} \sum_{m=1}^{\infty} \sum_{n=1}^{\infty} \frac{q_{mn}}{\left(\frac{m^2}{a^2} + \frac{n^2}{b^2}\right)^2} \sin\frac{m\pi x}{a} \sin\frac{n\pi y}{b}$$

(3.3.10)

where q_{mn} are given by equation (3.3.4).

For the case of a uniformly distributed force of intensity $q(x,y)=q$ the integral in equation (3.3.4) can be evaluated.

$$q_{mn} = \frac{4q}{ab} \int_0^a \int_0^b \sin\frac{m\pi x}{a} \sin\frac{n\pi y}{b} dx\,dy$$

$$= \frac{4q}{mn\pi^2} \left\{\cos\frac{m\pi x}{a}\right\}_0^a \times \left\{\cos\frac{n\pi y}{b}\right\}_0^b$$

$$= \frac{16q}{mn\pi^2} \text{ for } m \text{ and } n \text{ odd}$$

$$= 0 \text{ for } m \text{ and } n \text{ even}$$

(3.3.4a)

Therefore, the deformation w in (3.3.10) is given by

$$w = \frac{16qa^4b^4}{D\pi^6} \sum_{m=1,3..}^{\infty} \sum_{n=1,3..}^{\infty} \frac{1}{mn(m^2b^2+n^2a^2)^2} \sin\frac{m\pi x}{a} \sin\frac{n\pi y}{b}$$

(3.3.10a)

When the applied force q is uniform, it is obvious that the deformed surface is symmetrical with respect to the axes $x = \frac{1}{2}a$, $y = \frac{1}{2}b$. Therefore, the even numbers for m and n vanish in the solution since they give rise to unsymmetrical deformation.

The maximum deflection of the plate will be at center of the plate with $x = \frac{1}{2}a$, $y = \frac{1}{2}b$, which is obtained from (3.3.10a) as

70 Dynamics of Plates

$$w_{max} = \frac{16qa^4b^4}{D\pi^6} \sum_{m=1,3..}^{\infty} \sum_{n=1,3..}^{\infty} \frac{1}{mn(m^2b^2+n^2a^2)^2} \sin\frac{m\pi}{2}\sin\frac{n\pi}{2}$$

$$= \frac{16qa^4b^4}{D\pi^6} \sum_{m=1,3..}^{\infty} \sum_{n=1,3..}^{\infty} \frac{(-1)^{\frac{m+n}{2}-1}}{mn(m^2b^2+n^2a^2)^2}$$

(3.3.10b)

For a square plate, the above expression reduces to

$$w_{max} = \frac{16qa^4}{D\pi^6} \sum_{m=1,3..}^{\infty} \sum_{n=1,3..}^{\infty} \frac{(-1)^{\frac{m+n}{2}-1}}{mn(m^2+n^2)^2}$$

(3.3.10c)

which is a rapidly converging series. A one term approximation gives

$$w_{max} = \frac{4qa^4}{D\pi^6} = 0.00416\frac{qa^4}{D}$$

(3.3.10d)

Natural Frequencies and Mode Shapes

The simple solution given above for the static case can be easily extended to determine the natural frequencies and mode shapes of simply supported rectangular plates in lateral vibration. The differential equation is given by, see equation (3.2.54)

$$D\nabla^4 w + \rho h \ddot{w} = 0$$

(3.3.11)

Writing the solution of the above for harmonic motion as

$$w = W(x,y)\sin pt$$

(3.3.12)

equation (3.3.11) becomes

$$\nabla^4 W - \beta^2 W = 0$$

(3.3.13)

where

$$\beta^2 = \frac{\rho h p^2}{D}$$

(3.3.14)

We further write

$$W(x,y) = C\sin\frac{m\pi x}{a} \sin\frac{n\pi y}{b}$$

(3.3.7)

which makes (3.3.13) as

$$\beta^2 = \pi^4\left(\frac{m^2}{a^2} + \frac{n^2}{b^2}\right)^2$$

(3.3.15)

Substituting the above in (3.3.14)

$$\frac{\rho h p^2}{D} = \pi^4\left(\frac{m^2}{a^2} + \frac{n^2}{b^2}\right)^2$$

(3.3.16)

Since the above equation is valid for any m and n, the natural frequencies are given by

$$p = \sqrt{\frac{D}{\rho h}}\, \pi^2\left(\frac{m^2}{a^2} + \frac{n^2}{b^2}\right) \text{ rad/s}$$

(3.3.17)

with the corresponding mode shapes

$$W_{mn} = \sin\frac{m\pi x}{a} \sin\frac{n\pi y}{b}$$

(3.3.18)

Consider a rectangular plate with $\frac{a}{b} = k$ and define a frequency parameter

$$\lambda_{mn}^2 = \frac{\rho h a^4}{D} p_{mn}^2 = a^4 \beta_{mn}^2 = \pi^4(m^2 + n^2 k^2)^2$$

(3.3.18)

For a square plate, $k = 1$, the above equation simplifies to

$$\lambda_{mn}^2 = a^4 \beta_{mn}^2 = \pi^4(m^2 + n^2)^2$$

(3.3.19)

The non dimensional frequency parameters and the mode shapes are tabulated in Table 3.3.1 below.

Table 3.3.1 Frequency Parameters λ_{mn}^2 for a Square Plate

m	n	λ_{mn}^2	Mode Shape
1	1	389.64	$\sin\frac{\pi x}{a}\sin\frac{\pi y}{b}$
1	2	2,435.23	$\sin\frac{\pi x}{a}\sin\frac{2\pi y}{b}$
2	1	2,435.23	$\sin\frac{2\pi x}{a}\sin\frac{\pi y}{b}$
2	2	6,234.18	$\sin\frac{2\pi x}{a}\sin\frac{2\pi y}{b}$
1	3	9,740.91	$\sin\frac{\pi x}{a}\sin\frac{3\pi y}{b}$
3	1	9,740.91	$\sin\frac{3\pi x}{a}\sin\frac{\pi y}{b}$
2	3	16,462.14	$\sin\frac{2\pi x}{a}\sin\frac{3\pi y}{b}$
3	2	16,462.14	$\sin\frac{3\pi x}{a}\sin\frac{2\pi y}{b}$

The mode shapes are plotted in Fig. 3.3.1. It is interesting to note that the second and third modes have the same frequency. These two mode shapes can be superimposed to obtain

$$w_{23} = C\sin\frac{2\pi x}{a}\sin\frac{\pi y}{a} + D\sin\frac{\pi x}{a}\sin\frac{2\pi y}{a} \qquad (3.3.20)$$

where C and D are arbitrary constants. When $C = 0$, we get the mode $m = 1$ and $n = 2$ shown in Fig. 3.3.1. When $D = 0$, we get the mode $m = 2$ and $n = 1$. If $C = D$, equation (3.3.21) gives

$$w_{23} = 2C\sin\frac{\pi x}{a}\sin\frac{\pi y}{a}\left(\cos\frac{\pi x}{a} + \cos\frac{\pi y}{a}\right) \qquad (3.3.21)$$

The nodal points for this superimposed mode with the same frequency as $m = 1$ and $n = 2$ or $m = 2$ and $n = 1$ are defined by

$$\sin\frac{\pi x}{a} = 0$$
$$\sin\frac{\pi y}{a} = 0$$
$$\cos\frac{\pi x}{a} + \cos\frac{\pi y}{a} = 0$$

$$(3.3.22)$$

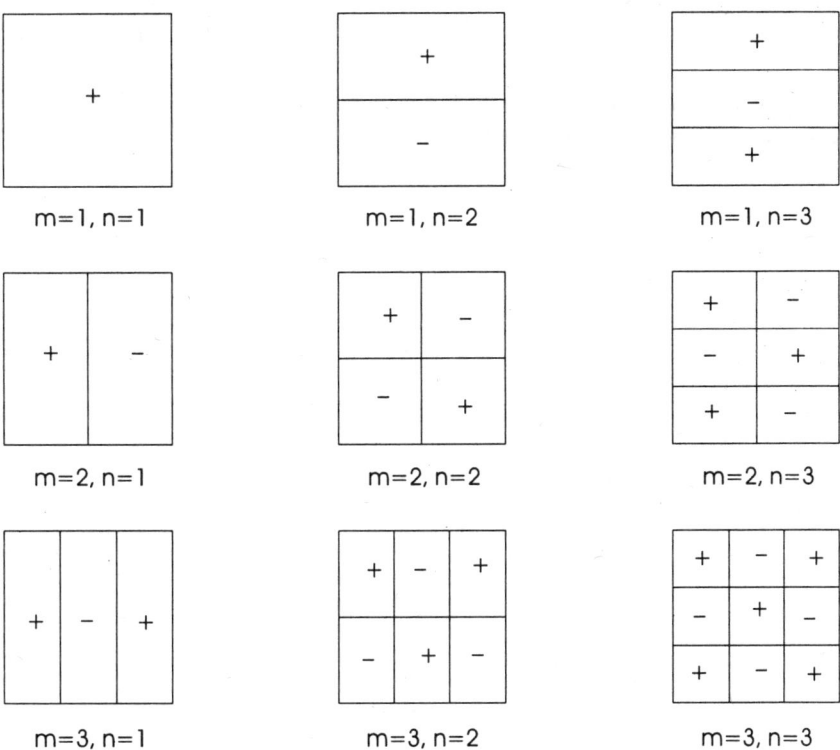

Fig. 3.3.1 Mode Shapes of a Square Plate

The first two equations of (3.3.22) above give us the four edges of the square plate. The third equation gives

$$\frac{\pi x}{a} = \pi - \frac{\pi y}{a} \tag{3.3.23}$$

i.e.,

$$x + y = a \tag{3.3.24}$$

Similarly, for $C = -D$, we get

$$x = y \tag{3.3.25}$$

These two mode shapes are shown in Fig. 3.3.2, which are diagonal lines of the plate. It is interesting to note that we can excite four different

74 Dynamics of Plates

modes with the same frequency. We can also deduce that a simply supported triangular plate of sides a has its fundamental frequency same as that of the second mode of the corresponding square plate. In this manner, one can extend the analysis for higher modes, e.g., $m = 2$ and $n = 3$ or $m = 3$ and $n = 2$.

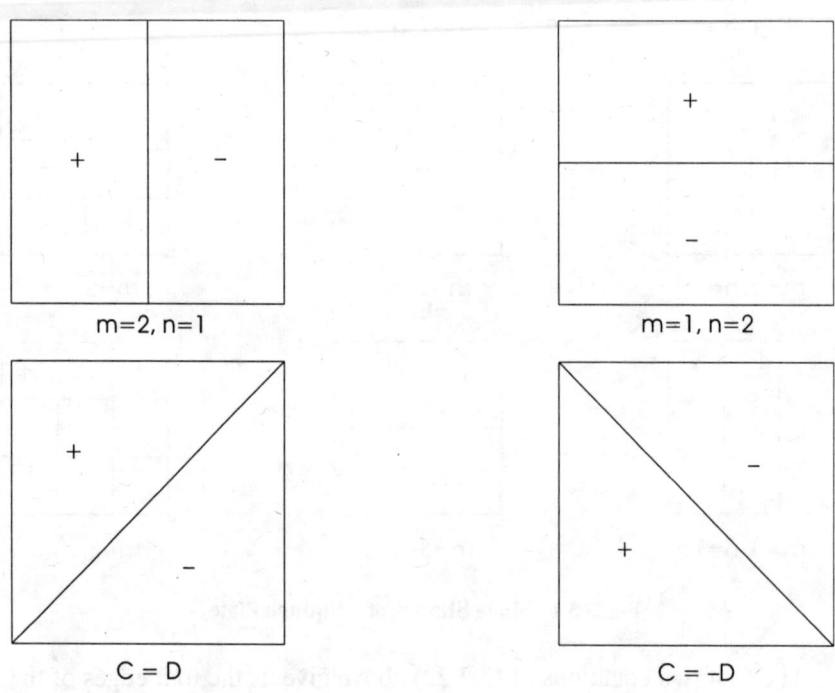

Fig. 3.3.2 **Different Mode Shapes with the Same Frequency**

3.4 FREE VIBRATION OF SIMPLY SUPPORTED RECTANGULAR PLATES

The differential equation for this problem is given by equation (3.3.11) and boundary conditions (3.2.38).

$$D\nabla^4 w + \rho h \, \ddot{w} = 0 \qquad (3.3.11)$$

$$w_{,xx} = 0 \text{ on the edge } x = 0 \text{ and } x = a$$

$$w = 0 \text{ on the edge } x = 0 \text{ and } x = a$$

and

$$w_{,yy} = 0 \text{ on the edge } y = 0 \text{ and } y = b$$

$w = 0$ on the edge $y = 0$ and $y = b$

(3.2.38)

Let the plate be acted on by initial displacement and velocity given by

$$w(x, y, 0) = w_0(x, y)$$
$$\dot{w}(x, y, 0) = \dot{w}_0(x, y)$$

(3.4.1)

The free vibration solution can be assumed by using modal analysis. The natural frequencies are given by equation (3.3.17)

$$p_{mn} = \sqrt{\frac{D}{\rho h}} \pi^2 \left(\frac{m^2}{a^2} + \frac{n^2}{b^2} \right)$$

(3.3.17)

The mode shapes are given in equation (3.3.18)

$$W_{mn} = \sin \frac{m\pi x}{a} \sin \frac{n\pi y}{b}$$

(3.3.18)

Hence

$$w = \sum_{m=1}^{\infty} \sum_{n=1}^{\infty} (A_{mn} \sin p_{mn} t + B_{mn} \cos p_{mn} t) \sin \frac{m\pi x}{a} \sin \frac{n\pi y}{b}$$

(3.4.2)

where A_{mn}, B_{mn} are constants of integration to be determined by using the initial conditions.

The constants of integration are determined using the orthogonality property of the mode shapes (sine functions)

$$\int_0^b \sin \frac{n\pi y}{b} \sin \frac{k\pi y}{b} dy = \frac{b}{2} \delta_{nk}$$

(3.3.2a)

$$\int_0^a \sin \frac{n\pi x}{a} \sin \frac{k\pi x}{a} dx = \frac{a}{2} \delta_{nk}$$

(3.3.2b)

Substituting the initial conditions in (3.4.1) in the modal solution (3.4.2), we have

76 Dynamics of Plates

Substituting the initial conditions in (3.4.1) in the modal solution (3.4.2), we have

$$w_0(x,y) = \sum_{m=1}^{\infty} \sum_{n=1}^{\infty} B_{mn} \sin \frac{m\pi x}{a} \sin \frac{n\pi y}{b} \qquad (3.4.3a)$$

$$\dot{w}_0(x,y) = \sum_{m=1}^{\infty} \sum_{n=1}^{\infty} A_{mn} p_{mn} \sin \frac{m\pi x}{a} \sin \frac{n\pi y}{b} \qquad (3.4.3b)$$

Multiplying both sides of equation (3.4.3a) by $\sin \frac{m'\pi x}{a} \sin \frac{n'\pi y}{b}$ and integrating, we get

$$\int_0^a \int_0^b w_0(x,y) \sin \frac{m'\pi x}{a} \sin \frac{n'\pi y}{b} dx\,dy$$

$$= \int_0^a \int_0^b \left(\sum_{m=1}^{\infty} \sum_{n=1}^{\infty} B_{mn} \sin \frac{m\pi x}{a} \sin \frac{n\pi y}{b} \right) \sin \frac{m'\pi x}{a} \sin \frac{n'\pi y}{b} dx\,dy$$

$$= \frac{ab}{4} B_{mn} \qquad (3.4.4)$$

Therefore

$$B_{mn} = \frac{4}{ab} \int_0^a \int_0^b w_0(x,y) \sin \frac{m\pi x}{a} \sin \frac{n\pi y}{b} dx\,dy \qquad (3.4.5)$$

Now, multiplying both sides of equation (3.4.3b) by $\sin \frac{m'\pi x}{a} \sin \frac{n'\pi y}{b}$ and integrating, we get

$$\int_0^a \int_0^b \dot{w}_0(x,y) \sin \frac{m'\pi x}{a} \sin \frac{n'\pi y}{b} dx\,dy$$

$$= \int_0^a \int_0^b \left(\sum_{m=1}^{\infty} \sum_{n=1}^{\infty} A_{mn} p_{mn} \sin \frac{m\pi x}{a} \sin \frac{n\pi y}{b} \right) \sin \frac{m'\pi x}{a} \sin \frac{n'\pi y}{b} dx\,dy$$

$$= \frac{ab}{4} p_{mn} A_{mn} \qquad (3.4.6)$$

Therefore

$$A_{mn} = \frac{4}{ab p_{mn}} \int_0^a \int_0^b \dot{w}_0(x,y) \sin \frac{m\pi x}{a} \sin \frac{n\pi y}{b} dx\,dy \qquad (3.4.7)$$

3.5 FORCED VIBRATION OF SIMPLY SUPPORTED RECTANGULAR PLATES

The differential equation for this problem is given by equation (3.2.54) and boundary conditions (3.2.38). Considering the excitation force to be harmonic with frequency ω, we have

$$D\nabla^4 w + \rho h \ddot{w} = F(x,y)\cos\omega t \tag{3.5.1}$$

The solution for steady forced vibration can be obtained either by direct solution or by modal analysis.

Direct method
The solution is written as

$$w = W(x,y)\cos\omega t \tag{3.5.2}$$

Let the force $F(x,y)$ be represented in the form of (3.3.1)

$$F(x,y) = \sum_{m=1}^{\infty}\sum_{n=1}^{\infty} F_{mn} \sin\frac{m\pi x}{a} \sin\frac{n\pi y}{b} \tag{3.5.3}$$

where F_{mn} can be obtained from (3.3.4)

$$F_{mn} = \frac{4}{ab}\int_0^a\int_0^b F(x,y)\sin\frac{m\pi x}{a}\sin\frac{n\pi y}{b}dx\,dy \tag{3.5.4}$$

Substituting (3.5.2) in equation (3.5.1), we get

$$D\nabla^4 W - \rho h\omega^2 W = \sum_{m=1}^{\infty}\sum_{n=1}^{\infty} F_{mn}\sin\frac{m\pi x}{a}\sin\frac{n\pi y}{b} \tag{3.5.5}$$

We can solve the above equation for each term on the right hand side and sum up all the terms to obtain the response of the system. For the mn term, we can write

$$D\nabla^4 W_{mn} - \rho h\omega^2 W_{mn} = F_{mn}\sin\frac{m\pi x}{a}\sin\frac{n\pi y}{b} \tag{3.5.5a}$$

78 Dynamics of Plates

The solution of the above can be written as

$$W_{mn} = C \sin \frac{m\pi x}{a} \sin \frac{n\pi y}{b}$$

(3.5.6)

where C is a constant to be determined. Substituting, we get

$$CD\pi^4 \left(\frac{m^2}{a^2} + \frac{n^2}{b^2}\right)^2 - \rho h \omega^2 C = F_{mn}$$

(3.5.7)

With the help of equation (3.3.16), the above equation is

$$C\rho h(p_{mn}^2 - \omega^2) = F_{mn}$$

(3.5.8)

Therefore, the constant C is

$$C = \frac{F_{mn}}{\rho h(p_{mn}^2 - \omega^2)}$$

(3.5.9)

The forced vibration response for the mn term is then given by

$$W_{mn} = \frac{F_{mn}}{\rho h p_{mn}^2 (1-r^2)} \sin \frac{m\pi x}{a} \sin \frac{n\pi y}{b} \cos \omega t$$

(3.5.10)

The total response is

$$w = \sum_{m=1}^{\infty} \sum_{n=1}^{\infty} \frac{F_{mn}}{\rho h p_{mn}^2 (1-r^2)} \sin \frac{m\pi x}{a} \sin \frac{n\pi y}{b} \cos \omega t$$

(3.5.11)

For a uniformly distributed force $F(x, y) = F$, equation (3.3.4a) gives

$$F_{mn} = \frac{16F}{mn\pi^2} \text{ for } m \text{ and } n \text{ odd}$$

$$= 0 \text{ for } m \text{ and } n \text{ even}$$

(3.5.12)

Therefore,

$$w = \frac{16F}{mn\pi^2 \rho h} \sum_{m=1,3,\ldots}^{\infty} \sum_{n=1,3,\ldots}^{\infty} \frac{1}{p_{mn}^2(1-r^2)} \sin\frac{m\pi x}{a} \sin\frac{n\pi y}{b} \cos\omega t$$

(3.5.13)

Modal method

In the modal method, we write the solution for (3.5.1) as

$$w = \sum_{m=1}^{\infty} \sum_{n=1}^{\infty} W_{mn} \eta_{mn}(t)$$

(3.5.14)

Substituting the above in (3.5.1) and using (3.5.3) for the force, we have

$$\rho h \sum_{m=1}^{\infty} \sum_{n=1}^{\infty} W_{mn} \ddot{\eta}_{mn} + D\nabla^4 \left(\sum_{m=1}^{\infty} \sum_{n=1}^{\infty} W_{mn} \eta_{mn} \right)$$

$$= \sum_{m=1}^{\infty} \sum_{n=1}^{\infty} F_{mn} \sin\frac{m\pi x}{a} \sin\frac{n\pi y}{b} \cos\omega t$$

(3.5.15)

From equations (3.3.13) and (3.3.14), we can write

$$\nabla^4 W_{mn} = \beta_{mn}^2 W_{mn}$$

$$= \frac{\rho h p_{mn}^2}{D} W_{mn}$$

(3.5.16)

Hence, equation (3.5.15) can be simplified as

$$\sum_{m=1}^{\infty} \sum_{n=1}^{\infty} W_{mn} \ddot{\eta}_{mn} + \sum_{m=1}^{\infty} \sum_{n=1}^{\infty} W_{mn} p_{mn}^2 \eta_{mn}$$

$$= \frac{1}{\rho h} \sum_{m=1}^{\infty} \sum_{n=1}^{\infty} F_{mn} \sin\frac{m\pi x}{a} \sin\frac{n\pi y}{b} \cos\omega t$$

(3.5.17)

Multiplying the above equation by W_{kl} and integrating over the region, we get

80 Dynamics of Plates

$$\int_0^a \int_0^b W_{kl}\left(\sum_{m=1}^{\infty}\sum_{n=1}^{\infty} W_{mn}\ddot{\eta}_{mn}\right)dx\,dy$$

$$+\int_0^a \int_0^b W_{kl}\left(\sum_{m=1}^{\infty}\sum_{n=1}^{\infty} W_{mn}p_{mn}^2\eta_{mn}\right)dx\,dy$$

$$=\frac{1}{\rho h}\int_0^a \int_0^b W_{kl}\sum_{m=1}^{\infty}\sum_{n=1}^{\infty} F_{mn}\sin\frac{m\pi x}{a}\sin\frac{n\pi y}{b}\cos\omega t\,dx\,dy$$

(3.5.18)

Making use of the results in (3.4.4) and (3.3.4a), the above simplify to the following uncoupled equations with m and n odd integers.

$$\frac{ab}{4}\ddot{\eta}_{mn}+\frac{ab}{4}p_{mn}^2\eta_{mn}=\frac{4ab}{mn\pi^2}\frac{1}{\rho h}F_{mn}\cos\omega t$$

$$\ddot{\eta}_{mn}+p_{mn}^2\eta_{mn}=\frac{16F_{mn}}{mn\pi^2\rho h}\cos\omega t$$

(3.5.19)

The solution of the above equation is

$$\eta_{mn}=\frac{16F_{mn}}{mn\pi^2\rho h}\left(\frac{1}{p_{mn}^2(1-r^2)}\right)\cos\omega t$$

(3.5.20)

Therefore, the total response is given by

$$w=\frac{16}{mn\pi^2\rho h}\sum_{m=1,3,\ldots}^{\infty}\sum_{n=1,3,\ldots}^{\infty}\frac{F_{mn}}{p_{mn}^2(1-r^2)}\cos\omega t$$

(3.5.21)

For a uniformly distributed force, the above equation becomes

$$w=\frac{16F}{mn\pi^2\rho h}\sum_{m=1,3,\ldots}^{\infty}\sum_{n=1,3,\ldots}^{\infty}\frac{1}{p_{mn}^2(1-r^2)}\sin\frac{m\pi x}{a}\sin\frac{n\pi y}{b}\cos\omega t$$

(3.5.13)

3.6 NADAI SOLUTION FOR SIMPLY SUPPORTED PLATES

It is convenient to locate the coordinate system at the center of the left edge of the plate with width in x direction as a and in y direction as b as shown in Fig. 3.6.1. The differential equation is

$$\nabla^4 w=\frac{q}{D}$$

(3.6.1)

Rectangular Plates   81

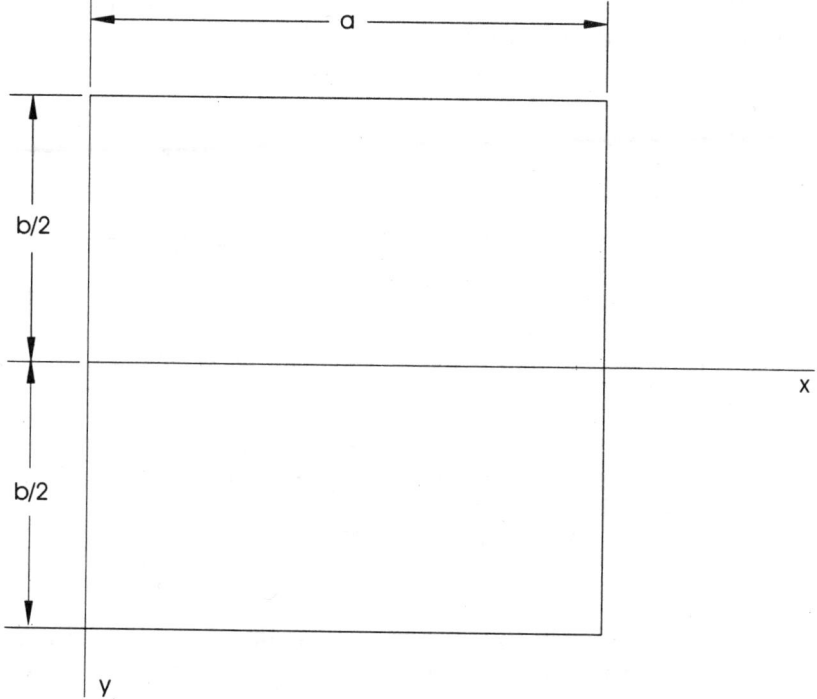

Fig. 3.6.1 Rectangular Plate

The solution for the deflected surface is taken in two parts. The first part is taken to represent the deflection of a uniformly loaded strip parallel to x axis

$$w_1 = \frac{q}{24D}(x^4 - 2ax^3 + a^3x)$$

(3.6.2)

which satisfies the differential equation and also the boundary conditions at $x=0$ and $x=a$. The second part of the solution should satisfy

$$\nabla^4 w_2 = 0$$

(3.6.3)

Let us consider the solution as in section 2.7 for a symmetrical deflection about x axis

$$w_2(x,y) = \sum_{m=1,3,5\cdots}^{\infty} Y_m(y) \sin\frac{m\pi x}{a}$$

(3.6.4)

which satisfies the boundary conditions $w = 0$ and $w_{,xx} = 0$ on the edges $x = 0$ and a. Y_m is determined to satisfy the boundary conditions on the other two edges $y = \pm \frac{b}{2}$. Substituting (3.6.4) in (3.6.3), we get

$$\sum_{m=1,3,5\cdots}^{\infty} \left[Y_m'''' - 2\left(\frac{m\pi}{a}\right)^2 Y_m'' + \left(\frac{m\pi}{a}\right)^4 Y_m \right] \sin \frac{m\pi x}{a} = 0$$

i.e., $Y_m'''' - 2\left(\frac{m\pi}{a}\right)^2 Y_m'' + \left(\frac{m\pi}{a}\right)^4 Y_m = 0$

(3.6.5)

The general solution of the above equation (3.6.5) can be written in the form

$$Y_m = \frac{qa^4}{D} \left(\begin{array}{l} A_m \cosh \frac{m\pi y}{a} + B_m \frac{m\pi y}{a} \sinh \frac{m\pi y}{a} \\ + C_m \sinh \frac{m\pi y}{a} + D_m \frac{m\pi y}{a} \cosh \frac{m\pi y}{a} \end{array} \right)$$

(3.6.6)

The constants C_m and D_m are taken zero so that deflection surface is symmetrical with respect to x axis for a uniformly distributed force q. Therefore, the total solution of (3.6.1) is given by

$$w = w_1 + w_2$$
$$= \frac{q}{24D}(x^4 - 2ax^3 + a^3 x)$$
$$+ \frac{qa^4}{D} \sum_{m=1,3,5\cdots}^{\infty} \left(A_m \cosh \frac{m\pi y}{a} + B_m \frac{m\pi y}{a} \sinh \frac{m\pi y}{a} \right) \sin \frac{m\pi x}{a}$$

(3.6.7)

The first term in the above equation is expressed in series form

$$\frac{q}{24D}(x^4 - 2ax^3 + a^3 x) = \frac{4qa^4}{\pi^5 D} \sum_{m=1,3,5\cdots}^{\infty} \frac{1}{m^5} \sin \frac{m\pi x}{a}$$

(3.6.8)

so that

$$w = \frac{qa^4}{D} \sum_{m=1,3,5\cdots}^{\infty} \left(\frac{4}{\pi^5 m^5} + A_m \cosh \frac{m\pi y}{a} + B_m \frac{m\pi y}{a} \sinh \frac{m\pi y}{a} \right) \sin \frac{m\pi x}{a}$$

(3.6.9)

From the boundary conditions at $y = \pm\frac{b}{2}$

$$\frac{4}{\pi^5 m^5} + A_m \cosh a_m + B_m a_m \sinh a_m = 0$$

$$(A_m + 2B_m)\cosh a_m + a_m B_m \sinh a_m = 0 \qquad (3.6.10)$$

where

$$a_m = \frac{m\pi b}{2a} \qquad (3.6.11)$$

Solving, the constants of integration can be obtained as

$$A_m = -\frac{2(a_m \tanh a_m + 2)}{\pi^5 m^5 \cosh a_m}$$

$$B_m = \frac{2}{\pi^5 m^5 \cosh a_m} \qquad (3.6.12)$$

The deflection surface of the plate is then given by

$$w = \frac{4qa^4}{\pi^5 D} \sum_{m=1,3,5\cdots}^{\infty} \frac{1}{m^5}\left(1 - \frac{a_m \tanh a_m + 2}{2\cosh a_m}\cosh\frac{m\pi y}{a} + \frac{a_m}{2\cosh a_m}\frac{2y}{b}\sinh\frac{m\pi y}{a}\right)\sin\frac{m\pi x}{a}$$

$$(3.6.13)$$

The maximum deflection of the plate at the center $x = \frac{a}{2}, y = 0$ is

$$w_{max} = \frac{4qa^4}{\pi^5 D}\sum_{m=1,3,5\cdots}^{\infty}\frac{(-1)^{\frac{m-1}{2}}}{m^5}\left(1 - \frac{a_m \tanh a_m + 2}{2\cosh a_m}\right) \qquad (3.6.14)$$

The first term on the right hand side represents the deflection of a uniformly loaded strip, hence

$$w_{max} = \frac{5}{384}\frac{qa^4}{D} - \frac{4qa^4}{\pi^5 D}\sum_{m=1,3,5\cdots}^{\infty}\frac{(-1)^{\frac{m-1}{2}}}{m^5}\frac{a_m \tanh a_m + 2}{2\cosh a_m} \qquad (3.6.15)$$

84 Dynamics of Plates

For a square plate, $a_m = \frac{m\pi}{2}$ and the above gives

$$w_{max} = \frac{5}{384}\frac{qa^4}{D} - \frac{4qa^4}{\pi^5 D}(0.6856 - 0.00025 + \cdots)$$

$$= 0.00406\frac{qa^4}{D}$$

(3.6.16)

The above compares well with Navier's solution in equation (3.3.10d).

3.7 CLAMPED PLATES

Let us consider the expression for the strain energy given in equation (3.2.33)

$$U = \iint \frac{D}{2}\left[(\nabla^2 w)^2 - 2(1-v)\{w_{,xx}w_{,yy} - w_{,xy}^2\}\right]dx\,dy$$

(3.2.33)

The integral of the curved bracket term can be evaluated by parts

$$I = \iint \{w_{,xx}w_{,yy} - w_{,xy}^2\}dx\,dy$$

(3.7.1)

The first integral in the above is integrated by parts with respect to y and using (2.2.27), we get

$$I_1 = \iint w_{,xx}(w_{,y})_{,y}\,dA$$

$$= \oint w_{,xx}w_{,y}a_{vy}dl - \iint_R w_{,xxy}w_{,y}dA$$

$$= -\oint w_{,xx}w_{,y}dx - \iint_R w_{,xxy}w_{,y}dA$$

(3.7.2)

The second integral in (3.7.1) is integrated by parts with respect to x and using (2.2.27), we get

$$I_2 = \iint w_{,xy}(w_{,y})_{,x}\,dA$$

$$= \oint w_{,xy}w_{,y}a_{vx}dl - \iint_R w_{,xxy}w_{,y}dA$$

$$= \oint w_{,xy}w_{,y}dy - \iint_R w_{,xxy}w_{,y}dA$$

(3.7.3)

The integral in (3.7.1) therefore reduces to a line integral

$$I = -\oint w_{,y}[w_{,xx}dx + w_{,xy}dy]$$
$$= -\oint w_{,y}\,d(w_{,x})$$
$$= -\oint w_{,y}(w_{,x})_{,l}\,dl$$

(3.7.4)

For a clamped plate with the coordinate system at the center, $w_{,y} = 0$ on $x = \pm a/2$ and $w_{,x} = 0$ on $y = \pm b/2$, therefore the above integral vanishes. The expression $\{w_{,xy}^2 - w_{,xx}w_{,yy}\}$ represents the Gaussian curvature of the deformed surface and for plates with clamped edges, this curvature becomes zero. As a result, the strain energy expression for clamped plates simplifies to

$$U = \iint \frac{D}{2}(\nabla^2 w)^2\,dx\,dy$$

(3.7.5)

The potential functional for the plate under a uniformly distributed force q can be written as

$$I = \frac{D}{2}\iint\left[(\nabla^2 w)^2 - \frac{2qw}{D}\right]dx\,dy$$

(3.7.6)

Let the plate of width a and length b with thickness h be placed in a coordinate system with the origin at the center of the plate and the edge with width a be parallel to x axis. Let us consider an approximate solution with one term

$$w = C_1(X^2 - 1)^2(Y^2 - 1)^2$$

(3.7.7)

satisfying the boundary conditions $w = w_{,X} = 0$ on $X = \pm 1$ and $w = w_{,Y} = 0$ on $Y = \pm 1$, where $X = \frac{2x}{a}$ and $Y = \frac{2y}{b}$. Taking the necessary derivatives,

$$w_{,xx} = \frac{16}{a^2}C_1(3X^2 - 1)(Y^2 - 1)^2$$

$$w_{,yy} = \frac{16}{b^2}C_1(X^2 - 1)^2(3Y^2 - 1)$$

$$\nabla^2 w = 16C_1\left[\frac{1}{a^2}(3X^2-1)(Y^2-1)^2 + \frac{1}{b^2}(X^2-1)^2(3Y^2-1)\right]$$

(3.7.8)

Upon substitution, (3.7.6) becomes

$$I = \frac{D}{2}\frac{ab}{4}\iint 256C_1^2\left\{\begin{array}{c}\frac{1}{a^2}(3X^2-1)(Y^2-1)^2 \\ +\frac{1}{b^2}(X^2-1)^2(3Y^2-1)\end{array}\right\}^2 dX\,dY$$

$$-\frac{qC_1 ab}{4}\iint(X^2-1)^2(Y^2-1)^2 dX\,dY$$

(3.7.9)

Upon extremizing the above with respect to C_1, we have

$$C_1 = \frac{q\int_{-1}^{1}\int_{-1}^{1}(X^2-1)^2(Y^2-1)^2 dX\,dY}{D\int_{-1}^{1}\int_{-1}^{1}256\left\{\begin{array}{c}\frac{1}{a^2}(3X^2-1)(Y^2-1)^2 \\ +\frac{1}{b^2}(X^2-1)^2(3Y^2-1)\end{array}\right\}^2 dX\,dY}$$

(3.7.10)

The integral in the numerator of equation (3.7.10) gives

$$\int_{-1}^{1}\int_{-1}^{1}(X^4-2X^2+1)(Y^4-2Y^2+1)dX\,dY$$

$$= \int_{-1}^{1}\left(\frac{X^5}{5}-\frac{2X^3}{3}+X\right)_{-1}^{1}(Y^4-2Y^2+1)dY$$

$$= \frac{16}{15}\left(\frac{Y^5}{5}-\frac{2Y^3}{3}+Y\right)_{-1}^{1} = \frac{256}{225}$$

(3.7.11)

One integral in the denominator of (2.3.10) is

$$\int_{-1}^{1}\int_{-1}^{1}(3X^2-1)^2(Y^2-1)^4 dX\,dY$$

$$= \int_{-1}^{1}\int_{-1}^{1}(9X^4-6X^2+1)(Y^8-4Y^6+6Y^4-4Y^2+1)dX\,dY$$

$$= \left(\frac{9X^5}{5}-2X^3+X\right)_{-1}^{1}\left(\frac{Y^9}{9}-\frac{4Y^7}{7}+\frac{6Y^5}{5}-\frac{4Y^3}{3}+Y\right)_{-1}^{1}$$

$$= \frac{128\times 16}{9\times 25\times 7}$$

(3.7.12)

Another integral in the denominator of (3.7.10) can be identified as

$$\int_{-1}^{1}\int_{-1}^{1}(3Y^2-1)^2(X^2-1)^4\,dX\,dY = \frac{128\times 16}{9\times 25\times 7}$$
(3.7.13)

The third and the last integral in the denominator of (3.7.10) is

$$\int_{-1}^{1}\int_{-1}^{1} 2(3X^2-1)(X^4-2X^2+1)(3Y^2-1)(Y^4-2Y^2+1)\,dX\,dY$$

$$=\int_{-1}^{1}\int_{-1}^{1} 2(3X^6-7X^4+5X^2-1)(3Y^6-7Y^4+5Y^2-1)\,dX\,dY$$

$$= 2\left(\frac{3X^7}{7}-\frac{7X^5}{5}+\frac{5X^3}{3}-X\right)_{-1}^{1}\left(\frac{3Y^7}{7}-\frac{7Y^5}{5}+\frac{5Y^3}{3}-Y\right)_{-1}^{1}$$

$$= \frac{2\times 64^2}{105^2}$$
(3.7.14)

With the help of the integrals in (3.7.11) to (3.7.14), equation (3.7.10) gives

$$C_1 = \frac{7q}{2048D\left(\dfrac{1}{a^4}+\dfrac{4}{7a^2b^2}+\dfrac{1}{b^4}\right)}$$
(3.7.15)

Hence, the deflection of the plate is

$$w = \frac{7q}{2048D\left(\dfrac{1}{a^4}+\dfrac{4}{7a^2b^2}+\dfrac{1}{b^4}\right)}(X^2-1)^2(Y^2-1)^2$$
(3.7.16)

For a square plate, the above equation gives

$$w = 0.001329\frac{qa^4}{D}(X^2-1)^2(Y^2-1)^2$$
(3.7.17)

The maximum deflection occurs at the center of the plate

$$w_{max} = 0.001329\frac{qa^4}{D}$$
(3.7.18)

The solution obtained by Kantorovich method for this problem is given in equations (2.8.11) and (2.8.12). The maximum deflection obtained here by Ritz method is 2.48% more than the Kantorovich solution.

Timoshenko considered this problem in a different way to obtain the solution in series form. His method consists of using the solution of a simply supported plate, for which a simple and exact solution can be obtained in series form, and then superimposing on to this the deflection of the plate by moments distributed along the edges, which are adjusted to satisfy the slope condition of fixed edges. The solution is lengthy, however, it gives the solution to the above problem upto five digits accuracy

$$w_{\max \text{ Timoshenko}} = 0.00126 \frac{qa^4}{D}$$

(3.7.19)

The solution given here with one term approximation is 5.19% more than Timoshenko series solution. The result by Ritz method given here or by Kantorovich method considered in section 2.8 will be considerably improved with two term or more approximation in the solution. However, a computer is required to perform such calculations. This is much easier way than obtaining a Timoshenko type solution, particularly for vibration problems.

Fundamental Natural Frequency

The functional concerned with vibration problem is

$$I = \iint \left[\frac{1}{2} m \, \dot{w}^2 - \frac{D}{2} (\nabla^2 w)^2 \right] dx \, dy$$

(3.7.20)

where m is the mass per unit area (ρh). Let us consider a one term approximation

$$w = C_1 (X^2 - 1)^2 (Y^2 - 1)^2 \sin pt$$

(3.7.21)

and use maximum energies in formulating the Ritz problem. We then have

$$\hat{w} = C_1 p (X^2 - 1)^2 (Y^2 - 1)^2$$

(3.7.22)

Substituting the above in (3.7.20) and noting that the strain energy term is already dealt with in equation (3.7.9), we have

$$I = -\frac{D}{2}\frac{ab}{4} \iint 256 C_1^2 \left\{ \begin{array}{l} \frac{1}{a^2}(3X^2-1)(Y^2-1)^2 \\ +\frac{1}{b^2}(X^2-1)^2(3Y^2-1) \end{array} \right\}^2 dX\,dY$$

$$+ \frac{m}{2}\frac{ab}{4} \iint p^2 C_1^2 (X^2-1)^4 (Y^2-1)^4 dX\,dY$$

(3.7.23)

Applying Ritz minimization process with respect to C_1, we get

$$\frac{\partial I}{\partial C_1} = 0$$

i.e., $$D \iint 256 \left\{ \begin{array}{l} \frac{1}{a^2}(3X^2-1)(Y^2-1)^2 \\ +\frac{1}{b^2}(X^2-1)^2(3Y^2-1) \end{array} \right\}^2 dX\,dY$$

$$= mp^2 \iint (X^2-1)^4 (Y^2-1)^4 dX\,dY$$

(3.7.24)

The left hand side integral has been already evaluated in equations (3.7.12) to (3.7.14). The right hand side integral is

$$\int_{-1}^{1}\int_{-1}^{1} (X^2-1)^4 (Y^2-1)^4 dX\,dY$$

$$= \int_{-1}^{1}\int_{-1}^{1} (X^8 - 4X^6 + 6X^4 - 4X^2 + 1)(Y^8 - 4Y^6 + 6Y^4 - 4Y^2 + 1)\,dX\,dY$$

$$= \left(\frac{X^9}{9} - \frac{4X^7}{7} + \frac{6X^5}{5} - \frac{4X^3}{3} + X\right)_{-1}^{1} \left(\frac{Y^9}{9} - \frac{4Y^7}{7} + \frac{6Y^5}{5} - \frac{4Y^3}{3} + Y\right)_{-1}^{1}$$

$$= \left(\frac{384 \times 2}{9 \times 5 \times 7 \times 3}\right)^2$$

(3.7.25)

The first mode natural frequency of the fixed rectangular plate is therefore

$$p^2 = \frac{256D\left[\dfrac{1}{a^4}\dfrac{128\times 16}{63\times 25} + \dfrac{1}{a^2b^2}\dfrac{8\times 32^2}{105^2} + \dfrac{1}{b^4}\dfrac{128\times 16}{63\times 25}\right]}{m\left(\dfrac{384\times 2}{9\times 5\times 7\times 3}\right)^2}$$

$$= \frac{504D}{m}\left(\frac{1}{a^4} + \frac{4}{7a^2b^2} + \frac{1}{b^4}\right)$$

(3.7.26)

For a square plate, the above gives

$$p = 36\sqrt{\frac{D}{ma^4}} \text{ rad/s}$$

(3.7.27)

General Eigenvalue Problem

To determine higher modes, we can assume an N term approximation for the shape function of the plate clamped on all edges. The functional for minimization is

$$I = \iint\left[\frac{1}{2}m\dot{w}^2 - \frac{D}{2}(\nabla^2 w)^2\right]dx\,dy$$

(3.7.20)

The N term approximation can be expressed as

$$\tilde{w} = \sum_{i=1}^{N} C_i\phi_i(x,y)\sin pt$$

(3.7.28)

where $\phi_i(x,y)$ satisfy at least the kinematic boundary conditions of the plate. We use maximum energies in formulating the Ritz problem, therefore

$$\hat{\tilde{w}} = \sum_{i=1}^{N} C_i\phi_i(x,y)$$

$$\hat{\dot{\tilde{w}}} = \sum_{i=1}^{N} C_i p\phi_i(x,y)$$

(3.7.29)

Substituting the above in (3.7.20), we have

$$I = \iint \left[\frac{1}{2} m \left(\sum_{i=1}^{N} C_i p \phi_i(x,y) \right)^2 - \frac{D}{2} \left[\sum_{i=1}^{N} C_i \nabla^2 \phi_i(x,y) \right]^2 \right] dx\, dy$$

(3.7.30)

We can rewrite the above

$$I = \frac{1}{2} p^2 \sum_{i=1}^{N} \sum_{j=1}^{N} C_i C_j Q_{ij} - \frac{1}{2} \sum_{i=1}^{N} \sum_{j=1}^{N} C_i C_j P_{ij}$$

(3.7.31)

where

$$Q_{ij} = m \iint \phi_i \phi_j\, dA$$

$$P_{ij} = \sum_{q=1}^{2} \sum_{p=1}^{2} D \iint \left(\frac{\partial^2 \phi_i}{\partial x_p^2} \right)\left(\frac{\partial^2 \phi_j}{\partial x_q^2} \right) dA$$

$$i, j = 1, 2, \ldots N$$

$$p, q = 1, 2$$

(3.7.32)

in which x and y are replaced by x_1 and x_2 respectively. Now, applying Ritz minimization process with respect to C_i, we get

$$\frac{\partial I}{\partial C_i} = 0 \quad i = 1, 2, \ldots N$$

(3.7.33)

i.e.,

$$[[P_{ij}] - p^2 [Q_{ij}]]\{C_i\} = 0$$

(3.7.34)

A computer program can be written to solve the above eigen value problem to determine the natural frequencies and mode shapes.

3.8 MINDLIN'S PLATE THEORY

The classical plate theory considered so far is similar to Euler-Bernoulli's beam theory in the assumption that the shear and rotary inertia effects are neglected. It may be recalled that Rayleigh introduced the effect of axial inertia (more popularly known as rotary inertia) and Timosheno introduced the effect of shear in bending of beams. Mindlin, likewise introduced the effects of shear and rotary inertia to the classical plate theory.

Displacement Field
We modify the displacement field of (3.2.15), thus

$$u_x = -z\psi_x$$
$$u_y = -z\psi_y$$
$$u_z = w$$

(3.8.1)

where ψ_x and ψ_y are the slopes due to bending alone in the respective planes.

Strain Field
The strain field is

$$\varepsilon_{xx} = -z\psi_{x,x}$$
$$\varepsilon_{yy} = -z\psi_{y,y}$$
$$\varepsilon_{xy} = -\frac{1}{2}z(\psi_{x,y} + \psi_{y,x})$$
$$\varepsilon_{xz} = \frac{1}{2}(w_{,x} - \psi_x)$$
$$\varepsilon_{yz} = \frac{1}{2}(w_{,y} - \psi_y)$$

(3.8.2)

Compared to the classical plate theory, we have now introduced all the shear strains in the above field.

Stress Field
The stress field is determined by applying Hooke's law for plane stress condition (*xy* plane), which gives

$$\tau_{xx} = \frac{E}{(1-v^2)}(\varepsilon_{xx} + v\varepsilon_{yy})$$
$$\tau_{yy} = \frac{E}{(1-v^2)}(\varepsilon_{yy} + v\varepsilon_{xx})$$

$$\tau_{xy} = 2G\varepsilon_{xy} = \frac{E}{(1+v)}\varepsilon_{xy} = -\frac{Ez}{2(1+v)}(\psi_{x,y} + \psi_{y,x})$$

$$\tau_{xz} = 2G\varepsilon_{xz} = \frac{E}{(1+v)}\varepsilon_{xz} = G(w_{,x} - \psi_x)$$

$$\tau_{yz} = 2G\varepsilon_{yz} = \frac{E}{(1+v)}\varepsilon_{yz} = G(w_{,y} - \psi_y)$$

(3.8.3)

Bending Moments
Bending moments (per unit length) M_x and M_y are given by

$$M_x = \int_{-h/2}^{h/2} \tau_{xx} z\, dz$$

$$= -\int_{-h/2}^{h/2} \frac{Ez^2}{(1-v^2)}(\psi_{x,x} + v\psi_{y,y})\, dz$$

$$= -D(\psi_{x,x} + v\psi_{y,y})$$

$$M_y = \int_{-h/2}^{h/2} \tau_{yy} z\, dz$$

$$= -\int_{-h/2}^{h/2} \frac{Ez^2}{(1-v^2)}(\psi_{y,y} + v\psi_{x,x})\, dz$$

$$= -D(\psi_{y,y} + v\psi_{x,x})$$

(3.8.4)

Twisting Moment
The twisting moment is given by

$$M_{xy} = -\int_{-h/2}^{h/2} \tau_{xy} z\, dz$$

$$= \int_{-h/2}^{h/2} \frac{E}{2(1+v)}(\psi_{x,y} + \psi_{y,x})z^2\, dz$$

$$= \frac{1}{2}D(1-v)(\psi_{x,y} + \psi_{y,x})$$

(3.8.5)

Shear Forces
The shear forces V_x and V_y per unit length due to the shear stresses in the transverse direction τ_{xz} and τ_{yz} are

$$V_x = \int_{-h/2}^{h/2} 2G\varepsilon_{xz}dz$$

$$= \int_{-h/2}^{h/2} G(w_{,x} - \psi_x)dz$$

$$= khG(w_{,x} - \psi_x)$$

$$V_y = \int_{-h/2}^{h/2} 2G\varepsilon_{yz}dz$$

$$= \int_{-h/2}^{h/2} G(w_{,y} - \psi_y)dz$$

$$= khG(w_{,y} - \psi_y)$$

(3.8.6)

where k is the shear correction factor. We can now formulate the Hamilton's principle as follows.

Kinetic Energy

$$T = \frac{1}{2} \iint_R \int_{-h/2}^{h/2} \rho(\dot{\psi}_x^2 z^2 + \dot{\psi}_y^2 z^2 + \dot{w}^2)dz\,dx\,dy$$

$$= \frac{1}{2} \iint_R \rho\left(\dot{\psi}_x^2 \frac{h^3}{12} + \dot{\psi}_y^2 \frac{h^3}{12} + h\dot{w}^2\right)dx\,dy$$

(3.8.7)

Strain Energy
Under plane stress condition, the strain energy is given by

$$U = \frac{1}{2} \iint_R \int_{-h/2}^{h/2} (\tau_{xx}\varepsilon_{xx} + \tau_{yy}\varepsilon_{yy} + 2\tau_{xy}\varepsilon_{xy} + 2\tau_{xz}\varepsilon_{xz} + 2\tau_{yz}\varepsilon_{yz})dz\,dx\,dy$$

$$= \frac{1}{2} \iint_R \int_{-h/2}^{h/2} \left[\frac{E}{(1-v^2)}(\varepsilon_{xx} + v\varepsilon_{yy})\varepsilon_{xx} + \frac{E}{(1-v^2)}(\varepsilon_{yy} + v\varepsilon_{xx})\varepsilon_{yy}\right]dz\,dx\,dy$$

$$+ \iint_R \int_{-h/2}^{h/2} \left[2G\varepsilon_{xy}^2 + \frac{1}{2}\tau_{xz}(w_{,x} - \psi_x) + \frac{1}{2}\tau_{yz}(w_{,y} - \psi_y)\right]dz\,dx\,dy$$

(3.8.8)

Substituting for the strains from equation (3.8.2), and integrating with respect to z, the above equation (3.8.8) becomes

$$U = \frac{1}{2}\iint_R \int_{-h/2}^{h/2} \frac{Ez^2}{1-v^2}\left(\psi_{x,x}^2 + \psi_{y,y}^2 + 2v\psi_{x,x}\psi_{y,y}\right)dz\,dx\,dy$$

$$+ \frac{1}{2}\iint_R \int_{-h/2}^{h/2}\left[\begin{array}{c}Gz^2(\psi_{x,y}+\psi_{y,x})^2 + \tau_{xz}(w_{,x}-\psi_x) \\ +\tau_{yz}(w_{,y}-\psi_y)\end{array}\right]dz\,dx\,dy$$

$$= \frac{1}{2}\iint_R \frac{Eh^3}{12(1-v^2)}\left(\psi_{x,x}^2 + \psi_{y,y}^2 + 2v\psi_{x,x}\psi_{y,y}\right)dx\,dy$$

$$+ \frac{1}{2}\iint_R \left[\begin{array}{c}\frac{Gh^3}{12}(\psi_{x,y}+\psi_{y,x})^2 + kh\tau_{xz}(w_{,x}-\psi_x) \\ +kh\tau_{yz}(w_{,y}-\psi_y)\end{array}\right]dx\,dy$$

(3.8.9)

With the help of (3.8.3) and using the expression for the plate rigidity, D, the above equation (3.8.9) yields

$$U = \frac{1}{2}\iint_R D\left(\psi_{x,x}^2 + \psi_{y,y}^2 + 2v\psi_{x,x}\psi_{y,y}\right)dx\,dy$$

$$+ \frac{1}{2}\iint_R \left[\frac{Gh^3}{12}(\psi_{x,y}+\psi_{y,x})^2 + khG\left\{\begin{array}{c}(w_{,x}-\psi_x)^2 \\ +(w_{,y}-\psi_y)^2\end{array}\right\}\right]dx\,dy$$

(3.8.10)

External Work

$$W = -V = \iint_R qw\,dx\,dy$$

(3.8.11)

Hamilton's Principle

$$\delta \int_{t_1}^{t_2}\left[\begin{array}{c}\frac{1}{2}\iint_R \rho\left(\dot{\psi}_x^2 \frac{h^3}{12} + \dot{\psi}_y^2 \frac{h^3}{12} + h\dot{w}^2\right)dx\,dy \\ -\frac{1}{2}\iint_R D\left(\psi_{x,x}^2 + \psi_{y,y}^2 + 2v\psi_{x,x}\psi_{y,y}\right)dx\,dy \\ -\frac{1}{2}\iint_R \left[\begin{array}{c}\frac{Gh^3}{12}(\psi_{x,y}+\psi_{y,x})^2 \\ +khG\{(w_{,x}-\psi_x)^2+(w_{,y}-\psi_y)^2\}\end{array}\right]dx\,dy \\ + \iint_R qw\,dx\,dy\end{array}\right]dt = 0$$

(3.8.12)

96 Dynamics of Plates

Carrying out the variation in the above equation, we get

$$\int_{t_1}^{t_2} \left[\begin{array}{c} \rho \iint_R \left[\dfrac{h^3}{12}(\dot{\psi}_x \delta \dot{\psi}_x + \dot{\psi}_y \delta \dot{\psi}_y) + h\dot{w}\delta\dot{w} \right] dx\,dy \\ -D \iint_R (\psi_{x,x}\delta\psi_{x,x} + \psi_{y,y}\delta\psi_{y,y} + v\psi_{x,x}\delta\psi_{y,y} + v\psi_{y,y}\delta\psi_{x,x})dx\,dy \\ -\dfrac{Gh^3}{12} \iint_R (\psi_{x,y} + \psi_{y,x})(\delta\psi_{x,y} + \delta\psi_{y,x})dx\,dy \\ -khG \iint_R \left\{ \begin{array}{c} (w_{,x} - \psi_x)(\delta w_{,x} - \delta\psi_x) \\ +(w_{,y} - \psi_y)(\delta w_{,y} - \delta\psi_y) \end{array} \right\} dx\,dy \\ + \iint_R q\delta w\,dx\,dy \end{array} \right] dt = 0$$

(3.8.13)

We can make use of equations (2.1.11) and (2.1.12) and integrate each of the terms in the above equation (3.8.13) by parts to give

$$\int_{t_1}^{t_2} \left[\begin{array}{c} \iint_R \left[-\dfrac{\rho h^3}{12}(\ddot{\psi}_x \delta\psi_x + \ddot{\psi}_y \delta\psi_y) - \rho h \ddot{w}\delta w \right] dx\,dy \\ -D \oint_\Gamma \psi_{x,x}\delta\psi_x\,dy + D \oint_\Gamma \psi_{y,y}\delta\psi_y\,dx \\ +Dv \oint_\Gamma \psi_{x,x}\delta\psi_y\,dx - Dv \oint_\Gamma \psi_{y,y}\delta\psi_x\,dy \\ +D \iint_R (\psi_{x,xx}\delta\psi_x + \psi_{y,yy}\delta\psi_y + v\psi_{x,xy}\delta\psi_y + v\psi_{y,xy}\delta\psi_x)dx\,dy \\ +\dfrac{Gh^3}{12} \oint_\Gamma \psi_{x,y}\delta\psi_x\,dx - \dfrac{Gh^3}{12} \oint_\Gamma \psi_{x,y}\delta\psi_y\,dy \\ +\dfrac{Gh^3}{12} \oint_\Gamma \psi_{y,x}\delta\psi_x\,dx - \dfrac{Gh^3}{12} \oint_\Gamma \psi_{y,x}\delta\psi_y\,dy \\ +\dfrac{Gh^3}{12} \iint_R (\psi_{x,yy}\delta\psi_x + \psi_{x,xy}\delta\psi_x + \psi_{y,xy}\delta\psi_x + \psi_{y,xx}\delta\psi_y)dx\,dy \\ -khG \oint_\Gamma w_{,x}\delta w\,dy + khG \oint_\Gamma \psi_x\delta w\,dy \\ -khG \oint_\Gamma \psi_y\delta w\,dx + khG \oint_\Gamma w_{,y}\delta w\,dx \\ +khG \iint_R [(w_{,xx} - \psi_{x,x})\delta w + (w_{,x} - \psi_x)\delta\psi_x]dx\,dy \\ +khG \iint_R [(w_{,yy} - \psi_{y,y})\delta w + (w_{,y} - \psi_y)\delta\psi_y]dx\,dy \\ + \iint_R q\delta w\,dx\,dy \end{array} \right] dt = 0$$

(3.8.13)

Separating the terms of $\delta w, \delta\psi_x$ and $\delta\psi_y$ in the double integral terms of the above equation (3.8.13) and substituting for $2G = \dfrac{E}{(1+v)}$, the differential equations for the plate are obtained.

Rectangular Plates

$$\rho h \ddot{w} - khG(w_{,xx} + w_{,yy} - \psi_{x,x} - \psi_{y,y}) = q$$

$$D(\psi_{x,xx} + \tfrac{1-\nu}{2}\psi_{x,yy} + \tfrac{1+\nu}{2}\psi_{y,xy}) + khG(w_{,x} - \psi_x) - \tfrac{\rho h^3}{12}\ddot{\psi}_x = 0$$

$$D(\psi_{y,yy} + \tfrac{1-\nu}{2}\psi_{y,xx} + \tfrac{1+\nu}{2}\psi_{x,xy}) + khG(w_{,y} - \psi_y) - \tfrac{\rho h^3}{12}\ddot{\psi}_y = 0$$

(3.8.14)

The boundary conditions for the Mindlin plate are

$$-khG\left[\oint_\Gamma w_{,x}\,dy - \oint_\Gamma \psi_x\,dy + \oint_\Gamma \psi_y\,dx - \oint_\Gamma w_{,y}\,dx\right]\delta w = 0$$

$$\left[-D\left\{\oint_\Gamma (\psi_{x,x} + \nu\psi_{y,y})dy\right\} + \tfrac{Gh^3}{12}\left\{\oint_\Gamma (\psi_{x,y} + \psi_{y,x})dx\right\}\right]\delta\psi_x = 0$$

$$\left[D\left\{\oint_\Gamma (\psi_{y,y} + \nu\psi_{x,x})dx\right\} - \tfrac{Gh^3}{12}\left\{\oint_\Gamma (\psi_{x,y} + \psi_{y,x})dy\right\}\right]\delta\psi_y = 0$$

(3.8.15)

As in Timoshenko beam theory, we can eliminate ψ_x, ψ_y in equations (3.8.14) to obtain a single differential equation for the plate in w. For this purpose, we can write equations (3.8.14) in operator form

$$\rho h \partial_t^2 w - khG(\partial_x^2 w + \partial_y^2 w - \partial_x \psi_x - \partial_y \psi_y) = q$$

$$D\left(\partial_x^2 \psi_x + \tfrac{1-\nu}{2}\partial_y^2 \psi_x + \tfrac{1+\nu}{2}\partial_x\partial_y \psi_y\right) + khG(\partial_x w - \psi_x) - \tfrac{\rho h^3}{12}\partial_t^2 \psi_x = 0$$

$$D\left(\partial_y^2 \psi_y + \tfrac{1-\nu}{2}\partial_x^2 \psi_y + \tfrac{1+\nu}{2}\partial_x\partial_y \psi_x\right) + khG(\partial_y w - \psi_y) - \tfrac{\rho h^3}{12}\partial_t^2 \psi_y = 0$$

(3.8.16)

The above can be written in matrix form as

$$\begin{bmatrix} \rho h \partial_t^2 - khG(\partial_x^2 + \partial_y^2) & khG\partial_x & khG\partial_y \\ khG\partial_x & D(\partial_x^2 + \tfrac{1-\nu}{2}\partial_y^2) - khG & D\tfrac{1+\nu}{2}\partial_x\partial_y \\ khG\partial_y & D\tfrac{1+\nu}{2}\partial_x\partial_y & D(\partial_y^2 + \tfrac{1-\nu}{2}\partial_x^2) - khG \end{bmatrix} \begin{Bmatrix} w \\ \psi_x \\ \psi_y \end{Bmatrix}$$

$$= \begin{Bmatrix} q \\ 0 \\ 0 \end{Bmatrix}$$

(3.8.17)

Solving for w in the above equation, we get a single differential equation for the plate,

$$\left(\nabla^2 - \frac{\rho}{kG}\partial_t^2\right)\left(D\nabla^2 - \frac{\rho h^3}{12}\partial_t^2\right)w + \rho h \partial_t^2 w = \left(1 - \frac{D}{kGh}\nabla^2 + \frac{\rho h^2}{12kG}\partial_t^2\right)q$$

(3.8.18)

The boundary conditions in equation (3.8.15) are rewritten with the help of (3.8.4) to (3.8.6) and substituting for $2G = \frac{E}{(1+v)}$, as

$$\left[-\oint_\Gamma V_x\,dy + \oint_\Gamma V_y\,dx\right]\delta w = 0$$
$$\left[\oint_\Gamma M_x\,dy + \oint_\Gamma M_{xy}\,dx\right]\delta\psi_x = 0$$
$$\left[-\oint_\Gamma M_y\,dx - \oint_\Gamma M_{xy}\,dy\right]\delta\psi_y = 0$$

(3.8.19)

With the help of equations (2.2.27), we can rewrite the above

$$-\left[\oint_\Gamma (V_x a_{vx} + V_y a_{vy})dl\right]\delta w = 0$$
$$\left[\oint_\Gamma (M_x a_{vx} - M_{xy} a_{vy})dl\right]\delta\psi_x = 0$$
$$\left[\oint_\Gamma (M_y a_{vy} - M_{xy} a_{vx})dl\right]\delta\psi_y = 0$$

(3.8.20)

To write the above boundary conditions in terms of the normal and tangential coordinates for an edge, we can use

$$\psi_x = \psi_v a_{vx} - \psi_l a_{vy}$$
$$\psi_y = \psi_v a_{vy} + \psi_l a_{vx}$$

(3.8.21)

Equation (3.8.20) can now be written as

$$-\left[\oint_\Gamma (V_x a_{vx} + V_y a_{vy})dl\right]\delta w = 0$$
$$\left[\oint_\Gamma (M_x a_{vx} - M_{xy} a_{vy})dl\right](\delta\psi_v a_{vx} - \delta\psi_l a_{vy}) = 0$$
$$\left[\oint_\Gamma (M_y a_{vy} - M_{xy} a_{vx})dl\right](\delta\psi_v a_{vy} + \delta\psi_l a_{vx}) = 0$$

(3.8.22)

Making use of (2.2.30) and expanding the second and third equations in (3.8.22) above, we get

$$-\left[\oint_\Gamma V_\nu dl\right]\delta w = 0$$
$$\left[\oint_\Gamma \{(M_x a_{vx}^2 - M_{xy}a_{vx}a_{vy})\delta\psi_\nu - (M_x a_{vx}a_{vy} - M_{xy}a_{vy}^2)\delta\psi_l\}dl\right] = 0$$
$$\left[\oint_\Gamma \{(M_y a_{vy}^2 - M_{xy}a_{vx}a_{vy})\delta\psi_\nu + (M_y a_{vx}a_{vy} - M_{xy}a_{vx}^2)\delta\psi_l\}dl\right] = 0$$

(3.8.23)

We can regroup the terms in the second and third equations of the above to give

$$-\left[\oint_\Gamma V_\nu dl\right]\delta w = 0$$
$$\left[\oint_\Gamma (M_x a_{vx}^2 + M_y a_{vy}^2 - 2M_{xy}a_{vx}a_{vy})\delta\psi_\nu dl\right] = 0$$
$$\left[\oint_\Gamma \{(M_y - M_x)a_{vx}a_{vy} - M_{xy}(a_{vx}^2 - a_{vy}^2)\}\delta\psi_l dl\right] = 0$$

(3.8.23a)

Using (2.2.30), once again, the above simplifies to

$$-\left[\oint_\Gamma V_\nu dl\right]\delta w = 0$$
$$\left[\oint_\Gamma M_\nu dl\right]\delta\psi_\nu = 0$$
$$\left[\oint_\Gamma M_{\nu l} dl\right]\delta\psi_l = 0$$

(3.8.24)

Therefore, in Mindlin plate theory, we have three conditions along the boundary

$V_\nu = 0$ or w specified
$M_\nu = 0$ or ψ_ν specified
$M_{\nu l} = 0$ or ψ_l specified

(3.8.25)

Simply supported rectangular plate

To determine the influence of transverse shear and rotary inertia, let us consider the free vibration of a simply supported rectangular plate of width a in x direction and length b in y direction with thickness h. Let us consider the differential equation (3.8.18) for this purpose, instead of dealing with three separate equations, with the external force set to zero

$$\left(\nabla^2 - \frac{\rho}{kG}\partial_t^2\right)\left(D\nabla^2 - \frac{\rho h^3}{12}\partial_t^2\right)w + \rho h \partial_t^2 w = 0$$

(3.8.26)

The free vibration solution for this can be written as

$$w = W(x,y) \sin pt \qquad (3.8.27)$$

Substituting in (3.8.26) above, we get

$$\left(\nabla^2 + \frac{\rho}{kG} p^2\right)\left(D\nabla^2 + \frac{\rho h^3}{12} p^2\right)W - \rho h p^2 W = 0 \qquad (3.8.28)$$

As in section 3.3, we can assume a series solution to the above, the mn term being

$$W(x,y) = C_{mn} \sin \frac{m\pi x}{a} \sin \frac{n\pi y}{b} \qquad (3.8.29)$$

Substituting the above in (3.8.28), we get

$$\left[\frac{\rho}{kG} p_{mn}^2 - \left\{\left(\frac{m\pi}{a}\right)^2 + \left(\frac{n\pi}{b}\right)^2\right\}\right]$$
$$\times \left[\frac{\rho h^3}{12} p_{mn}^2 - D\left\{\left(\frac{m\pi}{a}\right)^2 + \left(\frac{n\pi}{b}\right)^2\right\}\right] - \rho h p_{mn}^2$$
$$= 0 \qquad (3.8.30)$$

The above can be rewritten to give the frequency equation

$$\frac{\rho^2 h^3}{12kG} p_{mn}^4 - p_{mn}^2\left[\rho h + \left\{\left(\frac{m\pi}{a}\right)^2 + \left(\frac{n\pi}{b}\right)^2\right\}\left(\frac{\rho}{kG} D + \frac{\rho h^3}{12}\right)\right]$$
$$+ D\left\{\left(\frac{m\pi}{a}\right)^2 + \left(\frac{n\pi}{b}\right)^2\right\}^2$$
$$= 0 \qquad (3.8.31)$$

The classical plate theory relation (3.3.16) is obtained from the above by setting $\frac{\rho h^3}{12}$ and $\frac{1}{kGh}$ equal to zero.

$$p_{mn}^2 = \frac{D}{ph}\left\{\left(\frac{m\pi}{a}\right)^2 + \left(\frac{n\pi}{b}\right)^2\right\}^2$$

(3.3.16)

Let

$$\lambda_{mn}^2 = \frac{ph^2(1-v^2)}{12Es^4}p_{mn}^2$$

$$s^2 = \frac{1}{12}\left(\frac{h}{a}\right)^2$$

$$\chi_{mn} = \left(\frac{m}{a}\right) \div \left(\frac{n}{b}\right)$$

(3.8.32)

and substituting for $D = \frac{Eh^3}{12(1-v^2)}$, equation (3.3.16) becomes

$$p_{mn}^2 = \frac{Eh^3 m^4 \pi^4}{12(1-v^2)pha^4}\left\{1 + \frac{1}{\chi_{mn}^2}\right\}^2$$

$$[\lambda_{mn}^2]_{classical} = m^4 \pi^4 \Gamma_{mn}^4$$

(3.8.33)

where

$$\Gamma_{mn}^2 = 1 + \frac{1}{\chi_{mn}^2}$$

(3.8.34)

Now substituting (3.8.32) in (3.8.30) and rearranging, we get

$$\lambda_{mn}^4 - \lambda_{mn}^2\left[\frac{k'}{s^4} + \Gamma_{mn}^2\left(\frac{m\pi}{s}\right)^2(1+k')\right] + k'\left(\frac{m\pi}{s}\right)^4 \Gamma_{mn}^4 = 0$$

(3.8.35)

where

$$k' = \frac{1}{2}k(1-v)$$

(3.8.36)

The lower root of the above quadratic equation (3.8.35) gives the first group of non dimensional natural frequencies

102 Dynamics of Plates

$$\lambda_{mn}^2 = \frac{1}{2}\left[\frac{k'}{s^4} + \Gamma_{mn}^2\left(\frac{m\pi}{s}\right)^2(1+k')\right]$$

$$-\sqrt{\frac{1}{4}\left[\frac{k'}{s^4} + \Gamma_{mn}^2\left(\frac{m\pi}{s}\right)^2(1+k')\right]^2 - k'\left(\frac{m\pi}{s}\right)^4\Gamma_{mn}^4}$$

(3.8.37)

Let us consider two cases, a thin square plate and a thick square plate and determine the first mode frequency and compare the results with classical plate theory.;

Thin plate: Let $k' = 0.1, s^2 = 10^{-5}$
From (3.8.34)

$$\Gamma_{11}^2 = 1 + \frac{1}{\chi_{11}^2} = 2$$

From equation (3.8.37)

$$\lambda_{11}^2 = \frac{1}{2}[10^9 + 2 \times 10^5 \times \pi^2 \times 1.1]$$

$$- \sqrt{\frac{1}{4}[10^9 + 2 \times 10^5 \times \pi^2 \times 1.1]^2 - 0.1 \times \pi^4 \times 10^{10} \times 4}$$

$$= 5.01085656.4 \times 10^8 - \sqrt{2.51087 \times 10^{17} - 38.9636 \times 10^{10}}$$

$$= 5.01085656.4 \times 10^8 - 5.01085267.6 \times 10^8 = 388.8$$

Compared with the classical value, 389.64, given in Table 3.3.1, the frequency parameter decreased by only 0.22%.

Thick plate: Let $k' = 0.1, s^2 = 10^{-3}$
From equation (3.8.37)

$$\lambda_{11}^2 = \frac{1}{2}[10^5 + 2 \times 10^3 \times \pi^2 \times 1.1]$$

$$- \sqrt{\frac{1}{4}[10^5 + 2 \times 10^3 \times \pi^2 \times 1.1]^2 - 0.1 \times \pi^4 \times 10^6 \times 4}$$

$$= 6.085656484 \times 10^4 - \sqrt{3.70352 \times 10^9 - 38.9636 \times 10^6}$$

$$= 60856.56484 - 60535.59158 = 320.97$$

Compared with the classical value, 389.64, given in Table 3.3.1, the frequency parameter for the thick plate decreased by 17.62%. So we note that the correction due to shear and rotary inertia becomes important for thick plates and for thin plates, the correction is negligible.

3.9 PLATE ON VISCOELASTIC FOUNDATION

Consider a finite rectangular plate of dimensions a in x direction and b in y direction with h as its thickness. The plate is assumed to rest on a Kelvin type viscoelastic foundation having a stiffness coefficient k and damping coefficient c per unit area, with a shear layer of thickness H superimposed between the plate and the foundation as shown in Fig. 3.9.1. Let p be the excitation force acting on the plate and if q represents the foundation reaction on the plate, the equation of motion for the plate in accordance to the classical plate theory is given by

$$D\nabla^4 w(x,y,t) + \rho h \ddot{w} = p(x,y,t) - q(x,y,t) \tag{3.9.1}$$

The equilibrium relation for the shear layer are shown in Fig. 3.9.2, from which we can write

$$[-q(x,y,t) + q_s(x,y,t)]dx\,dy = \left(Q_x + \frac{\partial Q_x}{\partial x} - Q_x\right)dx\,dy$$
$$+ \left(Q_y + \frac{\partial Q_y}{\partial y} - Q_y\right)dx\,dy$$

i.e., $\quad \dfrac{\partial Q_x}{\partial x} + \dfrac{\partial Q_y}{\partial y} = q(x,y,t) + q_s(x,y,t)$

$$\tag{3.9.2}$$

where Q_x and Q_y are the shear forces per unit length of the shear layer in the direction of x and y axes respectively, given by

$$Q_x = \int_0^H \tau_{xz}\,dz$$

$$Q_y = \int_0^H \tau_{yz}\,dz$$

$$\tag{3.9.3}$$

104 Dynamics of Plates

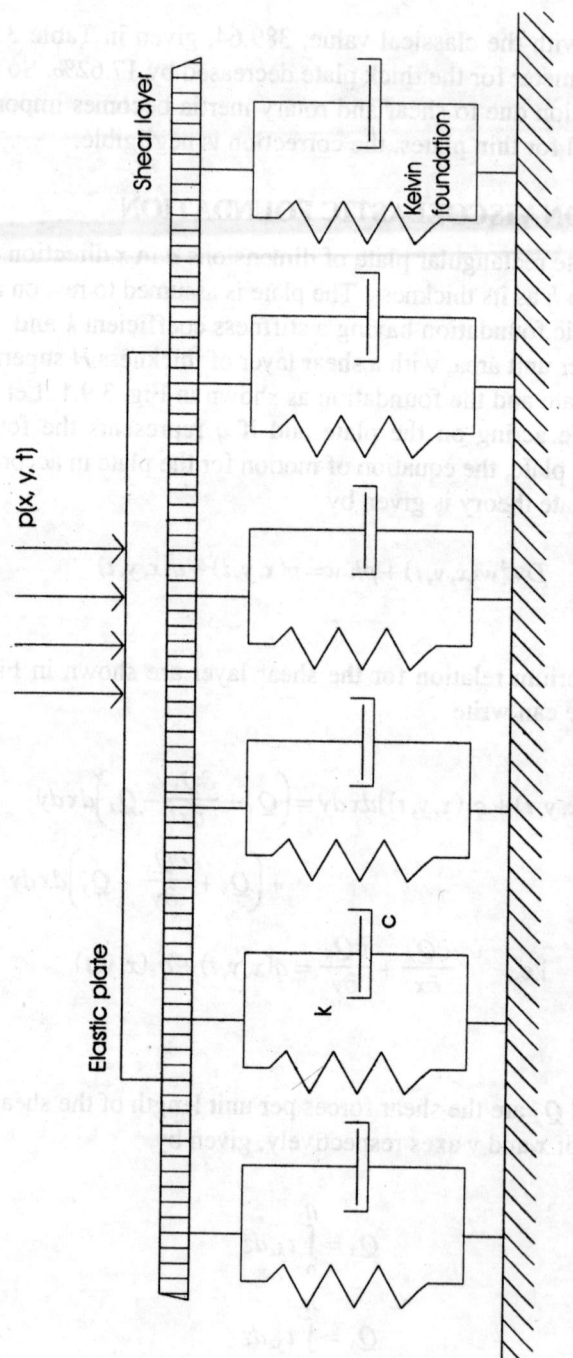

Fig. 3.9.1 Rectangular Elastic Plate on Viscoelastic Foundation

Writing the shear stresses as

$$\tau_{xz} = w_{,x}$$
$$\tau_{yz} = w_{,y}$$

equations (3.9.3) give

$$Q_x = GHw_{,x}$$
$$Q_y = GHw_{,y}$$
(3.9.4)

For a rate sensitive shear layer, the above equations are modified as

$$Q_x = GHw_{,xt}$$
$$Q_y = GHw_{,yt}$$
(3.9.5)

The reaction force between the shear layer and the viscoelastic foundation q_s is given by

$$q_s(x,y,t) = kw(x,y,t) + c\,\dot{w}(x,y,t)$$
(3.9.6)

Substituting (3.9.5) in (3.9.2), we get

$$GH\frac{\partial}{\partial t}[\nabla^2 w(x,y,t)] = -q(x,y,t) + q_s(x,y,t)$$
(3.9.7)

Substituting for q_s from (3.9.6), the above equation becomes

$$GH\frac{\partial}{\partial t}[\nabla^2 w(x,y,t)] = -q(x,y,t) + kw(x,y,t) + c\,\dot{w}(x,y,t)$$
(3.9.8)

With the help of the above equation (3.9.8), the equation of motion in (3.9.1) can be written as

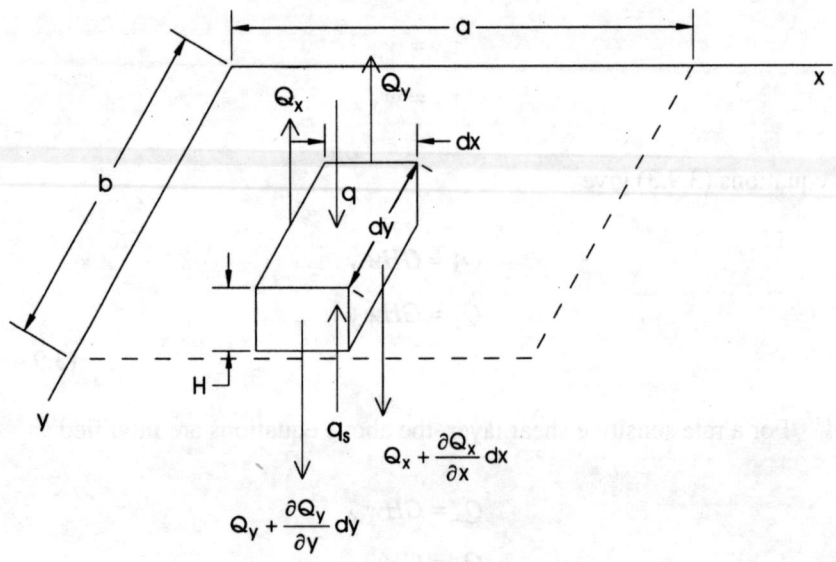

Fig. 3.9.2 Element of Shear Layer

$$D\nabla^4 w - GH\frac{\partial}{\partial t}[\nabla^2 w] + \rho h \ddot{w} + c\dot{w} + kw = p(x,y,t)$$

(3.9.9)

The boundary conditions for the plate are

$$w = w_{,xx} = 0 \quad \text{on } x = 0 \text{ and } x = a$$
$$w = w_{,yy} = 0 \quad \text{on } y = 0 \text{ and } y = b$$

(3.9.10)

The solution of the differential equation (3.9.9) can be obtained by Galerkin method. The solution is assumed as

$$w(x,y,t) = \sum_{m=1}^{\infty}\sum_{n=1}^{\infty} \phi_{mn}(x,y) f_{mn}(t)$$

(3.9.11)

where the shape functions ϕ_{mn} are assumed to satisfy the differential equation (3.3.13) and the above boundary conditions (3.9.10)

$$\nabla^4 \phi_{mn}(x,y) = \beta_{mn}^2 \phi_{mn}(x,y)$$

(3.3.13)

In the above

$$\phi_{mn}(x,y) = \sin\frac{m\pi x}{a} \sin\frac{n\pi y}{b}$$

(3.3.18)

$$\beta_{mn}^2 = \frac{ph}{D}p_{mn}^2$$

(3.3.14)

and

$$p_{mn}^2 = \frac{\pi^4 D}{ph}\left(\frac{m^2}{a^2} + \frac{n^2}{b^2}\right)^2$$

(3.3.17)

Substituting equation (3.9.11) in (3.9.9), we get the error in the differential equation as

$$\epsilon = D\sum_{m=1}^{\infty}\sum_{n=1}^{\infty} \nabla^4\phi_{mn}(x,y)f_{mn}(t) - GH\sum_{m=1}^{\infty}\sum_{n=1}^{\infty} \nabla^2\phi_{mn}(x,y)\dot{f}_{mn}(t)$$
$$+ ph\sum_{m=1}^{\infty}\sum_{n=1}^{\infty} \phi_{mn}(x,y)\ddot{f}_{mn}(t) + c\sum_{m=1}^{\infty}\sum_{n=1}^{\infty} \phi_{mn}(x,y)\dot{f}_{mn}(t)$$
$$+ k\sum_{m=1}^{\infty}\sum_{n=1}^{\infty} \phi_{mn}(x,y)f_{mn}(t) - p(x,y,t)$$

(3.9.12)

Substituting (3.3.13) in the above equation (3.9.12), we get

$$\epsilon = \sum_{m=1}^{\infty}\sum_{n=1}^{\infty}\left[\begin{array}{c} f_{mn}(t)\{D\beta_{mn}^2 + k\}\phi_{mn}(x,y) - GH\dot{f}_{mn}(t)\nabla^2\phi_{mn}(x,y) \\ ph\ddot{f}_{mn}(t)\phi_{mn}(x,y) + c\dot{f}_{mn}(t)\phi_{mn}(x,y) \end{array}\right]$$
$$- p(x,y,t)$$

(3.9.13)

According to Galerkin's method the error in the above equation is orthogonalized with the assumed mode shapes, i.e.,

108 Dynamics of Plates

$$\iint \left[\sum_{m=1}^{\infty} \sum_{n=1}^{\infty} \left\{ \begin{array}{c} f_{mn}(t)\{D\beta_{mn}^2 + k\}\phi_{mn}(x,y) \\ -GH f_{mn}(t)\nabla^2 \phi_{mn}(x,y) \\ \rho h \ddot{f}_{mn}(t)\phi_{mn}(x,y) + c\dot{f}_{mn}(t)\phi_{mn}(x,y) \end{array} \right\} - p(x,y,t) \right]$$

$$\times \phi_{m'n'}(x,y)dx\,dy$$

$$= 0$$

(3.9.14)

Using the result in equation (3.3.3) and rearranging, the above integral yields

$$[\rho h \ddot{f}_{mn}(t) + c\dot{f}_{mn}(t) + \{D\beta_{mn}^2 + k\}f_{mn}(t)]\frac{ab}{4}$$

$$+ GH \sum_{i=1}^{\infty} f_i(t)\frac{ab}{4}\left[\left(\frac{m\pi}{a}\right)^2 + \left(\frac{n\pi}{b}\right)^2\right]$$

$$- \iint p(x,y,t)\phi_{m'n'}(x,y)dx\,dy$$

$$= 0$$

(3.9.15)

Let $p(x,y,t)$ be a harmonic exciting force with frequency ω, expressed in the form of (3.3.1) and (3.3.4a) so that

$$p(x,y,t) = \sum_{m=1}^{\infty} \sum_{n=1}^{\infty} \frac{16P}{mn\pi^2} \sin\frac{m\pi x}{a} \sin\frac{n\pi y}{b} \sin\omega t \quad m \text{ and } n \text{ both odd}$$

(3.9.16)

Substituting the above in (3.9.15) and using the result of (3.3.3), we get

$$\ddot{f}_{mn}(t) + \frac{1}{\rho h}\left[c + GH\left\{\left(\frac{m\pi}{a}\right)^2 + \left(\frac{n\pi}{b}\right)^2\right\}\right]\dot{f}_{mn}(t)$$

$$+ \frac{D}{\rho h}\left\{\beta_{mn}^2 + \frac{k}{D}\right\}f_{mn}(t)$$

$$= \frac{16P}{\rho h mn\pi^2} \sin\omega t$$

(3.9.17)

Equation (3.9.17) can now be written as

$$\ddot{f}_{mn}(t) + 2r_{mn}\dot{f}_{mn}(t) + \lambda_{mn}^2 f_{mn}(t) = F_{mn}\sin\omega t \qquad (3.9.18)$$

where

$$2r_{mn} = \frac{1}{\rho h}\left[c + GH\left\{\left(\frac{m\pi}{a}\right)^2 + \left(\frac{n\pi}{b}\right)^2\right\}\right]$$

$$\lambda_{mn}^2 = \frac{D}{\rho h}\left\{\beta_{mn}^2 + \frac{k}{D}\right\} = p_{mn}^2\left[1 + \frac{k}{\pi^4 D\left(\frac{m^2}{a^2} + \frac{n^2}{b^2}\right)^2}\right]$$

$$F_{mn} = \frac{16P}{\rho h m n \pi^2} \qquad (3.9.19)$$

The natural frequencies of the plate on the viscoelastic foundation are higher than those of the classical frequencies of the simply supported rectangular plate given by

$$\lambda_{mn} = p_{mn}\sqrt{\left[1 + \frac{k}{\pi^4 D\left(\frac{m^2}{a^2} + \frac{n^2}{b^2}\right)^2}\right]} \qquad (3.9.20)$$

The solution of (3.9.18) is

$$f_{mn}(t) = \frac{F_{mn}\sin(\omega t - \psi_{mn})}{\sqrt{(\lambda_{mn}^2 - \omega^2)^2 + 4r_{mn}^2\omega^2}} \qquad (3.9.21)$$

where

$$\psi_{mn} = \tan^{-1}\left[\frac{2r_{mn}\omega}{\lambda_{mn}^2 - \omega^2}\right] \qquad (3.9.22)$$

Therefore, the response of the plate from (3.9.11) is given by

$$w(x,y,t) = \sum_{m=1}^{\infty}\sum_{n=1}^{\infty} \frac{F_{mn}\sin\frac{m\pi x}{a}\sin\frac{n\pi y}{b}}{\sqrt{(\lambda_{mn}^2 - \omega^2)^2 + 4r_{mn}^2\omega^2}}\sin(\omega t - \psi_j) \qquad (3.9.23)$$

4

Circular Plates

In chapter 2, we considered potential functionals with two independent space coordinates (for static deformations) or three independent parameters, two space coordinates and time parameter (for dynamical problems). These procedures are applied in chapter 3 to understand the behavior of rectangular shaped plates. Here, we consider circular plates, e.g., turbine discs. Evidently, it is convenient that the differential equation and boundary conditions for circular plates are treated in polar coordinates. We begin with this task first.

4.1 EQUATIONS OF BENDING OF PLATES IN POLAR COORDINATES

An element of a circular plate of thickness h and width dr subtending an angle $d\theta$ located at radius r and angle θ plate from x axis is shown in Fig. 4.1.1. The relations between Cartesian and polar coordinates are given by

$$r^2 = x^2 + y^2$$
$$\theta = \tan^{-1}\frac{y}{x}$$

(4.1.1)

The differentials in Cartesian coordinates can be converted into polar coordinates by using

$$\frac{\partial}{\partial x} = \frac{\partial}{\partial r}\frac{\partial r}{\partial x} + \frac{\partial}{\partial \theta}\frac{\partial \theta}{\partial x}$$
$$\frac{\partial}{\partial y} = \frac{\partial}{\partial r}\frac{\partial r}{\partial y} + \frac{\partial\partial\theta}{\partial\theta}\frac{\partial \theta}{\partial y}$$

(4.1.2)

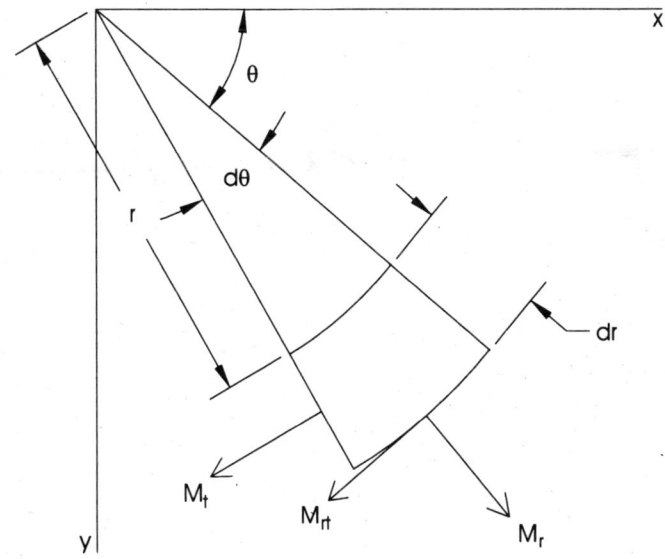

Fig. 4.1.1 Plate Element in Polar Coordinates

From (4.1.1), we get

$$\frac{\partial r}{\partial x} = \frac{x}{r} = \cos\theta$$

$$\frac{\partial r}{\partial y} = \frac{y}{r} = \sin\theta$$

(4.1.3a)

$$\frac{\partial \theta}{\partial x} = -\frac{y}{r^2} = -\frac{\sin\theta}{r}$$

$$\frac{\partial \theta}{\partial y} = \frac{x}{r^2} = \frac{\cos\theta}{r}$$

(4.1.3b)

Equation (4.1.2) can now be written as

$$\frac{\partial}{\partial x} = \frac{\partial}{\partial r}\cos\theta - \frac{1}{r}\sin\theta\frac{\partial}{\partial \theta}$$

$$\frac{\partial}{\partial y} = \frac{\partial}{\partial r}\sin\theta + \frac{1}{r}\cos\theta\frac{\partial}{\partial \theta}$$

(4.1.4)

Another differentiation yields

$$\frac{\partial^2}{\partial x^2} = \cos\theta \frac{\partial}{\partial r}\left(\frac{\partial}{\partial r}\cos\theta - \frac{1}{r}\sin\theta\frac{\partial}{\partial \theta}\right) - \frac{1}{r}\sin\theta\frac{\partial}{\partial \theta}\left(\frac{\partial}{\partial r}\cos\theta - \frac{1}{r}\sin\theta\frac{\partial}{\partial \theta}\right)$$

$$= \frac{\partial^2}{\partial r^2}\cos^2\theta + \frac{2}{r^2}\sin\theta\cos\theta\frac{\partial}{\partial \theta} - \frac{2}{r}\sin\theta\cos\theta\frac{\partial^2}{\partial r\partial \theta} + \frac{1}{r}\sin^2\theta\frac{\partial}{\partial r}$$

$$+ \frac{1}{r^2}\sin^2\theta\frac{\partial^2}{\partial \theta^2}$$

$$\frac{\partial^2}{\partial y^2} = \sin\theta\frac{\partial}{\partial r}\left(\frac{\partial}{\partial r}\sin\theta + \frac{1}{r}\cos\theta\frac{\partial}{\partial \theta}\right) + \frac{1}{r}\cos\theta\frac{\partial}{\partial \theta}\left(\frac{\partial}{\partial r}\sin\theta + \frac{1}{r}\cos\theta\frac{\partial}{\partial \theta}\right)$$

$$= \frac{\partial^2}{\partial r^2}\sin^2\theta - \frac{2}{r^2}\sin\theta\cos\theta\frac{\partial}{\partial \theta} + \frac{2}{r}\sin\theta\cos\theta\frac{\partial^2}{\partial r\partial \theta} + \frac{1}{r}\cos^2\theta\frac{\partial}{\partial r}$$

$$+ \frac{1}{r^2}\cos^2\theta\frac{\partial^2}{\partial \theta^2}$$

(4.1.5)

In a similar manner, we can obtain

$$\frac{\partial}{\partial x \partial y} = \frac{\partial^2}{\partial r^2}\sin\theta\cos\theta - \frac{1}{r^2}\cos 2\theta\frac{\partial}{\partial \theta} + \frac{1}{r}\cos 2\theta\frac{\partial^2}{\partial r\partial \theta}$$

$$- \frac{1}{r}\sin\theta\cos\theta\frac{\partial}{\partial r} - \frac{1}{r^2}\sin\theta\cos\theta\frac{\partial^2}{\partial \theta^2}$$

(4.1.6)

The Laplacian operator $\frac{\partial^2}{\partial x^2} + \frac{\partial^2}{\partial x^2}$ in polar coordinates can be obtained from (4.1.5) as

$$\nabla^2(r,\theta) = \frac{\partial^2}{\partial r^2} + \frac{1}{r}\frac{\partial}{\partial r} + \frac{1}{r^2}\frac{\partial^2}{\partial \theta^2}$$

(4.1.7)

Therefore the differential equation (3.2.36) for circular plates is

$$\left(\frac{\partial^2}{\partial r^2} + \frac{1}{r}\frac{\partial}{\partial r} + \frac{1}{r^2}\frac{\partial^2}{\partial \theta^2}\right)\left(\frac{\partial^2 w}{\partial r^2} + \frac{1}{r}\frac{\partial w}{\partial r} + \frac{1}{r^2}\frac{\partial^2 w}{\partial \theta^2}\right) = \frac{q}{D}$$

(4.1.8)

The bending and twisting moments shown in Fig. 4.1.1 can be obtained by considering the radial direction to coincide with the x axis, i.e., $\theta = 0$. Then, with the help of (3.2.45), (4.1.5) and (4.1.6) we get

$$M_r = -D(w_{,xx} + \nu w_{,yy})_{\theta=0}$$

$$= -D\left[w_{,rr} + \nu\left(\frac{1}{r}w_{,r} + \frac{1}{r^2}w_{,\theta\theta}\right)\right]$$

$$M_t = -D(w_{,yy} + \nu w_{,xx})_{\theta=0}$$

$$= -D\left[\frac{1}{r}w_{,r} + \frac{1}{r^2}w_{,\theta\theta} + \nu w_{,rr}\right]$$

$$M_{rt} = D(1-\nu)(w_{,xy})_{\theta=0}$$

$$= D(1-\nu)\left(\frac{1}{r}w_{,r\theta} - \frac{1}{r^2}w_{,\theta}\right)$$

(4.1.9)

In a similar manner, the shear forces in (3.2.47) and (3.2.48) are

$$V_r = \left[-D\frac{\partial}{\partial x}(\nabla^2 w)\right]_{\theta=0}$$

$$= -D\frac{\partial}{\partial r}(\nabla^2 w)$$

$$V_t = \left[-D\frac{\partial}{\partial y}(\nabla^2 w)\right]_{\theta=0}$$

$$= -\frac{D}{r}\frac{\partial}{\partial \theta}(\nabla^2 w)$$

(4.1.10)

The boundary conditions for a plate of radius a are

Fixed Edge

$$w(r=a) = 0$$
$$w_{,r}(r=a) = 0$$

Simply Supported Edge

$$w(r=a) = 0$$
$$M_r(r=a) = 0$$

Free Edge (Equation 3.2.51)

$$M_r(r = a) = 0$$

$$V_{\text{eff}} = \left[V_r - \frac{1}{r}M_{rt,\theta}\right](r = a) = 0$$

(4.1.11)

4.2 STRAIN ENERGY EXPRESSION IN POLAR COORDINATES

The strain energy for the plate in Cartesian coordinates is given by equation (3.2.33)

$$U = \iint \frac{D}{2}\left[(\nabla^2 w)^2 - 2(1-v)\{w_{,xx}w_{,yy} - w_{,xy}^2\}\right]dx\,dy$$

(3.2.33)

Using (4.1.5) and (4.1.6), the curved bracket term in the equation (3.2.33) above is

$$\begin{pmatrix} \frac{\partial^2 w}{\partial r^2}\cos^2\theta + \frac{2}{r^2}\sin\theta\cos\theta\frac{\partial w}{\partial \theta} - \frac{2}{r}\sin\theta\cos\theta\frac{\partial^2 w}{\partial r \partial \theta} \\ + \frac{1}{r}\sin^2\theta\frac{\partial w}{\partial r} + \frac{1}{r^2}\sin^2\theta\frac{\partial^2 w}{\partial \theta^2} \end{pmatrix}$$
$$\times \begin{pmatrix} \frac{\partial^2 w}{\partial r^2}\sin^2\theta - \frac{2}{r^2}\sin\theta\cos\theta\frac{\partial w}{\partial \theta} + \frac{2}{r}\sin\theta\cos\theta\frac{\partial^2 w}{\partial r \partial \theta} \\ + \frac{1}{r}\cos^2\theta\frac{\partial w}{\partial r} + \frac{1}{r^2}\cos^2\theta\frac{\partial^2 w}{\partial \theta^2} \end{pmatrix}$$
$$- \begin{pmatrix} \frac{\partial^2 w}{\partial r^2}\sin\theta\cos\theta - \frac{1}{r^2}\cos 2\theta\frac{\partial w}{\partial \theta} + \frac{1}{r}\cos 2\theta\frac{\partial^2 w}{\partial r \partial \theta} \\ - \frac{1}{r}\sin\theta\cos\theta\frac{\partial w}{\partial r} - \frac{1}{r^2}\sin\theta\cos\theta\frac{\partial^2 w}{\partial \theta^2} \end{pmatrix}^2$$
$$= \frac{\partial^2 w}{\partial r^2}\left(\frac{1}{r}\frac{\partial w}{\partial r} + \frac{1}{r^2}\frac{\partial^2 w}{\partial \theta^2}\right) - \left\{\frac{\partial}{\partial r}\left(\frac{1}{r}\frac{\partial w}{\partial \theta}\right)\right\}^2$$

(4.2.1)

Using (4.1.7) and the above expression in (4.2.1), the strain energy for a circular plate of radius a in (3.2.33) can be written as

$$U = \frac{D}{2}\int_0^{2\pi}\int_0^a \left[\begin{array}{c}\left(\frac{\partial^2 w}{\partial r^2} + \frac{1}{r}\frac{\partial w}{\partial r} + \frac{1}{r^2}\frac{\partial^2 w}{\partial \theta^2}\right)^2 - 2(1-v) \\ \left\{\frac{\partial^2 w}{\partial r^2}\left(\frac{1}{r}\frac{\partial w}{\partial r} + \frac{1}{r^2}\frac{\partial^2 w}{\partial \theta^2}\right) - \left[\frac{\partial}{\partial r}\left(\frac{1}{r}\frac{\partial w}{\partial \theta}\right)\right]^2\right\}\end{array}\right]r\,d\theta\,dr$$

(4.2.2)

For symmetric deflection of the plate, w is a function of r only and the above expression simplifies to

$$U = \pi D \int_0^a \left[\left(\frac{\partial^2 w}{\partial r^2} + \frac{1}{r}\frac{\partial w}{\partial r}\right)^2 - 2(1-\nu)\frac{\partial^2 w}{\partial r^2}\frac{1}{r}\frac{\partial w}{\partial r}\right] r\, dr$$

(4.2.3)

For a clamped edge as shown in section 3.6, equation (4.2.2) above simplifies to

$$U = \frac{D}{2} \int_0^{2\pi} \int_0^a \left(\frac{\partial^2 w}{\partial r^2} + \frac{1}{r}\frac{\partial w}{\partial r} + \frac{1}{r^2}\frac{\partial^2 w}{\partial \theta^2}\right)^2 r\, d\theta\, dr$$

(4.2.4)

For symmetric deflection of a clamped circular plate, the above simplifies further to

$$U = \pi D \int_0^a \left(\frac{\partial^2 w}{\partial r^2} + \frac{1}{r}\frac{\partial w}{\partial r}\right)^2 r\, dr$$

(4.2.5)

4.3 CLAMPED CIRCULAR PLATES

Static Deformation of Symmetrically Loaded Plate

Consider a circular plate of thickness h and radius a subjected to axially symmetric lateral force, $q(r)$. Evidently, the deflection surface will also be symmetric and independent of the angular coordinate θ. Therefore, the differential equation in (4.1.8) is

$$\left(\frac{\partial^2}{\partial r^2} + \frac{1}{r}\frac{\partial}{\partial r}\right)\left(\frac{\partial^2 w}{\partial r^2} + \frac{1}{r}\frac{\partial w}{\partial r}\right) = \frac{q(r)}{D}$$

$$\frac{1}{r}\frac{d}{dr}\left\{r\frac{d}{dr}\left[\frac{1}{r}\frac{d}{dr}\left(r\frac{dw}{dr}\right)\right]\right\} = \frac{q(r)}{D}$$

(4.3.1)

where

$$\nabla^2 w = \left(\frac{\partial^2 w}{\partial r^2} + \frac{1}{r}\frac{\partial w}{\partial r}\right)$$

$$= \frac{1}{r}\frac{d}{dr}\left(r\frac{dw}{dr}\right)$$

(4.3.2)

116 Dynamics of Plates

For simplicity, let us consider a uniformly distributed load, $q(r) = q$, equation (4.3.1) then becomes

$$\frac{d}{dr}\left\{r\frac{d}{dr}\nabla^2 w\right\} = \frac{qr}{D}$$

(4.3.3)

Integrating once with respect to r

$$\frac{d}{dr}(\nabla^2 w) = \frac{qr}{2D} + \frac{A_1}{r}$$

(4.3.4)

One more integration yields

$$\nabla^2 w = \frac{1}{r}\frac{d}{dr}\left(r\frac{dw}{dr}\right) = \frac{qr^2}{4D} + A_1 \ln r + B_1$$

(4.3.5)

We can integrate further to give

$$\frac{d}{dr}\left(r\frac{dw}{dr}\right) = \frac{qr^3}{4D} + A_1 r \ln r + B_1 r$$

$$r\frac{dw}{dr} = \frac{qr^4}{16D} + A_1\left(\frac{r^2}{2}\ln r - \frac{r^2}{4}\right) + B_1\frac{r^2}{2} + C_1$$

$$w = \frac{qr^4}{64D} + A_1\left(\frac{r^2}{4}\ln r - \frac{r^2}{4}\right) + B_1\frac{r^2}{4} + C_1 \ln r + E_1$$

(4.3.6)

We can redefine the constants of integration in the above equation and write the solution for the deflection surface as

$$w = \frac{qr^4}{64D} + Ar^2 \ln r + Br^2 + C \ln r + E$$

(4.3.7)

Since at $r=0$, the deflection is a finite value, the constants of integration $A = C = 0$. Therefore,

$$w = \frac{qr^4}{64D} + Br^2 + E$$

(4.3.8)

At $r = a$, we have

$$w(a) = \frac{qa^4}{64D} + Ba^2 + E = 0$$

$$w_{,r}(a) = \frac{qa^3}{16D} + 2Ba = 0$$

(4.3.9)

from which

$$B = -\frac{qa^2}{32D}$$

$$E = \frac{qa^4}{64D}$$

(4.3.10)

The deflection as a function of radial distance r is therefore

$$w(r) = \frac{q}{64D}(a^2 - r^2)^2$$

(4.3.11)

The maximum deflection occurs at the center of the plate

$$w_{max} = \frac{qa^4}{64D}$$

(4.3.12)

If the force applied is concentrated at the center of the plate and is of magnitude P, we can use the shear force relation in equation (4.1.10)

$$V_r = -D\frac{\partial}{\partial r}(\nabla^2 w)$$

(4.1.10a)

From (4.3.4), the above gives

$$\frac{d}{dr}(\nabla^2 w) = \frac{A_1}{r}$$

$$V_r = -D\frac{\partial}{\partial r}(\nabla^2 w)$$

$$= -D\frac{A_1}{r}$$

(4.3.13)

Force equilibrium of the plate demands that at any radius r

$$-2\pi r V_r = P \qquad (4.3.14)$$

Substituting in (4.3.13), we have

$$A_1 = \frac{P}{2\pi D} \qquad (4.3.15)$$

Noting that the constant of integration in (4.3.7) $A = \frac{A_1}{4}$ in (4.3.6), we have the deflection of the plate due to the concentrated force P at the center

$$w = \frac{P}{8\pi D} r^2 \ln r + B r^2 + E \qquad (4.3.16)$$

The first term on the right hand side of the above equation (4.3.16) represents a singularity solution. Applying the boundary condition $w_{,r} = 0$ at $r = a$,

$$B = -(1 + 2\ln a)\frac{P}{16\pi D} \qquad (4.3.17)$$

The second boundary condition $w = 0$ gives

$$E = \frac{Pa^2}{16\pi D} \qquad (4.3.18)$$

and

$$w = \frac{Pa^2}{8\pi D} \frac{r^2}{a^2} \ln \frac{r}{a} + \frac{Pa^2}{16\pi D}\left(1 - \frac{r^2}{a^2}\right) \qquad (4.3.19)$$

The maximum deflection occurs at the center of the plate

$$w_{max} = \frac{Pa^2}{16\pi D} \qquad (4.3.20)$$

Free Vibrations

For free vibrations, the differential equation is given by

$$D\nabla^4 w + \rho h \ddot{w} = 0 \tag{3.2.54}$$

For harmonic motion, $w(r,\theta,t) = W(r,\theta)\sin pt$ and the above can be written as

$$\nabla^4 W - \frac{\rho h}{D} p^2 W$$
$$= (\nabla^2 + \beta^2)(\nabla^2 - \beta^2)W = 0 \tag{4.3.21}$$

The complete solution to the above equation can be obtained by superimposing the solutions of the two following Bessel equations,

$$(\nabla^2 + \beta^2)W_1 = 0$$
$$(\nabla^2 - \beta^2)W_2 = 0 \tag{4.3.21a}$$

where $\nabla^2 = \frac{\partial^2}{\partial r^2} + \frac{1}{r}\frac{\partial}{\partial r} + \frac{1}{r^2}\frac{\partial^2}{\partial \theta^2}$ and the frequency parameter β is

$$\beta^2 = \sqrt{\frac{\rho h}{D}} p \tag{4.3.22}$$

The solution of (4.3.21) in terms of Fourier components can be assumed as

$$W(r,\theta) = \sum_{n=0}^{\infty} W_n(r)\cos n\theta + \sum_{n=0}^{\infty} W_n(r)\sin n\theta \tag{4.3.23}$$

Substituting the above in (4.3.21a), we can write the solution from Bessel equations as

$$W_n(r) = W_{n1}(r) + W_{n2}(r)$$
$$= A_n J_n(\beta r) + E_n Y_n(\beta r) + B_n I_n(\beta r) + F_n K_n(\beta r) \tag{4.3.23a}$$

where J_n, Y_n are Bessel functions of order n and I_n, K_n are modified Bessel functions of order n. Since Y and K tend to infinity as r tends to zero and that W is finite at the center of the plate, we take $E = F = 0$ and equation (4.3.23) gives

$$W_n = A_n J_n(\beta r) + B_n I_n(\beta r)$$
$$W'_n = A_n J'_n(\beta r) + B_n I'_n(\beta r)$$

(4.3.24)

Applying the boundary conditions

$$W_n(r=a) = A_n J_n(\beta a) + B_n I_n(\beta a) = 0$$
$$W'_n(r=a) = A_n J'_n(\beta a) + B_n I'_n(\beta a) = 0$$

(4.3.25)

Therefore

$$\frac{B_n}{A_n} = -\frac{J_n(\beta a)}{I_n(\beta a)} = -\frac{J'_n(\beta a)}{I'_n(\beta a)}$$

(4.3.26)

Making use of the recursive relation

$$\beta a J'_n(\beta a) = n J_n(\beta a) - \beta a J_{n+1}(\beta a)$$
$$\beta a I'_n(\beta a) = n I_n(\beta a) + \beta a I_{n+1}(\beta a)$$

(4.3.26a)

the frequency equation is obtained as

$$J_n(\beta a) I_{n+1}(\beta a) + I_n(\beta a) J_{n+1}(\beta a) = 0$$

(4.3.27)

Axisymmetric Vibration: From equation (4.3.23), we find that $n = 0$ represents the case of zero nodal diameters and the solution gives axisymmetric shape function for W. Equation (4.3.27) reduces for this case to

$$J_0(\beta a)I_1(\beta a) + I_0(\beta a)J_1(\beta a) = 0$$

(4.3.27a)

The first five roots of the above transcendental equation are

$$\beta_0 a = 3.1961$$
$$\beta_1 a = 6.3064$$
$$\beta_2 a = 9.4395$$
$$\beta_3 a = 12.577$$
$$\beta_4 a = 15.716$$

(4.3.28)

The natural frequencies are given by

$$p_i = \frac{(\beta_i a)^2}{a^2}\sqrt{\frac{D}{\rho h}} \text{ rad/s}$$

(4.3.29)

The corresponding mode shapes are given from equations (4.3.24) and (4.3.26)

$$W_i = C_i[I_0(\beta_i a)J_0(\beta_i r) - J_0(\beta_i a)I_0(\beta_i r)]$$
$$\text{or} \quad W_i = D_i[I_1(\beta_i a)J_0(\beta_i r) + J_1(\beta_i a)I_0(\beta_i r)]$$

(4.3.30)

where C_i and D_i are new constants. We can obtain orthonormal mode shapes by using

$$\iint_A \rho h \, W_i W_j \, dA = \delta_{ij}$$

(4.3.31)

From the above, using the first relation in (4.3.24), we have

$$\int_0^a \int_0^{2\pi} \rho h \, [AJ_0(\beta r) + BI_0(\beta r)]^2 r \, dr \, d\theta = 1$$

(4.3.32)

122 Dynamics of Plates

Integrating with respect to θ and expanding we get

$$A_i^2 \int_0^a rJ_0^2(\beta_i r)dr + B_i^2 \int_0^a rI_0^2(\beta_i r)dr$$
$$+ 2A_iB_i \int_0^a rJ_0(\beta_i r)I_0(\beta_i r)dr$$
$$= \frac{1}{2\pi\rho h}$$

(4.3.33)

The integrals in the above equation (4.3.33) are

$$\int_0^a rJ_0^2(\beta_i r)dr = \frac{a^2}{2}[J_0^2(\beta_i a) + J_1^2(\beta_i a)]$$

$$\int_0^a rI_0^2(\beta_i r)dr = \frac{a^2}{2}[I_0^2(\beta_i a) - I_1^2(\beta_i a)]$$

$$\int_0^a rJ_0(\beta_i r)I_0(\beta_i r)dr = \frac{a}{2\beta_i}[J_0(\beta_i a)I_1(\beta_i a) + I_0(\beta_i a)J_1(\beta_i a)]$$

$$= 0$$

(4.3.33a)

Equation (4.3.33) therefore becomes

$$A_i^2[J_0^2(\beta_i a) + J_1^2(\beta_i a)] + B_i^2[I_0^2(\beta_i a) - I_1^2(\beta_i a)] = \frac{1}{a^2\pi\rho h}$$

(4.3.34)

Upon rearrangement of the above

$$A_i\sqrt{[J_0^2(\beta_i a) + J_1^2(\beta_i a)] + \frac{B_i^2}{A_i^2}[I_0^2(\beta_i a) - I_1^2(\beta_i a)]} = \frac{1}{\sqrt{a^2\pi\rho h}}$$

(4.3.35)

Making use of (4.3.26), the above equation gives

$$A_i\sqrt{[J_0^2(\beta_i a) + J_1^2(\beta_i a)] + \frac{J_0^2(\beta_i a)}{I_0^2(\beta_i a)}[I_0^2(\beta_i a) - I_1^2(\beta_i a)]}$$

$$= A_i\sqrt{2J_0^2(\beta_i a)I_0^2(\beta_i a) + J_1^2(\beta_i a)I_0^2(\beta_i a) - J_0^2(\beta_i a)I_1^2(\beta_i a)}$$

$$= \frac{1}{\sqrt{a^2\pi\rho h}}$$

(4.3.36)

Now, making use of the frequency equation (4.3.27), the above yields

$$A_i = \frac{1}{J_0(\beta_i a) a \sqrt{2\pi \rho h}}$$

(4.3.37)

In a similar manner, we can obtain

$$B_i = -\frac{1}{I_0(\beta_i a) a \sqrt{2\pi \rho h}}$$

(4.3.38)

Therefore, the orthonormal mode shape of a clamped circular plate in axisymmetric motion is given by

$$W_i = \frac{1}{a}\sqrt{\frac{1}{2\pi \rho h}} \left[\frac{J_0(\beta_i r)}{J_0(\beta_i a)} - \frac{I_0(\beta_i r)}{I_0(\beta_i a)} \right]$$

(4.3.39)

The location of nodal circles is obtained from equation (4.3.30)

$$I_0(\beta_i a) J_0\left(\beta_i a \frac{r}{a}\right) = J_0(\beta_i a) I_0\left(\beta_i a \frac{r}{a}\right)$$

(4.3.40)

Using the values of $\beta_i a$ from (4.3.28), we get the locations of nodal circles for different modes in axisymmetric motion, which are:

$i = 0, \frac{r}{a} = 1$, represents no nodal circles other than the clamped boundary

$i = 1, \frac{r}{a} = 1, 0.379$ represents one nodal circle

$i = 2, \frac{r}{a} = 1, 0.583, 0.255$ represents two nodal circles

$i = 3, \frac{r}{a} = 1, 0.688, 0.439, 0.191$ represents three nodal circles

$i = 4, \frac{r}{a} = 1, 0.749, 0.55, 0.351, 0.153$ represents four nodal circles etc.

Free Vibration with Nodal Diameters For $n \geq 1$, the roots of equation (4.3.27) are given below.

Table 4.3.1 Values of $\beta_{ns}a$ for a clamped circular plate with $n \geq 1$

No. of Nodal Circles s	One Nodal Diameter $n = 1$	Two Nodal Diameters $n = 2$	Three Nodal Diameters $n = 3$
0	4.61	5.91	7.14
1	7.8	9.2	10.54
2	10.96	12.4	13.79
3	14.11	15.58	17.01
4	17.26	18.76	20.19

The nodal diameters are equiangularly spaced and the mode shapes as a function of the radial distance for different values of n and s are obtained from equations (4.3.24) and (4.3.26).

$$W_{ns} = \left[I_n(\beta_{ns}a) J_n\!\left(\beta_{ns}a\tfrac{r}{a}\right) - J_n(\beta_{ns}a) I_n\!\left(\beta_{ns}a\tfrac{r}{a}\right) \right]$$

(4.3.41)

The nodal circle locations are obtained from

$$I_n(\beta_{ns}a) J_n\!\left(\beta_{ns}a\tfrac{r}{a}\right) = J_n(\beta_{ns}a) I_n\!\left(\beta_{ns}a\tfrac{r}{a}\right)$$

(4.3.42)

These values are given in the table below.

Table 4.3.2 Location of nodal circles r/a for $n \geq 1$

No. of Nodal Circles s	One Nodal Diameter $n = 1$	Two Nodal Diameters $n = 2$	Three Nodal Diameters $n = 3$
1	1.000	1.000	1.000
	0.4899	0.559	0.606
2	1.000	1.000	1.000
	0.64	0.679	0.708
	0.35	0.414	0.462
3	1.000	1.000	1.000
	0.721	0.746	0.765
	0.497	0.540	0.574
	0.272	0.330	0.375
4	1.000	1.000	1.000
	0.767	0.789	0.803
	0.589	0.620	0.645
	0.407	0.449	0.488
	0.222	0.274	0.316

Response by Modal Analysis

The differential equation for forced vibration response of a circular plate of radius a and thickness h is given by

$$D\nabla^4 w + \rho h \ddot{w} = q(r,t) \tag{4.3.43}$$

where the density of the plate material is ρ and the flexural rigidity is D. For the present we restrict ourselves to axisymmetric vibrations with a clamped edge. Let the initial disturbance be given by

$$w(r,0) = w_0(r)$$
$$\dot{w}(r,0) = \dot{w}_0(r) \tag{4.3.44}$$

The general response is expressed in modal series

$$w = \sum_{i=1}^{\infty} W_i(r)\eta_i(t) \tag{4.3.45}$$

where W_i are orthonormal mode shapes given in equation (4.3.39) for the circular plate with fixed edge executing axisymmetric vibrations. Substituting (4.3.45) in (4.3.43), we get

$$D\nabla^4\left(\sum_{i=1}^{\infty} W_i(r)\eta_i(t)\right)w + \rho h\left(\sum_{i=1}^{\infty} W_i(r)\ddot{\eta}_i(t)\right) = q(r,t) \tag{4.3.46}$$

From (4.3.21), we can write

$$\nabla^4 W_i = \frac{\rho h}{D} p_i^2 W_i \tag{4.3.47}$$

and rewrite equation (4.3.46) as

$$\sum_{i=1}^{\infty} \rho h W_i(r)\left(\ddot{\eta}_i(t) + p_i^2 \eta_i(t)\right) = q(r,t) \tag{4.3.48}$$

Multiplying both sides of the above equation with $W_j(r)$ and integrating over the area of the plate, we get

$$\sum_{i=1}^{\infty}\left(\ddot{\eta}_i(t)+p_i^2\eta_i(t)\right)\iint_A \rho h W_i(r)W_j(r)\,dA = \iint_A q(r,t)W_j(r)\,dA$$
(4.3.49)

Using the orthonormal modes property in (4.3.31), the above equation simplifies to

$$\ddot{\eta}_j(t)+p_j^2\eta_j(t) = F_j(t)$$
(4.3.50)

where the generalized force $F_j(t)$ is

$$F_j(t) = \iint_A q(r,t)W_j(r)\,dA$$
(4.3.51)

The solution of (4.3.50) is

$$\eta_i(t) = \eta_i(0)\cos p_i t + \frac{\dot{\eta}_i(0)}{p_i}\sin p_i t + \frac{1}{p_i}\int_0^t F_i(\tau)\sin p_i(t-\tau)\,d\tau$$
(4.3.52)

To determine the $\eta_i(0)$ and $\dot{\eta}_i(0)$, we write

$$w_0 = \sum_{i=1}^{\infty} W_i(r)\eta_i(0)$$
(4.3.53)

and multiply both sides by $\rho h W_j(r)$ and integrate over the plate to give

$$\iint w_0 \rho h W_j(r)\,dA = \sum_{i=1}^{\infty} \eta_i(0)\iint \rho h W_j(r)W_i(r)\,dA$$
(4.3.54)

Therefore

$$\eta_j(0) = \iint w_0 \rho h W_j(r)\,dA$$
(4.3.55)

In a similar manner, we can obtain

$$\dot{\eta}_j(0) = \iint \dot{w}_0 \rho h W_j(r)\,dA$$
(4.3.56)

Consider the case where the plate is subjected to a suddenly applied force of magnitude q over the entire plate, then the generalized force from (4.3.53) and (4.3.39) is

$$F_j(t) = 2\pi \int_0^a q W_j(r) r \, dr$$

$$= \frac{2\pi q}{a} \sqrt{\frac{1}{2\pi \rho h}} \int_0^a \left[\frac{J_0(\beta_i r)}{J_0(\beta_i a)} - \frac{I_0(\beta_i r)}{I_0(\beta_i a)} \right] r \, dr$$

$$= \frac{2\pi q a}{\beta_i a} \sqrt{\frac{1}{2\pi \rho h}} \left[\frac{J_1(\beta_i a)}{J_0(\beta_i a)} - \frac{I_1(\beta_i a)}{I_0(\beta_i a)} \right]$$

(4.3.57)

Since the plate is at rest initially, $\eta_i(0)$ and $\dot{\eta}_i(0)$ are both zero, consequently, from (4.3.52) we get

$$\eta_i(t) = \frac{1}{p_i} \int_0^t F_i(\tau) \sin p_i(t-\tau) \, d\tau$$

$$= \frac{1}{p_i^2} \frac{2\pi q a}{\beta_i a} \sqrt{\frac{1}{2\pi \rho h}} \left[\frac{J_1(\beta_i a)}{J_0(\beta_i a)} - \frac{I_1(\beta_i a)}{I_0(\beta_i a)} \right] (1 - \cos p_i t)$$

(4.3.58)

Finally, from equations (4.3.45), (4.3.22) and (4.3.39), we get the response as

$$w = \frac{q a^4}{D} \sum_{i=1}^{\infty} \frac{1}{(\beta_i a)^5} \left[\frac{J_0(\beta_i r)}{J_0(\beta_i a)} - \frac{I_0(\beta_i r)}{I_0(\beta_i a)} \right] \left[\frac{J_1(\beta_i a)}{J_0(\beta_i a)} - \frac{I_1(\beta_i a)}{I_0(\beta_i a)} \right]$$
$$\times (1 - \cos p_i t)$$

(4.3.59)

4.4 SIMPLY SUPPORTED CIRCULAR PLATES

In the previous section, we considered clamped circular plates and determined the static deformation of symmetrically loaded plates, natural frequencies and mode shapes for a given nodal diameters and circles. We also discussed the modal analysis of axisymmetric clamped circular plates. In this section, we will study the natural frequencies and mode shapes of simply supported plates. The forced vibration analysis can be conducted in a similar manner as that of the previous section and is not dealt with here.

For free vibrations, the differential equation is given by

$$D\nabla^4 w + \rho h \ddot{w} = 0$$

(3.2.54)

For harmonic motion, $w(r, \theta, t) = W(r, \theta) \sin pt$ the above becomes

$$\nabla^4 W - \frac{\rho h}{D} p^2 W =$$
$$(\nabla^2 + \beta^2)(\nabla^2 - \beta^2) W = 0$$

(4.3.21)

As in the previous section, the complete solution to the above equation is

$$(\nabla^2 + \beta^2) W_1 = 0$$
$$(\nabla^2 - \beta^2) W_2 = 0$$

(4.3.21a)

where $\nabla^2 = \frac{\partial^2}{\partial r^2} + \frac{1}{r}\frac{\partial}{\partial r} + \frac{1}{r^2}\frac{\partial^2}{\partial \theta^2}$ and the frequency parameter β is

$$\beta^2 = \sqrt{\frac{\rho h}{D}} p$$

(4.3.22)

Expressing the solution of (4.3.21) in terms of Fourier components, see (4.3.23) and following the arguments given in the previous section, we get

$$W_n = A_n J_n(\beta r) + B_n I_n(\beta r)$$
$$W_n' = A_n J_n'(\beta r) + B_n I_n'(\beta r)$$

(4.3.24)

Circular Plates

The deflection boundary condition at the support gives

$$W_n(r=a) = A_n J_n(\beta a) + B_n I_n(\beta a) = 0$$

(4.4.1)

The bending moment boundary condition from (4.1.9) with $w_{,\theta\theta} = 0$ at the support gives

$$M_r(r=a) = \left[w_{,rr} + \frac{v}{r}w_{,r}\right]_{r=a}$$

$$= A_n\left[J_n''(\beta a) + \frac{v}{\beta a}J_n'(\beta a)\right] + B_n\left[I_n''(\beta a) + \frac{v}{\beta a}I_n'(\beta a)\right]$$

$$= 0$$

(4.4.2)

The frequency equation is therefore given by

$$\begin{vmatrix} J_n(\beta a) & I_n(\beta a) \\ \left[J_n''(\beta a) + \frac{v}{\beta a}J_n'(\beta a)\right] & \left[I_n''(\beta a) + \frac{v}{\beta a}I_n'(\beta a)\right] \end{vmatrix} = 0$$

i.e.,

$$J_n(\beta a)[\beta a I_n''(\beta a) + v I_n'(\beta a)] - I_n(\beta a)[\beta a J_n''(\beta a) + v J_n'(\beta a)] = 0$$

(4.4.3)

Making use of the recursive relations

$$\beta a J_n'(\beta a) = n J_n(\beta a) - \beta a J_{n+1}(\beta a)$$

$$\beta a I_n'(\beta a) = n I_n(\beta a) + \beta a I_{n+1}(\beta a)$$

(4.3.26a)

equation (4.4.3) gives

$$J_n(\beta a)\beta a\{n I_n'(\beta a) + \beta a I_{n+1}'(\beta a)\}$$
$$+ v J_n(\beta a)\{n I_n(\beta a) + \beta a I_{n+1}(\beta a)\}$$
$$- I_n(\beta a)\beta a\{n J_n'(\beta a) - \beta a J_{n+1}'(\beta a)\}$$
$$- v I_n(\beta a)\{n J_n(\beta a) - \beta a J_{n+1}(\beta a)\}$$
$$= 0$$

$$n\beta a I'_n(\beta a)J_n(\beta a) + \beta^2 a^2 I'_{n+1}(\beta a)J_n(\beta a)$$
$$- n\beta a I_n(\beta a)J'_n(\beta a) + \beta^2 a^2 I_n(\beta a)J'_{n+1}(\beta a)$$
$$+ v\beta a [J_n(\beta a)I_{n+1}(\beta a) + I_n(\beta a)J_{n+1}(\beta a)]$$
$$= 0$$

$$(2n+1)\{I_{n+1}(\beta a)J_n(\beta a) + J_{n+1}(\beta a)I_n(\beta a)\}$$
$$+ \beta a\{I_{n+2}(\beta a)J_n(\beta a) - J_{n+2}(\beta a)I_n(\beta a)\}$$
$$+ v[J_n(\beta a)I_{n+1}(\beta a) + I_n(\beta a)J_{n+1}(\beta a)]$$
$$= 0$$

(4.4.4)

Using the following recursive relations of Bessel functions

$$J_{n+2}(\beta a) = \frac{2}{\beta a}(n+1)J_{n+1}(\beta a) - J_n(\beta a)$$

$$I_{n+2}(\beta a) = -\frac{2}{\beta a}(n+1)I_{n+1}(\beta a) + I_n(\beta a)$$

(4.4.5)

equation (4.4.4) can be written as

$$2\beta a I_n(\beta a)J_n(\beta a) + (v-1)\{J_{n+1}(\beta a)I_n(\beta a) + I_{n+1}(\beta a)J_n(\beta a)\} = 0$$

Finally the frequency equation simplifies to

$$\left\{\frac{J_{n+1}(\beta a)}{J_n(\beta a)} + \frac{I_{n+1}(\beta a)}{I_n(\beta a)}\right\} = \frac{2\beta a}{1-v}$$

(4.4.6)

The roots of above frequency equation (4.4.6) are given in Table 4.4.1 for different combinations of n and s representing the number of nodal diameters and nodal circles respectively and Poisson's ratio $v = 0.3$.

Table 4.4.1 Values of $\beta_{ns}a$ for a simply supported circular plate for $v = 0.3$

No. of Nodal Circles s	Zero Nodal Diameters $n = 0$	One Nodal Diameter $n = 1$	Two Nodal Diameters $n = 2$
0	2.23	3.73	5.07
1	5.46	6.97	8.38
2	8.61	10.14	11.59
3	11.76	13.3	14.77

The nodal diameters are equiangularly spaced and the mode shapes as a function of the radial distance for different values of n and s are obtained from equation (4.4.1) which gives

$$\frac{A_{ns}}{B_{ns}} = -\frac{I_n(\beta_{ns}a)}{J_n(\beta_{ns}a)}$$

(4.4.7)

With the help of the above value the mode shape is given from (4.3.24) as

$$W_{ns} = \frac{A_{ns}}{B_{ns}} J_n\left(\beta_{ns}a\frac{r}{a}\right) + I_n\left(\beta_{ns}a\frac{r}{a}\right)$$

$$= -I_n(\beta_{ns}a)J_n\left(\beta_{ns}a\frac{r}{a}\right) + J_n(\beta_{ns}a)I_n\left(\beta_{ns}a\frac{r}{a}\right)$$

(4.4.8)

The locations of the nodal circles is obtained from the roots of the following equation.

$$I_n(\beta_{ns}a)J_n(\beta_{ns}a\tfrac{r}{a}) = J_n(\beta_{ns}a)I_n(\beta_{ns}a\tfrac{r}{a})$$

(4.4.8)

These values are given in Table 4.4.2.

4.5 NATURAL FREQUENCIES OF A CIRCULAR PLATE WITH FREE BOUNDARY

Consider a free circular plate of thickness h and radius a executing lateral vibrations. Such a plate may represent a turbine disc which is likely to be excited with nodal diameters and circles, because of excitation from steam or gas medium on the blades. Therefore, the deflection surface will be a function of the radial coordinate r as well as the angular coordinate θ. The differential equation in (4.1.8) is

Table 4.4.2 Location of nodal circles r/a for a simply supported circular plate

No. of Nodal Circles s	Zero Nodal Diameters $n = 0$	One Nodal Diameter $n = 1$	Two Nodal Diameters $n = 2$
0	1	1	1
1	1.000	1.000	1.000
	0.441	0.550	0.613
2	1.000	1.000	1.000
	0.644	0.692	0.726
	0.279	0.378	0.443
3	1.000	1.000	1.000
	0.736	0.765	0.787
	0.469	0.528	0.570
	0.204	0.288	0.348

$$D\left(\frac{\partial^2}{\partial r^2} + \frac{1}{r}\frac{\partial}{\partial r} + \frac{1}{r^2}\frac{\partial^2}{\partial \theta^2}\right)\left(\frac{\partial^2 w}{\partial r^2} + \frac{1}{r}\frac{\partial w}{\partial r} + \frac{1}{r^2}\frac{\partial^2}{\partial \theta^2}\right) + \rho h \ddot{w} = 0$$

(4.5.1)

For harmonic motion, $w(r,\theta,t) = W(r,\theta)\sin pt$, the above can be written as

$$\nabla^4 W - \frac{\rho h}{D}p^2 W =$$
$$(\nabla^2 + \beta^2)(\nabla^2 - \beta^2)W = 0$$

(4.5.2)

where the frequency parameter β is

$$\beta^2 = \sqrt{\frac{\rho h}{D}}\, p$$

(4.5.3)

The solution of (4.5.2) is written as

$$W(r,\theta) = [A_{1n}J_n(\beta r) + A_{3n}Y_n(\beta r) + B_{1n}I_n(\beta r) + B_{3n}K_n(\beta r)]\sin n\theta$$
$$+ [A_{2n}J_n(\beta r) + A_{4n}Y_n(\beta r) + B_{2n}I_n(\beta r) + B_{4n}K_n(\beta r)]\cos n\theta$$

(4.5.4)

Circular Plates 133

where J_n, Y_n are Bessel functions of order n and I_n, K_n are modified Bessel functions of order n. Since Y and K tend to infinity as r tends to zero and that W is finite at the center of the plate, we take the solution above as

$$W(r, \theta) = [A_{1n}J_n(\beta r) + B_{1n}I_n(\beta r)] \sin n\theta$$
$$+ [A_{2n}J_n(\beta r) + B_{2n}I_n(\beta r)] \cos n\theta$$
$$= W_n[J_n(\beta r) + \mu_n I_n(\beta r)] \cos(n\theta - \epsilon_n)$$

(4.5.5)

where

$$\cos(n\epsilon_n) = \frac{A_{1n}}{W_n} = \frac{B_{1n}}{\mu_n W_n}$$

$$\sin(n\epsilon_n) = \frac{A_{2n}}{W_n} = \frac{B_{2n}}{\mu_n W_n}$$

(4.5.6)

It is evident in the above solution that n represents the number of nodal diameters that are present in the mode shape. The boundary conditions are

$$M_r(r = a) = 0$$
$$V_{\text{eff}} = \left[V_r - \frac{1}{r}M_{rt,\theta}\right](r = a) = 0$$

(4.1.11)

i.e.,

$$\left[w_{,rr} + \nu\left(\frac{1}{r}w_{,r} + \frac{1}{r^2}w_{,\theta\theta}\right)\right]_{r=a} = 0$$

$$\left[\frac{\partial}{\partial r}(\nabla^2 w) + (1-\nu)\frac{1}{r}\left(\frac{1}{r}w_{,r\theta\theta} - \frac{1}{r^2}w_{,\theta\theta}\right)\right]_{r=a} = 0$$

(4.5.7)

Applying the above boundary conditions (4.5.7), equation (4.5.5) gives the frequency equation as

$$\frac{\beta^2 a^2}{2n} \left[\begin{array}{l} \{\beta^4 a^4 + (1-v)^2 n^2(n^2-1)\} \\ \times \{J_{n-1}(\beta a)I_{n+1}(\beta a) + J_{n+1}(\beta a)I_{n-1}(\beta a)\} \\ -2\beta^2 a^2 (1-v) n \left\{ \begin{array}{l} (n-1)I_{n-1}(\beta a)J_{n-1}(\beta a) \\ +(n+1)I_{n+1}(\beta a)J_{n+1}(\beta a) \end{array} \right\} \end{array} \right] = 0$$

(4.5.8)

The above equation gives the natural frequencies for different values of n. For example, with $n=1$, one nodal diameter, the frequency equation is

$$\frac{\beta^4 a^4}{2} \left[\begin{array}{l} \beta^2 a^2 \{J_0(\beta a)I_2(\beta a) + J_2(\beta a)I_0(\beta a)\} \\ -4(1-v)I_2(\beta a)J_2(\beta a) \end{array} \right] = 0$$

(4.5.9)

The natural frequencies are

$$p_{ns} = \frac{(\beta_{ns} a)^2}{a^2} \sqrt{\frac{D}{\rho h}} \text{ rad/s}$$

(4.5.10)

where β_{ns} are the roots of the frequency equation (4.5.8) for different values of n. $s = 1, 2, 3$, etc., give the number of nodal circles for each number of nodal diameters n.

Note that the above expressions are not valid for zero nodal diameters, i.e., $n=0$. For zero nodal diameters, the solution is to be obtained as in section 4.3 and for the free edge, the frequency equation can be shown to be

$$\beta a[J_0(\beta a)I_1(\beta a) + I_0(\beta a)J_1(\beta a)] - 2(1-v)I_1(\beta a)J_1(\beta a) = 0$$

(4.5.11)

The corresponding mode shapes are

$$\phi_{ns}(r,\theta) = [J_n(\beta_{ns}r) + \mu_{ns}I_n(\beta_{ns}r)]\cos(n\theta - \epsilon_n)$$

(4.5.12)

where

$$\mu_{ns} = \frac{(\beta_{ns}a)^3 J'_n(\beta_{ns}r) + (1-v)n^2[\beta_{ns}aJ'_n(\beta_{ns}r) - J_n(\beta_{ns}r)]}{(\beta_{ns}a)^3 I'_n(\beta_{ns}r) - (1-v)n^2[\beta_{ns}aI'_n(\beta_{ns}r) - I_n(\beta_{ns}r)]}$$

for $n \neq 0$

$$\mu_{ns} = \frac{(\beta_{ns}a)^2 J_n(\beta_{ns}r) + (1-v)\beta_{ns}aJ'_n(\beta_{ns}r)}{(\beta_{ns}a)^2 I_n(\beta_{ns}r) - (1-v)\beta_{ns}aI'_n(\beta_{ns}r)} = \frac{(\beta_{ns}a)^3 J'_n(\beta_{ns}r)}{(\beta_{ns}a)^3 I'_n(\beta_{ns}r)}$$

for $n = 0$

(4.5.13)

where J' and I' are derivatives of J and I with respect to r.

The mode shape in equation (4.5.12) can be expressed as

$$\phi_{ns}(r,\theta) = f_{ns}(r)\cos(n\theta - \epsilon_n)$$

(4.5.14)

where

$$f_{ns}(r) = [J_n(\beta_{ns}r) + \mu_{ns}I_n(\beta_{ns}r)]$$

(4.5.15)

defines the location of nodal circles for each nodal diameter family. The natural frequencies are obtained from the roots $\beta_{ns}a$ of the frequency equations (4.5.11) and (4.5.9). Here, the non dimensional frequency parameter is dependent on the Poisson's ratio v. Table 4.5.1 gives the roots of the frequency equations for different combinations of the number of nodal circles s and nodal diameters n for $v = 0.33$.

Table 4.5.1 Values of $\beta_{ns}a$ for a circular plate with free boundary for $v = 0.33$

No. of Nodal Circles s	Zero Nodal Diameters $n = 0$	One Nodal Diameter $n = 1$	Two Nodal Diameters $n = 2$	Three Nodal Diameters $n = 3$
0	0	0	2.29	3.5
1	3.01	4.53	5.94	7.27
2	6.21	7.74	9.16	10.55
3	9.37	10.91	12.41	13.86
4	12.53	14.08	15.58	17.05

136 Dynamics of Plates

The nodal circle locations are obtained from equation (4.5.15) which gives

$$J_n\!\left(\beta_{ns}a\tfrac{r}{a}\right)+\mu_{ns}I_n\!\left(\beta_{ns}a\tfrac{r}{a}\right)=0$$

(4.5.16)

The roots of the above equation are given in the table below.

Table 4.5.2 Location of nodal circles r/a for free circular plates

No. of Nodal Circles s	Zero Nodal Diameters $n=0$	One Nodal Diameter $n=1$	Two Nodal Diameters $n=2$	Three Nodal Diameters $n=3$
1	0.68	0.78	0.82	0.85
2	0.840	0.871	0.890	0.925
	0.391	0.497	0.562	0.605
3	0.893	0.932	0.936	0.930
	0.591	0.643	0.678	0.704
	0.257	0.351	0.414	0.460
4	0.941	0.946	0.950	0.951
	0.691	0.723	0.746	0.763
	0.441	0.498	0.540	0.572
	0.192	0.272	0.330	0.374

4.6 MINDLIN'S THEORY FOR AXISYMMETRIC PLATES

Here, we will develop the required equations for a circular plate with axisymmetric motion in a similar manner to that of rectangular plates in section 3.8.

Displacement Field

The displacement field in polar coordinates is assumed as

$$u_r = -z\psi(r)$$
$$u_\theta = 0$$
$$u_z = w(r)$$

(4.6.1)

where $\psi(r)$ is the slope in radial direction due to bending only.

Strain Field

The strain field is obtained as follows.

$$\varepsilon_{rr} = u_{r,r} = -z\psi_{,r}$$

$$\varepsilon_{\theta\theta} = \frac{1}{r}u_{\theta,\theta} + \frac{u_r}{r} = -\frac{z}{r}\psi$$

$$\varepsilon_{zz} = u_{z,z} = 0$$

$$\varepsilon_{r\theta} = \frac{1}{2}\left(\frac{1}{r}u_{r,\theta} + u_{\theta,r} - \frac{u_\theta}{r}\right) = 0$$

$$\varepsilon_{rz} = \frac{1}{2}(u_{z,r} + u_{r,z}) = \frac{1}{2}(w_{,r} - \psi)$$

$$\varepsilon_{\theta z} = \frac{1}{2}\left(u_{\theta,r} + \frac{1}{r}u_{z,\theta}\right) = 0$$

(4.6.2)

Stress Field

We use plane stress condition and obtain the stress field as

$$\tau_{rr} = \frac{E}{1-v^2}(\varepsilon_{rr} + v\varepsilon_{\theta\theta})$$

$$\tau_{\theta\theta} = \frac{E}{1-v^2}(\varepsilon_{\theta\theta} + v\varepsilon_{rr})$$

$$\tau_{rz} = 2G\varepsilon_{rz}$$

(4.6.3)

Strain Energy

The strain energy expression consistent with plane stress assumption is

$$U = \frac{1}{2}\int_0^{2\pi}\int_0^a\int_{-h/2}^{h/2}(\tau_{rr}\varepsilon_{rr} + \tau_{\theta\theta}\varepsilon_{\theta\theta} + 2\tau_{rz}\varepsilon_{rz})d\theta\,dr\,dz$$

(4.6.4)

Integrating with respect to θ, the above becomes

$$U = \pi\int_0^a\int_{-h/2}^{h/2}(\tau_{rr}\varepsilon_{rr} + \tau_{\theta\theta}\varepsilon_{\theta\theta} + 2\tau_{rz}\varepsilon_{rz})\,rdr\,dz$$

(4.6.5)

Substituting for the strains and stresses from equations (4.6.2) and (4.6.3), we get

$$U = \pi\int_0^a\int_{-h/2}^{h/2}\left[\frac{E}{1-v^2}(\varepsilon_{rr}^2 + \varepsilon_{\theta\theta}^2 + 2v\varepsilon_{\theta\theta}\varepsilon_{rr}) + 4G\varepsilon_{rz}^2\right]rdr\,dz$$

$$= \pi\int_0^a\int_{-h/2}^{h/2}\left[\frac{Ez^2}{1-v^2}\left(\psi_{,r}^2 + \frac{\psi^2}{r^2} + \frac{2v}{r}\psi\psi_{,r}\right) + G(w_{,r} - \psi)^2\right]rdr\,dz$$

(4.6.6)

Kinetic Energy

$$T = \frac{1}{2}\int_0^{2\pi}\int_0^a\int_{-h/2}^{h/2}\rho[z^2\dot{\psi}^2 + \dot{w}^2]\,r\,dr\,d\theta\,dz$$

$$= \pi\int_0^a\int_{-h/2}^{h/2}\rho[z^2\dot{\psi}^2 + \dot{w}^2]\,r\,dr\,dz$$

(4.6.7)

Work of External Force

$$W = -V = \int_0^{2\pi}\int_0^a qw\,r\,dr\,d\theta$$

$$= 2\pi\int_0^a qw\,r\,dr$$

(4.6.8)

Hamilton's Principle

$$\delta\int_{t_1}^{t_2}\left[\begin{array}{c}\pi\int_0^a\int_{-h/2}^{h/2}\rho[z^2\dot{\psi}^2 + \dot{w}^2]\,r\,dr\,dz + \pi\int_0^a 2qw\,r\,dr \\ -\pi\int_0^a\int_{-h/2}^{h/2}\left[\frac{Ez^2}{1-v^2}\left(\psi_{,r}^2 + \frac{\psi^2}{r^2} + \frac{2v}{r}\psi\psi_{,r}\right)\right.\\ \left.+G(w_{,r}-\psi)^2\right]r\,dr\,dz\end{array}\right]dt = 0$$

(4.6.9)

Carrying out the integration with respect to z, the above equation can be simplified as

$$\delta\int_{t_1}^{t_2}\left[\begin{array}{c}\int_0^a\rho\left[\frac{h^3}{12}\dot{\psi}^2 + h\dot{w}^2\right]r\,dr + \int_0^a 2qw\,r\,dr \\ -\int_0^a\left[D\left(\psi_{,r}^2 + \frac{\psi^2}{r^2} + \frac{2v}{r}\psi\psi_{,r}\right) + Gh(w_{,r}-\psi)^2\right]r\,dr\end{array}\right]dt = 0$$

(4.6.10)

Now, taking the variation in the above equation (4.6.10)

$$\int_{t_1}^{t_2}\left[\begin{array}{c}\int_0^a\rho\left[\frac{h^3}{12}\dot{\psi}\,\delta\dot{\psi} + h\dot{w}\,\delta\dot{w}\right]r\,dr + \int_0^a q\delta w\,r\,dr \\ -\int_0^a\left[D\left(\psi_{,r}\delta\psi_{,r} + \frac{\psi}{r^2}\delta\psi + \frac{v}{r}\psi\delta\psi_{,r} + \frac{v}{r}\psi_{,r}\delta\psi\right)\right.\\ \left.-Gh(w_{,r}-\psi)\delta\psi + Gh(w_{,r}-\psi)\delta w_{,r}\right]r\,dr\end{array}\right]dt = 0$$

(4.6.11)

Circular Plates 139

First let us consider the integral

$$I_1 = \int_{t_1}^{t_2} \int_0^a \rho\left[\frac{h^3}{12} \dot\psi \, \delta\dot\psi + h\dot w \, \delta\dot w\right] r \, dr \, dt$$

$$= \int_0^a \left\{\int_{t_1}^{t_2} \rho r\left[\frac{h^3}{12}\dot\psi\,\delta\dot\psi + h\dot w\,\delta\dot w\right] dt\right\} dr$$

$$= \int_0^a \left[\left\{\rho r \frac{h^3}{12}\dot\psi\,\delta\psi + \rho r h\,\dot w\,\delta w\right\}_{t_1}^{t_2} - \int_{t_1}^{t_2} \rho r\left\{\frac{h^3}{12}\ddot\psi\,\delta\psi + h\ddot w\,\delta w\right\}\cdot dt\right] dr$$

$$= -\int_{t_1}^{t_2} \int_0^a \rho r\left[\frac{h^3}{12}\ddot\psi\,\delta\psi + h\ddot w\,\delta w\right] dr\,dt$$

(4.6.12)

Next consider the following integral from (4.6.11)

$$I_2 = -\int_{t_1}^{t_2}\int_0^a D(r\psi_{,r} + v\psi)\delta\psi_{,r}\,dr\,dt$$

$$= -\int_{t_1}^{t_2}\left[\{D(r\psi_{,r} + v\psi)\delta\psi\}_0^a - \int_0^a D\{r\psi_{,rr} + \psi_{,r}(1+v)\}\delta\psi\,dr\right]dt$$

(4.6.13)

The last integral term in (4.6.11) is

$$I_3 = -\int_{t_1}^{t_2}\int_0^a Gh\,r(w_{,r} - \psi)\delta w_{,r}\,dr\,dt$$

$$= -\int_{t_1}^{t_2}\left[\{Gh\,r(w_{,r} - \psi)\delta w\}_0^a - \int_0^a Gh(rw_{,rr} + w_{,r} - \psi - r\psi_{,r})\delta w\,dr\right]dt$$

(4.6.14)

Substituting equations (4.6.12) to (4.6.14) in (4.6.11), we get

$$\int_{t_1}^{t_2}\begin{bmatrix} -\int_0^a \rho r\left[\frac{h^3}{12}\ddot\psi\,\delta\psi + h\ddot w\,\delta w\right]dr + \int_0^a q\delta w\,r\,dr \\ -\int_0^a D(\frac{\psi}{r^2} + \frac{v}{r}\psi_{,r})\delta\psi\,r\,dr \\ -\{D(r\psi_{,r} + v\psi)\delta\psi\}_0^a + \int_0^a D\{r\psi_{,rr} + \psi_{,r}(1+v)\}\delta\psi\,dr \\ \int_0^a Gh(w_{,r} - \psi)\delta\psi\,r\,dr \\ -\{Gh\,r(w_{,r} - \psi)\delta w\}_0^a + \int_0^a Gh(rw_{,rr} + w_{,r} - \psi - r\psi_{,r})\delta w\,dr \end{bmatrix} dt = 0$$

(4.6.15)

Regrouping the terms in the above equation (4.6.15), we can obtain the two differential equations and the boundary conditions

$$D\left\{\psi_{,rr} + \frac{1}{r}\psi_{,r}(1+v)\right\} + Gh(w_{,r} - \psi) - \rho\frac{h^3}{12}\ddot{\psi} - D\left(\frac{\psi}{r^2} + \frac{v}{r}\psi_{,r}\right) = 0$$

$$\frac{1}{r}Gh(rw_{,rr} + w_{,r} - \psi - r\psi_{,r}) - \rho h \ddot{w} + q = 0$$

$$\{D(r\psi_{,r} + v\psi)\delta\psi\}_0^a = 0$$

$$\{Gh\,r(w_{,r} - \psi)\delta w\}_0^a = 0$$

(4.6.16)

Introducing the shear correction factor and rewriting the first two equations in (4.6.16), we get

$$D\left\{\psi_{,rr} + \frac{1}{r}\psi_{,r} - \frac{\psi}{r^2}\right\} + kGh(w_{,r} - \psi) - \rho\frac{h^3}{12}\ddot{\psi} = 0$$

$$\frac{1}{r}\frac{\partial}{\partial r}[kGhr(w_{,r} - \psi)] - \rho h \ddot{w} + q = 0$$

(4.6.17)

Noting that $\psi = 0$ at the center of the plate, the boundary conditions at the edge become

$$(a\psi_{,r} + v\psi) = 0 \quad \text{or} \quad \text{Specify } \psi$$
$$(w_{,r} - \psi) = 0 \quad \text{or} \quad \text{Specify } w$$

(4.6.18)

In order to obtain a single differential equation, instead of the two coupled equations in (4.6.17), we rewrite the first two equations in (4.6.16) in operator form as

$$\left\{D\nabla^2 - \frac{D}{r^2} - kGh - \rho\frac{h^3}{12}\partial_t^2\right\}\psi + kGh\partial_r w = 0$$

$$-\frac{1}{r}kGh(1 + r\partial_r)\psi + \left\{kGh\partial_r^2 + \frac{1}{r}kGh\partial_r - \rho h\partial_t^2\right\}w = -q$$

(4.6.19)

i.e.,

$$\begin{bmatrix} \left\{D\nabla^2 - \dfrac{D}{r^2} - kGh - \rho\dfrac{h^3}{12}\partial_t^2\right\} & kGh\partial_r \\ -\dfrac{1}{r}kGh(1+r\partial_r) & \{kGh\nabla^2 - \rho h\partial_t^2\} \end{bmatrix} \begin{Bmatrix} \psi \\ w \end{Bmatrix} = \begin{Bmatrix} 0 \\ -q \end{Bmatrix}$$

(4.6.20)

We can now eliminate ψ in the above equation (4.6.20) to give

$$w = \dfrac{\begin{vmatrix} \left\{D\nabla^2 - \dfrac{D}{r^2} - kGh - \rho\dfrac{h^3}{12}\partial_t^2\right\} & 0 \\ -\dfrac{1}{r}kGh(1+r\partial_r) & -q \end{vmatrix}}{\begin{vmatrix} \left\{D\nabla^2 - \dfrac{D}{r^2} - kGh - \rho\dfrac{h^3}{12}\partial_t^2\right\} & kGh\partial_r \\ -\dfrac{1}{r}kGh(1+r\partial_r) & \{kGh\nabla^2 - \rho h\partial_t^2\} \end{vmatrix}}$$

(4.6.21)

Cross multiplying and simplifying

$$D\nabla^4 w - \left(\dfrac{D\rho}{kG} + \rho\dfrac{h^3}{12}\right)\nabla^2\partial_t^2 w - \left(\dfrac{D}{r^2} + kGh\right)\nabla^2 w +$$

$$\left(\dfrac{D}{r^2}\dfrac{\rho}{kG} + \rho h\right)\partial_t^2 w + \dfrac{\rho^2 h^3}{12kG}\partial_t^4 w$$

$$= -\dfrac{D\nabla^2 q}{kGh} + \dfrac{Dq}{r^2 kGh} + q + \dfrac{\rho h^2}{12kG}\partial_t^2 q$$

(4.6.22)

For free vibrations, the above equation simplifies to

$$D\nabla^4 w - \left(\dfrac{D\rho}{kG} + \rho\dfrac{h^3}{12}\right)\nabla^2\partial_t^2 w - \left(\dfrac{D}{r^2} + kGh\right)\nabla^2 w$$

$$+ \left(\dfrac{D}{r^2}\dfrac{\rho}{kG} + \rho h\right)\partial_t^2 w + \dfrac{\rho^2 h^3}{12kG}\partial_t^4 w$$

$$= 0$$

(4.6.23)

4.7 CIRCULAR PLATES ON VISCOELASTIC SUPPORT

Consider a circular plate of sufficiently large radius such that it can be considered as infinity and thickness h, supported on a uniformly distributed standard linear viscoelastic foundation as shown in Fig. 4.7.1. The standard linear solid is taken as three element system consisting of a linear elastic stiffness k_1 (stiffness per unit area N/m/m^2) connected in series with a parallel combination of a linear spring k_2(stiffness per unit area N/m/m^2) and a viscous dashpot having a damping constant c (damping per unit area Ns/m/m^2). Consider an axisymmetric uniformly distributed load $p(r,t)$ acting at the center of the plate over a radius $\bar{r}$ as shown in the figure. The dynamic foundation reaction is denoted by $q(r,t)$, then the equation of motion for axisymmetric motion of this plate is

$$D\nabla^4 w + \rho h \ddot{w} = p(r,t) - q(r,t) \qquad (4.7.1)$$

Fig. 4.7.2 shows the foundation element. Let w_1 represent the displacement at the junction point, then

$$k_2 w_1 + c \dot{w}_1 = q$$
$$k_1(w - w_1) = q \qquad (4.7.2)$$

From the second equation in the above, we get

$$w_1 = w - \frac{q}{w_1} \qquad (4.7.3)$$

Substituting for $w_1, \dot{w}_1$ from the above equation (4.7.3) in the first equation of (4.7.2)

$$k_2 w + c \dot{w} = \frac{k_2 + k_1}{k_1} q + \frac{c}{k_1} \dot{q} \qquad (4.7.4)$$

For the plate under consideration, the boundary conditions can be written as

$$w = w' = w'' = w''' = w'''' = 0 \text{ at } r \to \infty$$
$$w \neq \infty \text{ at } r = 0 \qquad (4.7.5)$$

Circular Plates 143

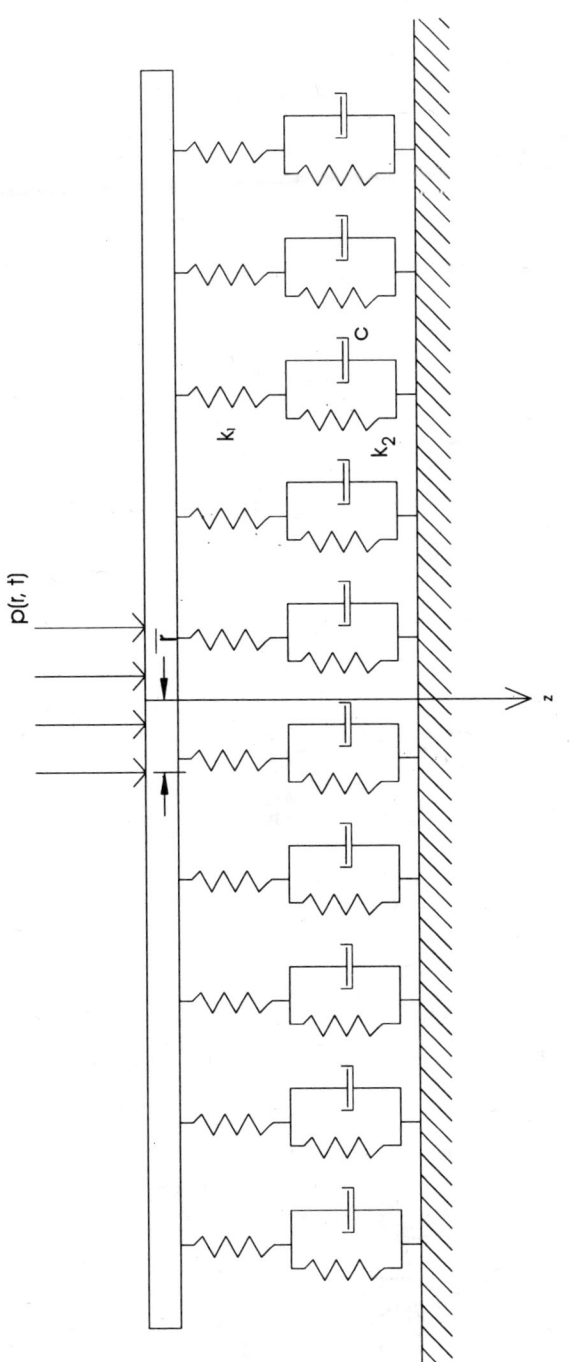

Fig. 4.7.1 Circular Plate of Infinite Radius on Standard Linear Viscoelastic Solid

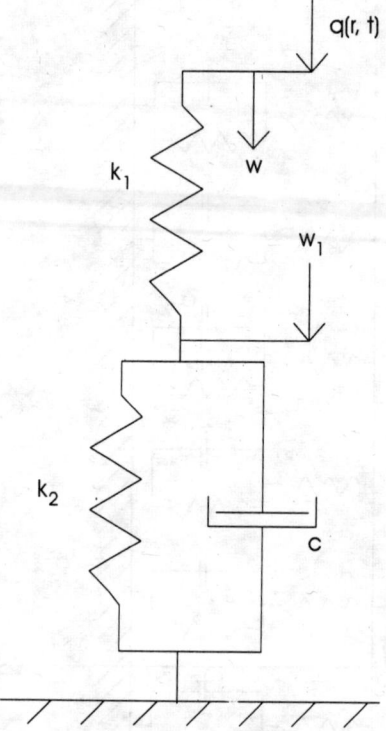

Fig. 4.7.2 Foundation Element

Let the initial conditions of the plate be

$$w = \dot{w} = 0 \text{ at } t = 0 \qquad (4.7.6)$$

The following non dimensional parameters are defined.

$$R = \frac{r}{h}$$

$$T = t\sqrt{\frac{E}{12(1-v^2)\rho h^2}} = t\sqrt{\frac{D}{\rho h^5}}$$

$$C = \sqrt{12(1-v^2)\frac{c^2}{\rho E}}$$

$$K_{1,2} = \sqrt{12(1-v^2)\frac{hk_{1,2}}{E}}$$

$$W = \frac{w}{h}$$

$$Q = 12\frac{(1-v^2)}{E}q$$

$$P = 12\frac{(1-v^2)}{E}p$$

(4.7.7)

then equations (4.7.1), (4.7.4) to (4.7.6) become $\left(\text{with } \cdot = \frac{d}{dt} \text{ and } ' = \frac{d}{dR}\right)$

$$\nabla^4 W + \ddot{W} = P(R,T) - Q(R,T)$$

(4.7.1a)

$$K_2^2 W + C\,\dot{W} = \frac{K_2^2 + K_1^2}{K_1^2}Q + \frac{C}{K_1^2}\dot{Q}$$

(4.7.4a)

$$W = W' = W'' = W''' = W'''' = 0 \text{ at } R \to \infty$$
$$W \neq \infty \text{ at } R = 0$$

(4.7.5a)

Let the initial conditions of the plate be

$$W = \dot{W} = 0 \text{ at } T = 0$$

(4.7.6a)

The above set of equations are solved by using integral transforms, see Sneddon (1951); Hankel transform for the space coordinate R and Laplace transform for time coordinate T. Hankel transform $f^*(\xi,T)$ of a function $f(R,T)$ is given by

$$f^*(\xi,T) = \int_0^\infty R J_0(\xi R) f(R,T)\, dR$$

(4.7.8)

Its inverse is given by

$$f(R,T) = \int_0^\infty \xi J_0(\xi R) f^*(\xi,T)\, d\xi$$

(4.7.9)

where ξ is Hankel transform parameter and J_0 is the Bessel function of the first kind and zero order. The Laplace transform $\overline{f}(R,s)$ of a function $f(R,T)$ is

$$\overline{f}(R,s) = \int_0^\infty e^{-sT} f(R,T)\, dT \tag{4.7.10}$$

with its inverse as

$$f(R,T) = \frac{1}{2\pi i} \int_{\gamma-i\infty}^{\gamma+i\infty} e^{sT} \overline{f}(R,s)\, ds \tag{4.7.11}$$

where s is the Laplace transform parameter and its integration in (4.7.11) is carried in the complex plane along Bromwich contour. Applying the Hankel transform to equations (4.7.1a), (4.7.4a) and and (4.7.6a) we get

$$\xi^4 W^*(\xi,T) + \frac{d^2 W^*(\xi,T)}{dT^2} = P^*(\xi,T) - Q^*(\xi,T) \tag{4.7.12}$$

$$K_2^2 W^*(\xi,T) + C \frac{dW^*(\xi,T)}{dT} = \frac{K_2^2 + K_1^2}{K_1^2} Q^*(\xi,T) + \frac{C}{K_1^2} \frac{dQ^*(\xi,T)}{dT} \tag{4.7.13}$$

$$W^*(\xi,0) = \frac{dW^*(\xi,0)}{dT} = 0 \tag{4.7.14}$$

Now, applying Laplace transform to equations (4.7.12) and (4.7.13) gives

$$(\xi^4 + s^2) W^*(\xi,s) + \overline{Q}^*(\xi,s) = \overline{P}^*(\xi,s) \tag{4.7.15}$$

$$(K_2^2 + Cs) W^*(\xi,s) - \left(\frac{K_2^2 + K_1^2}{K_1^2} + \frac{C}{K_1^2} s \right) \overline{Q}^*(\xi,s) = 0 \tag{4.7.16}$$

The above two equations can be expressed as

$$\begin{bmatrix} (\xi^4 + s^2) & 1 \\ (K_2^2 + Cs) & -\left(\dfrac{K_2^2 + K_1^2}{K_1^2} + \dfrac{C}{K_1^2}s\right) \end{bmatrix} \begin{Bmatrix} W^*(\xi,s) \\ \overline{Q}^*(\xi,s) \end{Bmatrix} = \begin{Bmatrix} \overline{P}^*(\xi,s) \\ 0 \end{Bmatrix}$$

(4.7.16a)

Solving the above matrix equation, we get

$$W^*(\xi,s) = \dfrac{\overline{P}^*(\xi,s)}{\xi^4 + s^2 + \dfrac{K_2^2 + Cs}{\dfrac{K_2^2 + K_1^2}{K_1^2} + \dfrac{C}{K_1^2}s}}$$

$$Q^*(\xi,s) = \dfrac{\overline{P}^*(\xi,s) \dfrac{K_2^2 + Cs}{\dfrac{K_2^2 + K_1^2}{K_1^2} + \dfrac{C}{K_1^2}s}}{\xi^4 + s^2 + \dfrac{K_2^2 + Cs}{\dfrac{K_2^2 + K_1^2}{K_1^2} + \dfrac{C}{K_1^2}s}}$$

(4.7.17)

Let the plate be subjected to an impulsive force $P(R,T)$ defined by

$$P(R,T) = P_1(R)\delta(T)$$

(4.7.18)

where δ is the Dirac delta function satisfying

$$\int_{-\infty}^{\infty} \delta(T)\,dT = 1$$

$$\int_{0}^{\infty} e^{-sT}\delta(T)\,dT = 1$$

(4.7.19)

148 Dynamics of Plates

Applying Laplace and Hankel transform to the applied impulsive force in (4.7.18), we get

$$\overline{P}^*(\xi,s) = \overline{P}_1^*(\xi)$$

(4.7.20)

Let $P_1(R)$ in (4.7.18) be

$$P_1(R) = \begin{cases} \dfrac{Ph^2}{\pi \overline{r}^2} & \text{for } R \leq \dfrac{\overline{r}}{h} \\ 0 & \text{for } R > \dfrac{\overline{r}}{h} \end{cases}$$

(4.7.21)

where $\underline{P}$ is the total non dimensional force given by

$$\underline{P} = \frac{12(1-v^2)}{Eh^2} p$$

(4.7.22)

Hankel transform of (4.7.21) is

$$P_1^*(\xi) = \int_0^\infty RJ_0(\xi R)\frac{Ph^2}{\pi \overline{r}^2} dR$$

$$= \frac{\overline{R}}{\xi} \frac{P}{\pi \overline{R}^2} J_1(\xi \overline{R})$$

(4.7.23)

where $\overline{R} = \dfrac{\overline{r}}{h}$ and J_1 is the Bessel function of the first kind and first order. If the force is concentrated at the center of the plate, then

$$\lim_{\overline{R}\to 0} \frac{J_1(\xi \overline{R})}{(\xi \overline{R})} = \frac{1}{2}$$

and the above equation (4.7.23) becomes

$$P_1^*(\xi) = \frac{P}{2\pi}$$

(4.7.24)

Equation (4.7.20) for a concentrated force at the center of the plate is then given by

$$\overline{P}^*(\xi,s) = \frac{P}{2\pi}$$

(4.7.20a)

Substituting the above result in (4.7.17), we get

$$W^*(\xi,s) = \frac{P}{2\pi} \frac{1}{\xi^4 + d_1^4}$$

(4.7.25)

where

$$d_1^4 = \left[s^2 + \frac{K_2^2 + Cs}{\frac{K_2^2 + K_1^2}{K_1^2} + \frac{C}{K_1^2}s} \right]$$

(4.7.26)

Now, applying inverse Hankel transform, see Erdelyi (1954), to the above equation (4.7.25), we get

$$W(R,s) = -\frac{P}{2\pi} \frac{1}{d_1^2} \Im[\text{ke}_0(R d_1)]$$

(4.7.27)

where ke_0 is the Kelvin function of zero order. Applying inverse Laplace transform to equation (4.7.27), we get

$$W(R,T) = -\frac{P}{2\pi} \frac{1}{2\pi i} \int_{Y-i\infty}^{Y+i\infty} e^{sT} \frac{1}{d_1^2} \Im[\text{ke}_0(R d_1)] \, ds$$

(4.7.28)

The maximum dynamic response occurs at the center of the plate. The Kelvin function here becomes

$$\Im[\text{ke}_0(R d_1)]_{R=0} = -\frac{\pi}{4}$$

therefore

$$W(0,T) = \frac{P}{8} \frac{1}{2\pi i} \int_{Y-i\infty}^{Y+i\infty} \frac{e^{sT}}{d_1^2} ds$$

$$= \frac{P}{8} \frac{1}{2\pi i} \int_{Y-i\infty}^{Y+i\infty} \frac{e^{sT} \sqrt{\frac{K_2^2 + K_1^2}{K_1^2} + \frac{C}{K_1^2} s}}{\sqrt{\frac{C}{K_1^2} s^3 + \frac{K_2^2 + K_1^2}{K_1^2} s^2 + Cs + K_2^2}} ds$$

(4.7.29)

Kelvin Foundation A special case of the three element standard linear viscoelastic solid is Kelvin type foundation which is obtained by letting $k_1 \to \infty$.

$$W(0,T) = \frac{P}{8} \frac{1}{2\pi i} \int_{Y-i\infty}^{Y+i\infty} \frac{e^{sT}}{\sqrt{\left(s + \frac{1}{2}C\right)^2 + \left(K_2^2 - \frac{1}{4}C^2\right)}} ds$$

$$= \frac{P}{8} e^{-\frac{1}{2}CT} J_0\left(\sqrt{K_2^2 - \frac{1}{4}C^2}\; T\right)$$

(4.7.30)

Let

$$C_c = 2K_2$$
$$\zeta = \frac{C}{C_c}$$
$$T^* = K_2(1 - \zeta^2)T$$

(4.7.31)

then

$$W(0,T) = \frac{P}{8} e^{-\frac{\zeta}{\sqrt{1-\zeta^2}} T^*} J_0(T^*)$$

(4.7.32)

Steady State Response Let the plate on a Kelvin foundation be subjected to a harmonic force with frequency ω instead of the impulsive force considered earlier. The non dimensional frequency of excitation from equation (4.7.7) is

$$\Omega = \omega\sqrt{\frac{ph^5}{D}}$$

(4.7.33)

With the help of Duhamel's integral, the response in equation (4.7.30) for the present case becomes

$$W(0,T) = \frac{P}{8}\int_0^T e^{-\frac{1}{2}C\tau}J_0\left(\sqrt{K_2^2 - \frac{1}{4}C^2}\,T\right)\Omega(T-\tau)\,d\tau$$

(4.7.34)

The steady state response can be obtained by letting $T \to \infty$

$$W(0,T) = \frac{P}{8}\int_0^\infty e^{-\frac{1}{2}C\tau}J_0\left(\sqrt{K_2^2 - \frac{1}{4}C^2}\,T\right)\Omega(T-\tau)\,d\tau$$

$$= \frac{P}{8}\Im\left[\int_0^\infty e^{-\frac{1}{2}C\tau}J_0\left(\sqrt{K_2^2 - \frac{1}{4}C^2}\,T\right)e^{i\Omega(T-\tau)}\,d\tau\right]$$

(4.7.34)

Carrying out the integration and taking the imaginary part, we get the response as

$$W(0,T) = \frac{P}{8}\frac{\sin(\Omega T - \Theta)}{[(K_2^2 - \Omega^2)^2 + (C\Omega)^2]^{1/4}}$$

(4.7.35)

where

$$\Theta = \frac{1}{02}\tan^{-1}\left[\frac{C\Omega}{(K_2^2 - \Omega^2)}\right]$$

(4.7.36)

5

Finite Element Methods for Thin Plates

In chapter 2 we studied the variational principles which established basic procedures for the derivation of governing differential equations and boundary conditions for various types of plates. In this chapter, we also established basic solution procedures for different types of plates under static as well as dynamic conditions. In chapter 3, we applied the principles of chapter 2 for rectangular thin plates and in chapter 4, circular plates were studied. In both the chapters 3 and 4, we used variational methods to solve the static as well as dynamic problems treating the entire plate as one region for the purpose of numerical integration that is involved in the numerical procedures of Lagrange, Ritz, Galerkin, Levy or Kantorovich methods. As long as the plate is rectangular or circular (well defined geometry) and of uniformly varying thickness (or as a special case, constant thickness), these procedures of chapters 3 and 4 are not impractical. However, when the plate is of arbitrary shape and thickness, the procedures outlined in chapters 3 and 4 become very unwieldy for practical purposes and we need finite element techniques to come to our aid in solving such problems. A special case is that of plates of airfoil cross-section as employed in turbine and compressor blading, where thickness in the region of the plate cannot be defined as a suitable mathematical function.

Finite Element method for a plate basically consists of dividing its region into suitable elements of triangular or rectangular shape. These elements are not infinitely small as in the case of derivation of system equations, they are chosen as small as possible depending on the accuracy desired and the need to idealize a given plate with elements of uniform thickness. Each element is a plate by itself and the deformation behavior is assumed in a suitable manner to satisfy the extremal property of the plate functional as defined earlier by a suitable method such as Ritz. Here, the

Finite Element Methods for Thin Plates 153

reader is assumed to have a basic understanding of finite element procedures, such as one dimensional structures of bars, rods and beams.

There are several elements developed since early 1960's and we describe here two different elements, a triangular element and a rectangular element.

5.1 TOCHER'S NONCONFORMING TRIANGULAR ELEMENT

Select Element Type The element is chosen of triangular shape as shown in Fig. 5.1.1. The element has obviously three nodes 1, 2 and 3. The nodal coordinate system $\hat{x}\hat{y}$ is located with its origin at the node 1 and the elemental $\hat{x}$ coordinate is chosen to lie on the nodal line 12. The global coordinate system is denoted XY as shown in the figure. The nodal coordinates for the points 1, 2 and 3 respectively are $(0,0)$, $(x_2, 0)$ and (x_3, y_3). In accordance to plate theory, we will assign three degrees of freedom to each node, viz., w, $w_{,y}$, $w_{,x}$, so that each element has nine degrees of freedom given below.

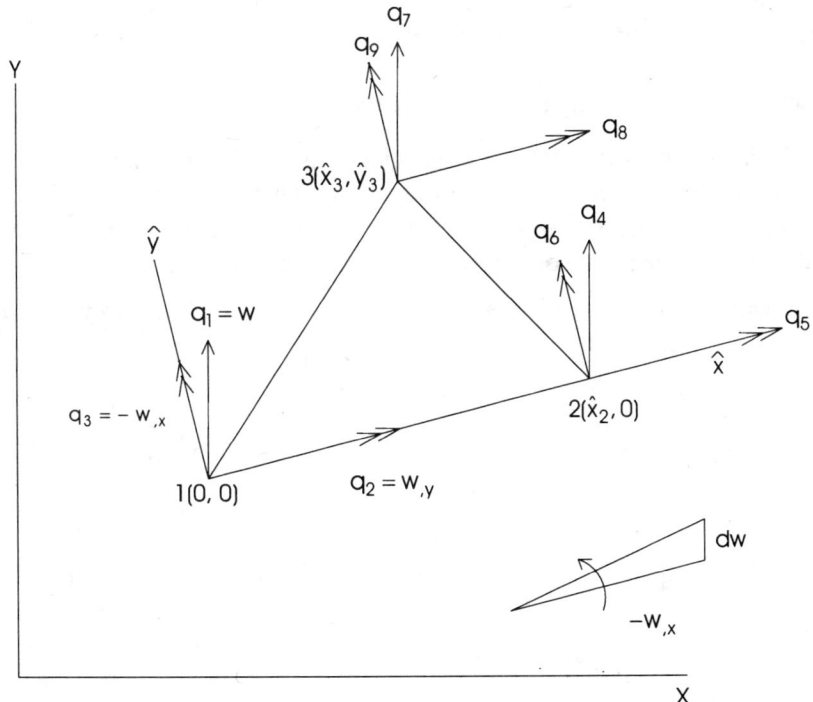

Fig. 5.1.1 Triangular Plate Element

$$q_1 = w(0,0)$$
$$q_2 = w_{,y}(0,0)$$
$$q_3 = -w_{,x}(0,0)$$
$$q_4 = w(x_2,0)$$
$$q_5 = w_{,y}(x_2,0)$$
$$q_6 = -w_{,x}(x_2,0)$$
$$q_7 = w(x_3,y_3)$$
$$q_8 = w_{,y}(x_3,y_3)$$
$$q_9 = -w_{,x}(x_3,y_3)$$

(5.1.1)

The nodal degrees q_3, q_6 and q_9 are taken negative, since for a positive displacement dw, the slope is negative as shown in the figure. The nodal degrees of freedom of the element are expressed as a vector

$$\{\hat{d}\} = \{\ \hat{q}_1\ \hat{q}_2\ \hat{q}_3\ \hat{q}_4\ \hat{q}_5\ \hat{q}_6\ \hat{q}_7\ \hat{q}_8\ \hat{q}_9\ \}^T$$

(5.1.2)

where a hat above any quantity $\widehat{}$ refers to the elemental values.

Displacement Function The displacement within the element at any point $(\hat{x}, \hat{y})$ is represented by a polynomial with nine terms, so that the nine coefficients in the polynomial can be evaluated to satisfy the nodal degrees of freedom. Tocher used the following polynomial

$$w(\hat{x},\hat{y}) = a_1 + a_2\hat{x} + a_3\hat{y} + a_4\hat{x}^2 + a_5\hat{x}\hat{y} + a_6\hat{y}^2 + a_7\hat{x}^3 + a_8(\hat{x}^2\hat{y} + \hat{x}\hat{y}^2)$$
$$+ a_9\hat{y}^3$$

$$= \begin{bmatrix} 1 & \hat{x} & \hat{y} & \hat{x}^2 & \hat{x}\hat{y} & \hat{y}^2 & \hat{x}^3 & (\hat{x}^2\hat{y}+\hat{x}\hat{y}^2) & \hat{y}^3 \end{bmatrix} \begin{Bmatrix} a_1 \\ a_2 \\ a_3 \\ a_4 \\ a_5 \\ a_6 \\ a_7 \\ a_8 \\ a_9 \end{Bmatrix} = [N]\{a\}$$

(5.1.3)

The above shape function does not represent a complete third degree polynomial as it has ten terms. Tocher combined two of the terms in the third degree polynomial to obtain the above function. Alternatively, one can take a ten term third degree polynomial and suppress one of them in the Ritz minimization process. Another way is to neglect $\hat{y}^3$ term as used by Adini. In any of these cases, we find that the deflection as well as slopes are not continuous across the elements through a common node. Therefore, the shape function above produces a *non conformal element*. Cooper later developed a conforming element, however, it consists of a shape function with 21 terms.

Differentiating equation (5.1.3)

$$\begin{Bmatrix} w \\ w_{,y} \\ -w_{,x} \end{Bmatrix} = \begin{bmatrix} 1 & \hat{x} & \hat{y} & \hat{x}^2 & \hat{x}\hat{y} & \hat{y}^2 & \hat{x}^3 & (\hat{x}^2\hat{y}+\hat{x}\hat{y}^2) & \hat{y}^3 \\ 0 & 0 & 1 & 0 & \hat{x} & 2\hat{y} & 0 & (\hat{x}^2+2\hat{x}\hat{y}) & 3\hat{y}^2 \\ 0 & -1 & 0 & -2\hat{x} & -\hat{y} & 0 & -3\hat{x}^2 & -(2\hat{x}\hat{y}+\hat{y}^2) & 0 \end{bmatrix} \{a\}$$

(5.1.4)

With the help of the above definition of shape function, equations (5.1.1) give

$$\begin{bmatrix} 1 & 0 & 0 & 0 & 0 & 0 & 0 & 0 & 0 \\ 0 & 0 & 1 & 0 & 0 & 0 & 0 & 0 & 0 \\ 0 & -1 & 0 & 0 & 0 & 0 & 0 & 0 & 0 \\ 1 & x_2 & 0 & x_2^2 & 0 & 0 & x_2^3 & 0 & 0 \\ 0 & 0 & 1 & 0 & x_2 & 0 & 0 & x_2^2 & 0 \\ 0 & -1 & 0 & -2x_2 & 0 & 0 & -3x_2^2 & 0 & 0 \\ 1 & x_3 & y_3 & x_3^2 & x_3 y_3 & y_3^2 & x_3^3 & (x_3^2 y_3 + x_3 y_3^2) & y_3^3 \\ 0 & 0 & 1 & 0 & x_3 & 2y_3 & 0 & (x_3^2 + 2x_3 y_3) & 3y_3^2 \\ 0 & -1 & 0 & -2x_3 & -y_3 & 0 & -3x_3^2 & -(2x_3 y_3 + y_3^2) & 0 \end{bmatrix} \begin{Bmatrix} a_1 \\ a_2 \\ a_3 \\ a_4 \\ a_5 \\ a_6 \\ a_7 \\ a_8 \\ a_9 \end{Bmatrix} = \begin{Bmatrix} \hat{q}_1 \\ \hat{q}_2 \\ \hat{q}_3 \\ \hat{q}_4 \\ \hat{q}_5 \\ \hat{q}_6 \\ \hat{q}_7 \\ \hat{q}_8 \\ \hat{q}_9 \end{Bmatrix}$$

i.e., $[\hat{A}]\{a\} = \{\hat{q}\}$

(5.1.5)

Therefore

$$\{a\} = [\hat{A}]^{-1}\{\hat{q}\}$$

(5.1.6)

From equation (5.1.3)

$$w = [N][\hat{A}]^{-1}\{\hat{q}\}$$

(5.1.7)

The elemental matrix $[\hat{A}]$ is singular whenever

$$x_2 - 2x_3 - y_3 = 0$$

(5.1.8)

therefore, the elements should be so chosen by altering the nodal positions to avaoid the above condition.

Strain Displacement Relations   The strain displacement relations for thin plate theory are given by equation (3.2.16)

$$\varepsilon_{xx} = -zw_{,xx}$$
$$\varepsilon_{yy} = -zw_{,yy}$$
$$\gamma_{xy} = -2zw_{,xy}$$

(3.2.16)

Differentiating the displacement function of (5.1.3),

$$\begin{Bmatrix} w_{,xx} \\ w_{,yy} \\ -w_{,xy} \end{Bmatrix} = \begin{bmatrix} 0 & 0 & 0 & 2 & 0 & 0 & 6\hat{x} & 2\hat{y} & 0 \\ 0 & 0 & 0 & 0 & 0 & 2 & 0 & 2\hat{x} & 6\hat{y} \\ 0 & 0 & 0 & 0 & 1 & 0 & 0 & 2(\hat{x}+\hat{y}) & 0 \end{bmatrix}\{a\}$$

(5.1.9)

Therefore, equations (3.2.16) can be expressed as

$$\{\hat{\varepsilon}\} = -z \begin{bmatrix} 0 & 0 & 0 & 2 & 0 & 0 & 6\hat{x} & 2\hat{y} & 0 \\ 0 & 0 & 0 & 0 & 0 & 2 & 0 & 2\hat{x} & 6\hat{y} \\ 0 & 0 & 0 & 0 & 2 & 0 & 0 & 4(\hat{x}+\hat{y}) & 0 \end{bmatrix}\{a\}$$

$$= [\overline{B}]\{a\}$$

(5.1.10)

where

$$[\bar{B}] = -z \begin{bmatrix} 0 & 0 & 0 & 2 & 0 & 0 & 6\hat{x} & 2\hat{y} & 0 \\ 0 & 0 & 0 & 0 & 0 & 2 & 0 & 2\hat{x} & 6\hat{y} \\ 0 & 0 & 0 & 0 & 2 & 0 & 0 & 4(\hat{x}+\hat{y}) & 0 \end{bmatrix}$$

(5.1.12)

With the help of (5.1.6), (5.1.10) becomes

$$\{\hat{\varepsilon}\} = [\bar{B}][\hat{A}]^{-1}\{\hat{q}\}$$
$$= [B]\{\hat{q}\}$$

(5.1.13)

where

$$[B] = [\bar{B}][\hat{A}]^{-1}$$

(5.1.14)

Stress Strain Relations The stress strain relations for the plate are given by (3.2.18)

$$\tau_{xx} = \frac{E}{(1-v^2)}(\varepsilon_{xx} + v\varepsilon_{yy})$$

$$\tau_{yy} = \frac{E}{(1-v^2)}(\varepsilon_{yy} + v\varepsilon_{xx})$$

$$\tau_{xy} = G\gamma_{xy} = \frac{E(1-v)}{2(1-v^2)}\gamma_{xy}$$

(3.2.18)

which are expressed as

$$\{\hat{\tau}\} = \frac{E}{1-v^2} \begin{bmatrix} 1 & v & 0 \\ v & 1 & 0 \\ 0 & 0 & \frac{1-v}{2} \end{bmatrix} \{\hat{\varepsilon}\}$$

$$= [D]\{\hat{\varepsilon}\}$$

(5.1.15)

where

$$[D] = \frac{E}{1-v^2} \begin{bmatrix} 1 & v & 0 \\ v & 1 & 0 \\ 0 & 0 & \frac{1-v}{2} \end{bmatrix}$$

(5.1.16)

Potential Functional The strain energy of the plate is

$$\hat{U} = \frac{1}{2} \iiint_V \{\hat{\varepsilon}\}^T [\hat{\tau}] dV$$

$$= \frac{1}{2} \iiint_V \{\hat{\varepsilon}\}^T [D] \{\hat{\varepsilon}\} dV$$

$$= \frac{1}{2} \iiint_V \{\hat{q}\}^T [B]^T [D][B] \{\hat{q}\} dV$$

(5.1.17)

Let $Q(x,y)$ be the lateral force applied on the surface of the plate, then the work done by this force is

$$W = \iint_A w Q(x,y) dA$$

$$= \iint_A \{\hat{q}\}^T [N]^T \left([\hat{A}]^{-1}\right)^T Q(x,y) dA$$

(5.1.18)

The potential functional for the plate problem is then written as

$$\Pi = \frac{1}{2} \{\hat{q}\}^T \left[\iiint_V [B]^T [D][B] dV\right] \{\hat{q}\} - \iint_A \{\hat{q}\}^T [N]^T \left([\hat{A}]^{-1}\right)^T Q(x,y) dA$$

(5.1.19)

Elemental Equation We now apply Ritz method to the potential functional in (5.1.19)

$$\frac{\partial \Pi}{\partial \{q\}^T} = 0$$

(5.1.20)

which gives

$$\left[\iiint_V [B]^T[D][B]dV\right]\{\hat{q}\} = \iint_A [N]^T\left([\hat{A}]^{-1}\right)^T Q(x,y)dA$$

$$[\hat{K}]\{\hat{q}\} = \{\hat{Q}\}$$

(5.1.21)

where $[\hat{K}]$ is the elemental stiffness matrix given by

$$[\hat{K}] = \left[\iiint_V [B]^T[D][B]dV\right]$$

(5.1.22)

and the nodal force vector is

$$\{\hat{Q}\} = \left([\hat{A}]^{-1}\right)^T \iint_A [N]^T Q(x,y)dA$$

(5.1.23)

Stiffness Matrix The stiffness matrix in (5.1.22) can be expanded by using equation (5.1.14) as

$$[\hat{K}] = \left([\hat{A}]^{-1}\right)^T \left[\iint_A dA \int_{-h/2}^{h/2} [\overline{B}]^T[D][\overline{B}]dz\right][\hat{A}]^{-1}$$

$$= \left([\hat{A}]^{-1}\right)^T [\overline{K}][\hat{A}]^{-1}$$

(5.1.24)

where

$$[\overline{K}] = \iint_A dA \int_{-h/2}^{h/2} [\overline{B}]^T[D][\overline{B}]dz$$

$$= \frac{Eh^3}{12(1-v^2)} \iint_A \begin{bmatrix} 0 & 0 & 0 \\ 0 & 0 & 0 \\ 0 & 0 & 0 \\ 2 & 0 & 0 \\ 0 & 0 & 2 \\ 0 & 2 & 0 \\ 6\hat{x} & 0 & 0 \\ 2\hat{y} & 2\hat{x} & 4(\hat{x}+\hat{y}) \\ 0 & 6\hat{y} & 0 \end{bmatrix} \begin{bmatrix} 1 & v & 0 \\ v & 1 & 0 \\ 0 & 0 & \frac{1-v}{2} \end{bmatrix}$$

$$\begin{bmatrix} 0 & 0 & 0 & 2 & 0 & 0 & 6\hat{x} & 2\hat{y} & 0 \\ 0 & 0 & 0 & 0 & 0 & 2 & 0 & 2\hat{x} & 6\hat{y} \\ 0 & 0 & 0 & 0 & 2 & 0 & 0 & 4(\hat{x}+\hat{y}) & 0 \end{bmatrix} d\hat{x}\,d\hat{y}$$

(5.1.25)

160 Dynamics of Plates

Carrying the matrix multiplication in the above, we get

$$[\bar{K}] = \frac{Eh^3}{12(1-v^2)} \iint_A \begin{bmatrix} 0 & 0 & 0 \\ 0 & 0 & 0 \\ 0 & 0 & 0 \\ 2 & 0 & 0 \\ 0 & 0 & 2 \\ 0 & 2 & 0 \\ 6\hat{x} & 0 & 0 \\ 2\hat{y} & 2\hat{x} & 4(\hat{x}+\hat{y}) \\ 0 & 6\hat{y} & 0 \end{bmatrix}$$

$$\begin{bmatrix} 0 & 0 & 0 & 2 & 0 & 2v & 6\hat{x} & 2(v\hat{x}+\hat{y}) & 6v\hat{y} \\ 0 & 0 & 0 & 2v & 0 & 2 & 6v\hat{x} & 2(\hat{x}+v\hat{y}) & 6\hat{y} \\ 0 & 0 & 0 & 0 & (1-v) & 0 & 0 & 2(1-v)(\hat{x}+\hat{y}) & 0 \end{bmatrix} d\hat{x}\,d\hat{y}$$

(5.1.26)

Finally, $[\bar{K}]$ matrix is given by

$$[\bar{K}] = \frac{Eh^3}{12(1-v^2)} \iint_A \begin{bmatrix} 0 & 0 & 0 & 0 & 0 & 0 & 0 & 0 & 0 \\ 0 & 0 & 0 & 0 & 0 & 0 & 0 & 0 & 0 \\ & 0 & 0 & 0 & 0 & 0 & 0 & 0 & 0 \\ & & 4 & 0 & 4v & 12\hat{x} & 4(v\hat{x}+\hat{y}) & 12v\hat{y} \\ & & & 2(1-v) & 0 & 0 & 4(1-v)(\hat{x}+\hat{y}) & 0 \\ & & & & 4 & 12v\hat{x} & 4(\hat{x}+v\hat{y}) & 12\hat{y} \\ & & & & & 36\hat{x}^2 & 12(v\hat{x}^2+\hat{x}\hat{y}) & 36v\hat{x}\hat{y} \\ & & \text{Sym} & & & & \begin{matrix}(12-8v)\\ \times(\hat{x}+\hat{y})^2 \\ -8(1-v)\hat{x}\hat{y}\end{matrix} & 12(v\hat{y}^2+\hat{x}\hat{y}) \\ & & & & & & & 36\hat{y}^2 \end{bmatrix} d\hat{x}\,d\hat{y}$$

(5.1.27)

Finite Element Methods for Thin Plates

The integrals in the above can be evaluated by using the following formulae,

$$\iint_A x^m y^n dA = \sum_{r=0}^{m+1} \sum_{s=0}^{r} \frac{(-1)^{r+s} m!}{(m+1-r)!(r-s)!s!(n+r+1)} x_2^{m+1-s} x_3^s y_3^{n+1}$$

$$- \frac{x_3^{m+1} y_3^{n+1}}{(m+1)(m+n+2)}$$

for $x_3 \neq 0, x_3 \neq x_2$

(5.1.28)

$$\iint_A x^m y^n dA = \sum_{r=0}^{m+1} \frac{(-1)^r m!}{(m+1-r)!r!(n+r+1)} x_2^{m+1} y_3^{n+1}$$

for $x_3 = 0$

(5.1.29)

$$\iint_A x^m y^n dA = \sum_{r=0}^{m+1} \frac{1}{(n+1)(m+n+2)} x_2^{m+1} y_3^{n+1}$$

for $x_3 = x_2$

(5.1.30)

Global Equation The stiffness matrix and the forcing vector are now to be transformed into the global system of Fig. 5.1.1. The elemental degrees of freedom can be expressed in global coordinates as

$$\begin{Bmatrix} w \\ w_{,y} \\ w_{,x} \end{Bmatrix} = \begin{bmatrix} 1 & 0 & 0 \\ 0 & \cos(x,X) & \cos(x,Y) \\ 0 & \cos(y,X) & \cos(y,Y) \end{bmatrix} \begin{Bmatrix} w \\ w_{,Y} \\ w_{,X} \end{Bmatrix}$$

$$= [L] \begin{Bmatrix} w \\ w_{,Y} \\ w_{,X} \end{Bmatrix}$$

(5.1.31)

where

$$[L] = \begin{bmatrix} 1 & 0 & 0 \\ 0 & \cos(x,X) & \cos(x,Y) \\ 0 & \cos(y,X) & \cos(y,Y) \end{bmatrix}$$

The directional cosines can be obtained from

$$\cos(x, X) = \frac{X_2 - X_1}{\sqrt{(X_2 - X_1)^2 + (Y_2 - Y_1)^2}}$$

$$\cos(x, Y) = \frac{Y_2 - Y_1}{\sqrt{(X_2 - X_1)^2 + (Y_2 - Y_1)^2}}$$

$$\cos(y, X) = -\frac{Y_2 - Y_1}{\sqrt{(X_2 - X_1)^2 + (Y_2 - Y_1)^2}}$$

$$\cos(y, Y) = \frac{X_2 - X_1}{\sqrt{(X_2 - X_1)^2 + (Y_2 - Y_1)^2}}$$

(5.1.32)

The transformation matrix is therefore given by

$$\{\hat{q}\} = [R]\{q\} \tag{5.1.33}$$

where

$$[R] = \begin{bmatrix} [L] & 0 & 0 \\ 0 & [L] & 0 \\ 0 & 0 & [L] \end{bmatrix}$$

(5.1.33a)

The potential functional of (5.1.19) in global coordinates becomes

$$\Pi = \frac{1}{2}[R]^T\{q\}^T \left[\iiint_V [B]^T[D][B] dV \right] [R]\{q\}$$

$$- \iint_A [R]^T\{q\}^T [N]^T \left([\hat{A}]^{-1}\right)^T Q(x,y) dA$$

(5.1.19a)

Hence, the stiffness matrix and forcing vector in global coordinates are given by

$$[K] = [R]^T [\hat{K}][R] \tag{5.1.34}$$

$$\{Q\} = [R]^T \{\hat{Q}\} \tag{5.1.35}$$

From (5.1.21), we now have the elemental equation in global coordinates

$$[K]\{q\} = \{Q\}$$

(5.1.36)

Assembly of Elemental Stiffness Matrices in Global Coordinates

In accordance to the direct stiffness method, the stiffness matrix and forcing vector for each element in global coordinates are assembled by using superposition principle. Then, the boundary conditions are applied for the displacement and the corresponding slope on the respective outer edges. The resulting equations give rise to the system equations

$$[K]_G\{q\}_G = \{Q\}_G$$

(5.1.36)

These equations are solved to determine the displacements, slopes and thus the bending moments and stresses.

Example

Consider a square clamped plate subjected to uniformly distributed force. Typical finite element idealization is shown in Fig. 5.1.2 for $n=1$, $n=2$, $n=4$ etc. for a quarter of plate. Because of symmetry it is sufficient to consider quarter of a plate, with fixed boundary conditions on the two edges and the slopes $w_{,x}$ and $w_{,y}$ zero on the edges 2-4 and 3-4 respectively. The deflection coefficient $\frac{10^3 D}{qa^4}$ is plotted in Fig. 5.1.3 as a function of the mesh size used in Fig. 5.1.2. It should be noted that monotonic convergency is not always obtained with nonconforming elements.

Elemental Consistent Mass Matrix

The kinetic energy of the plate element is given by (3.2.52)

$$T = \tfrac{1}{2} \iint_A \rho h \, \dot{w}^2 \, d\hat{x} \, d\hat{y}$$

(3.2.52)

From (5.1.7)

$$\dot{w} = [N][\hat{A}]^{-1}\{\dot{\hat{q}}\}$$

(5.1.7)

164 *Dynamics of Plates*

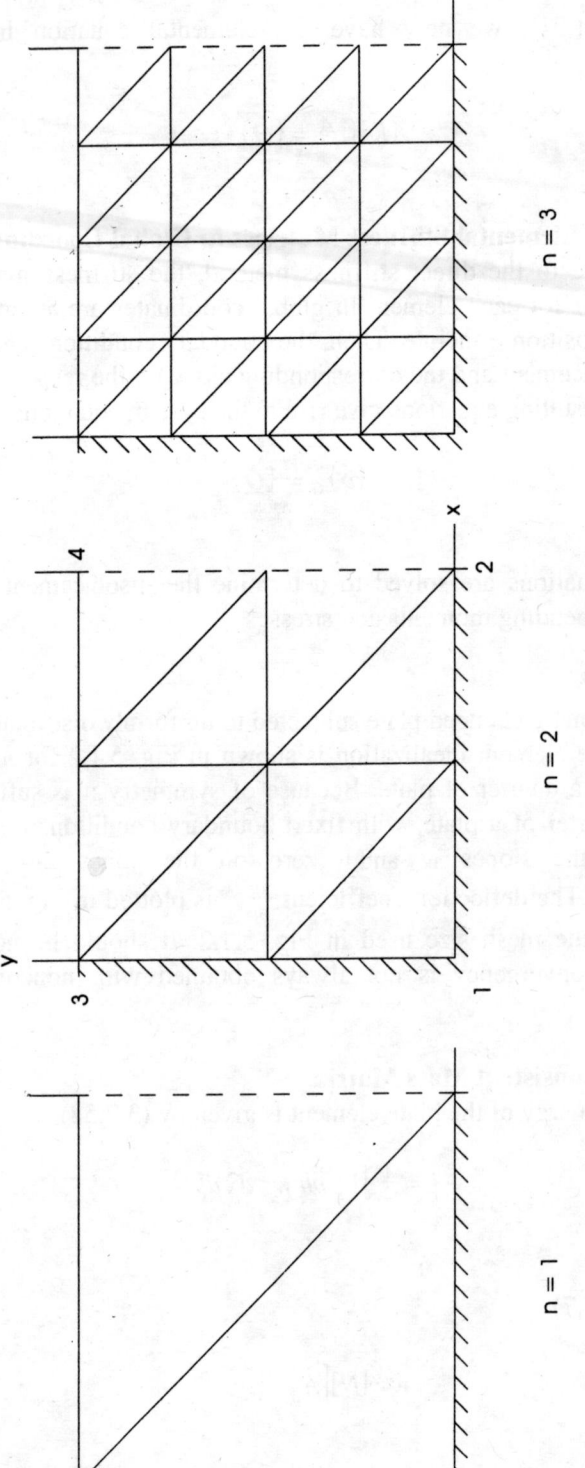

Fig. 5.1.2 Typical Mesh for a Square Plate

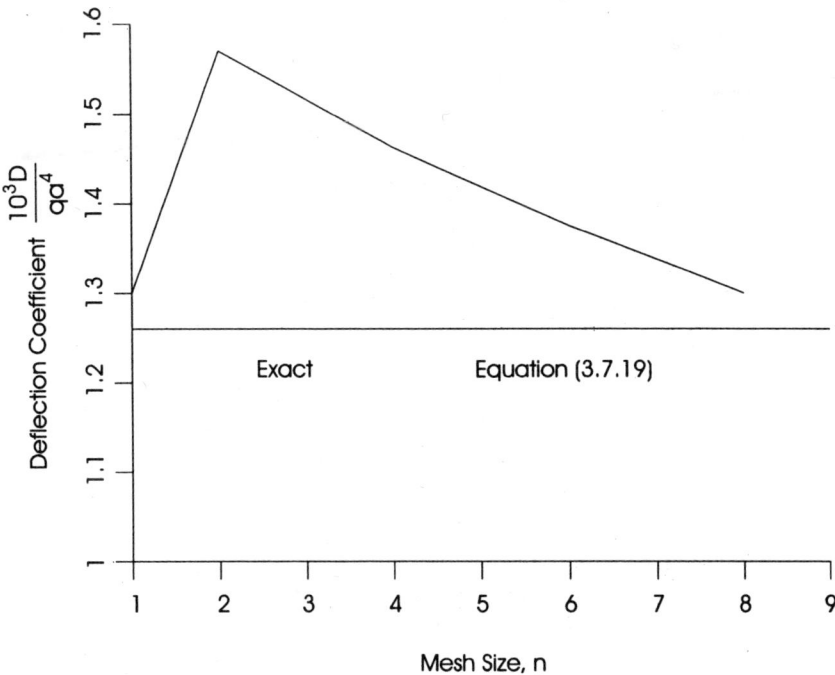

Fig. 5.1.3 Convergency Characteristic of Tocher's Element

in the above equation (3.2.52), we get

$$T = \tfrac{1}{2}\left([\hat{A}]^{-1}\right)^T \{\hat{q}\}^T \iint_A \rho h[N]^T dA [N][\hat{A}]^{-1}\{\hat{q}\} \tag{5.1.37}$$

To derive the governing equation of motion for the vibrating plate, we can use Lagrangian equations or Hamilton's principle directly. The Lagrangian equation is

$$\frac{d}{dt}\left(\frac{\partial L}{\partial \{\hat{\dot{q}}\}}\right) - \left(\frac{\partial L}{\partial \{\hat{q}\}}\right) = 0 \tag{5.1.38}$$

where

$$L = T - \Pi \tag{5.1.39}$$

Applying the first term of (5.1.38) to (5.1.37) gives

$$\left([\hat{A}]^{-1}\right)^T\left[\iint_A \rho h[N]^T[N]dA\right][\hat{A}]^{-1}\{\hat{q}\}$$
$$= [\widehat{M}]\{\hat{q}\} \tag{5.1.40}$$

The elemental mass matrix in the above is

$$[\widehat{M}] = \rho h\left([\hat{A}]^{-1}\right)^T\left[\iint_A [N]^T[N]dA\right][\hat{A}]^{-1}$$
$$= \rho h\left([\hat{A}]^{-1}\right)^T[\overline{M}][\hat{A}]^{-1} \tag{5.1.41}$$

where

$$[\overline{M}] = \iint_A [N]^T[N]\,d\hat{x}\,d\hat{y} \tag{5.1.42}$$

Substituting for $[N]$ from (5.1.3), the integrand $[N]^T[N]$ in the above equation is

$$\begin{bmatrix}
1 & & & & & & & & \\
\hat{x} & \hat{x}^2 & & & & & & & \\
\hat{y} & \hat{x}\hat{y} & \hat{y}^2 & & & \text{sym} & & & \\
\hat{x}^2 & \hat{x}^3 & \hat{x}^2\hat{y} & \hat{x}^4 & & & & & \\
\hat{x}\hat{y} & \hat{x}^2\hat{y} & \hat{x}\hat{y}^2 & \hat{x}^3\hat{y} & \hat{x}^2\hat{y}^2 & & & & \\
\hat{y}^2 & \hat{x}\hat{y}^2 & \hat{y}^3 & \hat{x}^2\hat{y}^2 & \hat{x}\hat{y}^3 & \hat{y}^4 & & & \\
\hat{x}^3 & \hat{x}^4 & \hat{x}^3\hat{y} & \hat{x}^5 & \hat{x}^4\hat{y} & \hat{x}^3\hat{y}^2 & \hat{x}^6 & & \\
\hat{x}^2\hat{y}+ & \hat{x}^2\hat{y}^2+ & \hat{x}\hat{y}^3+ & \hat{x}^3\hat{y}^2+ & \hat{x}^2\hat{y}^3+ & \hat{x}\hat{y}^4+ & \hat{x}^4\hat{y}^2+ & \left(\hat{x}\hat{y}^2 \atop +\hat{x}^2\hat{y}\right)^2 & \\
\hat{x}\hat{y}^2 & \hat{x}^3\hat{y} & \hat{x}^2\hat{y}^2 & \hat{x}^4\hat{y} & \hat{x}^3\hat{y}^2 & \hat{x}^2\hat{y}^3 & \hat{x}^5\hat{y} & & \\
\hat{y}^3 & \hat{x}\hat{y}^3 & \hat{y}^4 & \hat{x}^2\hat{y}^3 & \hat{x}\hat{y}^4 & \hat{y}^5 & \hat{x}^2\hat{y}^3 & \hat{x}\hat{y}^5+ \atop \hat{x}^3\hat{y}^3 & \hat{y}^6
\end{bmatrix}$$

$$\tag{5.1.43}$$

The integrals in the above are determined using the formulae given in equation (5.1.28) to (5.1.30). The elemental mass matrix $[\widehat{M}]$ in global coordinates is obtained in a similar manner to that of the stiffness matrix in equation (5.1.34)

$$[M] = [R]^T[\widehat{M}][R]$$

(5.1.44)

Global Equation of Motion
The elemental mass matrices in global coordinates are assembled using the principle of superposition of direct stiffness method to obtain the system mass matrix. Combining the mass matrix thus obtained with equation (5.1.36), we get the system equation in global coordinates as

$$[M]_G\{\ddot{q}\}_G + [K]_G\{q\}_G = \{Q\}_G$$

(5.1.45)

Consider as an example a square cantilever plate shown in Fig. 5.1.4 discretized into 50 triangular elements with 36 nodes. Each node has 3 degrees of freedom and six nodes are fixed. Therefore, there are 90 degrees of freedom in all for the plate. The natural frequencies obtained by Anderson, Irons and Zienkiewicz are

$$p_1 = 3.469\sqrt{\frac{D}{\rho h a^4}} \quad p_2 = 8.535\sqrt{\frac{D}{\rho h a^4}}$$

$$p_3 = 21.450\sqrt{\frac{D}{\rho h a^4}} \quad p_4 = 27.059\sqrt{\frac{D}{\rho h a^4}}$$

5.2 ACM NONCONFORMING RECTANGULAR ELEMENT

Select Element Type This element is commonly called Adini, Clough and Melosh element with a rectangular shape as shown in Fig. 5.2.1. The element has four nodes 1, 2, 3 and 4. The nodal coordinate system $\hat{x}\hat{y}$ is located with its origin at the center of the element of width $2a$ in $\hat{x}$ direction and $2b$ in $\hat{y}$ direction as shown. Because of the rectangular geometry, it is convenient to use non dimensional elemental coordinates as

$$\xi = \frac{x}{a} \quad \eta = \frac{y}{b}$$

(5.2.1)

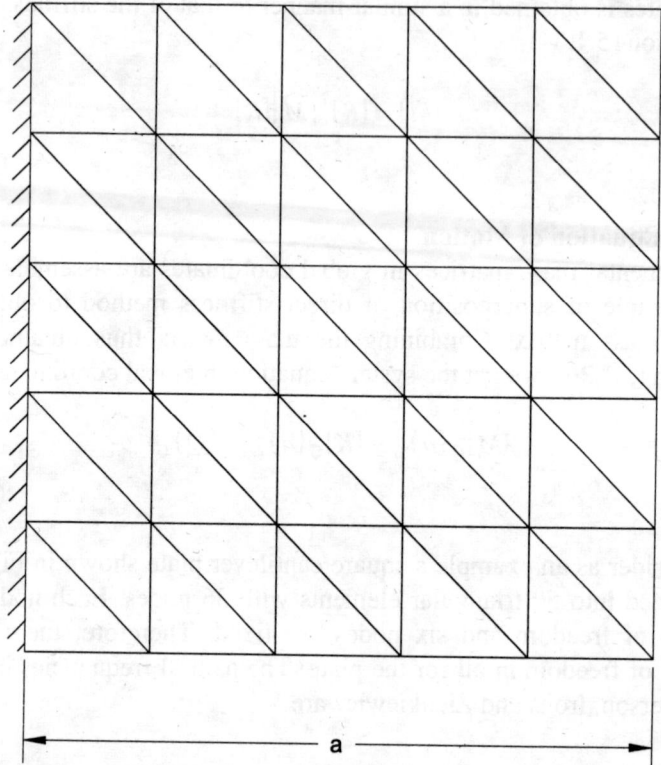

Fig. 5.1.4 A Square Cantilever Plate 50 Elements, 36 Nodes and 90 Degrees of Freedom.

In accordance to plate theory, we will assign three degrees of freedom to each node, viz., w, $w_{,y}, w_{,x}$, so that each element has twelve degrees of freedom given below.

$$q_1 = w_1$$
$$q_2 = (w_{,\eta})_1$$
$$q_3 = -(w_{,\xi})_1$$
$$q_4 = w_2$$
$$q_5 = (w_{,\eta})_2$$
$$q_6 = -(w_{,\xi})_2 \quad \cdots$$
$$q_{10} = w_4$$
$$q_{11} = (w_{,\eta})_4$$

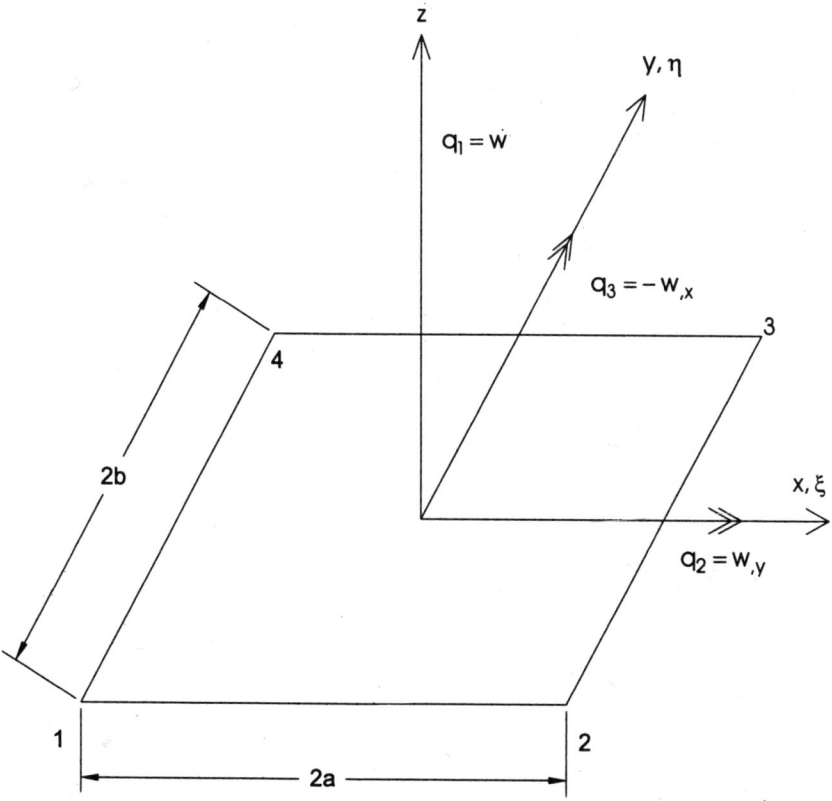

Fig. 5.2.1 Rectangular Plate Element

$$q_{12} = -(w_{,\xi})_4 \qquad (5.2.2)$$

The nodal degrees of freedom of the element are expressed as a vector

$$\{\hat{d}\} = \{\, \hat{q}_1\ \hat{q}_2\ \hat{q}_3\ \hat{q}_4\ \hat{q}_5\ \hat{q}_6\ \hat{q}_7\ \cdots\ \hat{q}_{12}\,\}^T \qquad (5.2.3)$$

where a hat above any quantity $\widehat{}$ refers to the elemental values.

Displacement Function The displacement within the element at any point $(\hat{x}, \hat{y})$ is represented by a polynomial with twelve terms, so that the coefficients in the polynomial can be evaluated to satisfy the nodal degrees of freedom.

$w(\xi, \eta) =$

$$= \begin{bmatrix} 1 & \xi & \eta & \xi^2 & \xi\eta & \eta^2 & \xi^3 & \xi^2\eta & \xi\eta^2 & \eta^3 & \xi^3\eta & \xi\eta^3 \end{bmatrix} \begin{Bmatrix} a_1 \\ a_2 \\ a_3 \\ a_4 \\ a_5 \\ a_6 \\ a_7 \\ a_8 \\ a_9 \\ a_{10} \\ a_{11} \\ a_{12} \end{Bmatrix}$$

$$= [N]\{a\}$$

(5.2.4)

The above shape function is a complete third degree polynomial with two additional quartic terms. As in the case of Tocher's triangular element, we find that the deflection as well as slopes are not continuous across the elements through a common node. Therefore, the shape function above produces a non conformal element. Differentiating equation (5.2.4)

$$w_{,\xi} = \begin{bmatrix} 0 & 1 & 0 & 2\xi & \eta & 0 & 3\xi^2 & 2\xi\eta & \eta^2 & 0 & 3\xi^2\eta & \eta^3 \end{bmatrix}\{a\}$$

$$w_{,\eta} = \begin{bmatrix} 0 & 0 & 1 & 0 & \xi & 2\eta & 0 & \xi^2 & 2\xi\eta & 3\eta^2 & \xi^3 & 3\xi\eta^2 \end{bmatrix}\{a\}$$

(5.2.5)

The shape function and its derivatives are evaluated at the four nodes of the element $(\xi, \eta) = (-1, -1), (1, -1), (1, 1)$ and $(-1, 1)$ and substituting in (5.2.2), we get

$$[\hat{A}]\{a\} = \{\hat{q}\}$$

(5.2.6)

where

$$[\hat{A}] = \begin{bmatrix} 1 & -1 & -1 & 1 & 1 & 1 & -1 & -1 & -1 & -1 & 1 & 1 \\ 0 & 0 & 1 & 0 & -1 & -2 & 0 & 1 & 2 & 3 & -1 & -3 \\ 0 & -1 & 0 & 2 & 1 & 0 & -3 & -2 & -1 & 0 & 3 & 1 \\ 1 & 1 & -1 & 1 & -1 & 1 & 1 & -1 & 1 & -1 & -1 & -1 \\ 0 & 0 & 1 & 0 & 1 & -2 & 0 & 1 & -2 & 3 & 1 & 3 \\ 0 & -1 & 0 & -2 & 1 & 0 & -3 & 2 & -1 & 0 & 3 & 1 \\ 1 & 1 & 1 & 1 & 1 & 1 & 1 & 1 & 1 & 1 & 1 & 1 \\ 0 & 0 & 1 & 0 & 1 & 2 & 0 & 1 & 2 & 3 & 1 & 3 \\ 0 & -1 & 0 & -2 & -1 & 0 & -3 & -2 & -1 & 0 & -3 & -1 \\ 1 & -1 & 1 & 1 & -1 & 1 & -1 & 1 & -1 & 1 & -1 & -1 \\ 0 & 0 & 1 & 0 & -1 & 2 & 0 & 1 & -2 & 3 & -1 & -3 \\ 0 & -1 & 0 & 2 & -1 & 0 & -3 & 2 & -1 & 0 & -3 & -1 \end{bmatrix}$$

(5.2.7)

Therefore

$$\{a\} = [\hat{A}]^{-1}\{\hat{q}\}$$

(5.2.8)

where

$$[\hat{A}]^{-1} = \frac{1}{8}\begin{bmatrix} 2 & 1 & -1 & 2 & 1 & 1 & 2 & -1 & 1 & 2 & -1 & -1 \\ -3 & -1 & 1 & 3 & 1 & 1 & 3 & -1 & 1 & -3 & 1 & 1 \\ -3 & -1 & 1 & -3 & -1 & -1 & 3 & -1 & 1 & 3 & -1 & -1 \\ 0 & 0 & 1 & 0 & 0 & -1 & 0 & 0 & -1 & 0 & 0 & 1 \\ 4 & 1 & -1 & -4 & -1 & -1 & 4 & -1 & 1 & -4 & 1 & 1 \\ 0 & -1 & 0 & 0 & -1 & 0 & 0 & 1 & 0 & 0 & 1 & 0 \\ 1 & 0 & -1 & -1 & 0 & -1 & -1 & 0 & -1 & 1 & 0 & -1 \\ 0 & 0 & -1 & 0 & 0 & 1 & 0 & 0 & -1 & 0 & 0 & -1 \\ 0 & 1 & 0 & 0 & -1 & 0 & 0 & 1 & 0 & 0 & -1 & 0 \\ 1 & 1 & 0 & 1 & 1 & 0 & -1 & 1 & 0 & -1 & 1 & 0 \\ -1 & 0 & 1 & 1 & 0 & 1 & -1 & 0 & -1 & 1 & 0 & -1 \\ -1 & -1 & 0 & 1 & 1 & 0 & -1 & 1 & 0 & 1 & -1 & 0 \end{bmatrix}$$

(5.2.9)

From equation (5.2.4)

$$\begin{aligned} w &= [N][\hat{A}]^{-1}\{\hat{q}\} \\ &= \left[\{N_1(\xi,\eta)\} \ \{N_2(\xi,\eta)\} \ \{N_3(\xi,\eta)\} \ \{N_4(\xi,\eta)\} \right]\{\hat{q}\} \\ &= [\hat{N}]\{\hat{q}\} \end{aligned}$$

(5.2.10)

The shape function matrices $\{N_i(\xi,\eta)\}$ in the element shape function matrix $[\hat{N}]$ are given by

$$\{N_i(\xi,\eta)\}^T = \left\{ \begin{array}{l} \frac{1}{8}(1+\xi_i\xi)(1+\eta_i\eta)(2+\xi_i\xi+\eta_i\eta-\xi^2-\eta^2) \\ \frac{b}{8}(1+\xi_i\xi)(\eta_i+\eta)(\eta^2-1) \\ \frac{a}{8}(\xi_i+\xi)(1+\eta_i\eta)(\xi^2-1) \end{array} \right\}$$

(5.2.11)

where ξ_i, η_i are the coordinates of the node i.

Strain Displacement Relations The strain displacement relations for thin plate theory are given by equation (3.2.16)

$$\varepsilon_{xx} = -zw_{,xx}$$
$$\varepsilon_{yy} = -zw_{,yy}$$
$$\gamma_{xy} = -2zw_{,xy}$$

(3.2.16)

Differentiating the displacement function of (5.2.4),

$$w_{,\xi\xi} = \left[0 \ 0 \ 0 \ 2 \ 0 \ 0 \ 6\xi \ 2\eta \ 0 \ 0 \ 6\xi\eta \ 0 \right]\{a\}$$
$$w_{,\eta\eta} = \left[0 \ 0 \ 0 \ 0 \ 0 \ 2 \ 0 \ 0 \ 2\xi \ 6\eta \ 0 \ 6\xi\eta \right]\{a\}$$
$$w_{,\xi\eta} = \left[0 \ 0 \ 0 \ 0 \ 1 \ 0 \ 0 \ 2\xi \ 2\eta \ 0 \ 3\xi^2 \ 3\eta^2 \right]\{a\}$$

(5.2.12)

Therefore, equations (3.2.16) can be expressed as

$$\{\hat{\varepsilon}\} = -z \begin{bmatrix} 0 & 0 & 0 & \frac{2}{a^2} & 0 & 0 & \frac{6\xi}{a^2} & \frac{2\eta}{a^2} & 0 & 0 & \frac{6\xi\eta}{a^2} & 0 \\ 0 & 0 & 0 & 0 & 0 & \frac{2}{b^2} & 0 & 0 & \frac{2\xi}{b^2} & \frac{6\eta}{b^2} & 0 & \frac{6\xi\eta}{b^2} \\ 0 & 0 & 0 & 0 & 2 & 0 & 0 & \frac{4\xi}{ab} & \frac{4\eta}{ab} & 0 & \frac{6\xi^2}{ab} & \frac{6\eta^2}{ab} \end{bmatrix} \{a\}$$

$$= [\overline{B}]\{a\}$$

(5.2.13)

where

$$[\overline{B}] = -z \begin{bmatrix} 0 & 0 & 0 & \frac{2}{a^2} & 0 & 0 & \frac{6\xi}{a^2} & \frac{2\eta}{a^2} & 0 & 0 & \frac{6\xi\eta}{a^2} & 0 \\ 0 & 0 & 0 & 0 & 0 & \frac{2}{b^2} & 0 & 0 & \frac{2\xi}{b^2} & \frac{6\eta}{b^2} & 0 & \frac{6\xi\eta}{b^2} \\ 0 & 0 & 0 & 0 & 2 & 0 & 0 & \frac{4\xi}{ab} & \frac{4\eta}{ab} & 0 & \frac{6\xi^2}{ab} & \frac{6\eta^2}{ab} \end{bmatrix}$$

(5.2.14)

With the help of (5.2.8), (5.2.13) becomes

$$\{\hat{\varepsilon}\} = [\overline{B}][\hat{A}]^{-1}\{\hat{q}\}$$
$$= [B]\{\hat{q}\}$$

(5.2.15)

where

$$[B] = [\overline{B}][\hat{A}]^{-1}$$

(5.2.16)

Elemental Equation

Following the procedure in section 5.1, we get the elemental equation

$$[\hat{K}]\{\hat{q}\} = \{\hat{Q}\}$$

(5.2.17)

Stiffness Matrix

The elemental stiffness matrix $[\hat{K}] = \left[\iiint_V [B]^T[D][B] dV \right]$ as evaluated by Smith is

174 Dynamics of Plates

$$[\hat{K}] = \frac{Eh^3}{48(1-v^2)ab} \begin{bmatrix} [K_{11}] & & & \\ [K_{21}] & [K_{22}] & \text{Sym} & \\ [K_{31}] & [K_{32}] & [K_{33}] & \\ [K_{41}] & [K_{42}] & [K_{43}] & [K_{44}] \end{bmatrix}$$

(5.2.18)

The sub matrices in the above with $\alpha = \frac{a}{b}$, $\beta = \frac{b}{a}$ are given by

$$[K_{11}] = \begin{bmatrix} \begin{bmatrix} 4(\beta^2+\alpha^2) \\ +\frac{2}{5}(7-2v) \end{bmatrix} & & & \text{Sym} \\ b\begin{bmatrix} 2\alpha^2 \\ +\frac{1}{5}(1+4v) \end{bmatrix} & b^2\begin{bmatrix} \frac{4}{3}\alpha^2 \\ +\frac{4}{15}(1-v) \end{bmatrix} & & \\ -a\begin{bmatrix} 2\beta^2 \\ +\frac{1}{5}(1+4v) \end{bmatrix} & -vab & a^2\begin{bmatrix} \frac{4}{3}\beta^2 \\ +\frac{4}{15}(1-v) \end{bmatrix} \end{bmatrix}$$

$$[K_{21}] = \begin{bmatrix} \begin{bmatrix} -2(2\beta^2-\alpha^2) \\ -\frac{2}{5}(7-2v) \end{bmatrix} & b\begin{bmatrix} \alpha^2 \\ -\frac{1}{5}(1+4v) \end{bmatrix} & a\begin{bmatrix} 2\beta^2 \\ +\frac{1}{5}(1-v) \end{bmatrix} \\ b\begin{bmatrix} \alpha^2 \\ -\frac{1}{5}(1+4v) \end{bmatrix} & b^2\begin{bmatrix} \frac{2}{3}\alpha^2 \\ -\frac{4}{15}(1-v) \end{bmatrix} & 0 \\ -a\begin{bmatrix} 2\beta^2 \\ +\frac{1}{5}(1-v) \end{bmatrix} & 0 & a^2\begin{bmatrix} \frac{2}{3}\beta^2 \\ -\frac{1}{15}(1-v) \end{bmatrix} \end{bmatrix}$$

$$[K_{31}] = \begin{bmatrix} \begin{bmatrix} -2(2\beta^2+\alpha^2) \\ -\frac{2}{5}(7-2v) \end{bmatrix} & b\begin{bmatrix} -\alpha^2 \\ +\frac{1}{5}(1-v) \end{bmatrix} & a\begin{bmatrix} \beta^2 \\ -\frac{1}{5}(1-v) \end{bmatrix} \\ b\begin{bmatrix} \alpha^2 \\ -\frac{1}{5}(1-v) \end{bmatrix} & b^2\begin{bmatrix} \frac{1}{3}\alpha^2 \\ +\frac{1}{15}(1-v) \end{bmatrix} & 0 \\ -a\begin{bmatrix} \beta^2 \\ -\frac{1}{5}(1-v) \end{bmatrix} & 0 & a^2\begin{bmatrix} \frac{1}{3}\beta^2 \\ +\frac{1}{15}(1-v) \end{bmatrix} \end{bmatrix}$$

$$[K_{41}] = \begin{bmatrix} \begin{bmatrix} 2(\beta^2 - 2a^2) \\ -\frac{2}{5}(7-2v) \end{bmatrix} & -b\begin{bmatrix} 2a^2 \\ +\frac{1}{5}(1-v) \end{bmatrix} & a\begin{bmatrix} -\beta^2 \\ +\frac{1}{5}(1+4v) \end{bmatrix} \\ b\begin{bmatrix} 2a^2 \\ +\frac{1}{5}(1-v) \end{bmatrix} & b^2\begin{bmatrix} \frac{2}{3}a^2 \\ -\frac{1}{15}(1-v) \end{bmatrix} & 0 \\ -a\begin{bmatrix} \beta^2 \\ -\frac{1}{5}(1+4v) \end{bmatrix} & 0 & a^2\begin{bmatrix} \frac{2}{3}\beta^2 \\ -\frac{4}{15}(1-v) \end{bmatrix} \end{bmatrix}$$

$$[K_{22}] = [I_3]^T [K_{11}][I_3]$$
$$[K_{32}] = [I_3]^T [K_{41}][I_3]$$
$$[K_{42}] = [I_3]^T [K_{32}][I_3]$$

$$[K_{33}] = [I_1]^T [K_{11}][I_1]$$
$$[K_{43}] = [I_1]^T [K_{21}][I_1]$$

$$[K_{44}] = [I_2]^T [K_{11}][I_2] \tag{5.2.19}$$

where

$$[I_1] = \begin{bmatrix} -1 & & \\ & 1 & \\ & & 1 \end{bmatrix}$$

$$[I_2] = \begin{bmatrix} 1 & & \\ & -1 & \\ & & 1 \end{bmatrix}$$

$$[I_3] = \begin{bmatrix} 1 & & \\ & 1 & \\ & & -1 \end{bmatrix} \tag{5.2.20}$$

Nodal Force Vector

The nodal force vector is

$$\{\hat{Q}\} = \iint_A [\hat{N}]^T Q(x,y) dA \tag{5.2.21}$$

Assembly of Elemental Stiffness Matrices

In accordance to the direct stiffness method, the stiffness matrix and forcing vector for each element are assembled by using superposition principle. Then, the boundary conditions are applied for the displacement and the corresponding slope on the respective outer edges. The resulting equations give rise to the system equations

$$[K]_G \{q\}_G = \{Q\}_G$$

(5.2.23)

These equations are solved to determine the displacements, slopes and thus the bending moments and stresses.

Example

Consider the square clamped plate subjected to uniformly distributed force. Typical finite element idealization is shown in Fig. 5.1.2 for $n = 1$, $n = 2$, $n = 4$ etc. for a quarter of plate. Because of symmetry it is sufficient to consider quarter of a plate, with fixed boundary conditions on the two edges and the slopes $w_{,x}$ and $w_{,y}$ zero on the edges 2-4 and 3-4 respectively. The deflection coefficient $\frac{10^3 D}{qa^4}$ is plotted in Fig. 5.2.2 as a function of the mesh size used in Fig. 5.1.2.

Elemental Consistent Mass Matrix

Following the derivation in section 5.1, the elemental mass matrix is

$$\left[\widehat{M}\right] = \rho h \iint_A [\widehat{N}]^T [\widehat{N}] dA$$

$$= \rho h a b \int_{-1}^{1} \int_{-1}^{1} [\widehat{N}]^T [\widehat{N}] d\xi\, d\eta$$

(5.2.24)

Substituting for $[\widehat{N}]$ from (5.2.10) and evaluating the integrals in the above, we get the elemental mass matrix as

$$\left[\widehat{M}\right] = \frac{\rho h a b}{6300} \begin{bmatrix} [M_{11}] & [M_{21}]^T \\ [M_{21}] & [M_{22}] \end{bmatrix}$$

(5.2.25)

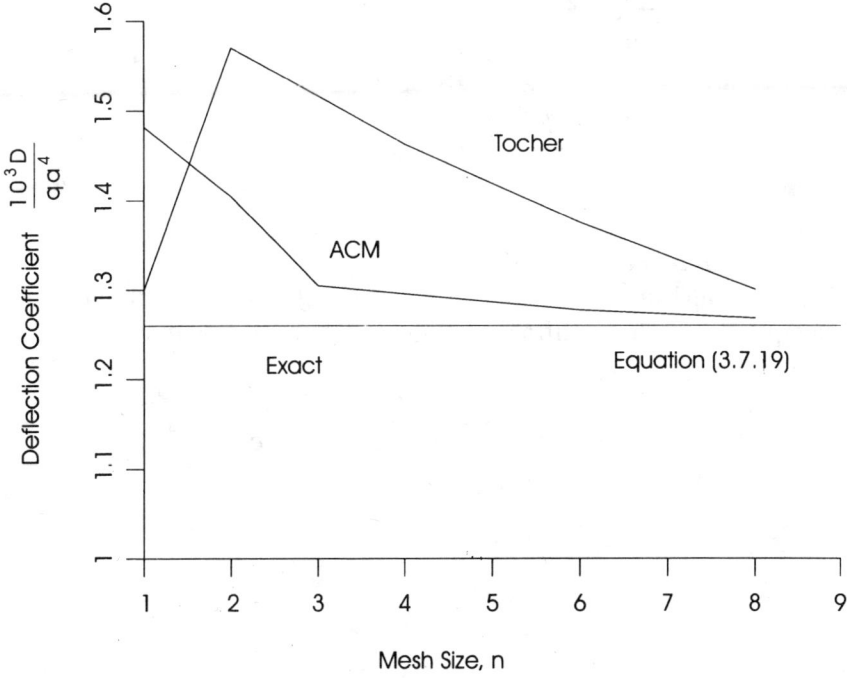

Fig. 5.2.2 Convergency Characteristic of ACM and Tocher Elements

where

$$[M_{11}] = \begin{bmatrix} 3454 & & & & & & \\ 922b & 320b^2 & & & \text{Sym} & & \\ -922a & -252ab & 320a^2 & & & & \\ 1226 & 398b & -548a & 3454 & & & \\ 398b & 160b^2 & -168ab & 922b & 320b^2 & & \\ 548a & 168ab & -240a^2 & 922a & 252ab & 320a^2 \end{bmatrix}$$

$$[M_{21}] = \begin{bmatrix} 394 & 232b & -232a & 1226 & 548b & 398a \\ -232b & -120b^2 & 112ab & -548b & -240b^2 & -168ab \\ 232a & 112ab & -120a^2 & 398a & 168ab & 160a^2 \\ 1226 & 548b & -398a & 394 & 232b & 232a \\ -548b & -240b^2 & 168ab & -232b & -120b^2 & -112ab \\ -398a & -168ab & 160a^2 & -232b & -112ab & -120a^2 \end{bmatrix}$$

178 Dynamics of Plates

$$[M_{22}] = \begin{bmatrix} 3454 & & & & & & \\ -922b & 320b^2 & & & \text{Sym} & & \\ 922a & -252ab & 320a^2 & & & & \\ 1226 & -398b & 548a & 3454 & & & \\ -398b & 160b^2 & -168ab & -922b & 320b^2 & & \\ -548a & 168ab & -240a^2 & -922a & 252ab & 320a^2 \end{bmatrix}$$

(5.2.26)

Global Equation of Motion

The elemental mass matrices are assembled using the principle of superposition of direct stiffness method to obtain the system mass matrix. Then,

$$[M]_G\{\ddot{q}\}_G + [K]_G\{q\}_G = \{Q\}_G$$

(5.2.27)

6

Plates with Combined Lateral and in-Plane Forces

So far we considered plates which are subjected to lateral forces only. There are several examples where the plates are subjected to forces not only lateral to its surface but also forces that are acting in the plane of the plate. Here, we consider thin plates as in section 3.2, subjected to lateral as well as in-plane forces.

6.1 VON KARMAN'S THEORY

Displacement Field
The displacement field of classical thin plate theory is modified to take into account the components of the displacement of the middle surface. Thus the displacements at any point can be expressed as

$$u_x = u(x,y) - zw_{,x}$$
$$u_y = v(x,y) - zw_{,y}$$
$$u_z = w(x,y)$$

(6.1.1)

Strain Field
The linear strain field (Kirchoff strain displacement relations) is

$$\varepsilon_{xx\text{linear}} = u_{,x} - zw_{,xx}$$
$$\varepsilon_{yy\text{linear}} = v_{,y} - zw_{,yy}$$
$$\varepsilon_{xy\text{linear}} = \frac{1}{2}(u_{,y} + v_{,x} - 2zw_{,xy})$$

(6.1.2)

We can deduce from the above strain field, that proceeding further to define stresses, strain energy etc. in the usual manner, we will arrive at uncoupled equations for the motion in the lateral and in-plane directions. In order to find a coupling mechanism between the lateral and in-plane motions, we should consider the components of the in-plane forces, which are tangential to the deformed middle surface of the plate, in the vertical direction. This will mean we should consider large deflections which are of the order of plate thickness. Therefore, we account for rotations in the strain displacement relations. For this purpose, we use the Green strain tensor (see Appendix 3 of *Advanced Theory of Vibration book*)

$$\varepsilon_{ij} = \frac{1}{2}(u_{i,j} + u_{j,i} + u_{k,i}u_{k,j})$$

(6.1.3)

This tensor can be approximated for our purpose in the following nonlinear form

$$\varepsilon_{ij} = \frac{1}{2}(u_{i,j} + u_{j,i}) + \frac{1}{2}\omega_{k,i}\omega_{kj}$$

(6.1.3a)

where

$$\omega_{ij} = \frac{1}{2}(u_{i,j} - u_{j,i})$$

Therefore, the nonlinear strain displacement relations are written as

$$\varepsilon_{xx} = u_{,x} + \frac{1}{2}w_{,x}^2 - zw_{,xx}$$

$$\varepsilon_{yy} = v_{,y} + \frac{1}{2}w_{,y}^2 - zw_{,yy}$$

$$\varepsilon_{zz} = \frac{1}{2}w_{,x}^2 + \frac{1}{2}w_{,y}^2$$

$$\varepsilon_{xy} = \frac{1}{2}(u_{,y} + v_{,x} - 2zw_{,xy}) + \frac{1}{2}w_{,x}w_{,y}$$

(6.1.4)

With reference to our plate problem, we can relate the strains above to an element dx shown in Fig. 6.1.1a. The axial displacements of this element are u_x and $u_x + u_{x,x}dx$. Therefore, the elongation in the x direction is $u_{x,x}dx$. The elongation of the same element due to lateral deflection is $dx - dx\cos w_x \simeq \frac{1}{2}w_{,x}^2 dx$. Therefore, the total elongation is

$u_{x,x}dx + \frac{1}{2}w_{,x}^2 dx$ and the strain in x direction $\varepsilon_{xx} = \left(u_{x,x}dx + \frac{1}{2}w_{,x}^2 dx\right) \div dx$
$= u_{x,x} + \frac{1}{2}w_{,x}^2$. The first equation in (6.1.4) is precisely this relation. In a similar manner the second strain relation can be explained.

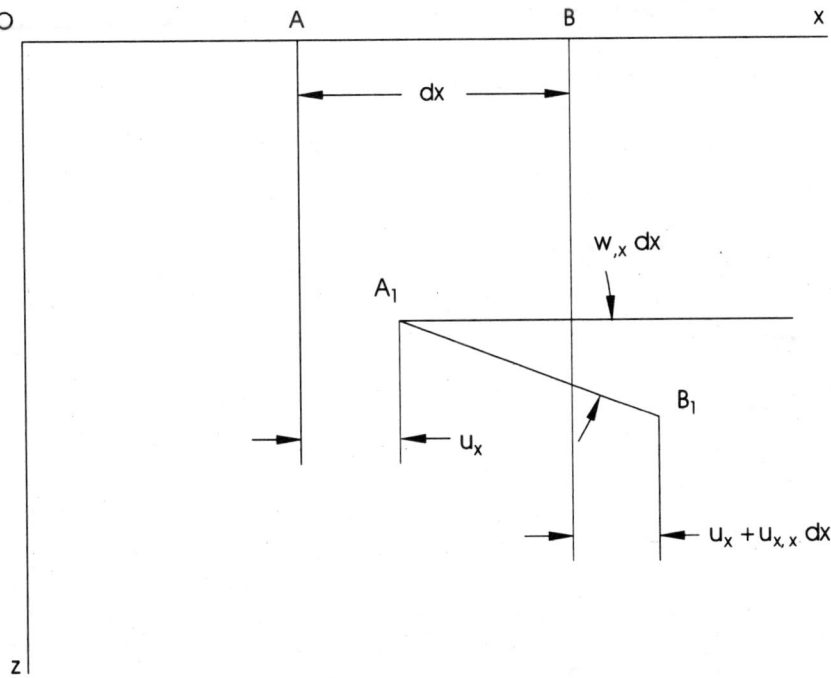

Fig. 6.1.1a Strain in x Direction

With regard to the shear strain, consider Fig. 6.1.1b of an element whose sides are dx and dy. The engineering shearing strain γ_{xy} due to displacements u_x and u_y is $u_{x,y} + u_{y,x}$. The shearing strain due to displacement w is the difference between $\frac{\pi}{2}$ and the angle $A_1O_1B_1$ as shown in the figure. Draw O_1B_2 parallel to OB and $A_1O_1B_2$ is a right angle. Let us now rotate the plane $A_1O_1B_2$ about A_1O_1 through an angle $w_{,y}$ to bring the plane $A_1O_1B_2$ to plane $A_1O_1B_1$. In this process, the point B_2 moves to the point C. The displacement B_2C is then given by $\overline{BC} = w_{,y}dy$ and is inclined at an angle $w_{,x}$ from B_2B_1. The angle CO_1B_1 represents the engineering shear strain, which is equal to
$\gamma_{xy} = \dfrac{\overline{B_1C}}{\overline{O_1B_1}} = \dfrac{\overline{B_2C} \times w_{,x}}{\overline{O_1B_1}} = \dfrac{w_{,y}dy \times w_{,x}}{dy} = w_{,x}w_{,y}$. Therefore, the total engineering shearing strain is $\gamma_{xy} = u_{x,y} + u_{y,x} + w_{,x}w_{,y}$ which corresponds to the nonlinear shearing strain in equation (6.1.4).

182 Dynamics of Plates

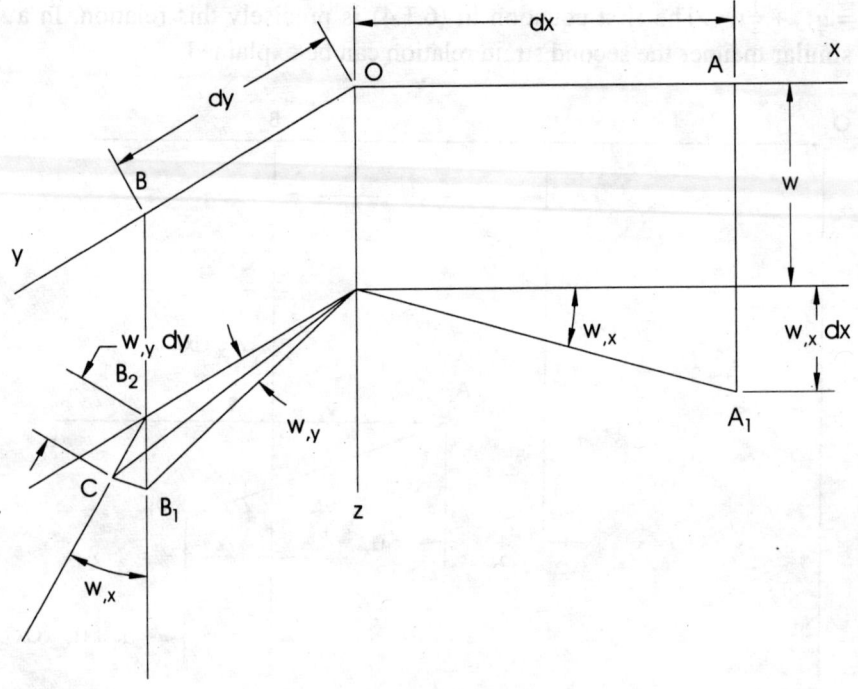

Fig. 6.1.1b Shear Strain Relation

Stress Field
The stress field is determined by applying Hooke's law for plane stress condition,

$$\tau_{xx} = \frac{E}{(1-\nu^2)}(\varepsilon_{xx} + \nu\varepsilon_{yy})$$

$$\tau_{yy} = \frac{E}{(1-\nu^2)}(\varepsilon_{yy} + \nu\varepsilon_{xx})$$

$$\tau_{xy} = 2G\varepsilon_{xy} = \frac{E}{(1+\nu)}\varepsilon_{xy}$$

(3.2.18)

Bending and Twisting Moments
As in section 3.2, we find

$$M_x = \int_{-h/2}^{h/2} \tau_{xx} z \, dz = -D(w_{,xx} + \nu w_{,yy})$$

$$M_y = \int_{-h/2}^{h/2} \tau_{yy} z \, dz = -D(w_{,yy} + v w_{,xx})$$

$$M_{xy} = -\int_{-h/2}^{h/2} \tau_{xy} z \, dz = D(1-v) w_{,xy}$$

(6.1.5)

Axial Stress Resultants

The *axial stress resultants* are defined as

$$N_x = \int_{-h/2}^{h/2} \tau_{xx} \, dz$$

$$N_y = \int_{-h/2}^{h/2} \tau_{yy} \, dz$$

$$N_{xy} = \int_{-h/2}^{h/2} \tau_{xy} \, dz$$

$$N_{yx} = \int_{-h/2}^{h/2} \tau_{yx} \, dz$$

(6.1.6)

Substituting for the stresses in equation (3.2.18) and making use of (6.1.4), the above stress resultants are

$$N_x = C\left[\left(u_{,x} + \tfrac{1}{2}w_{,x}^2\right) + v\left(v_{,y} + \tfrac{1}{2}w_{,y}^2\right)\right]$$

$$N_y = C\left[\left(v_{,y} + \tfrac{1}{2}w_{,y}^2\right) + v\left(u_{,x} + \tfrac{1}{2}w_{,x}^2\right)\right]$$

$$N_{xy} = \tfrac{1}{2}C(1-v)(u_{,y} + v_{,x} + w_{,x} w_{,y})$$

(6.1.7)

where C is the *extensional rigidity* given by

$$C = \frac{Eh}{1-v^2}$$

(6.1.7a)

Principle of Virtual Work

We can use the principle of virtual work directly, which is, see equation (5.70), *Advanced Theory of Vibration* book,

$$\iiint_V \tau_{ij} \delta \varepsilon_{ij} \, dv = \iiint_V B_i \delta u_i \, dv + \iint_S T_i^{(v)} \delta u_i \, dA$$

184 Dynamics of Plates

For the potential energy of the system considered above, we can write

$$\delta U = \iiint_V \tau_{ij}\delta\varepsilon_{ij} dv$$

$$= \iint_R \int_{-h/2}^{h/2} (\tau_{xx}\delta\varepsilon_{xx} + 2\tau_{xy}\delta\varepsilon_{xy} + \tau_{yy}\delta\varepsilon_{yy})dz\,dA$$

(6.1.8)

With the help of (6.1.4), the above equation (6.1.8) can be written as

$$\delta U = \iint_R \int_{-h/2}^{h/2} \left\{ \begin{array}{l} \tau_{xx}(\delta u_{,x} + w_{,x}\delta w_{,x} - z\delta w_{,xx}) \\ +\tau_{xy}(\delta u_{,y} + \delta v_{,x} - 2z\delta w_{,xy} + w_{,x}\delta w_{,y} + w_{,y}\delta w_{,x}) \\ +\tau_{yy}(\delta v_{,y} + w_{,y}\delta w_{,y} - z\delta w_{,yy}) \end{array} \right\} dz\,dA$$

(6.1.9)

Making use of the definitions of the bending and twisting moments and the axial stress resultants in equations (6.1.5) and (6.1.6), the above equation (6.1.9) can be rewritten as

$$\delta U = \iint_R \left\{ \begin{array}{l} N_x(\delta u_{,x} + w_{,x}\delta w_{,x}) - M_x\delta w_{,xx} \\ +N_{xy}(\delta u_{,y} + \delta v_{,x} + w_{,x}\delta w_{,y} + w_{,y}\delta w_{,x}) + 2M_{xy}\delta w_{,xy} \\ +N_y(\delta v_{,y} + w_{,y}\delta w_{,y}) - M_y\delta w_{,yy} \end{array} \right\} dA$$

(6.1.10)

Potential of the Applied Forces

Let the applied forces be q per unit area in the lateral direction acting on the plate top surface and $\overline{N}_x, \overline{N}_y$ the components of the *in-plane force* per unit length acting on the boundary of the plate as shown in Fig. 6.1.2. The in-plane force can also be expressed as

$$\overline{N}_v = \overline{N}_x a_{vx}^2 + \overline{N}_y a_{vy}^2$$

$$\overline{N}_t = \frac{1}{2}(\overline{N}_y - \overline{N}_x)a_{vx}a_{vy}$$

(6.1.11)

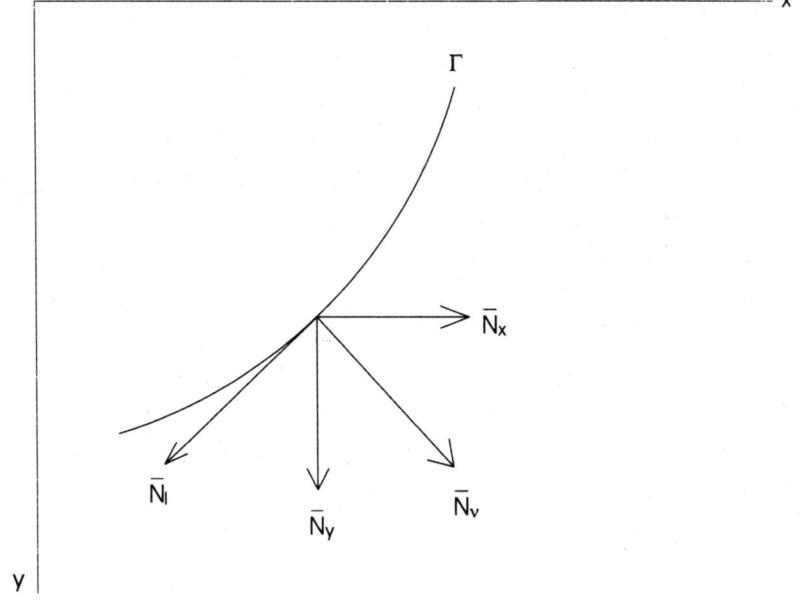

Fig. 6.1.2 Applied In-Plane Forces

Then

$$\delta V = -\iint_R q\delta w\, dA - \oint_\Gamma \overline{N}_v \delta u_v\, dl - \oint_\Gamma \overline{N}_l \delta u_l\, dl \tag{6.1.12}$$

Total Potential Energy Principle

Combining equations (6.1.10) and (6.1.12), we get

$$\delta(U+V) = 0$$

$$= \iint_R \left\{ \begin{array}{c} N_x(\delta u_{,x} + w_{,x}\delta w_{,x}) - M_x \delta w_{,xx} \\ +N_{xy}(\delta u_{,y} + \delta v_{,x} + w_{,x}\delta w_{,y} + w_{,y}\delta w_{,x}) + 2M_{xy}\delta w_{,xy} \\ +N_y(\delta v_{,y} + w_{,y}\delta w_{,y}) - M_y \delta w_{,yy} \end{array} \right\} dA$$

$$- \iint_R q\delta w\, dA - \oint_\Gamma \overline{N}_v \delta u_v\, dl - \oint_\Gamma \overline{N}_l \delta u_l\, dl$$

(6.1.13)

The double integrals in the above equation (6.1.13) are evaluated using Green's theorem, see equations (2.1.11) and (2.1.12), to give

$$\iint_R N_x \delta u_{,x}\, dx\, dy = \oint_\Gamma \delta u N_x\, dy - \iint_R N_{x,x} \delta u\, dx\, dy$$

$$\iint_R N_y \delta v_{,y}\, dx\, dy = -\oint_\Gamma \delta v N_y\, dx - \iint_R N_{y,y} \delta v\, dx\, dy$$

$$\iint_R N_x w_{,x} \delta w_{,x}\, dx\, dy = \oint_\Gamma \delta w N_x w_{,x}\, dy - \iint_R (N_x w_{,x})_{,x} \delta w\, dx\, dy$$

$$\iint_R N_y w_{,y} \delta w_{,y}\, dx\, dy = -\oint_\Gamma \delta w N_y w_{,y}\, dx - \iint_R (N_y w_{,y})_{,y} \delta w\, dx\, dy$$

$$-\iint_R M_x \delta w_{,xx}\, dx\, dy = -\oint_\Gamma \delta w_{,x} M_x\, dy + \oint_\Gamma \delta w M_{x,x}\, dy - \iint_R M_{x,xx} \delta w\, dx\, dy$$

$$-\iint_R M_y \delta w_{,yy}\, dx\, dy = \oint_\Gamma \delta w_{,y} M_y\, dx - \oint_\Gamma \delta w M_{y,y}\, dx - \iint_R M_{y,yy} \delta w\, dx\, dy$$

$$\iint_R N_{xy} \delta u_{,y}\, dx\, dy = -\oint_\Gamma \delta u N_{xy}\, dx - \iint_R N_{xy,y} \delta u\, dx\, dy$$

$$\iint_R N_{xy} \delta v_{,x}\, dx\, dy = \oint_\Gamma \delta v N_{xy}\, dy - \iint_R N_{xy,x} \delta v\, dx\, dy$$

$$\iint_R N_{xy} w_{,x} \delta w_{,y}\, dx\, dy = -\oint_\Gamma \delta w N_{xy} w_{,x}\, dx - \iint_R (N_{xy} w_{,x})_{,y} \delta w\, dx\, dy$$

$$\iint_R N_{xy} w_{,y} \delta w_{,x}\, dx\, dy = \oint_\Gamma \delta w N_{xy} w_{,y}\, dy - \iint_R (N_{xy} w_{,y})_{,x} \delta w\, dx\, dy$$

$$\iint_R M_{xy} \delta w_{,xy}\, dx\, dy = -\oint_\Gamma \delta w_{,x} M_{xy}\, dx - \oint_\Gamma \delta w M_{xy,y}\, dy + \iint_R M_{xy,xy} \delta w\, dx\, dy$$

$$\iint_R M_{xy} \delta w_{,xy}\, dx\, dy = \oint_\Gamma \delta w_{,y} M_{xy}\, dy + \oint_\Gamma \delta w M_{xy,x}\, dx + \iint_R M_{xy,xy} \delta w\, dx\, dy$$

(6.1.14)

The last two integrals in the above are same, the sum of which gives the twisting moment term in equation (6.1.13). These integrals are evaluated, first with respect to x and then with y, and vice versa. Now, substituting the above integrals in (6.1.13), we get

$$-\iint_R [N_{x,x} + N_{xy,y}] \delta u\, dx\, dy - \oint_\Gamma \delta u N_{xy}\, dx + \oint_\Gamma \delta u N_x\, dy$$

$$-\iint_R [N_{y,y} + N_{xy,x}] \delta v\, dx\, dy + \oint_\Gamma \delta v N_{xy}\, dy - \oint_\Gamma \delta v N_y\, dx$$

$$-\iint_R \left[\begin{array}{l} (N_x w_{,x})_{,x} + (N_y w_{,y})_{,y} + (N_{xy} w_{,x})_{,y} + (N_{xy} w_{,y})_{,x} \\ + M_{x,xx} + M_{y,yy} - 2 M_{xy,xy} + q \end{array} \right] \delta w\, dx\, dy$$

$$+ \oint_\Gamma [N_x w_{,x} + M_{x,x} + N_{xy} w_{,y} - M_{xy,y}] \delta w \, dy$$

$$- \oint_\Gamma [N_y w_{,y} + M_{y,y} + N_{xy} w_{,x} - M_{xy,x}] \delta w \, dx$$

$$- \oint_\Gamma \delta w_{,x} M_x dy + \oint_\Gamma \delta w_{,y} M_y dx - \oint_\Gamma \delta w_{,x} M_{xy} dx + \oint_\Gamma \delta w_{,y} M_{xy} dy$$

$$- \oint_\Gamma \overline{N}_v \delta u_v \, dl - \oint_\Gamma \overline{N}_l \delta u_l \, dl$$

$$= 0$$

(6.1.13a)

The differential equations for the plate from the above equation (6.1.13a) can be written as

$$N_{x,x} + N_{xy,y} = 0$$

$$N_{y,y} + N_{xy,x} = 0$$

$$(N_x w_{,x})_{,x} + (N_y w_{,y})_{,y} + (N_{xy} w_{,x})_{,y} + (N_{xy} w_{,y})_{,x} + M_{x,xx} + M_{y,yy}$$

$$- 2 M_{xy,xy} + q = 0$$

(6.1.15)

After integrating the first four terms in the third differential equation and making use of the first two differential equations of the above (6.1.15), we get the final form of the differential equation of the plate under combined lateral and in-plane forces.

$$M_{x,xx} + M_{y,yy} - 2 M_{xy,xy} = -(N_x w_{,xx} + N_y w_{,yy} + 2 N_{xy} w_{,xy} + q)$$

(6.1.16)

The terms $N_x w_{,xx} + N_y w_{,yy} + 2 N_{xy} w_{,xy}$ on the right hand side of the above equation (6.1.16) are referred as *Reduced Loads*. Making use of equations (3.2.47) and (3.2.48), the boundary condition terms in equation (6.1.13a) are

$$\left\{ \begin{array}{c} [-\oint_\Gamma N_{xy} dx + \oint_\Gamma N_x dy] \delta u + [\oint_\Gamma N_{xy} dy - \oint_\Gamma N_y dx] \delta v \\ - \oint_\Gamma \overline{N}_v \delta u_v \, dl - \oint_\Gamma \overline{N}_l \delta u_l \, dl \end{array} \right\} = 0$$

$$\oint_\Gamma [N_x w_{,x} + N_{xy} w_{,y} + V_x] \delta w \, dy - \oint_\Gamma [N_y w_{,y} + N_{xy} w_{,x} + V_y] \delta w \, dx = 0$$

$$- [\oint_\Gamma M_x dy + \oint_\Gamma M_{xy} dx] \delta w_{,x} + [\oint_\Gamma M_y dx + \oint_\Gamma M_{xy} dy] \delta w_{,y} = 0$$

(6.1.17)

188 Dynamics of Plates

Now, we make use of (2.2.26) and (2.2.27) along with

$$u = u_v a_{vx} - u_l a_{vy}$$
$$v = u_v a_{vy} + u_l a_{vx}$$

(6.1.18)

to rewrite equations (6.1.17) as

$$\left\{\begin{array}{c} [\oint_\Gamma N_{xy}a_{vy}dl + \oint_\Gamma N_x a_{vx} dl](\delta u_v a_{vx} - \delta u_l a_{vy}) \\ +[\oint_\Gamma N_{xy}a_{vx} dl + \oint_\Gamma N_y a_{vy}dl](\delta u_v a_{vy} + \delta u_l a_{vx}) \\ -\oint_\Gamma \overline{N}_v \delta u_v\, dl - \oint_\Gamma \overline{N}_l \delta u_l\, dl \end{array}\right\} = 0$$

$$\oint_\Gamma [N_x(a_{vx}w_{,v} - a_{vy}w_{,l}) + N_{xy}(a_{vx}w_{,l} + a_{vy}w_{,v}) + V_x]\delta w a_{vx}\, dl = 0$$

$$\oint_\Gamma [N_y(a_{vx}w_{,l} + a_{vy}w_{,v}) + N_{xy}(a_{vx}w_{,v} - a_{vy}w_{,l}) + V_y]\delta w a_{vy} dl = 0$$

$$-\left[\oint_\Gamma M_x a_{vx}\, dl - \oint_\Gamma M_{xy}a_{vy}dl\right](\delta a_{vx}w_{,v} - \delta a_{vy}w_{,l}) = 0$$

$$\left[-\oint_\Gamma M_y a_{vy}dl + \oint_\Gamma M_{xy}a_{vx}\, dl\right](\delta a_{vx}w_{,l} + \delta a_{vy}w_{,v}) = 0$$

(6.1.19)

On further simplification, we get

$$\left[\oint_\Gamma \{2N_{xy}a_{vx}a_{vy} + N_x a_{vx}^2 + N_y a_{vy}^2 - \overline{N}_v\}\, dl\right]\delta u_v = 0$$

$$\left[\oint_\Gamma \{N_{xy}(a_{vx}^2 - a_{vy}^2) + (N_y - N_x)a_{vx}a_{vy} - \overline{N}_l\}dl\right]\delta u_l = 0$$

$$\oint_\Gamma [N_x(a_{vx}^2 w_{,v} - a_{vx}a_{vy}w_{,l}) + N_{xy}(a_{vx}^2 w_{,l} + a_{vx}a_{vy}w_{,v}) + V_x a_{vx}]\delta w\, dl = 0$$

$$\oint_\Gamma [N_y(a_{vx}a_{vy}w_{,l} + a_{vy}^2 w_{,v}) + N_{xy}(a_{vx}a_{vy}w_{,v} - a_{vy}^2 w_{,l}) + V_y a_{vy}]\delta w\, dl = 0$$

$$-\left[\oint_\Gamma \{M_x a_{vx}^2 + M_y a_{vy}^2 - 2M_{xy}a_{vx}a_{vy}\}dl\right]\delta w_{,v} = 0$$

$$\left[\oint_\Gamma \{(M_x - M_y)a_{vx}a_{vy} + M_{xy}(a_{vx}^2 - a_{vy}^2)\}dl\right]\delta w_{,l} = 0$$

(6.1.20)

We can simplify further with the help of (2.2.30) and obtain

Plates with Combined Lateral and In-Plane Forces 189

$$[\oint_\Gamma (N_v - \overline{N}_v) dl] \delta u_v = 0$$

$$[\oint_\Gamma (N_{vl} - \overline{N}_l) dl] \delta u_l = 0$$

$$\oint_\Gamma [N_v w_{,v} + N_{vl} w_{,l} + V_v] \delta w \, dl = 0$$

$$-[\oint_\Gamma M_v dl] \delta w_{,v} = 0$$

$$-[\oint_\Gamma M_{vl} dl] \delta w_{,l} = 0$$

(6.1.21)

where

$$N_v = N_x a_{vx}^2 + N_y a_{vy}^2 + 2N_{xy} a_{vx} a_{vy}$$

$$N_{vl} = N_{xy}(a_{vx}^2 - a_{vy}^2) + (N_y - N_x) a_{vx} a_{vy}$$

$$V_v = V_x a_{vx} + V_y a_{vy}$$

(6.1.22)

Integrating the last term by parts, the boundary conditions in (6.1.21) can now be stated as

$$N_v = \overline{N}_v \quad \text{or specify } u_v$$

$$N_{vl} = \overline{N}_l \quad \text{or specify } u_l$$

$$N_v w_{,v} + N_{vl} w_{,l} + V_v + M_{vl,l} = 0 \quad \text{or specify } w$$

$$M_v = 0 \quad \text{or specify } w_{,v}$$

$$[M_{vl} \delta w]_\Gamma = 0 \quad \text{at discontinuities}$$

(6.1.23)

Now, making use of (6.1.5), the differential equation (6.1.16) for the Von Karman plate can be written down in terms of the displacements as

$$D\nabla^4 w = N_x w_{,xx} + N_y w_{,yy} + 2N_{xy} w_{,xy} + q$$

(6.1.24)

The forces N_x, N_y and N_{xy} depend not only on the external forces applied in the *xy* plane, but also on the strain of the middle surface of the

190 Dynamics of Plates

plate due to bending. The first two equations in (6.1.15) are not sufficient to determine all the three quantities N_x, N_y and N_{xy} and we need another equation for this purpose. This equation is obtained by considering the strain displacement relations in (6.1.4) for the middle plane, $z = 0$,

$$\varepsilon_{xx} = u_{,x} + \frac{1}{2}w_{,x}^2$$

$$\varepsilon_{yy} = v_{,y} + \frac{1}{2}w_{,y}^2$$

$$\gamma_{xy} = (u_{,y} + v_{,x}) + w_{,x}w_{,y}$$

(6.1.25)

Taking the second derivatives of the three equations above in (6.1.25)

$$\varepsilon_{xx,yy} = u_{,xyy} + w_{,xy}w_{,xy} + w_{,x}w_{,xyy}$$

$$\varepsilon_{yy,xx} = v_{,xxy} + w_{,xy}w_{,xy} + w_{,y}w_{,xxy}$$

$$\gamma_{xy,xy} = (u_{,xyy} + v_{,xxy}) + w_{,xxy}w_{,y} + w_{,xx}w_{,yy} + w_{,xy}w_{,xy} + w_{,x}w_{,xyy}$$

(6.1.26)

Combining the above three equations in (6.1.26), we get

$$\varepsilon_{xx,yy} + \varepsilon_{yy,xx} - \gamma_{xy,xy} = w_{,xy}^2 - w_{,xx}w_{,yy}$$

(6.1.27)

We can express the strain components in the above as

$$\varepsilon_{xx} = \frac{1}{hE}(N_x - \nu N_y)$$

$$\varepsilon_{yy} = \frac{1}{hE}(N_y - \nu N_x)$$

$$\gamma_{xy} = \frac{1}{hG}N_{xy}$$

(6.1.28)

and obtain the third equation required to determine N_x, N_y and N_{xy}. This procedure may prove to be cumbersome and therefore, we seek a simplified process which can be obtained by expressing N_x, N_y and N_{xy} as

$$N_x = F_{,yy}$$

$$N_y = F_{,xx}$$
$$N_{xy} = -F_{,xy}$$
(6.1.29)

where F is an Airy stress function. Then, equations (6.1.28) become

$$\varepsilon_{xx} = \frac{1}{hE}(F_{,yy} - vF_{,xx})$$
$$\varepsilon_{yy} = \frac{1}{hE}(F_{,xx} - vF_{,yy})$$
$$\gamma_{xy} = -\frac{2(1+v)}{hE}F_{,xy}$$
(6.1.30)

Using these strain expressions (6.1.30) in equation (6.1.27) and (6.1.29) in (6.1.24) give us the following two differential equations with w and F as the dependent variables.

$$\nabla^4 F = Eh\left(w_{,xy}^2 - w_{,xx}w_{,yy}\right)$$
(6.1.31)

$$D\nabla^4 w = F_{,yy}w_{,xx} + F_{,xx}w_{,yy} - 2F_{,xy}w_{,xy} + q$$
(6.1.32)

The above are Von Karman's plate equations. Using the following nonlinear operator L

$$L(p,q) = p_{,yy}q_{,xx} + p_{,xx}q_{,yy} - 2p_{,xy}q_{,xy}$$
(6.1.33)

equations (6.1.31) and (6.1.32) can be expressed in a simple form as

$$\nabla^4 F = -\frac{1}{2}Eh\,L(w,w)$$
$$D\nabla^4 w = L(F,w) + q$$
(6.1.34)

Before attempting a solution of the above coupled nonlinear differential equations, we will first consider membranes subjected to in plane forces, then proceed to the case of plates subjected to in plane forces with small deflections (no coupling between transverse and axial deformations). Finally, we will consider the solution of coupled equations (6.1.34) given above.

For circular plates, equation (6.1.33) is expressed in polar coordinates

$$L(p,q) = p_{,rr}\left(\frac{1}{r}q_{,r} + \frac{1}{r^2}q_{,\theta\theta}\right) + \left(\frac{1}{r}p_{,r} + \frac{1}{r^2}q_{,\theta\theta}\right)q_{,rr}$$
$$- 2\left(\frac{1}{r}q_{,\theta}\right)_{,r}\left(\frac{1}{r}p_{,\theta}\right)_{,r}$$

(6.1.35)

For an axisymmetric case, the above simplifies to

$$L(p,q) = p_{,rr}\left(\frac{1}{r}q_{,r}\right) + \left(\frac{1}{r}p_{,r}\right)q_{,rr}$$

(6.1.36)

6.2 RECTANGULAR MEMBRANES

A *membrane* is a member which cannot offer any resistance to vertical forces unless it is stretched initially by a tensile in plane force. The flexural rigidity D of a membrane is zero. We will consider first a rectangular membrane which is initially stressed by force N per unit length on the boundary as shown in Fig. 6.2.1. This force is constant throughout the membrane area and therefore we have simple linear equation (with $D = 0$) from (6.1.24)

$$w_{,xx} + w_{,yy} = -\frac{q}{N}$$

(6.2.1)

The solution of the above equation can be assumed by the following series

$$w = \sum_{m=1}^{\infty}\sum_{n=1}^{\infty} C_{mn} \sin\frac{m\pi x}{a} \sin\frac{n\pi y}{b}$$

(6.2.2)

which satisfies the boundary conditions that w is zero on all the four edges. We express the uniformly distributed force q in the form of (3.3.1) and (3.3.4a)

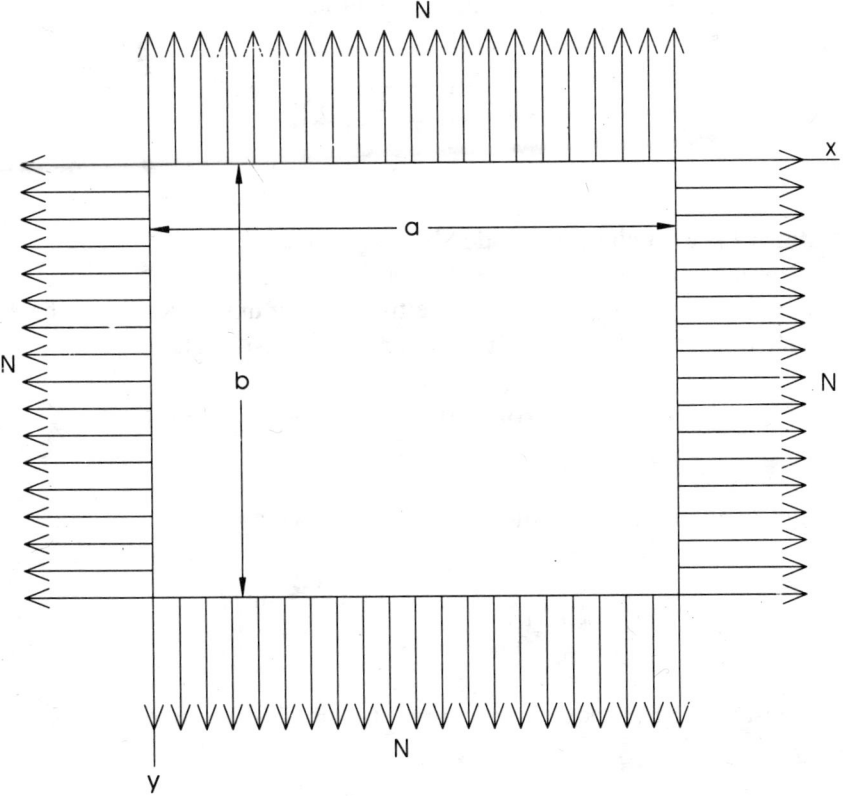

Fig. 6.2.1 A Rectangular Membrane

$$q(x,y) = \sum_{m=1}^{\infty} \sum_{n=1}^{\infty} q_{mn} \sin\frac{m\pi x}{a} \sin\frac{n\pi y}{b}$$

(3.3.1)

$$q_{mn} = \frac{16q}{mn\pi^2} \text{ for } m \text{ and } n \text{ odd}$$

$$= 0 \text{ for } m \text{ and } n \text{ even}$$

(3.3.4a)

Substituting the required derivatives of w from equation (6.2.2) and the above equation (3.3.1) for q in (6.1.1), we get

$$C_{mn} = \frac{16\,q}{Nmn\pi^2\left[\left(\frac{m\pi}{a}\right)^2 + \left(\frac{n\pi}{b}\right)^2\right]}$$

(6.2.3)

194 Dynamics of Plates

The deflection of the membrane is therefore given by

$$w = \frac{16q}{N\pi^4} \sum_{m=1,3,5...}^{\infty} \sum_{n=1,3,5...}^{\infty} \frac{1}{mn\left[\left(\frac{m}{a}\right)^2 + \left(\frac{n}{b}\right)^2\right]} \sin\frac{m\pi x}{a} \sin\frac{n\pi y}{b}$$

(6.2.4)

Natural Frequencies and Mode Shapes

To determine the free vibration behavior of the membrane, we replace the steady force q per unit area by the inertia force per unit area $-\rho h \ddot{w}$ where h is the thickness of the membrane and ρ is its density. Then,

$$N(w_{,xx} + w_{,yy}) = \rho h \ddot{w}$$

(6.2.5)

For free vibration, the solution can be assumed as

$$w = \sum_{m=1}^{\infty} \sum_{n=1}^{\infty} C_{mn} \sin\frac{m\pi x}{a} \sin\frac{n\pi y}{b} \sin pt$$

(6.2.6)

Substituting the above equation (6.2.6) and its derivatives in (6.2.5), we get

$$p_{mn}^2 = \frac{N\pi^2}{\rho h}\left[\left(\frac{m}{a}\right)^2 + \left(\frac{n}{b}\right)^2\right]$$

$$= \frac{N\pi^2}{\rho h a^2}(m^2 + n^2 k^2)$$

(6.2.7)

where k is the aspect ratio = a/b. The corresponding mode shapes are given by

$$W_{mn} = \sin\frac{m\pi x}{a} \sin\frac{n\pi y}{b}$$

(6.2.8)

Let

$$\lambda_{mn}^2 = \frac{\rho h a^2}{N} p_{mn}^2$$

$$= \pi^2(m^2 + n^2 k^2)$$

(6.2.9)

The non dimensional frequency parameter above for a square membrane $k = 1$ is given in Table 6.2.1 for different mode shapes. We see that the mode shapes are similar to that of a square plate dealt in section 3.3.

Table 6.2.1 Frequency Parameters λ_{mn}^2 for a Square Membrane

m	n	λ_{mn}^2	Mode Shape
1	1	19.74	$\sin\frac{\pi x}{a} \sin\frac{\pi y}{b}$
1	2	49.35	$\sin\frac{\pi x}{a} \sin\frac{2\pi y}{b}$
2	1	49.35	$\sin\frac{2\pi x}{a} \sin\frac{\pi y}{b}$
2	2	78.96	$\sin\frac{2\pi x}{a} \sin\frac{2\pi y}{b}$
1	3	98.7	$\sin\frac{\pi x}{a} \sin\frac{3\pi y}{b}$
3	1	98.7	$\sin\frac{3\pi x}{a} \sin\frac{\pi y}{b}$
2	3	128.31	$\sin\frac{2\pi x}{a} \sin\frac{3\pi y}{b}$
3	2	128.31	$\sin\frac{3\pi x}{a} \sin\frac{2\pi y}{b}$

6.3 CIRCULAR MEMBRANES

We will now consider a circular membrane of radius a which is initially stressed by force N per unit length on the boundary as shown in Fig. 6.3.1. This force is constant throughout the membrane area and therefore we have simple linear equation for axisymmetric case (with $D = 0$) from (6.1.24) in polar coordinates

$$\nabla^2 w = \frac{1}{r}\frac{d}{dr}\left(r\frac{dw}{dr}\right) = -\frac{q}{N} \tag{6.3.1}$$

The solution of the above equation can be obtained by direct integration as follows.

$$\frac{d}{dr}\left(r\frac{dw}{dr}\right) = -\frac{qr}{N}$$

$$r\frac{dw}{dr} = -\frac{qr^2}{2N} + A$$

$$\frac{dw}{dr} = -\frac{qr}{2N} + \frac{A}{r}$$

$$w = -\frac{qr^2}{4N} + A\ln r + B$$

$$\tag{6.3.2}$$

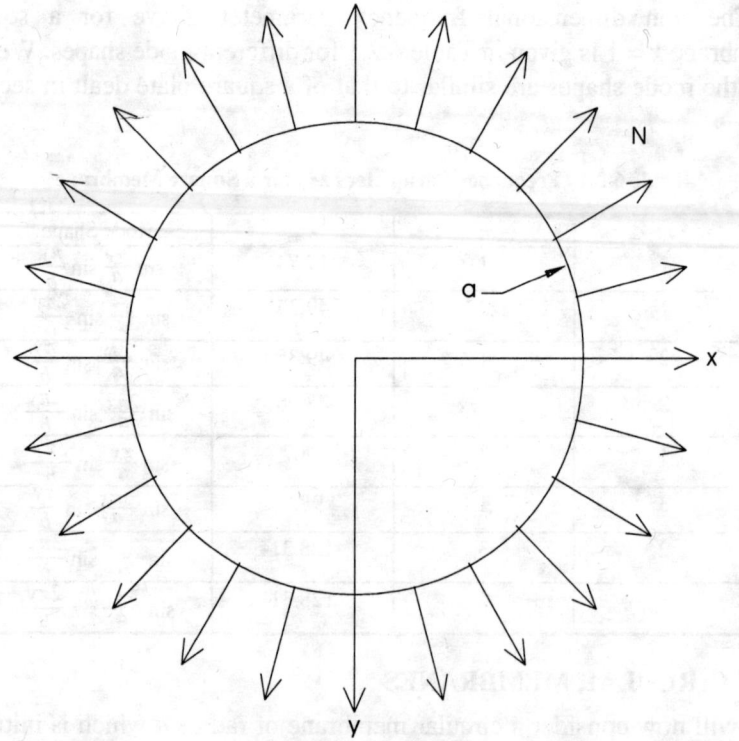

Fig. 6.3.1 Initially Stressed Circular Membrane

We first note that the membrane deformation is finite at $r = 0$, therefore, $A = 0$ in the above equation (6.3.2). The boundary condition at the radius $r = a$ gives

$$B = \frac{qa^2}{4N}$$

(6.3.3)

Therefore,

$$w = \frac{q}{4N}(a^2 - r^2)$$

(6.3.4)

The maximum deflection occurs at the center of the membrane,

$$w_{max} = \frac{qa^2}{4N}$$

(6.3.5)

Frequencies and Mode Shapes

The equation of motion for free vibration is

$$N\nabla^2 w = \rho h \ddot{w} \tag{6.3.6}$$

The solution for free harmonic vibration is

$$w = W \sin pt \tag{6.3.7}$$

which makes equation (6.3.6) to be

$$(\nabla^2 + \lambda^2)W = 0 \tag{6.3.8}$$

where

$$\lambda^2 = \left(\frac{\rho h}{N}\right)p^2 \tag{6.3.9}$$

The solution of the above equation (6.3.8) is given by

$$W = AJ_0(\lambda r) + BY_0(\lambda r) \tag{6.3.10}$$

where J_0 and Y_0 are Bessel functions of order 0. Since Y_0 tends to infinity as r tends to zero, the constant B is made zero. Then, the solution is

$$w = AJ_0(\lambda r) \sin pt \tag{6.3.11}$$

Applying the boundary condition that $w = 0$ at $r = a$, we get the frequency equation as

$$J_0(\lambda a) = 0 \tag{6.3.12}$$

The first five roots of the above equation (6.3.12) are

$$(\lambda a)_1 = 2.404$$

$$(\lambda a)_2 = 5.520$$
$$(\lambda a)_3 = 8.654$$
$$(\lambda a)_4 = 11.792$$
$$(\lambda a)_5 = 14.931$$

The natural frequencies are given by

$$p_i = (\lambda a)_i \sqrt{\frac{N}{\rho h a^2}} \text{ rad/s} \qquad (6.3.13)$$

with the corresponding mode shapes

$$W_i = J_0\left[(\lambda a)_i \frac{r}{a}\right] \qquad (6.3.14)$$

For the first mode, the nodal circle occurs at the boundary of the membrane where $r = a$. For the second mode, an additional nodal circle appears at

$$J_0\left[5.52\frac{r}{a}\right] = 0$$
$$5.52\frac{r}{a} = 2.404$$
$$\frac{r}{a} = 0.4355 \qquad (6.3.15)$$

For the third mode, we will have two nodal circles defined by

$$J_0\left[8.654\frac{r}{a}\right] = 0$$
$$8.654\left(\frac{r}{a}\right)_1 = 2.404$$
$$\left(\frac{r}{a}\right)_1 = 0.2778$$
$$8.654\left(\frac{r}{a}\right)_2 = 5.52$$
$$\left(\frac{r}{a}\right)_2 = 0.6379 \qquad (6.3.16)$$

Plates with Combined Lateral and In-Plane Forces 199

In a similar manner, we can show that as the mode number increases, the number of nodal circles also increase, one at a time, nth mode for example will have $n-1$ nodal circles, the locations of which can be determined as outlined above. The first three modes are shown in Fig. 6.3.2.

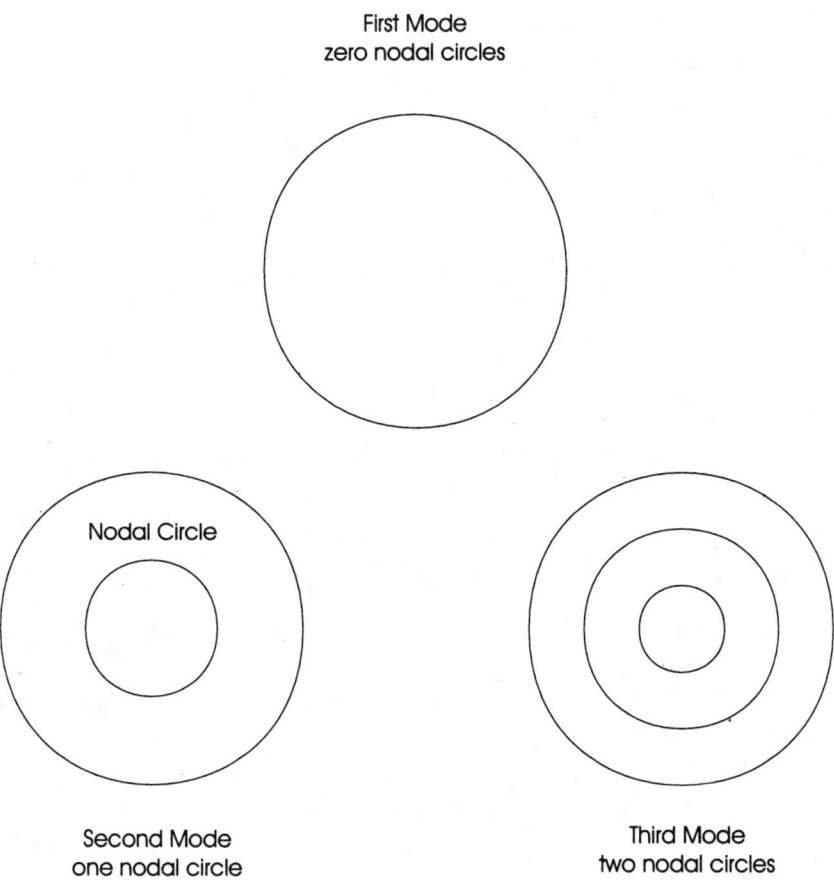

Fig. 6.3.2 First Three Modes of a Membrane

6.4 INITIALLY STRESSED RECTANGULAR PLATES

In section 6.2 we considered transversely loaded rectangular membranes. Here, we extend this analysis to plates. The deformations are considered small, therefore, there will be no coupling between extension and lateral deformations. Let the rectangular plate be initially stressed by a force N in x direction as shown in Fig. 6.4.1, then the equations governing such a plate can be deduced from equation (6.1.15) to be

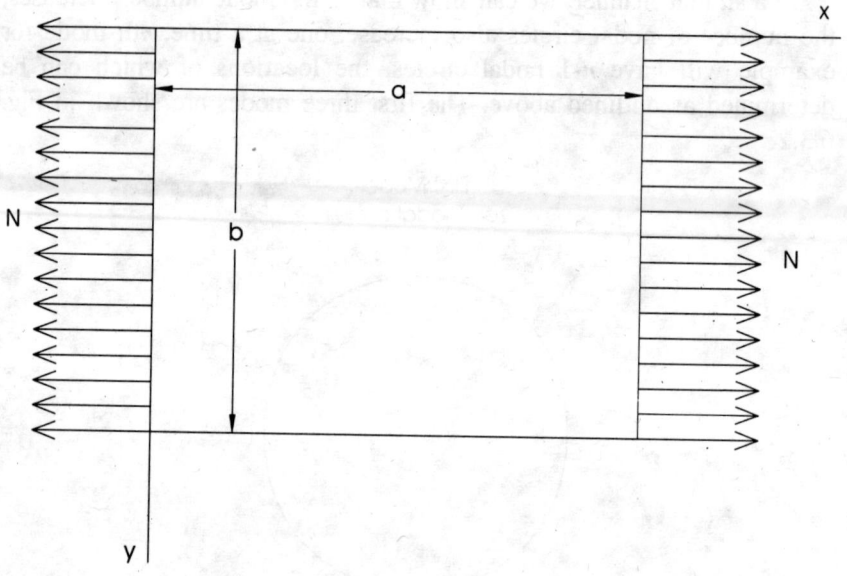

Fig. 6.4.1 A Rectangular Plate under Tension

$$N_{x,x} = 0$$
$$D\nabla^4 w - N_x w_{,xx} = q$$

(6.4.1)

The first equation in the above (6.4.1) gives us $N_x = N$ and the second equation becomes

$$D\nabla^4 w - N w_{,xx} = q$$

(6.4.2)

Representing the uniformly distributed force in the form of double Fourier series of (3.3.1) and (3.3.4a)

$$q(x,y) = \sum_{m=1}^{\infty} \sum_{n=1}^{\infty} q_{mn} \sin \frac{m\pi x}{a} \sin \frac{n\pi y}{b}$$

(3.3.1)

$$q_{mn} = \frac{16q}{mn\pi^2} \text{ for } m \text{ and } n \text{ odd}$$
$$= 0 \text{ for } m \text{ and } n \text{ even}$$

(3.3.4a)

Therefore, we seek the solution of

$$D\nabla^4 w - Nw_{,xx} = q_{mn} \sin\frac{m\pi x}{a} \sin\frac{n\pi y}{b} \qquad (6.4.3)$$

The solution of the above equation can be assumed as

$$w = C_{mn} \sin\frac{m\pi x}{a} \sin\frac{n\pi y}{b} \qquad (6.4.4)$$

Upon substitution of the above equation (6.4.4) and its derivatives, we can evaluate C_{mn} as

$$C_{mn} = \frac{q_{mn}}{D\left[\left(\frac{m\pi}{a}\right)^2 + \left(\frac{n\pi}{b}\right)^2\right]^2 + N\left(\frac{m\pi}{a}\right)^2} \qquad (6.4.5)$$

The solution of (6.4.3) is therefore given by

$$w_{mn} = \frac{q_{mn} \sin\frac{m\pi x}{a} \sin\frac{n\pi y}{b}}{D\left[\left(\frac{m\pi}{a}\right)^2 + \left(\frac{n\pi}{b}\right)^2\right]^2 + N\left(\frac{m\pi}{a}\right)^2} \qquad (6.4.6)$$

The solution of (6.4.2) is then given by

$$w = \frac{16q}{\pi^2} \sum_{m=1,3,5...}^{\infty} \sum_{n=1,3,5...}^{\infty} \frac{\sin\frac{m\pi x}{a} \sin\frac{n\pi y}{b}}{Dmn\left[\left(\frac{m\pi}{a}\right)^2 + \left(\frac{n\pi}{b}\right)^2\right]^2 + N\left(\frac{m\pi}{a}\right)^2} \qquad (6.4.6a)$$

So, we see that the axial tensile force stiffens the plate and the deflection of the plate decreases in its presence.

Buckling of the Plate

Let us consider the case when the in-plane force is compressive as shown in Fig. 6.4.2 and that the transverse force is zero. Then the governing equation for the system is

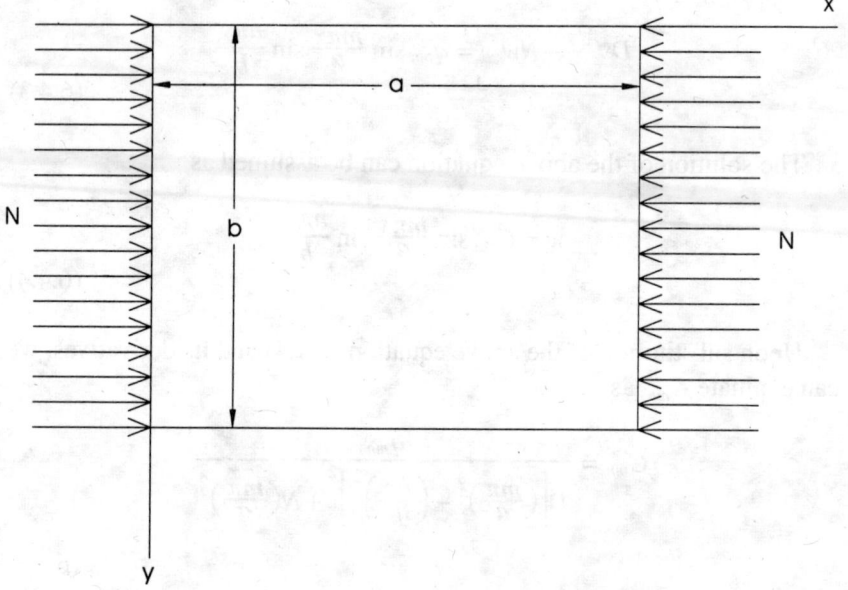

Fig. 6.4.2 A Rectangular Plate in Compression

$$D\nabla^4 w + N w_{,xx} = 0$$

(6.4.7)

For a solution given by (6.4.4), we get

$$D\left[\left(\frac{m\pi}{a}\right)^2 + \left(\frac{n\pi}{b}\right)^2\right]^2 w - N\left(\frac{m\pi}{a}\right)^2 w = 0$$

(6.4.8)

Therefore, we have either the deflection $w = 0$ or

$$N = \frac{D\pi^2}{b^2}\left(\frac{m}{k} + \frac{kn^2}{m}\right)^2$$

(6.4.9)

where k is the aspect ratio = a/b. Equation (6.4.9) represents the nontrivial solution for which the plate assumes an equilibrium configuration with $w \neq 0$ even when the transverse force $q = 0$. The axial force N assumes smallest value when $n = 1$. Therefore, the critical value of N is

$$N_{cr} = \frac{D\pi^2}{b^2} F$$

(6.4.9a)

where

$$F = \left(\frac{m}{k} + \frac{k}{m}\right)^2 \tag{6.4.10}$$

The factor F is plotted in Fig. 6.4.3 as a function of aspect ratio k for different values of m. The number of half waves in the buckling of plate can be obtained from (6.4.4) which are given by m. The minimum value of factor F, i.e., the buckling load N is determined by the aspect ratio and m. The intersection point for the curves $m = 1$ and $m = 2$ is given from (6.4.10) by

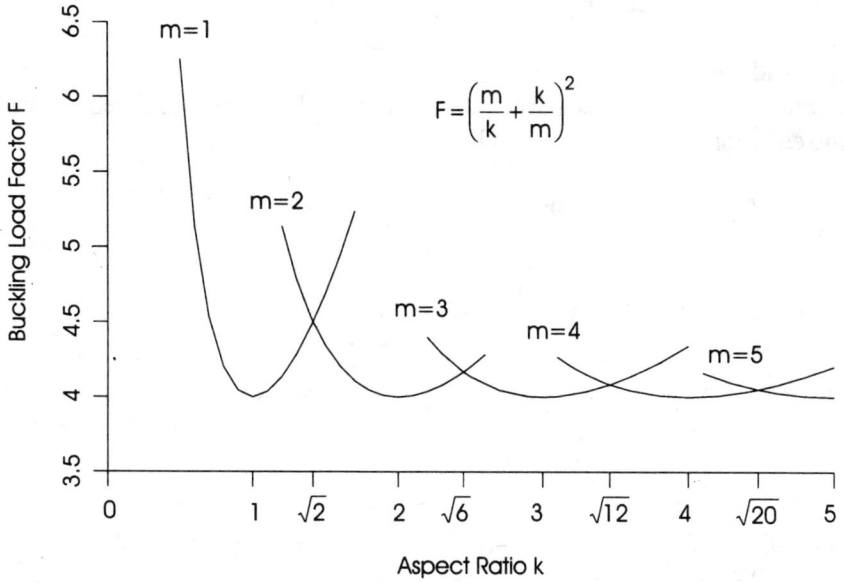

Fig. 6.4.3 Buckling Load vs Aspect Ratio

$$\left(\frac{1}{k} + \frac{k}{1}\right) = \left(\frac{2}{k} + \frac{k}{2}\right)$$

$$2(1 + k^2) = 4 + k^2$$

$$k = \sqrt{2} \tag{6.4.11}$$

Similarly the intersection points for the curves $m = 2$ and $m = 3$, etc., can be obtained as $\sqrt{6}, \sqrt{12}, \sqrt{20}$ etc. as shown in the figure. The minimum value of the factor F with respect to the parameter m/k, is obtained from

204 Dynamics of Plates

$$\frac{d}{d\left(\frac{m}{k}\right)}\left[\left(\frac{m}{k}+\frac{k}{m}\right)\right]=0$$

(6.4.12)

This gives $k = m$ and $F = 4$ at $k = 1, 2, 3, 4$ as shown in Fig. 6.4.3. Therefore, no buckling takes place when

$$N < \frac{4D\pi^2}{b^2}$$

(6.4.13)

for any given aspect ratio k.

Natural Frequencies
Let us now consider free harmonic vibration of the plate considered so far, the equation of motion is given by

$$D\nabla^4 w - N_x w_{,xx} + \rho h \ddot{w} = 0$$

(6.4.14)

The response of the plate can be assumed as

$$w = C_{mn} \sin\frac{m\pi x}{a} \sin\frac{n\pi y}{b} \sin pt$$

(6.4.15)

which satisfies the boundary conditions of the simply supported rectangular plate. Substituting the above equation (6.4.15) and its derivatives in (6.4.14), we get

$$p_{mn}^2 = \frac{D}{\rho h}\left[\left\{\left(\frac{m\pi}{a}\right)^2 + \left(\frac{n\pi}{b}\right)^2\right\}^2 + \frac{N}{D}\left(\frac{m\pi}{a}\right)^2\right]$$

$$= \frac{D\pi^4}{\rho h a^4}\{m^2+n^2k^2\}^2 + \frac{D}{\rho h}\frac{N}{D}\left(\frac{m\pi}{a}\right)^2$$

(6.4.16)

Let

$$\lambda_{mn}^2 = \frac{\rho h a^4}{D} p_{mnc}^2 = \pi^4(m^2+n^2k^2)^2$$

(3.3.18)

where p_{mnc} is the natural frequency of the plate without any initial stress. Then, equation (6.4.16) is

$$p_{mn}^2 = \frac{D}{\rho h a^4}\lambda_{mn}^2 + \frac{D}{\rho h}\frac{N}{D}\left(\frac{m\pi}{a}\right)^2$$

(6.4.17)

Further, let us define

$$\bar{\lambda}_{mn}^2 = \frac{\rho h a^4}{D}p_{mn}^2$$

(6.4.18)

then

$$\bar{\lambda}_{mn}^2 = \lambda_{mn}^2 + \frac{Na^2 m^2 \pi^2}{D}$$

(6.4.19)

which can be written as

$$\begin{aligned}\frac{\bar{\lambda}_{mn}^2}{\lambda_{mn}^2} &= 1 + \frac{Na^2 m^2 \pi^2}{D\lambda_{mn}^2} \\ &= 1 + \frac{Na^2 m^2}{D\pi^2(m^2 + n^2 k^2)^2} \\ &= 1 + \bar{N}\end{aligned}$$

(6.4.20)

where

$$\bar{N} = \frac{Na^2 m^2}{D\pi^2(m^2 + n^2 k^2)^2}$$

(6.4.21)

is the non dimensional preload. The relation (6.4.20) is plotted in Fig. 6.4.4. The non dimensional frequency parameter increases linearly with the preload parameter.

The non dimensional frequency parameter becomes zero when $\bar{N} = -1$, which corresponds to a compressive force given by the relation (6.4.9).

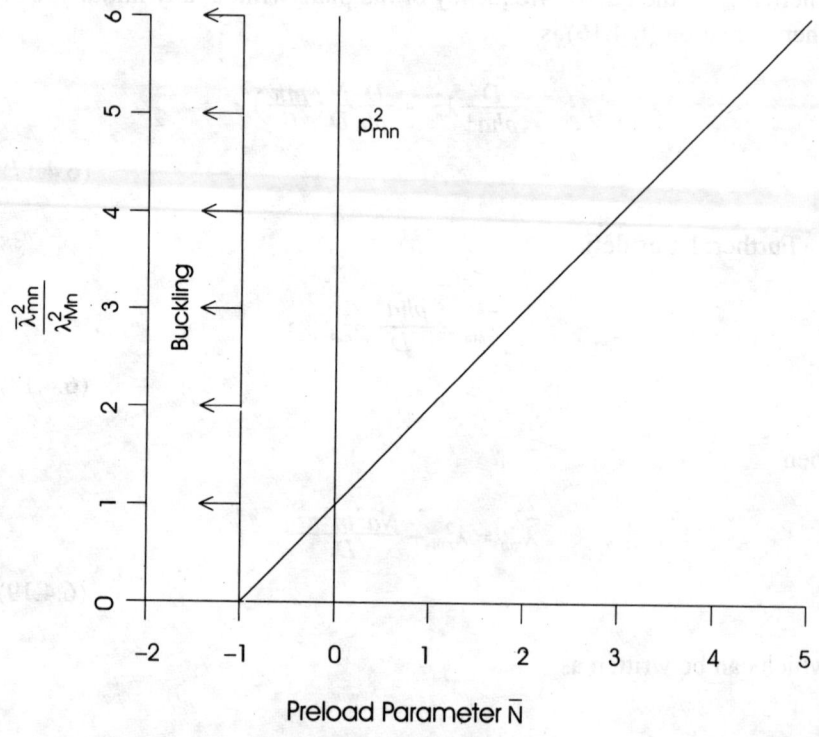

Fig. 6.4.4 Natural Frequency vs Preload

6.5 CLAMPED CIRCULAR PLATES

We will now consider a clamped circular plate of radius a which is initially stressed by force N per unit length on the boundary similar to that shown in Fig. 6.3.1. This force is constant throughout the plate area and therefore the governing equation for axisymmetric case from (6.1.24) in polar coordinates

$$D\nabla^4 w = q + N\nabla^2 w \qquad (6.5.1)$$

Let

$$\beta^2 = \frac{N}{D} \qquad (6.5.2)$$

then equation (6.5.1) can be written as

$$\nabla^2(\nabla^2 - \beta^2)w = \frac{q}{D} \qquad (6.5.3)$$

The solution of the above equation can be written in the form

$$w = A_1 I_0(\beta r) + A_2 + A_3 K_0(\beta r) + A_4 \ln\left(\frac{r}{a}\right) - \frac{qr^2}{4N}$$

(6.5.4)

where I and K are modified Bessel functions of the first and second kind respectively of order 0. We first note that the plate deformation is finite at $r = 0$, therefore, $A_3 = A_4 = 0$ in the above equation (6.5.4). With the help of the boundary conditions at the radius $r = a$ for a clamped plate and noting that $I_0'(x) = I_1(x)$, we get

$$A_2 = \frac{qa^2}{4N} - \frac{qa^2 I_0(\beta a)}{2N(\beta a) I_1(\beta a)}$$

$$A_1 = \frac{qa^2}{2N(\beta a) I_1(\beta a)}$$

(6.5.5)

Therefore,

$$w = \frac{qa^2}{2N}\left[\frac{I_0(\beta r) - I_0(\beta a)}{(\beta a) I_1(\beta a)} + \frac{1}{2}\left(1 - \frac{r^2}{a^2}\right)\right]$$

(6.5.6)

We can rewrite the above in a non dimensional form

$$\frac{2Dw}{qa^4} = \frac{1}{(\beta a)^2}\left[\frac{I_0(\beta r) - I_0(\beta a)}{(\beta a) I_1(\beta a)} + \frac{1}{2}\left(1 - \frac{r^2}{a^2}\right)\right]$$

(6.5.7)

Compressive Axial Force

For the case of a compressive axial force, let

$$\bar{\beta}^2 = -\beta^2 = -\frac{N}{D}$$

(6.5.8)

so that

$$\beta = i\bar{\beta}$$
$$\beta^2 a^2 = -\bar{\beta}^2 a^2$$
$$I_0(\beta r) = I_0(i\bar{\beta}r) = J_0(\bar{\beta}r)$$
$$I_1(\beta a) = I_1(i\bar{\beta}a) = -\frac{1}{i}J_1(\bar{\beta}a)$$

(6.5.9)

We can now write (6.5.7) for the axial compressive force case

$$\frac{2Dw}{qa^4} = \frac{1}{(\bar{\beta}a)^2}\left[\frac{J_0(\bar{\beta}r) - J_0(\bar{\beta}a)}{(\bar{\beta}a)J_1(\bar{\beta}a)} - \frac{1}{2}\left(1 - \frac{r^2}{a^2}\right)\right]$$

(6.5.10)

The plate deflection w becomes unbounded when

$$J_1(\bar{\beta}a) = 0$$

(6.5.11)

The lowest positive root of the above transcendental equation (6.5.11) is

$$(\bar{\beta}a)_1 = (\bar{\beta}a)_{cr} = 3.832$$

(6.5.12)

Consequently, the buckling load is given by

$$(\bar{\beta}a)^2_{cr} = -\beta^2_{cr}a^2 = N_{cr}\frac{a^2}{D} = 3.832^2$$

or

$$N_{cr} = 14.6842\frac{D}{a^2}$$

(6.5.13)

Frequencies and Mode Shapes
The equation of motion for free vibration is

$$D\nabla^4 w - N\nabla^2 w + \rho h \ddot{w} = 0 \tag{6.5.14}$$

The solution for free harmonic vibration is

$$w = W \sin pt \tag{6.5.15}$$

which makes equation (6.5.14) to be

$$(\nabla^4 - \beta^2 \nabla^2 - \lambda^2) W = 0 \tag{6.5.16}$$

where

$$\lambda^2 = \left(\frac{\rho h}{D}\right) p^2 \tag{6.5.17}$$

Equation (6.5.16) can be expressed as

$$(\nabla^2 + a_n^2)(\nabla^2 - \theta_n^2) W_n = 0 \tag{6.5.18}$$

where

$$a_n^2 a^2 = \frac{1}{2}\left(\sqrt{L^2 + \Gamma_n^2} - L\right)$$
$$\theta_n^2 a^2 = \frac{1}{2}\left(\sqrt{L^2 + \Gamma_n^2} + L\right)$$
$$L = \beta^2 a^2$$
$$\Gamma_n^2 = 4a^4 \lambda_n^2$$

$$\tag{6.5.19}$$

The solution of the above equation (6.5.18) is given by

$$W = A_n J_0(a_n r) + B_n Y_0(a_n r) + C_n I_0(\theta_n r) + D_n K_0(\theta_n r)$$

$$\tag{6.5.20}$$

where J_0 and Y_0 are Bessel functions and I_0 and K_0 are modified Bessel functions all of order 0. Since Y_0 and K_0 tend to infinity as r tends to zero, the constants B and D are made zero. Then, the solution is

$$W_n = A_n J_0(a_n r) + C_n I_0(\theta_n r)$$
$$W'_n = A_n a_n J'_0(a_n r) + C_n \theta_n I'_0(\theta_n r)$$

(6.5.21)

Applying the boundary condition that $w = w' = 0$ at $r = a$, we get

$$A_n J_0(a_n a) + C_n I_0(\theta_n a) = 0$$
$$A_n a_n J'_0(a_n a) + C_n \theta_n I'_0(\theta_n a) = 0$$

(6.5.22)

The second equation gives us

$$A_n a_n J_1(a_n a) = C_n \theta_n I_1(\theta_n a)$$

(6.5.23)

Substituting the above result in the first equation of (6.5.22), we get the frequency equation as

$$\theta_n I_1(\theta_n a) J_0(a_n a) + a_n J_1(a_n a) I_0(\theta_n a) = 0$$

(6.5.24)

The first three non dimensional frequencies Γ given by a solution of the transcendental equation (6.5.24) above, are given in Fig. 6.5.1 as a function of the non dimensional pre load L. Buckling takes place when the first frequency becomes zero, which corresponds to

$$a^2 a^2 = 0$$
$$\theta^2 a^2 = L = -14.682$$

(6.5.25)

This case of buckling is shown by the vertical line in Fig. 6.5.1.

When $L = 0$, the natural frequencies given in Fig. 6.5.1 correspond to the case of a clamped circular plate without any pre load.

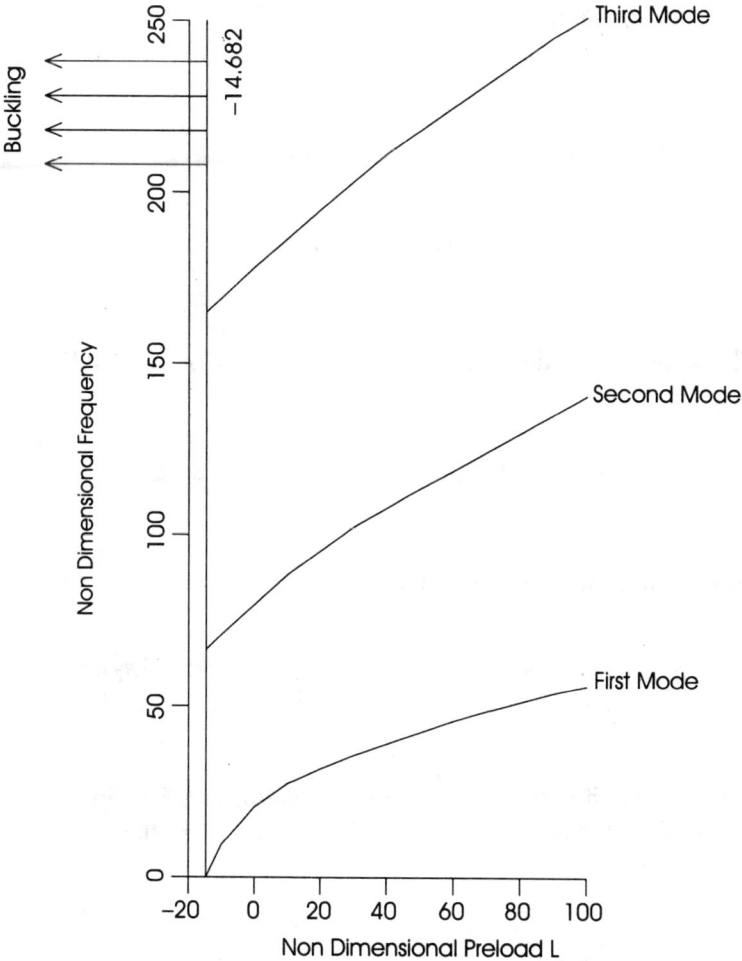

Fig. 6.5.1 **Natural Frequency of a Preloaded Clamped Circular Plate**

6.6 BERGER'S APPROXIMATION FOR RECTANGULAR PLATES

Here, we consider an approximation of Von Karman's equations as proposed by Berger. The nonlinear strain displacement relations of the plate are given by equation (6.1.4)

$$\varepsilon_{xx} = u_{,x} + \frac{1}{2}w_{,x}^2 - zw_{,xx}$$

$$\varepsilon_{yy} = v_{,y} + \frac{1}{2}w_{,y}^2 - zw_{,yy}$$

$$\gamma_{xy} = u_{,y} + v_{,x} - 2zw_{,xy} + w_{,x}w_{,y}$$

(6.6.1)

The corresponding stresses are given by (3.2.18)

$$\tau_{xx} = \frac{E}{(1-v^2)}(\varepsilon_{xx} + v\varepsilon_{yy})$$

$$\tau_{yy} = \frac{E}{(1-v^2)}(\varepsilon_{yy} + v\varepsilon_{xx})$$

$$\tau_{xy} = G\gamma_{xy}$$

(6.6.2)

The strain energy due to bending and stretching of the middle surface of the plate is given by

$$V = \frac{1}{2}\iiint(\tau_{xx}\varepsilon_{xx} + \tau_{yy}\varepsilon_{yy} + \tau_{xy}\gamma_{xy})dx\,dy\,dz$$

(6.6.3)

Substituting equations (6.6.2) in (6.6.3) above, we get

$$V = \frac{1}{2}\iint\left[\int_{-h/2}^{h/2}\left\{\frac{E}{(1-v^2)}(\varepsilon_{xx}^2 + \varepsilon_{yy}^2 + 2v\varepsilon_{yy}\varepsilon_{xx}) + G\gamma_{xy}^2\right\}dz\right]dx\,dy$$

(6.6.4)

Substituting the strain displacement relations (6.6.1) in the above equation (6.6.4) and integrating with respect to z, the following is obtained.

$$V = \frac{Gh}{2(1-v)}\iint\left[(\bar{\varepsilon}_{xx}^2 + \bar{\varepsilon}_{yy}^2 + 2v\bar{\varepsilon}_{yy}\bar{\varepsilon}_{xx}) + \frac{(1-v)}{2}\bar{\gamma}_{xy}^2\right]dx\,dy$$

$$+ \frac{D}{2}\iint\left[(\nabla^2 w)^2 - 2(1-v)(w_{,xx}w_{,yy} - w_{,xy}^2)\right]dx\,dy$$

(6.6.5)

where

$$\bar{\varepsilon}_{xx} = u_{,x} + \frac{1}{2}w_{,x}^2$$

$$\bar{\varepsilon}_{yy} = v_{,y} + \frac{1}{2}w_{,y}^2$$

$$\bar{\gamma}_{xy} = u_{,y} + v_{,x} + w_{,x}w_{,y}$$

(6.6.6)

are the strains of the middle surface of the plate.

The *first and second strain invariants* of the middle surface are given by

$$e_1 = \bar{\varepsilon}_{xx} + \bar{\varepsilon}_{yy}$$

$$e_2 = \bar{\varepsilon}_{xx}\bar{\varepsilon}_{yy} - \frac{1}{4}\bar{\gamma}_{xy}^2$$

(6.6.7)

In terms of the strain invariants above, the strain energy expression in (6.6.5) can be written as

$$V = \frac{D}{2} \iint \left[(\nabla^2 w)^2 + \frac{12}{h^2} e_1^2 - 2(1-v)\left(\frac{12}{h^2} e_2 + w_{,xx} w_{,yy} - w_{,xy}^2\right) \right] dx\, dy$$

(6.6.8)

Berger's approximation consists of neglecting the second strain invariant containing product and square terms of the strains, then equation (6.6.8) above is simplified as

$$V = \frac{D}{2} \iint \left[(\nabla^2 w)^2 + \frac{12}{h^2} e_1^2 - 2(1-v)\left(w_{,xx} w_{,yy} - w_{,xy}^2\right) \right] dx\, dy$$

(6.6.8a)

The kinetic energy of the plate is given by

$$T = \frac{1}{2}\rho h \iint [\dot{u}^2 + \dot{v}^2 + \dot{w}^2]\, dx\, dy$$

(6.6.9)

From Hamilton's principle, we get

$$\delta \int_{t_1}^{t_2} \left[\iint \frac{D}{2}\left\{ (\nabla^2 w)^2 + \frac{12}{h^2} e_1^2 - 2(1-v)(w_{,xx}w_{,yy} - w_{,xy}^2) \right\} dx\, dy \right. \\ \left. - \frac{1}{2}\rho h \iint \{\dot{u}^2 + \dot{v}^2 + \dot{w}^2\}\, dx\, dy \right] dt = 0$$

(6.6.10)

The above leads to

$$e_{1,x} = \frac{\rho h^3}{12D} \ddot{u}$$

$$e_{1,y} = \frac{\rho h^3}{12D} \ddot{v}$$

$$D\nabla^4 w - \frac{12D}{h^2}\left[(e_1 w_{,x})_{,x} + (e_1 w_{,y})_{,y}\right] + \rho h \ddot{w} = 0$$

(6.6.11)

Assuming the in-plane inertia effects to be negligible, the first two equations in the above give

$$e_{1,x} = 0$$
$$e_{1,y} = 0$$

(6.6.12)

As a result of the above, the third equation in (6.6.11) simplifies to

$$D\nabla^4 w - \frac{12D}{h^2} e_1 \nabla^2 w + \rho h \ddot{w} = 0$$

(6.6.13)

From (6.1.7) and (6.6.6), we find

$$N_x = \frac{Eh}{1-v^2}\left[\bar{\varepsilon}_{xx} + v\bar{\varepsilon}_{yy}\right]$$

$$N_y = \frac{Eh}{1-v^2}\left[\bar{\varepsilon}_{yy} + v\bar{\varepsilon}_{xx}\right]$$

(6.6.14)

Adding the above two equations in (6.6.14), we get

$$N_x + N_y = \frac{Eh}{1-v^2}[e_1 + ve_1]$$

$$= \frac{Eh}{1-v^2} e_1(1+v)$$

(6.6.15)

Let

$$N = \frac{N_x + N_y}{1+v}$$

(6.6.16)

then

$$e_1 = N\frac{(1-v^2)}{Eh}$$

$$= \frac{Nh^2}{12D}$$

(6.6.17)

Equation (6.6.13) now can be written as

$$D\nabla^4 w - N\nabla^2 w + ph\ddot{w} = 0 \qquad (6.6.18)$$

Elastic Plate on Kelvin Foundation Consider the case of the rectangular plate shown in Fig. 6.6.1. Equation (6.6.18) for this case modifies to

$$D\nabla^4 w - N\nabla^2 w + ph\ddot{w} + c\dot{w} + kw = p(x, y, t) \qquad (6.6.19)$$

From (6.6.17), (6.6.7) and (6.6.6), we have

$$\frac{Nh^2}{12D} = u_{,x} + v_{,y} + \frac{1}{2}\left(w_{,x}^2 + w_{,y}^2\right) \qquad (6.6.20)$$

Integrating the above equation over the area of the plate, we get

$$\frac{Nh^2}{12D}ab = \iint (u_{,x} + v_{,y})dx\,dy + \frac{1}{2}\iint \left(w_{,x}^2 + w_{,y}^2\right)dx\,dy \qquad (6.6.21)$$

The boundary conditions for the in-plane displacements are

$$\begin{aligned} u &= 0 \quad \text{at } x = 0, a \text{ and } 0 \le y \le b \\ v &= 0 \quad \text{at } y = 0, b \text{ and } 0 \le x \le a \end{aligned} \qquad (6.6.22)$$

In view of the above conditions, equation (6.6.21) reduces to

$$\frac{Nh^2}{12D}ab = \frac{1}{2}\iint \left(w_{,x}^2 + w_{,y}^2\right)dx\,dy \qquad (6.6.23)$$

Integrating by parts

$$\frac{Nh^2}{12D}ab = \frac{1}{2}\int [w_{,x}w]_{x=0}^{x=a}dy + \frac{1}{2}\int [w_{,y}w]_{y=0}^{y=b}dx$$
$$- \frac{1}{2}\iint (w\nabla^2 w)dx\,dy \qquad (6.6.24)$$

216 *Dynamics of Plates*

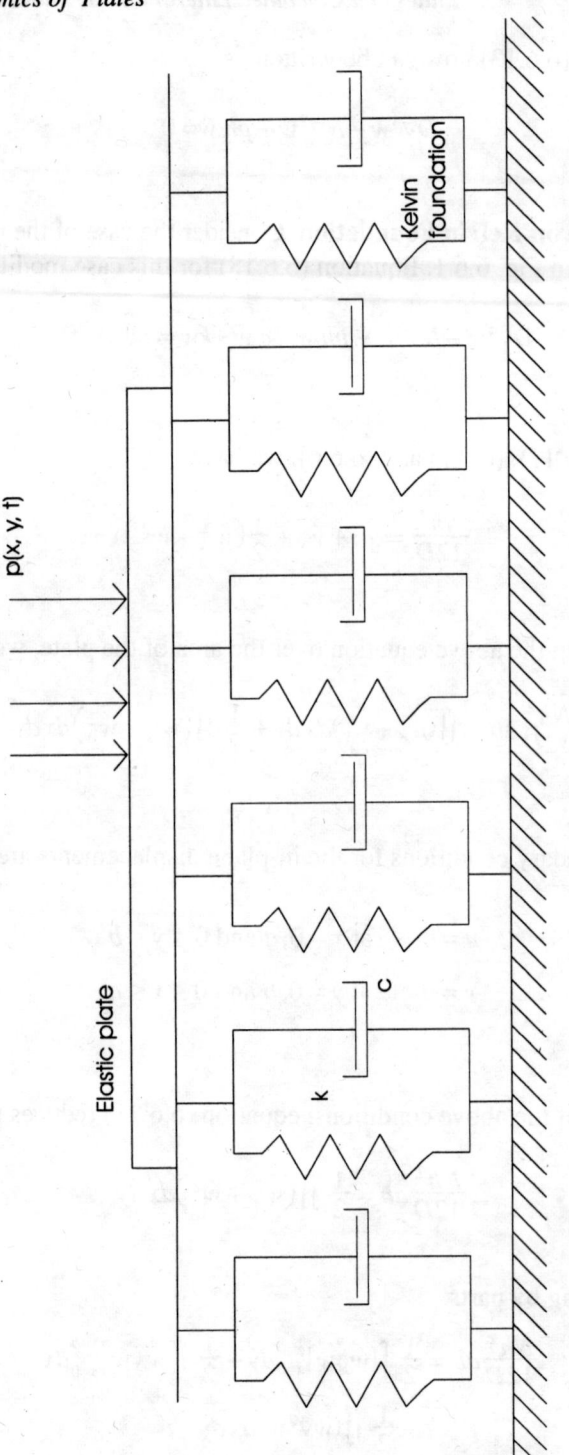

Figure 6.6.1 Elastic Plate on Kelvin Foundation

Plates and Combined Lateral and In-Plane Forces 217

The boundary conditions for the transverse deflection of the plate are

$$w = w_{,xx} = 0 \quad \text{at } x = 0, a \text{ and } 0 \leq y \leq b$$
$$w = w_{,yy} = 0 \quad \text{at } y = 0, b \text{ and } 0 \leq x \leq a$$

(6.6.25)

In view of the above, equation (6.6.24) becomes

$$\frac{N}{D} = -\frac{6}{abh^2} \iint (w\nabla^2 w) dx\, dy$$

(6.6.26)

Making use of the above result, equation (6.6.19) becomes

$$D\nabla^4 w + \left[\frac{6D}{abh^2} \iint (w\nabla^2 w) dx\, dy\right] \nabla^2 w + ph\ddot{w} + c\dot{w} + kw$$
$$= p(x, y, t)$$

(6.6.27)

To solve the above nonlinear equation, we can employ Galerkin's method. Here, we consider a one term approximation

$$w(x, y, t) = \phi(x, y) f(t)$$
$$= \sin \frac{\pi x}{a} \sin \frac{\pi y}{b} f(t)$$

(6.6.28)

Substituting the above in (6.6.27), and making use of (3.3.13) we get the error in the differential equation as

$$\epsilon = D\beta^2 \phi(x, y) f +$$
$$\left[\frac{6D}{abh^2} f^2 \iint \phi(x, y) \nabla^2 \phi(x, y) dx\, dy\right] \nabla^2 \phi(x, y) f$$
$$+ ph\phi(x, y) \ddot{f} + c\phi(x, y) \dot{f} + k\phi(x, y) f - p(x, y, t)$$

(6.6.29)

According to Galerkin's method,

$$f \iint D\beta^2 \phi^2(x,y)\,dx\,dy +$$

$$\frac{6D}{abh^2} f^3 \iint [\iint \phi(x,y)\nabla^2\phi(x,y)\,dx\,dy]\nabla^2\phi(x,y)\,\phi(x,y)\,dx\,dy$$

$$+ \rho h \ddot{f} \iint \phi^2(x,y)\,dx\,dy + c\dot{f} \iint \phi^2(x,y)\,dx\,dy$$

$$+ kf \iint \phi^2(x,y)\,dx\,dy - \iint p(x,y,t)\phi(x,y)\,dx\,dy$$

$$= 0$$

(6.6.30)

Noting that

$$\int_0^a \int_0^b \phi^2\,dx\,dy = \frac{ab}{4}$$

$$\int_0^a \int_0^b \phi\nabla^2\phi\,dx\,dy = -\frac{a^2+b^2}{4ab}\pi^2$$

(6.6.31)

equation (6.6.30) becomes

$$fD\beta^2 \frac{ab}{4} + \frac{6D}{abh^2} f^3 \left[\frac{a^2+b^2}{4ab}\pi^2\right]^2$$

$$+ \rho h \ddot{f} \frac{ab}{4} + c\dot{f} \frac{ab}{4} + kf\frac{ab}{4}$$

$$= \iint p(x,y,t)\phi(x,y)\,dx\,dy$$

(6.6.32)

Expressing the excitation force in the form

$$p(x,y,t) = p_1 \sin \omega t + p_2 \cos \omega t$$

(6.6.33)

and making use of the result in (3.3.3), the above equation (6.6.32) further reduces to

$$fD\beta^2 + \frac{3D\pi^4(a^2+b^2)^2}{2a^4b^4h^2}f^3$$
$$+ ph\ddot{f} + c\dot{f} + kf = p_1 \sin\omega t + p_2 \cos\omega t$$

(6.6.34)

The above equation can be rearranged as

$$\ddot{f} + \frac{c}{ph}\dot{f} + \frac{(k+D\beta^2)}{ph}f + \frac{3D\pi^4(a^2+b^2)^2}{2ph^3a^4b^4}f^3$$
$$= \frac{p_1}{ph}\sin\omega t + \frac{p_2}{ph}\cos\omega t$$

(6.6.34a)

Let us introduce a non dimensional time parameter

$$T = \frac{\pi^2}{ab}\sqrt{\frac{D}{ph}}\, t$$

(6.6.35)

then, equation (6.6.34) becomes

$$\ddot{f} + 2a\dot{f} + \omega_1^2 f + \gamma^2 f^3 = P_1 \sin\Omega T + P_2 \cos\Omega T$$

(6.6.36)

where

$$2a = \frac{cab}{\pi^2\sqrt{D\rho h}}$$

$$\omega_1^2 = \frac{(k+D\beta^2)a^2b^2}{\pi^4 D}$$

$$\gamma^2 = \frac{3}{2}\frac{(a^2+b^2)^2}{h^2a^2b^2}$$

$$P_1 = \frac{p_1 a^2 b^2}{\pi^4 D}$$

$$P_2 = \frac{p_2 a^2 b^2}{\pi^4 D}$$

$$\Omega = \frac{\omega ab}{\pi^2}\sqrt{\frac{\rho h}{D}}$$

(6.6.37)

220 Dynamics of Plates

Equation (6.6.36) is the familiar Duffing's equation, the solution of which can be obtained by using Harmonic Balancing method. Here, we seek the first harmonic solution and write this as

$$f(T) = A \sin \Omega T \tag{6.6.38}$$

Substituting the above in equation (6.6.36) we get

$$\left[(\omega_1^2 - \Omega^2)A + \frac{3}{4}\gamma^2 A^3\right] \sin \Omega T + 2a\Omega A \cos \Omega T$$
$$= P_1 \sin \Omega T + P_2 \cos \Omega T \tag{6.6.39}$$

From the above, the following two equations can be obtained.

$$\left[(\omega_1^2 - \Omega^2)A + \frac{3}{4}\gamma^2 A^3\right] = P_1$$
$$2a\Omega A = P_2 \tag{6.6.40}$$

Squaring and adding the above two equations, we get

$$\left[(\omega_1^2 - \Omega^2)A + \frac{3}{4}\gamma^2 A^3\right]^2 + [2a\Omega A]^2 = P^2 \tag{6.6.41}$$

where

$$P = \frac{a^2 b^2}{\pi^4 D} \sqrt{p_1^2 + p_2^2} \tag{6.6.42}$$

Equation (6.6.41) can be solved numerically by a trial and error method to determine the three roots of A^2. Depending on the value of the frequency ratio $\frac{\Omega}{\omega_1}$, there will be either one real root with two complex roots or all three real roots. A plot of these roots gives the familiar Duffing's equation solution exhibiting the jump phenomenon.

For a detailed discussion on the Duffing's equation, reference may be made to the book *Advanced Theory of Vibration* by the author.

6.7 BERGER'S APPROXIMATION FOR CIRCULAR PLATES

The equation of motion for the plate under Berger's approximation is given by equation (6.6.18)

$$D\nabla^4 w(r,\theta,t) - N\nabla^2 w(r,\theta,t) + ph\,\ddot{w}(r,\theta,t) = 0 \tag{6.7.1}$$

Elastic Plate on Kelvin Foundation Consider the case of a circular plate as shown in Fig. 6.7.1. Equation (6.7.1) for this case modifies to

$$D\nabla^4 w(r,\theta,t) - N\nabla^2 w(r,\theta,t) + ph\,\ddot{w}(r,\theta,t)$$
$$+ c\,\dot{w}(r,\theta,t) + kw(r,\theta,t)$$
$$= p(r,\theta,t) \tag{6.7.2}$$

The boundary conditions for transverse deflection of the plate are

$$w(r,\theta,t) = 0 \quad \text{at } r = a \text{ and } w \text{ is finite at } r = 0$$
$$w_{,rr} + \frac{\nu}{r} w_{,r} = 0 \tag{6.7.3}$$

The middle surface strains of a circular plate are

$$\bar{\varepsilon}_{rr} = u_{,r} + \frac{1}{2} w_{,r}^2$$
$$= \frac{1}{Eh}(N_r - \nu N_\theta)$$
$$\bar{\varepsilon}_{\theta\theta} = \frac{u}{r} + \frac{1}{r} v_{,\theta} + \frac{1}{2r^2} w_{,\theta}^2$$
$$= \frac{1}{Eh}(N_\theta - \nu N_r) \tag{6.7.4}$$

and the first strain invariant is

$$e_1 = \bar{\varepsilon}_{rr} + \bar{\varepsilon}_{\theta\theta}$$
$$= \frac{Nh^2}{12D} \tag{6.7.5}$$

222 Dynamics of Plates

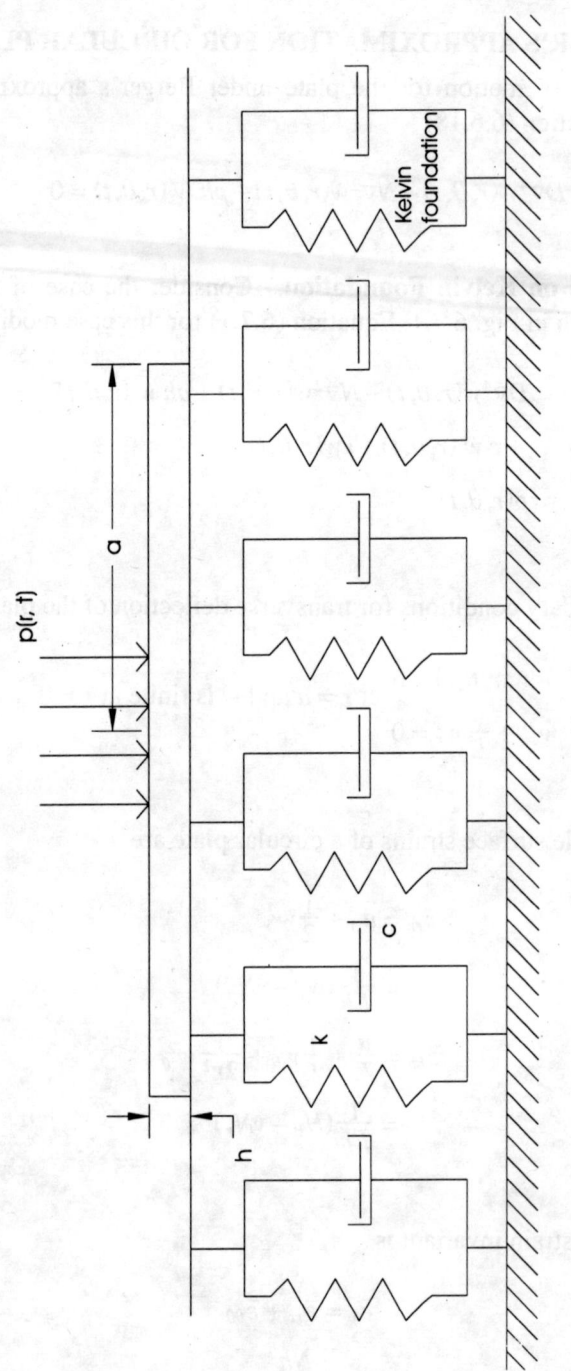

Fig. 6.7.1 Circular Plate on Viscoelastic Foundation

where

$$N = \frac{N_r + N_\theta}{1+v} \tag{6.7.6}$$

and N_r and N_θ are the stress resultants. From (6.7.4) and (6.7.5), we can write

$$\frac{Nh^2}{12D} = \frac{u}{r} + u_{,r} + \frac{1}{r}v_{,\theta} + \frac{1}{2}\left(w_{,r}^2 + \frac{1}{r^2}w_{,\theta}^2\right) \tag{6.7.7}$$

Integrating the above over the area of the plate

$$\iint \frac{Nh^2}{12D}r\,dr\,d\theta = \iint \left(\frac{u}{r} + u_{,r}\right)r\,dr\,d\theta + \iint \frac{1}{r}v_{,\theta}r\,dr\,d\theta$$
$$+ \frac{1}{2}\iint \left(w_{,r}^2 + \frac{1}{r^2}w_{,\theta}^2\right)r\,dr\,d\theta \tag{6.7.8}$$

Let the outer edge of the plate be immovably constrained against the radial displacement and further let us consider axisymmetric case, then

$$\int_0^a \frac{Nh^2}{12D}r\,dr = \frac{1}{2}\int_0^a w_{,r}^2 r\,dr \tag{6.7.9}$$

Integrating

$$\frac{1}{2}\frac{Nh^2}{12D}a^2 = \frac{1}{2}\left[\{ww_{,r}r\}_0^a - \int_0^a w\nabla^2 w\,r\,dr\right]$$
$$= -\frac{1}{2}\int_0^a w\nabla^2 w\,r\,dr \tag{6.7.10}$$

Therefore

$$N = -\frac{12D}{a^2h^2}\int_0^a w\nabla^2 w\,r\,dr \tag{6.7.11}$$

Substituting the above in equation (6.7.2), we get

$$D\nabla^4 w(r,t) + \left[\frac{12D}{a^2 h^2}\int_0^a w\nabla^2 w\, r\, dr\right]\nabla^2 w(r,t)$$
$$+ \rho h \ddot{w}(r,t) + c\dot{w}(r,t) + kw(r,t)$$
$$= p(r,t) \qquad (6.7.12)$$

The solution of the above equation can be obtained by Galerkin's method. The solution is assumed as

$$w = \phi(r)f(t) \qquad (6.7.13)$$

where $\phi(r)$ satisfies the boundary conditions of the plate. Substituting equation (6.7.13) in (6.7.12), we get the error in the differential equation as

$$\epsilon = D\nabla^4 \phi(r)f(t) + \left[\frac{12D}{a^2 h^2}f^2(t)\int_0^a \phi(r)\nabla^2\phi(r)\, r\, dr\right]\nabla^2\phi(r)f(t)$$
$$+ \rho h \ddot{f}(t)\phi(r) + c\dot{f}\phi(r) + k\phi(r)f(t) - p(r,t) \qquad (6.7.14)$$

According to Galerkin's process

$$f(t)\int_0^a \phi(r)D\nabla^4\phi(r)\, r\, dr$$
$$+ \frac{12D}{a^2 h^2}f^3(t)\int_0^a\left[\int_0^a \phi(r)\nabla^2\phi(r)\, r\, dr\right]\phi(r)\nabla^2\phi(r)\, r\, dr$$
$$+ \rho h \ddot{f}(t)\int_0^a \phi^2(r)\, r\, dr + c\dot{f}\int_0^a \phi^2(r)\, r\, dr + kf(t)\int_0^a \phi^2(r)\, r\, dr$$
$$- \int_0^a p(r,t)\phi(r)\, r\, dr$$
$$= 0 \qquad (6.7.15)$$

The following shape function is assumed

$$\phi(r) = 1 + C_1\left(\frac{r}{a}\right)^2 + C_2\left(\frac{r}{a}\right)^4$$

(6.7.16)

where

$$C_1 = -2\left(\frac{v+3}{v+5}\right)$$

$$C_2 = \left(\frac{v+1}{v+5}\right)$$

(6.7.17)

which satisfies the boundary conditions in (6.7.3) and also the differential equation (4.3.21).

$$\nabla^4 \phi - \beta^4 \phi = 0$$

(4.3.21)

Substituting (6.7.16) in (6.7.15) and rearranging, we get

$$\frac{D}{ph}\left(\beta^4 + \frac{k}{D}\right)f(t) + \frac{12D}{ph^3 a^4}\frac{\psi^2}{\mu}f^3(t) +$$

$$\ddot{f}(t) + \frac{c}{ph}\dot{f} - \frac{1}{\mu p h a^2}Q(t)$$

$$= 0$$

(6.7.18)

where

$$\psi_{11} = \int_0^a \phi(r)\nabla^2 \phi(r)\, r\, dr$$

$$\mu_{11} = \frac{1}{a^2}\int_0^a \phi^2(r)\, r\, dr$$

$$Q(t) = \int_0^a p(r,t)\phi(r)\, r\, dr$$

(6.7.19)

Let

$$T = \frac{t}{a^2}\sqrt{\frac{D}{ph}}$$

(6.7.20)

then, equation (6.7.18) becomes

$$\ddot{f}(T) + 2a\dot{f} + \omega_1^2 f(T) + \gamma_1^2 f^3(T)$$
$$= \frac{a^4}{\mu D} Q(T)$$

(6.7.21)

where

$$\cdot = \frac{d}{dT}$$

$$\omega_1^2 = a^4\left(\beta^4 + \frac{k}{D}\right)$$

$$\gamma_1^2 = \frac{12}{h^2} \frac{\psi^2}{\mu}$$

(6.7.22)

Let the forcing function $p(r, T)$ be

$$p(r, T) = P\phi(r)F(T)$$

(6.7.23)

then

$$Q(T) = PF(T) \int_0^a \phi^2(r) r \, dr$$
$$= \mu a^2 P f(T)$$

(6.7.24)

and equation (6.7.21) becomes

$$\ddot{f}(T) + 2a\dot{f} + \omega_1^2 f(T) + \gamma_1^2 f^3(T)$$
$$= \frac{Pa^6}{D} F(T)$$

(6.7.25)

This is the familiar Duffing's equation, see section 6.6 for its solution.

6.8 VON KARMAN RECTANGULAR PLATE ON A VISCOELASTIC FOUNDATION

Let us consider the rectangular plate of Fig. 6.6.1, this time under Von Karman theory, equations (6.1.32) and (6.1.31). These equations are modified to take into account the viscoelastic foundation to give

$$D\nabla^4 w + \rho h \ddot{w} = q(x,y,t) + F_{,yy}w_{,xx} + F_{,xx}w_{,yy} - 2F_{,xy}w_{,xy}$$

$$\nabla^4 F = Eh(w_{,xy}^2 - w_{,xx}w_{,yy})$$

$$q(x,y,t) = p(x,y,t) - kw(x,y,t) - c\dot{w}(x,y,t)$$

(6.8.1)

where $p(x,y,t)$ is the external force acting on the plate, $q(x,y,t)$ is the foundation force acting on the plate, k and c are the foundation parameters and F is the Airy stress function defined by

$$N_x = F_{,yy}$$

$$N_y = F_{,xx}$$

$$N_{xy} = -F_{,xy}$$

(6.1.29)

Making use of the third equation in (6.8.1), the first equation becomes

$$D\nabla^4 w + \rho h \ddot{w} + c\dot{w} + kw - (F_{,yy}w_{,xx} + F_{,xx}w_{,yy} - 2F_{,xy}w_{,xy})$$
$$= p(x,y,t)$$

(6.8.1a)

The in-plane displacements from (6.1.4) can be written as

$$u = \int_0^x \left(\varepsilon_{xx} - \frac{1}{2}w_{,x}^2\right)dx$$

(6.8.2)

$$v = \int_0^y \left(\varepsilon_{yy} - \frac{1}{2}w_{,y}^2\right)dy$$

(6.8.3)

Making use of equations (6.1.30), the above equations can written as

$$u = \int_0^x \left[\frac{1}{hE}(F_{,yy} - vF_{,xx}) - \frac{1}{2}w_{,x}^2 \right] dx$$

(6.8.4)

$$v = \int_0^y \left[\frac{1}{hE}(F_{,xx} - vF_{,yy}) - \frac{1}{2}w_{,y}^2 \right] dy$$

(6.8.5)

The boundary conditions for the transverse deflections of the plate are

$$w = w_{,xx} = 0 \quad \text{at } x = 0, a \text{ and } 0 \le y \le b$$
$$w = w_{,yy} = 0 \quad \text{at } y = 0, b \text{ and } 0 \le x \le a$$

(6.8.6)

The boundary conditions for the in-plane displacements are

$$u = 0 \text{ and } F_{,xy} = 0 \quad \text{at } x = 0, a \text{ and } 0 \le y \le b$$
$$v = 0 \text{ and } F_{,xy} = 0 \quad \text{at } y = 0, b \text{ and } 0 \le x \le a$$

(6.8.7)

For solving equations (6.8.1), we can use Galerkin's method and assume the shape function with only one term as

$$w(x, y, t) = \phi(x, y) f(t)$$
$$= \sin \frac{\pi x}{a} \sin \frac{\pi y}{b} f(t)$$

(6.8.8)

which satisfies the boundary conditions in (6.8.4) and the linear plate differential equation

$$\nabla^4 w - \beta^2 W = 0$$

(3.3.13)

Substituting (6.8.8) in the second equation of (6.8.1), we get

Plates and Combined Lateral and In-Plane Forces 229

$$\nabla^4 F = \frac{1}{2} Ehf^2(t) \frac{\pi^4}{a^2 b^2} \left(\cos \frac{2\pi x}{a} + \cos \frac{2\pi y}{b} \right)$$

(6.8.9)

The particular integral of the above differential equation is expressed as

$$F = f_1(t) \cos \frac{2\pi x}{a} + f_2(t) \cos \frac{2\pi y}{b}$$

(6.8.10)

where f_1 and f_2 are to be determined. Substituting the above F in (6.8.9), we have

$$16 f_1(t) \frac{\pi^4}{a^4} \cos \frac{2\pi x}{a} + 16 f_2(t) \frac{\pi^4}{b^4} \cos \frac{2\pi y}{b}$$
$$= \frac{1}{2} Ehf^2(t) \frac{\pi^4}{a^2 b^2} \left(\cos \frac{2\pi x}{a} + \cos \frac{2\pi y}{b} \right)$$

(6.8.11)

Equating the coefficients of cosine terms in the above, the following two relations can be obtained.

$$f_1(t) = \frac{1}{32} Ehf^2(t) \frac{a^2}{b^2}$$

$$f_2(t) = \frac{1}{32} Ehf^2(t) \frac{b^2}{a^2}$$

(6.8.12)

Now, equation (6.8.10) becomes

$$F = \frac{1}{32} Ehf^2(t) \left(\frac{a^2}{b^2} \cos \frac{2\pi x}{a} + \frac{b^2}{a^2} \cos \frac{2\pi y}{b} \right)$$

(6.8.13)

Including the complementary function, the solution of (6.8.9) is

$$F = \frac{1}{2} C_2(t) x^2 + \frac{1}{2} C_1(t) y^2 + \frac{1}{32} Ehf^2(t) \left(\frac{a^2}{b^2} \cos \frac{2\pi x}{a} + \frac{b^2}{a^2} \cos \frac{2\pi y}{b} \right)$$

$$F_{,xx} = C_2(t) - \frac{1}{8} Ehf^2(t) \frac{\pi^2}{b^2} \cos \frac{2\pi x}{a}$$

$$F_{,yy} = C_1(t) - \frac{1}{8} Ehf^2(t) \frac{\pi^2}{a^2} \cos \frac{2\pi y}{b}$$

(6.8.14)

230 Dynamics of Plates

The above equation satisfies the boundary conditions for the in-plane displacements in (6.8.4) and (6.8.5) for $x = 0$ and $y = 0$. For $x = a$ and $y = b$, we get

$$u = \int_0^a \frac{1}{hE} \left\{ \begin{array}{l} C_1(t) - \frac{1}{8}Ehf^2(t)\frac{\pi^2}{a^2}\cos\frac{2\pi y}{b} \\ -v\left(C_2(t) - \frac{1}{8}Ehf^2(t)\frac{\pi^2}{b^2}\cos\frac{2\pi x}{a}\right) \\ -\frac{1}{2}\frac{\pi^2}{a^2}\cos^2\frac{\pi x}{a}\sin^2\frac{\pi y}{b}f^2(t) \end{array} \right\} dx = 0$$

for $0 \leq y \leq b$

$$v = \int_0^b \frac{1}{hE} \left\{ \begin{array}{l} C_2(t) - \frac{1}{8}Ehf^2(t)\frac{\pi^2}{b^2}\cos\frac{2\pi x}{a} \\ -v\left(C_1(t) - \frac{1}{8}Ehf^2(t)\frac{\pi^2}{a^2}\cos\frac{2\pi y}{b}\right) \\ -\frac{1}{2}\frac{\pi^2}{b^2}\sin^2\frac{\pi x}{a}\cos^2\frac{\pi y}{b}f^2(t) \end{array} \right\} dy = 0$$

for $0 \leq x \leq a$

(6.8.15)

On evaluating the integrals, the above equations simplify to

$$C_1(t) - vC_2(t) = \frac{1}{8}hEf^2(t)\frac{\pi^2}{a^2}$$

$$C_2(t) - vC_1(t) = \frac{1}{8}hEf^2(t)\frac{\pi^2}{b^2}$$

(6.8.16)

Solving the above, we get

$$C_1(t) = \frac{hE}{8(1-v^2)}f^2(t)\pi^2\left(\frac{1}{a^2} + \frac{v}{b^2}\right)$$

$$C_2(t) = \frac{hE}{8(1-v^2)}f^2(t)\pi^2\left(\frac{1}{b^2} + \frac{v}{a^2}\right)$$

(6.8.17)

The stress function F is now obtained as

$$F = \frac{hE}{16(1-v^2)}f^2(t)\pi^2\left[\left(\frac{1}{b^2} + \frac{v}{a^2}\right)x^2 + \left(\frac{1}{a^2} + \frac{v}{b^2}\right)y^2\right]$$

$$+ \frac{1}{32}Ehf^2(t)\left(\frac{a^2}{b^2}\cos\frac{2\pi x}{a} + \frac{b^2}{a^2}\cos\frac{2\pi y}{b}\right)$$

(6.8.18)

The derivatives of the stress function above are

$$F_{,yy} = \frac{hE}{8a^2}f^2(t)\pi^2\left[\frac{1}{(1-v^2)}\left(1+\frac{va^2}{b^2}\right)-\cos\frac{2\pi y}{b}\right]$$

$$F_{,xx} = \frac{hE}{8b^2}f^2(t)\pi^2\left[\frac{1}{(1-v^2)}\left(1+\frac{vb^2}{a^2}\right)-\cos\frac{2\pi x}{a}\right]$$

$$F_{,xy} = 0$$

(6.8.19)

With the help of the above, equation (6.8.1a) can be written as

$$D\nabla^4 w + ph\ddot{w} + c\dot{w} + kw$$

$$-\frac{hE}{8a^2}f^2(t)\pi^2\left[\frac{1}{(1-v^2)}\left(1+\frac{va^2}{b^2}\right)-\cos\frac{2\pi y}{b}\right]w_{,xx}$$

$$-\frac{hE}{8b^2}f^2(t)\pi^2\left[\frac{1}{(1-v^2)}\left(1+\frac{vb^2}{a^2}\right)-\cos\frac{2\pi x}{a}\right]w_{,yy}$$

$$= p(x,y,t)$$

(6.8.20)

Substituting for the assumed solution (6.8.8) in the above, we get the error in the differential equation as

$$\epsilon = D\pi^4\left(\frac{1}{a^2}+\frac{1}{b^2}\right)^2\sin\frac{\pi x}{a}\sin\frac{\pi y}{b}f(t) + ph\ddot{f}(t)\sin\frac{\pi x}{a}\sin\frac{\pi y}{b}$$

$$+ c\dot{f}(t)\sin\frac{\pi x}{a}\sin\frac{\pi y}{b} + k\sin\frac{\pi x}{a}\sin\frac{\pi y}{b}f(t) - p(x,y,t)$$

$$+ \frac{hE}{8a^4}f^3(t)\pi^4\left[\frac{1}{(1-v^2)}\left(1+\frac{va^2}{b^2}\right)-\cos\frac{2\pi y}{b}\right]\sin\frac{\pi x}{a}\sin\frac{\pi y}{b}$$

$$+ \frac{hE}{8b^4}f^3(t)\pi^4\left[\frac{1}{(1-v^2)}\left(1+\frac{vb^2}{a^2}\right)-\cos\frac{2\pi x}{a}\right]\sin\frac{\pi x}{a}\sin\frac{\pi y}{b}$$

(6.8.21)

Orthogonolizing the above error with the assumed mode shape, we get

232 Dynamics of Plates

$$\left[\rho h \ddot{f}(t)+c\dot{f}(t)+D\left\{\left(\frac{\pi^2}{a^2}+\frac{\pi^2}{b^2}\right)^2+\frac{k}{D}\right\}f(t)\right]\iint \sin^2\frac{\pi x}{a}\sin^2\frac{\pi y}{b}dx\,dy$$

$$-\frac{hE}{8}\iint\left[\begin{array}{c}\frac{\pi^4}{a^4}\cos\frac{2\pi y}{b}+\frac{\pi^4}{b^4}\cos\frac{2\pi x}{a}\\ \dfrac{1}{(1-\nu^2)}\left\{\begin{array}{c}\frac{\pi^2}{a^2}\left(\frac{\pi^2}{a^2}+\nu\frac{\pi^2}{b^2}\right)\\ +\frac{\pi^2}{b^2}\left(\frac{\pi^2}{b^2}+\nu\frac{\pi^2}{a^2}\right)\end{array}\right\}\end{array}\right]f^3(t)\sin^2\frac{\pi x}{a}\sin^2\frac{\pi y}{b}dx\,dy$$

$$=\iint p(x,y,t)\sin\frac{\pi x}{a}\sin\frac{\pi y}{b}dx\,dy$$

(6.8.22)

Making use of the following results

$$\iint \sin^2\frac{\pi x}{a}\sin^2\frac{\pi y}{b}dx\,dy=\frac{ab}{4}$$

$$\iint \cos\frac{2\pi y}{b}\sin^2\frac{\pi x}{a}\sin^2\frac{\pi y}{b}dx\,dy=-\frac{ab}{8}$$

$$\iint \cos\frac{2\pi x}{a}\sin^2\frac{\pi x}{a}\sin^2\frac{\pi y}{b}dx\,dy=-\frac{ab}{8}$$

(6.8.23)

equation (6.8.22) becomes

$$\left[\ddot{f}(t)+\frac{c}{\rho h}\dot{f}(t)+\frac{D}{\rho h}\left\{\left(\frac{\pi^2}{a^2}+\frac{\pi^2}{b^2}\right)^2+\frac{k}{D}\right\}f(t)\right]$$

$$+\frac{E}{8\rho}\left[\frac{1}{2}\left(\frac{\pi^4}{a^4}+\frac{\pi^4}{b^4}\right)+\frac{1}{(1-\nu^2)}\left\{\frac{\pi^4}{a^4}+2\nu\frac{\pi^2}{a^2}\frac{\pi^2}{b^2}+\frac{\pi^4}{b^4}\right\}\right]f^3(t)$$

$$=\frac{4}{ab\rho h}Q_1(t)$$

(6.8.24)

where

$$Q_1(t)=\iint p(x,y,t)\sin\frac{\pi x}{a}\sin\frac{\pi y}{b}dx\,dy$$

(6.8.25)

Using the non dimensional time parameter

$$T=\frac{\pi^2}{ab}\sqrt{\frac{D}{\rho h}}\,t$$

(6.8.26)

we get

$$\ddot{f}(T) + 2a\dot{f} + \omega_1^2 f(T) + \gamma_1^2 f^3(T)$$
$$= \frac{4ab}{\pi^4 D} Q_1(T)$$

(6.8.27)

the familiar Duffing's equation for the dynamic behavior of the plate. In the above

$$2a = \frac{cab}{\pi^2 \sqrt{D\rho h}}$$

$$\omega_1^2 = \frac{ka^2 b^2}{\pi^4 D} + \left(\frac{b}{a}\right)^2 \left[1 + \left(\frac{a}{b}\right)^2\right]^2$$

$$\gamma_1^2 = \frac{3b^2}{2a^2 h^2} \left[\frac{1}{2}(1-v^2)\left\{1 + \left(\frac{a}{b}\right)^4\right\} + 1 + 2v\left(\frac{a}{b}\right)^2 + \left(\frac{a}{b}\right)^4\right]$$

(6.8.28)

Let

$$p(x, y, T) = P \sin \frac{\pi x}{a} \sin \frac{\pi y}{b} F(T)$$

(6.8.29)

then

$$Q_1(T) = PF(t) \iint \sin^2 \frac{\pi x}{a} \sin^2 \frac{\pi y}{b} dx\, dy$$
$$= PF(t) \frac{ab}{4}$$

(6.8.30)

The plate equation can then be written as

$$\ddot{f}(T) + 2a\dot{f} + \omega_1^2 f(T) + \gamma_1^2 f^3(T) = \frac{a^2 b^2}{\pi^4 D} PF(T)$$

(6.8.31)

For a solution of the above equation, see section 6.6.

6.9 VON KARMAN CIRCULAR PLATE ON A VISCOELASTIC FOUNDATION

Let us consider the axisymmetric circular plate of Fig. 6.7.1, this time under Von Karman theory, equations (6.1.34), (6.1.36). These equations are modified to take into account the viscoelastic foundation to give

$$D\nabla^4 w(r,t) + ph\,\ddot{w}(r,t) = q(r,t) + \frac{1}{r}(F_{,r}w_{,r})_{,r}$$

$$\nabla^4 F = -\frac{Eh}{r}w_{,r}w_{,rr}$$

$$q(r,t) = p(r,t) - kw(r,t) - c\,\dot{w}(r,t)$$

(6.9.1)

where $p(r,t)$ is the external force acting on the plate, $q(r,t)$ is the foundation force acting on the plate, k and c are the foundation parameters and F is the Airy stress function defined by the stress resultants

$$N_r = \frac{1}{r}F_{,r}$$

$$N_\theta = F_{,rr}$$

(6.9.2)

The stress resultants in the above equation (6.9.2) are given by equations (6.7.4).

$$N_r = \frac{Eh}{1-v^2}(\bar{\varepsilon}_r + v\bar{\varepsilon}_\theta)$$

$$N_\theta = \frac{Eh}{1-v^2}(\bar{\varepsilon}_\theta + v\bar{\varepsilon}_r)$$

(6.9.3)

Making use of the third equation in (6.9.1), the first equation becomes

$$D\nabla^4 w + ph\,\ddot{w} + c\,\dot{w} + kw - \frac{1}{r}(F_{,r}w_{,r})_{,r} = p(r,t)$$

(6.9.1a)

The boundary conditions for the transverse deflections of the plate are

$$w = 0$$
$$w_{,rr} + \frac{v}{r}w_{,r} = 0 \qquad \text{at } r = a$$

(6.9.4)

For the case of an immovably constrained edge against radial displacements u_r, the following boundary condition is also prescribed

$$F_{,rr} - \frac{v}{r}F_{,r} = 0 \quad \text{at } r = a$$

(6.9.5)

For solving equations (6.9.1), we can use Galerkin's method and assume a one term approximation

$$w(r,t) = \phi(r)f(t)$$

(6.9.6)

The shape function in the above is taken as

$$\phi(r) = 1 + C_1\left(\frac{r}{a}\right)^2 + C_2\left(\frac{r}{a}\right)^4$$

(6.7.16)

where

$$C_1 = -2\left(\frac{v+3}{v+5}\right)$$
$$C_2 = \left(\frac{v+1}{v+5}\right)$$

(6.7.17)

satisfying the boundary conditions in (6.9.4). Substituting (6.9.6) in the second equation of (6.9.1), we get

$$\frac{1}{r}\frac{\partial}{\partial r}\left[r\frac{\partial}{\partial r}\left\{\frac{1}{r}\frac{\partial}{\partial r}\left(r\frac{\partial}{\partial r}\right)\right\}\right]F$$
$$= -\frac{4f^2(t)Eh}{a^4}\left[C_1^2 + 8C_1C_2\left(\frac{r}{a}\right)^2 + 12C_2^2\left(\frac{r}{a}\right)^4\right]$$

(6.9.7)

Integrating the above equation (6.9.7)

$$F = -f^2(t)Eh\left[C_4\left(\frac{r}{a}\right)^2 + \frac{C_1^2}{16}\left(\frac{r}{a}\right)^4 + \frac{C_1C_2}{18}\left(\frac{r}{a}\right)^6 + \frac{C_2^2}{48}\left(\frac{r}{a}\right)^8\right]$$

(6.9.8)

236 Dynamics of Plates

where the constant C_4 is to be determined from the boundary condition in equation (6.9.5). Substituting F given by above equation (6.9.8) in (6.9.5) and solving for C_4, we get

$$C_4 = -\frac{1}{24(v-1)}[3(v-3)C_1^2 + 4(v-5)C_1C_2 + 2(v-7)C_2^2]$$

(6.9.9)

It may be noted that C_1, C_2 and C_4 are all functions of the Poisson's ratio of the plate material. Substituting for F from (6.9.8) and the assumed solution (6.9.6) in equation (6.9.1a), we get the error in the differential equation

$$\epsilon = Df(t)\nabla^4\phi(r) + \rho h \ddot{f}(t)\phi(r) + c\dot{f}(t)\phi(r) + kf(t)\phi(r)$$

$$+ \frac{Eh}{a^4}f^3(t)\begin{bmatrix} 8C_1C_4 + 2(C_1^3 + 16C_2C_4)(\frac{r}{a})^2 \\ +10C_1^2C_2(\frac{r}{a})^4 + \frac{40}{3}C_1C_2^2(\frac{r}{a})^6 \\ + \frac{20}{3}C_2^3(\frac{r}{a})^8 \end{bmatrix} - p(r,t)$$

(6.9.10)

Orthogonolizing the above error with the assumed mode shape, we get

$$Df(t)\int_0^a \phi(r)\nabla^4\phi(r)r\,dr + \rho h \ddot{f}(t)\int_0^a \phi^2(r)r\,dr +$$

$$c\dot{f}(t)\int_0^a \phi^2(r)r\,dr + kf(t)\int_0^a \phi^2(r)r\,dr$$

$$+ \frac{Eh}{a^4}f^3(t)\int_0^a \begin{bmatrix} 8C_1C_4 + 2(C_1^3 + 16C_2C_4)(\frac{r}{a})^2 \\ +10C_1^2C_2(\frac{r}{a})^4 + \frac{40}{3}C_1C_2^2(\frac{r}{a})^6 \\ + \frac{20}{3}C_2^3(\frac{r}{a})^8 \end{bmatrix}\phi(r)r\,dr$$

$$= \int_0^a p(r,t)\phi(r)r\,dr$$

(6.9.11)

Noting that the assumed shape function satisfies the linear plate equation

$$\nabla^4 \phi - \beta^4 \phi = 0 \tag{4.3.21}$$

equation (6.9.11) can be written as

$$D\mu\beta^4 f(t) + \rho h\mu \ddot{f}(t) + c\mu \dot{f}(t) + k\mu f(t) + \frac{Eh\psi}{a^4} f^3(t) = Q(t) \tag{6.9.12}$$

where

$$\mu = \int_0^a \phi^2(r) r\, dr$$

$$\psi = \int_0^a \left[\begin{array}{c} 8C_1 C_4 + 2(C_1^3 + 16C_2 C_4)\left(\frac{r}{a}\right)^2 + \\ 10 C_1^2 C_2 \left(\frac{r}{a}\right)^4 + \frac{40}{3} C_1 C_2^2 \left(\frac{r}{a}\right)^6 + \frac{20}{3} C_2^3 \left(\frac{r}{a}\right)^8 \end{array} \right] \phi(r) r\, dr$$

$$Q(t) = \int_0^a p(r,t) \phi(r) r\, dr \tag{6.9.13}$$

Using the non dimensional time parameter

$$T = \sqrt{\frac{D}{\rho h a^4}}\, t \tag{6.9.14}$$

equation (6.9.12) can be written as

$$\ddot{f}(T) + 2a\dot{f} + \omega_1^2 f(T) + \gamma_1^2 f^3(T) = \frac{a^4}{\mu D} Q(T)$$

$$\cdot = \frac{d}{dT} \tag{6.9.15}$$

which is the familiar Duffing's equation for the dynamic behavior of the plate. In the above

$$2a = \frac{ca^2}{\sqrt{D\rho h}}$$

238 Dynamics of Plates

$$\omega_1^2 = \frac{ka^4}{D} + \beta^4 a^4$$

$$\gamma_1^2 = \frac{12(1-v^2)}{h^2} \frac{\psi}{\mu}$$

(6.9.16)

Let

$$p(r,T) = P\phi(r)F(T)$$

(6.9.17)

then

$$Q(T) = PF(t) \int_0^a \phi^2(r) r\, dr$$

$$= PF(t)\frac{ab}{4}$$

(6.9.18)

The plate equation can then be written as

$$\ddot{f}(T) + 2a\dot{f} + \omega_1^2 f(T) + \gamma_1^2 f^3(T) = \frac{Pa^4}{D} F(T)$$

(6.9.19)

See section 6.6 for a solution of this equation.

7

Finite Element Methods for Initially Stressed Thin Plates

We begin formulating the finite element method for initially stressed plates with in plane stresses only (membranes) and then combine the same with plate elements considered in chapter 5.

7.1 CONSTANT STRAIN TRIANGULAR ELEMENT

As in chapter 5, we follow the familiar steps in deriving the desired finite element.

Select Element Type The element is chosen of triangular shape as shown in Fig. 7.1.1. The element has obviously three nodes 1, 2 and 3. The nodal coordinate system $\hat{x}\hat{y}$ is located with its origin at the node 1 and the elemental $\hat{x}$ coordinate is chosen to lie on the nodal line 12. The global coordinate system is denoted XY as shown in the figure. The nodal coordinates for the points 1, 2 and 3 respectively are $(0,0)$, $(\hat{x}_2, 0)$ and $(\hat{x}_3, \hat{y}_3)$. For in plane displacements, we consider two degrees of freedom for each node, $\hat{u}_i, \hat{v}_i$ so that the nodal degrees of freedom are given by

$$\{\hat{d}\} = \begin{Bmatrix} \hat{u}_1 \\ \hat{v}_1 \\ \hat{u}_2 \\ \hat{v}_2 \\ \hat{u}_3 \\ \hat{v}_3 \end{Bmatrix}$$

(7.1.1)

Displacement Functions The displacements within the triangular element at any point $(\hat{x}, \hat{y})$ are represented by

240 *Dynamics of Plates*

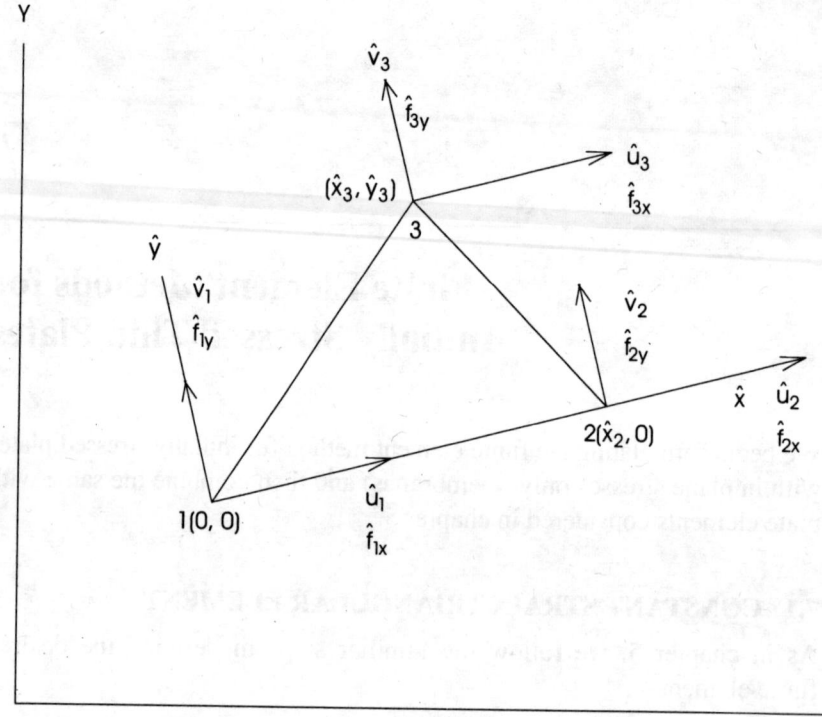

Fig. 7.1.1 A Triangular Membrane Element

$$\hat{u}(\hat{x},\hat{y}) = a_1 + a_2\hat{x} + a_3\hat{y}$$
$$\hat{v}(\hat{x},\hat{y}) = a_4 + a_5\hat{x} + a_6\hat{y}$$

(7.1.2)

which can be expressed as

$$\left\{\begin{array}{c} \hat{u} \\ \hat{v} \end{array}\right\} = \left[\begin{array}{cccccc} 1 & \hat{x} & \hat{y} & 0 & 0 & 0 \\ 0 & 0 & 0 & 1 & \hat{x} & \hat{y} \end{array}\right] \left\{\begin{array}{c} a_1 \\ a_2 \\ a_3 \\ a_4 \\ a_5 \\ a_6 \end{array}\right\}$$

(7.1.3)

where a are constants to be evaluated. The linear functions assumed for the displacements u and v ensures compatibility of displacements along the edge as well as at the nodes shared by adjacent elements. The nodal displacements are

$$\hat{u}_1 = a_1 + a_2\hat{x}_1 + a_3\hat{y}_1$$
$$\hat{u}_2 = a_1 + a_2\hat{x}_2 + a_3\hat{y}_2$$
$$\hat{u}_3 = a_1 + a_2\hat{x}_3 + a_3\hat{y}_3$$
$$\hat{v}_1 = a_4 + a_5\hat{x}_1 + a_6\hat{y}_1$$
$$\hat{v}_2 = a_4 + a_5\hat{x}_2 + a_6\hat{y}_2$$
$$\hat{v}_3 = a_4 + a_5\hat{x}_3 + a_6\hat{y}_3$$

(7.1.4)

We can express the above in equation (7.1.4) as

$$\{\hat{u}\} = \begin{bmatrix} 1 & \hat{x}_1 & \hat{y}_1 \\ 1 & \hat{x}_2 & \hat{y}_2 \\ 1 & \hat{x}_3 & \hat{y}_3 \end{bmatrix} \begin{Bmatrix} a_1 \\ a_2 \\ a_3 \end{Bmatrix}$$

$$\{\hat{v}\} = \begin{bmatrix} 1 & \hat{x}_1 & \hat{y}_1 \\ 1 & \hat{x}_2 & \hat{y}_2 \\ 1 & \hat{x}_3 & \hat{y}_3 \end{bmatrix} \begin{Bmatrix} a_4 \\ a_5 \\ a_6 \end{Bmatrix}$$

(7.1.5)

Therefore

$$\begin{Bmatrix} a_1 \\ a_2 \\ a_3 \end{Bmatrix} = \begin{bmatrix} 1 & \hat{x}_1 & \hat{y}_1 \\ 1 & \hat{x}_2 & \hat{y}_2 \\ 1 & \hat{x}_3 & \hat{y}_3 \end{bmatrix}^{-1} \{\hat{u}\}$$

$$\begin{Bmatrix} a_4 \\ a_5 \\ a_6 \end{Bmatrix} = \begin{bmatrix} 1 & \hat{x}_1 & \hat{y}_1 \\ 1 & \hat{x}_2 & \hat{y}_2 \\ 1 & \hat{x}_3 & \hat{y}_3 \end{bmatrix}^{-1} \{\hat{v}\}$$

(7.1.6)

242 Dynamics of Plates

Taking the inverse of the concerned matrices in the above equations (7.1.6), we get the arbitrary constants as

$$\begin{Bmatrix} a_1 \\ a_2 \\ a_3 \end{Bmatrix} = \frac{1}{2A} \begin{bmatrix} a_1 & a_2 & a_3 \\ \beta_1 & \beta_2 & \beta_3 \\ \gamma_1 & \gamma_2 & \gamma_3 \end{bmatrix} \{\hat{u}\}$$

$$\begin{Bmatrix} a_4 \\ a_5 \\ a_6 \end{Bmatrix} = \frac{1}{2A} \begin{bmatrix} a_1 & a_2 & a_3 \\ \beta_1 & \beta_2 & \beta_3 \\ \gamma_1 & \gamma_2 & \gamma_3 \end{bmatrix} \{\hat{v}\}$$

(7.1.7)

where A is the area of the triangle given by

$$A = \frac{1}{2} \begin{vmatrix} 1 & \hat{x}_1 & \hat{y}_1 \\ 1 & \hat{x}_2 & \hat{y}_2 \\ 1 & \hat{x}_3 & \hat{y}_3 \end{vmatrix} = \frac{1}{2}[\hat{x}_1(\hat{y}_2 - \hat{y}_3) + \hat{x}_2(\hat{y}_3 - \hat{y}_1) + \hat{x}_3(\hat{y}_1 - \hat{y}_2)]$$

(7.1.8)

and

$$a_1 = \hat{x}_2\hat{y}_3 - \hat{y}_2\hat{x}_3, \quad a_2 = \hat{x}_3\hat{y}_1 - \hat{y}_3\hat{x}_1 \quad \text{and} \quad a_3 = \hat{x}_1\hat{y}_2 - \hat{y}_1\hat{x}_2$$

$$\beta_1 = \hat{y}_2 - \hat{y}_3, \quad \beta_2 = \hat{y}_3 - \hat{y}_1 \quad \text{and} \quad \beta_3 = \hat{y}_1 - \hat{y}_2$$

$$\gamma_1 = \hat{x}_3 - \hat{x}_2, \quad \gamma_2 = \hat{x}_1 - \hat{x}_3 \quad \text{and} \quad \gamma_3 = \hat{x}_2 - \hat{x}_1$$

(7.1.9)

Substituting the first equation of (7.1.7) in the first equation of (7.1.3)

$$u(\hat{x}, \hat{y}) = \frac{1}{2A}\begin{bmatrix} 1 & \hat{x} & \hat{y} \end{bmatrix} \begin{bmatrix} a_1 & a_2 & a_3 \\ \beta_1 & \beta_2 & \beta_3 \\ \gamma_1 & \gamma_2 & \gamma_3 \end{bmatrix} \begin{Bmatrix} \hat{u}_1 \\ \hat{u}_2 \\ \hat{u}_3 \end{Bmatrix}$$

$$= \frac{1}{2A}\begin{bmatrix} 1 & \hat{x} & \hat{y} \end{bmatrix} \begin{Bmatrix} a_1\hat{u}_1 + a_2\hat{u}_2 + a_3\hat{u}_3 \\ \beta_1\hat{u}_1 + \beta_2\hat{u}_2 + \beta_3\hat{u}_3 \\ \gamma_1\hat{u}_1 + \gamma_2\hat{u}_2 + \gamma_3\hat{u}_3 \end{Bmatrix}$$

(7.1.10)

Expanding the above

$$u(\hat{x},\hat{y}) = \frac{1}{2A}\left[\begin{array}{c}(a_1+\beta_1\hat{x}+\gamma_1\hat{y})\hat{u}_1 + \\ (a_2+\beta_2\hat{x}+\gamma_2\hat{y})\hat{u}_2 + (a_3+\beta_3\hat{x}+\gamma_3\hat{y})\hat{u}_3\end{array}\right]$$

(7.1.11)

Similarly,

$$v(\hat{x},\hat{y}) = \frac{1}{2A}\left[\begin{array}{c}(a_1+\beta_1\hat{x}+\gamma_1\hat{y})\hat{v}_1 + (a_2+\beta_2\hat{x}+\gamma_2\hat{y})\hat{v}_2 \\ +(a_3+\beta_3\hat{x}+\gamma_3\hat{y})\hat{v}_3\end{array}\right]$$

(7.1.12)

The above equations (7.1.11) and (7.1.12) are now expressed as below to define the shape functions and the displacement function relation in terms of the nodal displacements.

$$u(\hat{x},\hat{y}) = [N_1\hat{u}_1 + N_2\hat{u}_2 + N_3\hat{u}_3]$$
$$v(\hat{x},\hat{y}) = [N_1\hat{v}_1 + N_2\hat{v}_2 + N_3\hat{v}_3]$$

(7.1.13)

where

$$N_1 = \frac{1}{2A}(a_1+\beta_1\hat{x}+\gamma_1\hat{y})$$
$$N_2 = \frac{1}{2A}(a_2+\beta_2\hat{x}+\gamma_2\hat{y})$$
$$N_3 = \frac{1}{2A}(a_3+\beta_3\hat{x}+\gamma_3\hat{y})$$

(7.1.14)

Finally, the displacement functions are expressed in matrix form as

$$\left\{\begin{array}{c}u(\hat{x},\hat{y})\\v(\hat{x},\hat{y})\end{array}\right\} = \left\{\begin{array}{c}N_1\hat{u}_1+N_2\hat{u}_2+N_3\hat{u}_3\\N_1\hat{v}_1+N_2\hat{v}_2+N_3\hat{v}_3\end{array}\right\}$$

$$= \begin{bmatrix} N_1 & 0 & N_2 & 0 & N_3 & 0 \\ 0 & N_1 & 0 & N_2 & 0 & N_3 \end{bmatrix} \begin{Bmatrix} \hat{u}_1 \\ \hat{v}_1 \\ \hat{u}_2 \\ \hat{v}_2 \\ \hat{u}_3 \\ \hat{v}_3 \end{Bmatrix}$$

$$= [N]\{\hat{d}\} \tag{7.1.15}$$

where

$$[N] = \begin{bmatrix} N_1 & 0 & N_2 & 0 & N_3 & 0 \\ 0 & N_1 & 0 & N_2 & 0 & N_3 \end{bmatrix} \tag{7.1.16}$$

Strain Displacement Relations The strain displacement relations are

$$\{\hat{\varepsilon}\} = \begin{Bmatrix} \varepsilon_{xx} \\ \varepsilon_{yy} \\ \gamma_{xy} \end{Bmatrix}$$

$$= \begin{Bmatrix} u_{,x} \\ v_{,y} \\ u_{,y} + v_{,x} \end{Bmatrix} \tag{7.1.17}$$

With the help of equation (7.1.15), the above equation (7.1.17) can be written as

$$\begin{Bmatrix} \varepsilon_{xx} \\ \varepsilon_{yy} \\ \gamma_{xy} \end{Bmatrix} = \begin{Bmatrix} N_{1,\hat{x}}\hat{u}_1 + N_{2,\hat{x}}\hat{u}_2 + N_{3,\hat{x}}\hat{u}_3 \\ N_{1,\hat{y}}\hat{v}_1 + N_{2,\hat{y}}\hat{v}_2 + N_{3,\hat{y}}\hat{v}_3 \\ \begin{bmatrix} (N_{1,\hat{y}}\hat{u}_1 + N_{2,\hat{y}}\hat{u}_2 + N_{3,\hat{y}}\hat{u}_3) \\ +(N_{1,\hat{x}}\hat{v}_1 + N_{2,\hat{x}}\hat{v}_2 + N_{3,\hat{x}}\hat{v}_3) \end{bmatrix} \end{Bmatrix} \tag{7.1.18}$$

Now, making use of (7.1.14), we can write the above as the following strain-displacement relations.

$$\left\{ \begin{array}{c} \varepsilon_{xx} \\ \varepsilon_{yy} \\ \gamma_{xy} \end{array} \right\} = \frac{1}{2A} \left\{ \begin{array}{c} \beta_1 \hat{u}_1 + \beta_2 \hat{u}_2 + \beta_3 \hat{u}_3 \\ \gamma_1 \hat{v}_1 + \gamma_2 \hat{v}_2 + \gamma_3 \hat{v}_3 \\ \left[\begin{array}{c} (\gamma_1 \hat{u}_1 + \gamma_2 \hat{u}_2 + \gamma_3 \hat{u}_3) \\ +(\beta_1 \hat{v}_1 + \beta_2 \hat{v}_2 + \beta_3 \hat{v}_3) \end{array} \right] \end{array} \right\}$$

$$= \frac{1}{2A} \begin{bmatrix} \beta_1 & 0 & \beta_2 & 0 & \beta_3 & 0 \\ 0 & \gamma_1 & 0 & \gamma_2 & 0 & \gamma_3 \\ \gamma_1 & \beta_1 & \gamma_2 & \beta_2 & \gamma_3 & \beta_3 \end{bmatrix} \left\{ \begin{array}{c} \hat{u}_1 \\ \hat{v}_1 \\ \hat{u}_2 \\ \hat{v}_2 \\ \hat{u}_3 \\ \hat{v}_3 \end{array} \right\}$$

(7.1.19)

The above equation is written in a simple form as

$$\{\hat{\varepsilon}\} = [B]\{\hat{d}\}$$

(7.1.20)

where

$$[B] = \frac{1}{2A} \begin{bmatrix} \beta_1 & 0 & \beta_2 & 0 & \beta_3 & 0 \\ 0 & \gamma_1 & 0 & \gamma_2 & 0 & \gamma_3 \\ \gamma_1 & \beta_1 & \gamma_2 & \beta_2 & \gamma_3 & \beta_3 \end{bmatrix}$$

(7.1.21)

Referring to equation (7.1.9), we find that β and γ are independent of the elemental coordinates $\hat{x}, \hat{y}$. They are solely dependent on the nodal coordinates. Therefore, the strains in equation (7.1.20) are constant throughout the element. Hence the name *Constant Strain Triangle* (CST) for this element.

Stress Strain Relations The stress strain relations (plane stress) for the element are given by

$$\tau_{xx} = \frac{E}{(1-v^2)}(\varepsilon_{xx} + v\varepsilon_{yy})$$

$$\tau_{yy} = \frac{E}{(1-v^2)}(\varepsilon_{yy} + v\varepsilon_{xx})$$

$$\tau_{xy} = G\gamma_{xy} = \frac{E(1-v)}{2(1-v^2)}\gamma_{xy}$$

(3.2.18)

which are expressed as

$$\{\hat{\tau}\} = \frac{E}{1-v^2} \begin{bmatrix} 1 & v & 0 \\ v & 1 & 0 \\ 0 & 0 & \frac{1-v}{2} \end{bmatrix} \{\hat{\varepsilon}\}$$

$$= [D]\{\hat{\varepsilon}\} \qquad (5.1.15)$$

where

$$[D] = \frac{E}{1-v^2} \begin{bmatrix} 1 & v & 0 \\ v & 1 & 0 \\ 0 & 0 & \frac{1-v}{2} \end{bmatrix} \qquad (5.1.16)$$

Potential Functional The strain energy of the membrane is

$$\hat{U} = \frac{1}{2} \iiint_V \{\hat{\varepsilon}\}^T [\hat{\tau}] dV$$

$$= \frac{1}{2} \iiint_V \{\hat{\varepsilon}\}^T [D]\{\hat{\varepsilon}\} dV$$

$$= \frac{1}{2} \iiint_V \{\hat{a}\}^T [B]^T [D][B]\{\hat{a}\} dV \qquad (7.1.22)$$

Let $\{\hat{P}\}$ be the concentrated external nodal forces, then the potential energy of the concentrated forces is

$$V_P = -\{\hat{a}\}^T \{\hat{P}\} \qquad (7.1.23)$$

Let $\{T\}$ represent the surface tractions acting over an area A, then the potential energy of this distributed force is

$$V_T = -\iint_A \{ u \ v \} \{T\} dA$$

$$= -\iint_A \{\hat{a}\}^T [N]^T \{T\} dA \qquad (7.1.24)$$

The potential functional for the membrane is then written as

$$\Pi = \frac{1}{2}\{\hat{a}\}^T\left[\iiint_V [B]^T[D][B]dV\right]\{\hat{a}\} - \{\hat{a}\}^T\{\hat{P}\} - \iint_A \{\hat{a}\}^T[N]^T\{T\}dA$$

$$= \frac{1}{2}\{\hat{a}\}^T\left[\iiint_V [B]^T[D][B]dV\right]\{\hat{a}\} - \{\hat{a}\}^T\{\hat{f}\}$$

(7.1.25)

where

$$\{\hat{f}\} = \{\hat{P}\} + \iint_A [N]^T\{T\}dA$$

(7.1.25a)

Elemental Equation We now apply Ritz method to the potential functional in (7.1.25)

$$\frac{\partial \Pi}{\partial \{q\}^T} = 0$$

which gives

$$\left[\iiint_V [B]^T[D][B]dV\right]\{\hat{a}\} = \{\hat{f}\}$$

$$[\hat{K}]\{\hat{a}\} = \{\hat{f}\}$$

(7.1.26)

where $[\hat{K}]$ is the elemental stiffness matrix given by

$$[\hat{K}] = \left[\iiint_V [B]^T[D][B]dV\right]$$

(7.1.27)

Since $[B]$ matrix is independent of the elemental coordinates, the above stiffness matrix can be simplified as

$$[\hat{K}] = hA[B]^T[D][B]$$

(7.1.28)

where h is the membrane thickness and A is the area of the element given in equation (7.1.8).

Expressing $[B]$ matrix in equation (7.1.21) as

$$[B] = \begin{bmatrix} [B_1] & [B_2] & [B_3] \end{bmatrix} \quad (7.1.29)$$

where

$$[B_1] = \frac{1}{2A} \begin{bmatrix} \beta_1 & 0 \\ 0 & \gamma_1 \\ \gamma_1 & \beta_1 \end{bmatrix}$$

$$[B_2] = \frac{1}{2A} \begin{bmatrix} \beta_2 & 0 \\ 0 & \gamma_2 \\ \gamma_2 & \beta_2 \end{bmatrix}$$

$$[B_3] = \frac{1}{2A} \begin{bmatrix} \beta_3 & 0 \\ 0 & \gamma_3 \\ \gamma_3 & \beta_3 \end{bmatrix}$$

$$(7.1.30)$$

the elemental stiffness matrix can be rewritten as

$$[\hat{K}] = \begin{bmatrix} [K_{11}] & [K_{12}] & [K_{13}] \\ [K_{21}] & [K_{22}] & [K_{23}] \\ [K_{31}] & [K_{32}] & [K_{33}] \end{bmatrix} \quad (7.1.31)$$

where $[K_{ij}]$ in the above equation are 2×2 sub matrices given by

$$[K_{ij}] = hA[B_i]^T[D][B_j] \quad (7.1.32)$$

Equation (7.1.26) can be rewritten as

$$[\hat{K}]\{\hat{a}\} = \{\hat{f}\}$$

$$\begin{bmatrix} [K_{11}] & [K_{12}] & [K_{13}] \\ [K_{21}] & [K_{22}] & [K_{23}] \\ [K_{31}] & [K_{32}] & [K_{33}] \end{bmatrix} \begin{Bmatrix} \{\hat{a}_1\} \\ \{\hat{a}_2\} \\ \{\hat{a}_3\} \end{Bmatrix} = \begin{Bmatrix} \{\hat{f}_1\} \\ \{\hat{f}_2\} \\ \{\hat{f}_3\} \end{Bmatrix}$$

$$(7.1.33)$$

where

$$\{\hat{d}_1\} = \begin{Bmatrix} \hat{u}_1 \\ \hat{v}_1 \end{Bmatrix}$$

$$\{\hat{f}_1\} = \begin{Bmatrix} \hat{f}_{x1} \\ \hat{f}_{y1} \end{Bmatrix}$$

...

(7.1.34)

The stiffness matrix in (7.1.33) can be expanded directly to give it in an explicit form, instead of involving matrix operations. This expression is

$$[\hat{K}] = [\widehat{K_n}] + [\widehat{K_s}]$$

(7.1.35)

where the stiffness matrix due to normal stresses is

$$[\widehat{K_n}] = \frac{hE}{4A(1-v^2)} \begin{bmatrix} \beta_1^2 & v\beta_1\gamma_1 & \beta_1\beta_2 & v\beta_1\gamma_2 & \beta_1\beta_3 & v\beta_1\gamma_3 \\ & \gamma_1^2 & v\beta_2\gamma_1 & \gamma_1\gamma_2 & v\beta_3\gamma_1 & \gamma_1\gamma_3 \\ & & \beta_2^2 & v\beta_2\gamma_2 & \beta_2\beta_3 & v\beta_2\gamma_3 \\ & & & \gamma_2^2 & v\beta_3\gamma_2 & \gamma_2\gamma_3 \\ & \text{sym} & & & \beta_3^2 & v\beta_3\gamma_3 \\ & & & & & \gamma_3^2 \end{bmatrix}$$

(7.1.36)

and the stiffness matrix due to shear stresses is

$$[\widehat{K_s}] = \frac{hE}{8A(1+v)} \begin{bmatrix} \gamma_1^2 & \beta_1\gamma_1 & \gamma_1\gamma_2 & \beta_2\gamma_1 & \gamma_1\gamma_3 & \beta_3\gamma_1 \\ & \beta_1^2 & \beta_1\gamma_2 & \beta_1\beta_2 & \beta_1\gamma_3 & \beta_1\beta_3 \\ & & \gamma_2^2 & \beta_2\gamma_2 & \gamma_2\gamma_3 & \beta_3\gamma_2 \\ & & & \beta_2^2 & \beta_2\gamma_3 & \beta_2\beta_3 \\ & \text{sym} & & & \gamma_3^2 & \beta_3\gamma_3 \\ & & & & & \beta_3^2 \end{bmatrix}$$

(7.1.37)

Global Equation The stiffness matrix and the forcing vector are now to be transformed into global system of Fig. 7.1.1. The transformation matrix can be written as

$$\{\hat{d}\} = [R]\{d\} \tag{7.1.38}$$

where

$$[R] = \begin{bmatrix} \cos(x,X) & \cos(x,Y) & 0 & 0 & 0 & 0 \\ \cos(y,X) & \cos(y,Y) & 0 & 0 & 0 & 0 \\ 0 & 0 & \cos(x,X) & \cos(x,Y) & 0 & 0 \\ 0 & 0 & \cos(y,X) & \cos(y,Y) & 0 & 0 \\ 0 & 0 & 0 & 0 & \cos(x,X) & \cos(x,Y) \\ 0 & 0 & 0 & 0 & \cos(y,X) & \cos(y,Y) \end{bmatrix}$$

$$\tag{7.1.39}$$

The directional cosines are determined from (5.1.32)

$$\cos(x,X) = \frac{X_2 - X_1}{\sqrt{(X_2 - X_1)^2 + (Y_2 - Y_1)^2}}$$

$$\cos(x,Y) = \frac{Y_2 - Y_1}{\sqrt{(X_2 - X_1)^2 + (Y_2 - Y_1)^2}}$$

$$\cos(y,X) = -\frac{Y_2 - Y_1}{\sqrt{(X_2 - X_1)^2 + (Y_2 - Y_1)^2}}$$

$$\cos(y,Y) = \frac{X_2 - X_1}{\sqrt{(X_2 - X_1)^2 + (Y_2 - Y_1)^2}}$$

$$\tag{5.1.32}$$

Hence, the stiffness matrix and forcing vector in global coordinates are given by

$$[K] = [R]^T [\hat{K}][R] \tag{7.1.40}$$

$$\{f\} = [R]^T \{\hat{f}\} \tag{7.1.41}$$

We now have the elemental equation in global coordinates

Finite Element Methods for Initially Stressed Thin Plates

$$[K]\{d\} = \{f\}$$

(7.1.42)

Assembly of Elemental Stiffness Matrices in Global Coordinates

In accordance to the direct stiffness method, the stiffness matrix and forcing vector for each element in global coordinates are assembled by using superposition principle. Then, the boundary conditions are applied for the displacement and the corresponding slope on the respective outer edges. The resulting equations give rise to the system equations

$$[K]_G\{d\}_G = \{f\}_G$$

(7.1.43)

These equations are solved to determine the nodal displacements.

Consistent Mass Matrix

The kinetic energy of the element is

$$T = \frac{1}{2} \iint_A \rho h (\dot{u}^2 + \dot{v}^2) dA$$

(7.1.44)

Using (7.1.15), we can write the above expression as

$$T = \frac{1}{2}\{\dot{d}\}^T \left[\iint_A \rho h [N]^T [N] dA \right] \{\dot{d}\}$$

(7.1.45)

Using Lagrangian approach, we obtain the elemental mass matrix as

$$[\widehat{M}] = \iint_A \rho h [N]^T [N] dA$$

(7.1.46)

Substituting equations (7.1.16) in the above equation (7.1.46), we get

$$[\widehat{M}] = \rho h \iint_A \begin{bmatrix} N_1 & \\ & N_1 \\ N_2 & \\ & N_2 \\ N_3 & \\ & N_3 \end{bmatrix} \begin{bmatrix} N_1 & 0 & N_2 & 0 & N_3 & 0 \\ 0 & N_1 & 0 & N_2 & 0 & N_3 \end{bmatrix} dA$$

$$= \rho h \iint_A \begin{bmatrix} N_1^2 & 0 & N_1N_2 & 0 & N_1N_3 & 0 \\ 0 & N_1^2 & 0 & N_1N_2 & 0 & N_1N_3 \\ N_1N_2 & 0 & N_2^2 & 0 & N_2N_3 & 0 \\ 0 & N_1N_2 & 0 & N_2^2 & 0 & N_2N_3 \\ N_1N_3 & 0 & N_2N_3 & 0 & N_3^2 & 0 \\ 0 & N_1N_3 & 0 & N_2N_3 & 0 & N_3^2 \end{bmatrix} dA$$

(7.1.47)

Making use of area coordinates, it can be shown that

$$\iint_A N_1^m N_2^n N_3^p \, dA = \frac{m!n!p!}{(m+n+p+2)!} 2A$$

(7.1.48)

With the help of the above equation (7.1.48), we can evaluate the integrals in (7.1.47) to give

$$[\widehat{M}] = \frac{1}{12}\rho h A \begin{bmatrix} 2 & 0 & 1 & 0 & 1 & 0 \\ 0 & 2 & 0 & 1 & 0 & 1 \\ 1 & 0 & 2 & 0 & 1 & 0 \\ 0 & 1 & 0 & 2 & 0 & 1 \\ 1 & 0 & 1 & 0 & 2 & 0 \\ 0 & 1 & 0 & 1 & 0 & 2 \end{bmatrix}$$

(7.1.49)

The elemental mass matrices are first transformed to global coordinates by using equation (7.1.38) and (7.1.39) to give

$$[M] = [R]^T [\widehat{M}][R]$$

(7.1.50)

System Equation for Vibration

For in-plane vibration of a membrane, the system equation of motion in global coordinates can be now written as

$$[M]_G \{\ddot{d}\}_G + [K]_G \{d\}_G = \{f(t)\}_G$$

(7.1.51)

For free vibrations, the equation of motion becomes

$$[M]_G\{\ddot{d}\}_G + [K]_G\{d\}_G = 0 \qquad (7.1.52)$$

Assuming a harmonic solution

$$\{d\}_G = \{\bar{d}\}_G \sin pt \qquad (7.1.53)$$

the eigen value problem is

$$[[K]_G - p^2[M]_G]\{\bar{d}\}_G = 0 \qquad (7.1.54)$$

The following simple example illustrates the application of static and dynamic problem discussed in this section.

Example
Consider an axially loaded plate (membrane) shown in Fig. 7.1.2. The plate is divided for simplicity into two elements. Young's modulus of the material is 2×10^7 N/cm^2 and the Poisson's ratio is 0.3.

Element 1
The element 1 in its own coordinates is shown in Fig. 7.1.3. The nodal coordinates of the element are:

$$\hat{x}_1 = \hat{y}_1 = 0$$
$$\hat{x}_2 = 20, \hat{y}_2 = 0$$
$$\hat{x}_3 = 20, \hat{y}_3 = 10$$

The area of the element is $A = 100$ cm^2. From equation (7.1.9)

$$a_1 = 200, \quad a_2 = 0 \quad \text{and} \quad a_3 = 0$$
$$\beta_1 = -10, \quad \beta_2 = 10 \quad \text{and} \quad \beta_3 = 0$$
$$\gamma_1 = 0, \quad \gamma_2 = -20 \quad \text{and} \quad \gamma_3 = 20$$

254 Dynamics of Plates

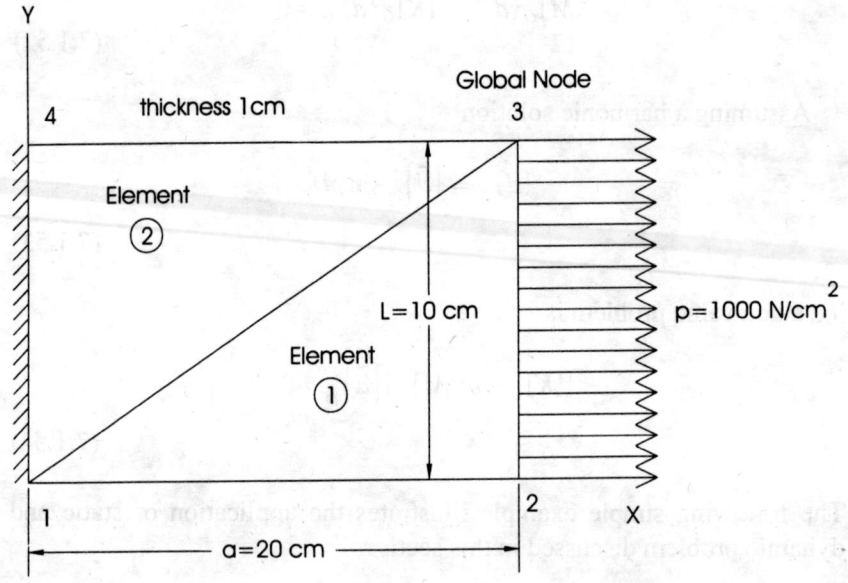

Fig. 7.1.2 Membrane under In Plane Force

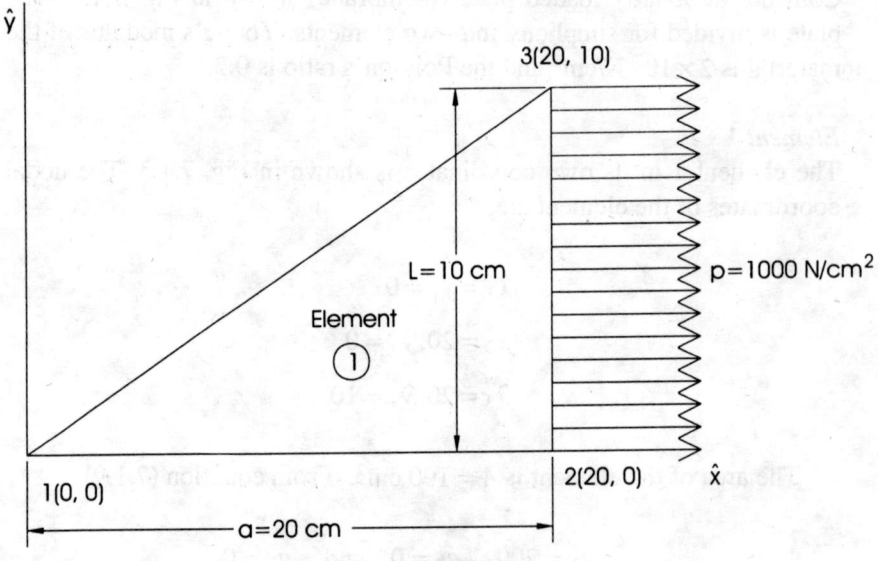

Fig. 7.1.3 Element 1

From (7.1.36)

$$[\widehat{K_n}] = 5.494505 \times 10^6 \begin{bmatrix} 1 & 0 & -1 & 0.6 & 0 & -0.6 \\ & 0 & 0 & 0 & 0 & 0 \\ & & 1 & -0.6 & 0 & 0.6 \\ & & & 4 & 0 & -4 \\ & \text{sym} & & & 0 & 0 \\ & & & & & 4 \end{bmatrix} \text{ N/cm}$$

From (7.1.37)

$$[\widehat{K_s}] = 1.923077 \times 10^6 \begin{bmatrix} 0 & 0 & 0 & 0 & 0 & 0 \\ & 1 & 2 & -1 & -2 & 0 \\ & & 4 & -2 & -4 & 0 \\ & & & 1 & 2 & 0 \\ & \text{sym} & & & 4 & 0 \\ & & & & & 0 \end{bmatrix} \text{ N/cm}$$

Stiffness matrix

$$[\widehat{K}] = 5.494505 \times 10^6 \begin{bmatrix} 1 & 0 & -1 & 0.6 & 0 & -0.6 \\ & 0.35 & 0.7 & -0.35 & -0.7 & 0 \\ & & 2.4 & -1.3 & -1.4 & 0.6 \\ & & & 4.35 & 0.7 & -4 \\ & \text{sym} & & & 1.4 & 0 \\ & & & & & 4 \end{bmatrix} \text{ N/cm}$$

Directional Cosines

$$\cos(x, X) = 1$$
$$\cos(x, Y) = 0$$
$$\cos(y, X) = 0$$
$$\cos(y, Y) = 1$$

Dynamics of Plates

Transformation Matrix

$$[R] = \begin{bmatrix} 1 & 0 & 0 & 0 & 0 & 0 \\ 0 & 1 & 0 & 0 & 0 & 0 \\ 0 & 0 & 1 & 0 & 0 & 0 \\ 0 & 0 & 0 & 1 & 0 & 0 \\ 0 & 0 & 0 & 0 & 1 & 0 \\ 0 & 0 & 0 & 0 & 0 & 1 \end{bmatrix}$$

Stiffness Matrix in Global Coordinates

$$\left[\widehat{K}_G^{(1)}\right] = 5.494505 \times 10^6 \begin{bmatrix} 1 & 0 & -1 & 0.6 & 0 & -0.6 \\ & 0.35 & 0.7 & -0.35 & -0.7 & 0 \\ & & 2.4 & -1.3 & -1.4 & 0.6 \\ & & & 4.35 & 0.7 & -4 \\ & \text{sym} & & & 1.4 & 0 \\ & & & & & 4 \end{bmatrix} \text{ N/cm}$$

Element 2

The element 2 in its own coordinates is shown in Fig. 7.1.4. The nodal coordinates of the element are:

$$\hat{x}_1 = \hat{y}_1 = 0$$
$$\hat{x}_2 = 10, \hat{y}_2 = 0$$
$$\hat{x}_3 = 0, \hat{y}_3 = 20$$

The area of the element is $A = 100$ cm². From equation (7.1.9)

$$a_1 = 200, \quad a_2 = 0 \quad \text{and} \quad a_3 = 0$$
$$\beta_1 = -20, \quad \beta_2 = 20 \quad \text{and} \quad \beta_3 = 0$$
$$\gamma_1 = -10, \quad \gamma_2 = 0 \quad \text{and} \quad \gamma_3 = 10$$

From (7.1.36)

$$\left[\widehat{K}_n\right] = 5.494505 \times 10^6 \begin{bmatrix} 4 & 0.6 & -4 & 0 & 0 & -0.6 \\ & 1 & -0.6 & 0 & 0 & -1 \\ & & 4 & 0 & 0 & 0.6 \\ & & & 0 & 0 & 0 \\ & \text{sym} & & & 0 & 0 \\ & & & & & 1 \end{bmatrix} \text{ N/cm}$$

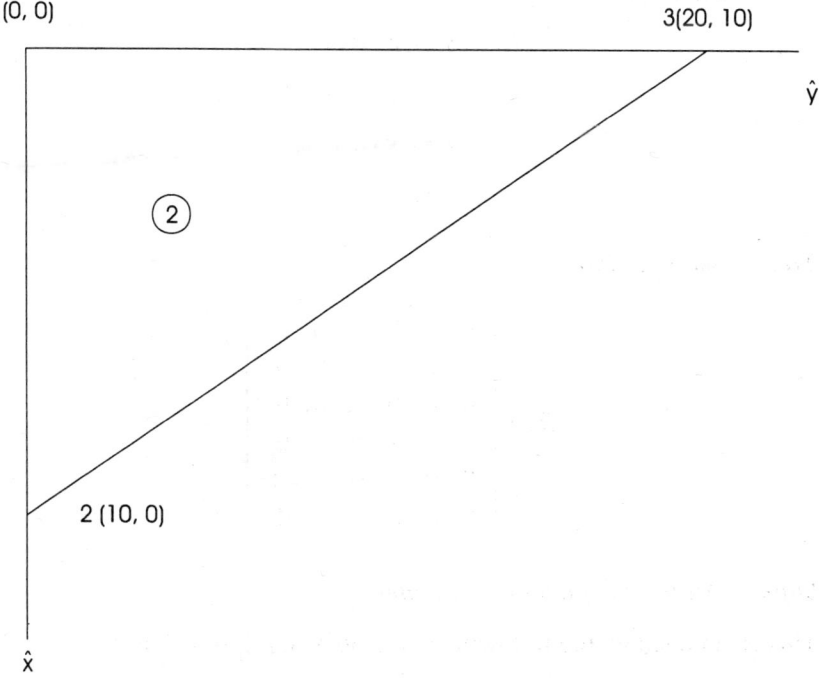

Fig. 7.1.4 Element 2

From (7.1.37)

$$[\widehat{K_s}] = 1.923077 \times 10^6 \begin{bmatrix} 1 & 2 & 0 & -2 & -1 & 0 \\ & 4 & 0 & -4 & -2 & 0 \\ & & 0 & 0 & 0 & 0 \\ & & & 4 & 2 & 0 \\ & \text{sym} & & & 1 & 0 \\ & & & & & 0 \end{bmatrix} \text{N/cm}$$

Stiffness Matrix

$$[\widehat{K}] = 5.494505 \times 10^6 \begin{bmatrix} 4.35 & 1.3 & -4 & -0.7 & -0.35 & -0.6 \\ & 2.4 & -0.6 & -1.4 & -0.7 & -1 \\ & & 4 & 0 & 0 & 0.6 \\ & & & 1.4 & 0.7 & 0 \\ & \text{sym} & & & 0.35 & 0 \\ & & & & & 1 \end{bmatrix} \text{N/cm}$$

258 Dynamics of Plates

Directional Cosines

$$\cos(x, X) = 0$$
$$\cos(x, Y) = -1$$
$$\cos(y, X) = 1$$
$$\cos(y, Y) = 0$$

Transformation Matrix

$$[R] = \begin{bmatrix} 0 & -1 & 0 & 0 & 0 & 0 \\ 1 & 0 & 0 & 0 & 0 & 0 \\ 0 & 0 & 0 & -1 & 0 & 0 \\ 0 & 0 & 1 & 0 & 0 & 0 \\ 0 & 0 & 0 & 0 & 0 & -1 \\ 0 & 0 & 0 & 0 & 1 & 0 \end{bmatrix}$$

Stiffness Matrix in Global Coordinates

This is obtained by the following two matrix multiplications:

$$[\hat{K}^{(2)}][R] = 5.494505 \times 10^6 \begin{bmatrix} 4.35 & 1.3 & -4 & -0.7 & -0.35 & -0.6 \\ & 2.4 & -0.6 & -1.4 & -0.7 & -1 \\ & & 4 & 0 & 0 & 0.6 \\ & & & 1.4 & 0.7 & 0 \\ & \text{sym} & & & 0.35 & 0 \\ & & & & & 1 \end{bmatrix}$$

$$\times \begin{bmatrix} 0 & -1 & 0 & 0 & 0 & 0 \\ 1 & 0 & 0 & 0 & 0 & 0 \\ 0 & 0 & 0 & -1 & 0 & 0 \\ 0 & 0 & 1 & 0 & 0 & 0 \\ 0 & 0 & 0 & 0 & 0 & -1 \\ 0 & 0 & 0 & 0 & 1 & 0 \end{bmatrix}$$

$$= 5.494505 \times 10^6 \begin{bmatrix} 1.3 & -4.35 & -0.7 & 4 & -0.6 & 0.35 \\ 2.4 & -1.3 & -1.4 & 0.6 & -1 & 0.7 \\ -0.6 & 4 & 0 & -4 & 0.6 & 0 \\ -1.4 & 0.7 & 1.4 & 0 & 0 & -0.7 \\ -0.7 & 0.35 & 0.7 & 0 & 0 & -0.35 \\ -1 & 0.6 & 0 & -0.6 & 1 & 0 \end{bmatrix}$$

$$[\overline{K_G}^{(2)}] = 5.494505 \times 10^6 \begin{bmatrix} 0 & 1 & 0 & 0 & 0 & 0 \\ -1 & 0 & 0 & 0 & 0 & 0 \\ 0 & 0 & 0 & 1 & 0 & 0 \\ 0 & 0 & -1 & 0 & 0 & 0 \\ 0 & 0 & 0 & 0 & 0 & 1 \\ 0 & 0 & 0 & 0 & -1 & 0 \end{bmatrix}$$

$$\times \begin{bmatrix} 1.3 & -4.35 & -0.7 & 4 & -0.6 & 0.35 \\ 2.4 & -1.3 & -1.4 & 0.6 & -1 & 0.7 \\ -0.6 & 4 & 0 & -4 & 0.6 & 0 \\ -1.4 & 0.7 & 1.4 & 0 & 0 & -0.7 \\ -0.7 & 0.35 & 0.7 & 0 & 0 & -0.35 \\ -1 & 0.6 & 0 & -0.6 & 1 & 0 \end{bmatrix}$$

$$= 5.494505 \times 10^6 \begin{bmatrix} 2.4 & -1.3 & -1.4 & 0.6 & -1 & 0.7 \\ & 4.35 & 0.7 & -4 & 0.6 & -0.35 \\ & & 1.4 & 0 & 0 & -0.7 \\ & & & 4 & -0.6 & 0 \\ & & & & 1 & 0 \\ & & & & & 0.35 \end{bmatrix} \text{N/cm}$$

System Global Stiffness Matrix

5.494505×10^6

$$\times \begin{bmatrix} u_1 & v_1 & u_2 & v_2 & u_3 & v_3 & u_4 & v_4 \\ (2.4) & (0) & -1 & 0.6 & (0) & (-1.3) & -1.4 & 0.7 \\ (0) & (4.35) & 0.7 & -0.35 & (-1.3) & (0) & 0.6 & -4 \\ -1 & 0.7 & 2.4 & -1.3 & -1.4 & 0.6 & & \\ 0.6 & -0.35 & -1.3 & 4.35 & 0.7 & -4 & & \\ (0) & (-1.3) & -1.4 & 0.7 & (2.4) & (0) & -1 & 0.6 \\ (-1.3) & 0 & 0.6 & -4 & (0) & (4.35) & 0.7 & -0.35 \\ -1.4 & 0.6 & & & -1 & 0.7 & 2.4 & -1.3 \\ 0.7 & -4 & & & 0.6 & -0.35 & -1.3 & 4.35 \end{bmatrix} \begin{matrix} u_1 \\ v_1 \\ u_2 \\ v_2 \\ u_3 \\ v_3 \\ u_4 \\ v_4 \end{matrix} \text{N/cm}$$

System Nodal Force Vector

This vector can be directly written down by splitting the edge force equally at the two nodes 2 and 3 in the global coordinates to give

260 Dynamics of Plates

$$\{f_G\} = \begin{Bmatrix} 0 \\ 0 \\ 5000 \\ 0 \\ 5000 \\ 0 \\ 0 \\ 0 \end{Bmatrix}$$

Alternatively, we can use equation (7.1.25a) for the distributed traction force on element 1 as

$$\{\hat{f}_T\} = \iint_A [N]^T \{T\} dA$$

Referring to the element 1 in Fig. 7.1.3, the traction force vector $\{T\} = \begin{Bmatrix} p \\ 0 \end{Bmatrix}$ where $p = 1000$ N/cm^2 and making use of (7.1.16), we get

$$\{\hat{f}_T\} = \iint_A \begin{bmatrix} N_1 & 0 \\ 0 & N_1 \\ N_2 & 0 \\ 0 & N_2 \\ N_3 & 0 \\ 0 & N_3 \end{bmatrix} \begin{Bmatrix} p \\ 0 \end{Bmatrix} dA$$

Let the length of the edge 1-2 be a which is equal to 20 cm and that of the edge on which the force is acting 2-3 be L which in the present case is 10 cm. The thickness h is 1 cm. The integration in the above is evaluated on the edge 2-3, which is $x = a$, from 0 to L and through the thickness h. Therefore,

$$\{\hat{f}_T\} = \int_0^h \int_0^L \begin{bmatrix} N_1 & 0 \\ 0 & N_1 \\ N_2 & 0 \\ 0 & N_2 \\ N_3 & 0 \\ 0 & N_3 \end{bmatrix} \begin{Bmatrix} p \\ 0 \end{Bmatrix} dz\, dy$$

$$= h \int_0^L \begin{bmatrix} N_1 p \\ 0 \\ N_2 p \\ 0 \\ N_3 p \\ 0 \end{bmatrix} dy$$

From (7.1.9), we have

$$a_1 = aL \quad a_2 = 0 \quad a_3 = 0$$
$$\beta_1 = -L \quad \beta_2 = L \quad \beta_3 = 0$$
$$\gamma_1 = 0 \quad \gamma_2 = -a \quad \gamma_3 = a$$

Therefore, from (7.1.14), we have

$$N_1 = \frac{L}{2A}(a - \hat{x})$$
$$N_2 = \frac{1}{2A}(L\hat{x} - a\hat{y})$$
$$N_3 = \frac{1}{2A}(a\hat{y})$$

Substituting the above in the nodal force vector and integrating with $\hat{x} = a$, we get

$$\{\hat{f}_T\} = \frac{h}{2A} \int_0^L \begin{Bmatrix} L(a-\hat{x})p \\ 0 \\ (L\hat{x} - a\hat{y})p \\ 0 \\ (a\hat{y})p \\ 0 \end{Bmatrix} dy$$

$$= \frac{h}{2A} \begin{Bmatrix} 0 \\ 0 \\ \frac{1}{2}L^2 ap \\ 0 \\ \frac{1}{2}L^2 ap \\ 0 \end{Bmatrix} = \begin{Bmatrix} 0 \\ 0 \\ \frac{1}{2}Lhp \\ 0 \\ \frac{1}{2}Lhp \\ 0 \end{Bmatrix}$$

262 Dynamics of Plates

The above is same as the nodal force vector written directly earlier.

System Equation
The system equation can now be written

$$5.494505 \times 10^6 \begin{bmatrix} 2.4 & 0 & -1 & 0.6 & 0 & -1.3 & -1.4 & 0.7 \\ 0 & 4.35 & 0.7 & -0.35 & -1.3 & 0 & 0.6 & -4 \\ -1 & 0.7 & 2.4 & -1.3 & -1.4 & 0.6 & & \\ 0.6 & -0.35 & -1.3 & 4.35 & 0.7 & -4 & & \\ 0 & -1.3 & -1.4 & 0.7 & 2.4 & 0 & -1 & 0.6 \\ -1.3 & 0 & 0.6 & -4 & 0 & 4.35 & 0.7 & -0.35 \\ -1.4 & 0.6 & & & -1 & 0.7 & 2.4 & -1.3 \\ 0.7 & -4 & & & 0.6 & -0.35 & -1.3 & 4.35 \end{bmatrix} \times \begin{Bmatrix} u_1 \\ v_1 \\ u_2 \\ v_2 \\ u_3 \\ v_3 \\ u_4 \\ v_4 \end{Bmatrix} = \begin{Bmatrix} 0 \\ 0 \\ 5000 \\ 0 \\ 5000 \\ 0 \\ 0 \\ 0 \end{Bmatrix}$$

Applying the boundary conditions, the above reduces to

$$5.494505 \times 10^6 \begin{bmatrix} 2.4 & -1.3 & -1.4 & 0.6 \\ -1.3 & 4.35 & 0.7 & -4 \\ -1.4 & 0.7 & 2.4 & 0 \\ 0.6 & -4 & 0 & 4.35 \end{bmatrix} \begin{Bmatrix} u_2 \\ v_2 \\ u_3 \\ v_3 \end{Bmatrix} = \begin{Bmatrix} 5000 \\ 0 \\ 5000 \\ 0 \end{Bmatrix}$$

Therefore,

$$u_2 = \frac{\begin{vmatrix} 5000 & -1.3 & -1.4 & 0.6 \\ 0 & 4.35 & 0.7 & -4 \\ 5000 & 0.7 & 2.4 & 0 \\ 0 & -4 & 0 & 4.35 \end{vmatrix}}{5.494505 \times 10^6 \begin{vmatrix} 2.4 & -1.3 & -1.4 & 0.6 \\ -1.3 & 4.35 & 0.7 & -4 \\ -1.4 & 0.7 & 2.4 & 0 \\ 0.6 & -4 & 0 & 4.35 \end{vmatrix}} = 0.0009955 \text{cm}$$

This value compares well with the simple analytical solution

$$\delta = \frac{FL}{AE} = \frac{10000 \times 20}{10 \times 2 \times 10^7} = 0.001 \text{ cm}$$

The other deformations can be similarly obtained as

$$v_2 = 0.1561 \times 10^{-3}$$
$$u_3 = 0.9144 \times 10^{-3} \text{ cm}$$
$$v_3 = 0.0062 \times 10^{-3}$$

Natural Frequencies

The elemental mass matrices are obtained from equation (7.1.49)

Element 1: Let the density of the plate material be $\rho = 0.008$ kg/cm^3

$$[\widehat{M}]^{(1)} = [M_G]^{(1)} = \frac{2}{15} \begin{bmatrix} 2 & 0 & 1 & 0 & 1 & 0 \\ 0 & 2 & 0 & 1 & 0 & 1 \\ 1 & 0 & 2 & 0 & 1 & 0 \\ 0 & 1 & 0 & 2 & 0 & 1 \\ 1 & 0 & 1 & 0 & 2 & 0 \\ 0 & 1 & 0 & 1 & 0 & 2 \end{bmatrix}$$

Element 2:

$$[\widehat{M}]^{(2)} = \frac{2}{15} \begin{bmatrix} 2 & 0 & 1 & 0 & 1 & 0 \\ 0 & 2 & 0 & 1 & 0 & 1 \\ 1 & 0 & 2 & 0 & 1 & 0 \\ 0 & 1 & 0 & 2 & 0 & 1 \\ 1 & 0 & 1 & 0 & 2 & 0 \\ 0 & 1 & 0 & 1 & 0 & 2 \end{bmatrix}$$

$$[M_G]^{(2)} = [R]^T [\widehat{M}]^{(2)} [R]$$

264 Dynamics of Plates

$$[\widehat{M}^{(2)}][R] = \frac{2}{15} \begin{bmatrix} 2 & 0 & 1 & 0 & 1 & 0 \\ 0 & 2 & 0 & 1 & 0 & 1 \\ 1 & 0 & 2 & 0 & 1 & 0 \\ 0 & 1 & 0 & 2 & 0 & 1 \\ 1 & 0 & 1 & 0 & 2 & 0 \\ 0 & 1 & 0 & 1 & 0 & 2 \end{bmatrix} \begin{bmatrix} 0 & -1 & 0 & 0 & 0 & 0 \\ 1 & 0 & 0 & 0 & 0 & 0 \\ 0 & 0 & 0 & -1 & 0 & 0 \\ 0 & 0 & 1 & 0 & 0 & 0 \\ 0 & 0 & 0 & 0 & 0 & -1 \\ 0 & 0 & 0 & 0 & 1 & 0 \end{bmatrix}$$

$$= \frac{2}{15} \begin{bmatrix} 0 & -2 & 0 & -1 & 0 & -1 \\ 2 & 0 & 1 & 0 & 1 & 0 \\ 0 & -1 & 0 & -2 & 0 & -1 \\ 1 & 0 & 2 & 0 & 1 & 0 \\ 0 & -1 & 0 & -1 & 0 & -2 \\ 1 & 0 & 1 & 0 & 2 & 0 \end{bmatrix}$$

$$[\widehat{M_G}^{(2)}] = \frac{2}{15} \begin{bmatrix} 0 & 1 & 0 & 0 & 0 & 0 \\ -1 & 0 & 0 & 0 & 0 & 0 \\ 0 & 0 & 0 & 1 & 0 & 0 \\ 0 & 0 & -1 & 0 & 0 & 0 \\ 0 & 0 & 0 & 0 & 0 & 1 \\ 0 & 0 & 0 & 0 & -1 & 0 \end{bmatrix} \begin{bmatrix} 0 & -2 & 0 & -1 & 0 & -1 \\ 2 & 0 & 1 & 0 & 1 & 0 \\ 0 & -1 & 0 & -2 & 0 & -1 \\ 1 & 0 & 2 & 0 & 1 & 0 \\ 0 & -1 & 0 & -1 & 0 & -2 \\ 1 & 0 & 1 & 0 & 2 & 0 \end{bmatrix}$$

$$= \frac{2}{15} \begin{bmatrix} 2 & 0 & 1 & 0 & 1 & 0 \\ & 2 & 0 & 1 & 0 & 1 \\ & & 2 & 0 & 1 & 0 \\ & & & 2 & 0 & 1 \\ & \text{sym} & & & 2 & 0 \\ & & & & & 2 \end{bmatrix}$$

System Global Mass Matrix

$$[M_G] = \frac{2}{15} \begin{bmatrix} u_1 & v_1 & u_2 & v_2 & u_3 & v_3 & u_4 & v_4 \\ (2+2) & (0+0) & 1 & 0 & (1+1) & (0+0) & 1 & 0 \\ (0+0) & (2+2) & 0 & 1 & (0+0) & (1+1) & 0 & 1 \\ 1 & 0 & 2 & 0 & 1 & 0 & & \\ 0 & 1 & 0 & 2 & 0 & 1 & & \\ (1+1) & (0+0) & 1 & 0 & (2+2) & (0+0) & 1 & 0 \\ (0+0) & (1+1) & 0 & 1 & (0+0) & (2+2) & 0 & 1 \\ 1 & 0 & & & 1 & 0 & 2. & 0 \\ 0 & 1 & & & 0 & 1 & 0 & 2 \end{bmatrix} \begin{matrix} u_1 \\ v_1 \\ u_2 \\ v_2 \\ u_3 \\ v_3 \\ u_4 \\ v_4 \end{matrix}$$

$$= \frac{2}{15}\begin{bmatrix} 4 & 0 & 1 & 0 & 2 & 0 & 1 & 0 \\ 0 & 4 & 0 & 1 & 0 & 2 & 0 & 1 \\ 1 & 0 & 2 & 0 & 1 & 0 & 0 & 0 \\ 0 & 1 & 0 & 2 & 0 & 1 & 0 & 0 \\ 2 & 0 & 1 & 0 & 4 & 0 & 1 & 0 \\ 0 & 2 & 0 & 1 & 0 & 4 & 0 & 1 \\ 1 & 0 & 0 & 0 & 1 & 0 & 2. & 0 \\ 0 & 1 & 0 & 0 & 0 & 1 & 0 & 2 \end{bmatrix}$$

Therefore, the eigen value problem for free vibration is

$$\left| 5.494505 \times 10^6 \begin{bmatrix} 2.4 & -1.3 & -1.4 & 0.6 \\ -1.3 & 4.35 & 0.7 & -4 \\ -1.4 & 0.7 & 2.4 & 0 \\ 0.6 & -4 & 0 & 4.35 \end{bmatrix} - \frac{2}{15}p^2 \begin{bmatrix} 2 & 0 & 1 & 0 \\ 0 & 2 & 0 & 1 \\ 1 & 0 & 4 & 0 \\ 0 & 1 & 0 & 4 \end{bmatrix} \right| = 0$$

The eigen values are

$$p^2 = 1 \times 10^8 \begin{Bmatrix} 2.1011 \\ 0.7991 \\ 0.0980 \\ 0.0219 \end{Bmatrix}$$

The natural frequencies are 1480, 3130, 8939 and 14495 rad/s. Because there are two triangular elements in one rectangle, the symmetry is lost and the natural frequencies will be very approximate. One needs more number of elements to obtain converged eigen values. In this respect, the rectangular element in the next section may be preferred.

7.2 RECTANGULAR MEMBRANE ELEMENT

Select Element Type We choose the element with four nodes 1, 2, 3 and 4 as shown in Fig. 7.2.1. The nodal coordinate system $\hat{x}\hat{y}$ is located with its origin at the center of the element of width $2a$ in $\hat{x}$ direction and $2b$ in $\hat{y}$ direction as shown. Because of the rectangular geometry, it is convenient to use non dimensional elemental coordinates as $\xi = \frac{x}{a}$, $\eta = \frac{y}{b}$, so that

$$\xi_1 = -1, \eta_1 = -1$$
$$\xi_2 = 1, \eta_2 = -1$$

266 Dynamics of Plates

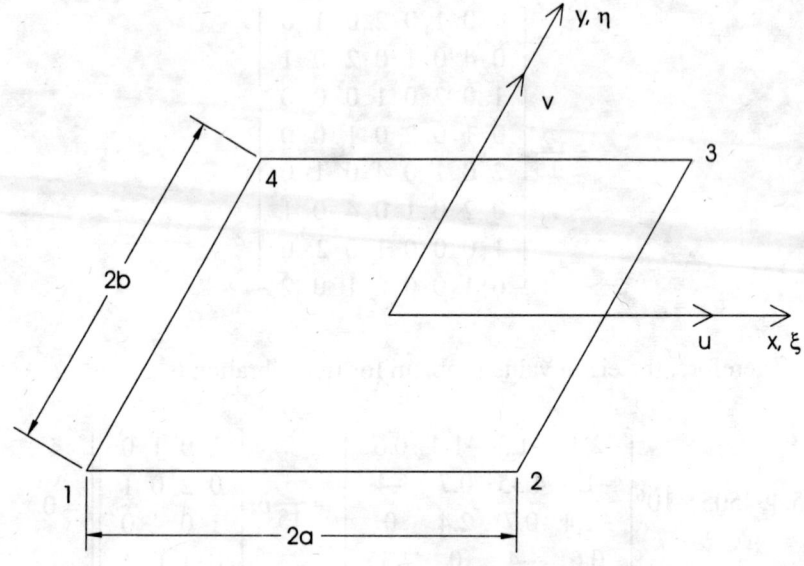

Fig. 7.2.1 Rectangular Membrane Element

$$\xi_3 = 1, \eta_3 = 1$$
$$\xi_4 = -1, \eta_4 = 1$$

(7.2.1)

For the membrane element, we will assign two degrees of freedom to each node, viz., u and v, so that each element has eight degrees of freedom given below.

$$\{\hat{d}\} = \begin{Bmatrix} \hat{u}_1 \\ \hat{v}_1 \\ \hat{u}_2 \\ \hat{v}_2 \\ \hat{u}_3 \\ \hat{v}_3 \\ \hat{u}_4 \\ \hat{v}_4 \end{Bmatrix}$$

(7.2.2)

Displacement Functions The displacements within the rectangular element at any point $(\hat{x}, \hat{y})$ are represented by

$$u(\hat{x},\hat{y}) = N_1(\xi,\eta)\hat{u}_1 + N_2(\xi,\eta)\hat{u}_2 + N_3(\xi,\eta)\hat{u}_3 + N_4(\xi,\eta)\hat{u}_4$$
$$v(\hat{x},\hat{y}) = N_1(\xi,\eta)\hat{v}_1 + N_2(\xi,\eta)\hat{v}_2 + N_3(\xi,\eta)\hat{v}_3 + N_4(\xi,\eta)\hat{v}_4$$

(7.2.3)

where

$$N_1 = \frac{1}{4}(1-\xi)(1-\eta)$$
$$N_2 = \frac{1}{4}(1+\xi)(1-\eta)$$
$$N_3 = \frac{1}{4}(1+\xi)(1+\eta)$$
$$N_4 = \frac{1}{4}(1-\xi)(1+\eta)$$

(7.2.4)

In terms of the nodal coordinates, the above shape functions are expressed as

$$N_i = \frac{1}{4}(1+\xi_i\xi)(1+\eta_i\eta)$$

(7.2.5)

As in the triangular element of section 7.1, the displacements vary linearly along each side of the element and are uniquely determined by the values of the nodal displacements. Therefore, the displacements are continuous between elements, i.e., the element is a *conformal element*. We can now express the displacement functions (7.2.3) in matrix form as

$$\left\{\begin{array}{c} u(\hat{x},\hat{y}) \\ v(\hat{x},\hat{y}) \end{array}\right\} = \begin{bmatrix} N_1 & 0 & N_2 & 0 & N_3 & 0 & N_4 & 0 \\ 0 & N_1 & 0 & N_2 & 0 & N_3 & 0 & N_4 \end{bmatrix} \left\{\begin{array}{c} \hat{u}_1 \\ \hat{v}_1 \\ \hat{u}_2 \\ \hat{v}_2 \\ \hat{u}_3 \\ \hat{v}_3 \\ \hat{u}_4 \\ \hat{v}_4 \end{array}\right\}$$

$$= [N]\{\hat{d}\}$$

(7.2.6)

where

$$[N] = \begin{bmatrix} N_1 & 0 & N_2 & 0 & N_3 & 0 & N_4 & 0 \\ 0 & N_1 & 0 & N_2 & 0 & N_3 & 0 & N_4 \end{bmatrix}$$

(7.2.7)

Strain Displacement Relations The strain displacement relations in (7.1.17) for the rectangular element can be expressed as

$$\left\{ \begin{array}{c} \varepsilon_{xx} \\ \varepsilon_{yy} \\ \gamma_{xy} \end{array} \right\} = \left\{ \begin{array}{c} N_{1,\hat{x}}\hat{u}_1 + N_{2,\hat{x}}\hat{u}_2 + N_{3,\hat{x}}\hat{u}_3 + N_{4,\hat{x}}\hat{u}_4 \\ N_{1,\hat{y}}\hat{v}_1 + N_{2,\hat{y}}\hat{v}_2 + N_{3,\hat{y}}\hat{v}_3 + N_{4,\hat{y}}\hat{v}_4 \\ \left[\begin{array}{c} (N_{1,\hat{y}}\hat{u}_1 + N_{2,\hat{y}}\hat{u}_2 + N_{3,\hat{y}}\hat{u}_3 + N_{4,\hat{y}}\hat{u}_4) \\ +(N_{1,\hat{x}}\hat{v}_1 + N_{2,\hat{x}}\hat{v}_2 + N_{3,\hat{x}}\hat{v}_3 + N_{4,\hat{x}}\hat{v}_4) \end{array} \right] \end{array} \right\}$$

(7.2.8)

Differentiating (7.2.5), we have

$$N_{i,\hat{x}} = \frac{1}{4}\left(\frac{\xi_i}{a}\right)(1 + \eta_i\eta)$$

$$N_{i,\hat{y}} = \frac{1}{4}\left(\frac{\eta_i}{b}\right)(1 + \xi_i\xi)$$

(7.2.9)

Making use of the above equation (7.2.9) and (7.2.1), equation (7.2.8) becomes

$$\{\varepsilon\} = \frac{1}{4} \begin{bmatrix} -\frac{1-\eta}{a} & 0 & \frac{1-\eta}{a} & 0 & \frac{1+\eta}{a} & 0 & -\frac{1+\eta}{a} & 0 \\ 0 & -\frac{1-\xi}{b} & 0 & -\frac{1+\xi}{b} & 0 & \frac{1+\xi}{b} & 0 & \frac{1-\xi}{b} \\ -\frac{1-\xi}{b} & -\frac{1-\eta}{a} & -\frac{1+\xi}{b} & \frac{1-\eta}{a} & \frac{1+\xi}{b} & \frac{1+\eta}{a} & \frac{1-\xi}{b} & -\frac{1+\eta}{a} \end{bmatrix}$$

$$\times \left\{ \begin{array}{c} \hat{u}_1 \\ \hat{v}_1 \\ \hat{u}_2 \\ \hat{v}_2 \\ \hat{u}_3 \\ \hat{v}_3 \\ \hat{u}_4 \\ \hat{v}_4 \end{array} \right\}$$

$$= [B]\{\hat{d}\}$$

(7.2.10)

where

$$[B] = \frac{1}{4}\begin{bmatrix} -\frac{1-\eta}{a} & 0 & \frac{1-\eta}{a} & 0 & \frac{1+\eta}{a} & 0 & -\frac{1+\eta}{a} & 0 \\ 0 & -\frac{1-\xi}{b} & 0 & -\frac{1+\xi}{b} & 0 & \frac{1+\xi}{b} & 0 & \frac{1-\xi}{b} \\ -\frac{1-\xi}{b} & -\frac{1-\eta}{a} & -\frac{1+\xi}{b} & \frac{1-\eta}{a} & \frac{1+\xi}{b} & \frac{1+\eta}{a} & \frac{1-\xi}{b} & -\frac{1+\eta}{a} \end{bmatrix}$$

(7.2.11)

A correction is applied to the above strain displacement relation for the following reason. Consider the element in Fig. 7.2.1 to be subjected to pure bending in ξ direction as in a simple Euler-Bernoulli beam. The nodal displacements can be expressed as

$$\hat{u}_1 = \hat{u}_3 = u$$

$$\hat{u}_2 = \hat{u}_4 = -u$$

$$\hat{v}_1 = \hat{v}_2 = \hat{v}_3 = \hat{v}_4 = 0$$

(7.2.12)

Substituting (7.2.12) in the strain-displacement relations, we get

$$\{\varepsilon\} = \frac{1}{4}\begin{bmatrix} -\frac{1-\eta}{a} & 0 & \frac{1-\eta}{a} & 0 & \frac{1+\eta}{a} & 0 & -\frac{1+\eta}{a} & 0 \\ 0 & -\frac{1-\xi}{b} & 0 & -\frac{1+\xi}{b} & 0 & \frac{1+\xi}{b} & 0 & \frac{1-\xi}{b} \\ -\frac{1-\xi}{b} & -\frac{1-\eta}{a} & -\frac{1+\xi}{b} & \frac{1-\eta}{a} & \frac{1+\xi}{b} & \frac{1+\eta}{a} & \frac{1-\xi}{b} & -\frac{1+\eta}{a} \end{bmatrix}$$

$$\times \begin{Bmatrix} u \\ 0 \\ -u \\ 0 \\ u \\ 0 \\ -u \\ 0 \end{Bmatrix}$$

$$\varepsilon_{xx} = \frac{u}{a}\eta$$

$$\varepsilon_{yy} = 0$$

$$\gamma_{xy} = \frac{u}{b}\xi$$

(7.2.13)

270 Dynamics of Plates

The normal strain is predicted exactly, however, we find that the shear strain is not zero as it should be in pure bending of a beam. The shear strain is zero only along $\xi = 0$. In a similar manner, we can consider pure bending along η direction and show that the shear strain will be zero along $\eta = 0$. Thus, we find that the shear strain is exact only when $\xi = \eta = 0$. To accommodate this, the strain displacement relation in (7.2.11) is modified by writing the $[B]$ matrix as

$$[B] = \frac{1}{4ab} \times$$

$$\begin{bmatrix} -(1-\eta)b & 0 & (1-\eta)b & 0 & (1+\eta)b & 0 & -(1+\eta)b & 0 \\ 0 & -(1-\xi)a & 0 & -(1+\xi)a & 0 & (1+\xi)a & 0 & (1-\xi)a \\ -a & -b & -a & b & a & b & a & -b \end{bmatrix}$$

(7.2.14)

The above modification alters the conformity of the element.

Elemental Stiffness Matrix

Following the procedure in section 7.1, the elemental stiffness matrix is given by

$$[\hat{K}] = \iiint_V [B]^T [D][B] dV$$

(7.1.27)

Substituting from equations (7.2.14) and (5.1.16), the above equation gives the elemental stiffness matrix

$$[\hat{K}] = \frac{Eh}{16ab(1-v^2)} \int_{-1}^{1} \int_{-1}^{1} \begin{bmatrix} -b+\eta b & 0 & -a \\ 0 & -a+\xi a & -b \\ b-\eta b & 0 & -a \\ 0 & -a-\xi a & b \\ b+\eta b & 0 & a \\ 0 & a+\xi a & b \\ -b-\eta b & 0 & a \\ 0 & a-\xi a & -b \end{bmatrix} \begin{bmatrix} 1 & v & 0 \\ v & 1 & 0 \\ 0 & 0 & \frac{1-v}{2} \end{bmatrix}$$

$$\begin{bmatrix} -b+\eta b & 0 & b-\eta b & 0 & b+\eta b & 0 & -b-\eta b & 0 \\ 0 & -a+\xi a & 0 & -a-\xi a & 0 & a+\xi a & 0 & a-\xi a \\ -a & -b & -a & b & a & b & a & -b \end{bmatrix} d\xi \, d\eta$$

(7.2.15)

Performing the matrix multiplications involved in the above equation and integrating, we get the stiffness matrix as

$$[\hat{k}] = [K_1] + [K_2] \tag{7.2.16}$$

where

$$[K_1] = \frac{Eh}{16ab(1-v^2)} \begin{bmatrix} 2\bar{b} & & & & & & & \\ \bar{v} & 2\bar{a} & & & & \text{sym} & & \\ -2\bar{b} & -\bar{v} & 2\bar{b} & & & & & \\ \bar{v} & \bar{a} & -\bar{v} & 2\bar{a} & & & & \\ -\bar{b} & -\bar{v} & \bar{b} & -\bar{v} & 2\bar{b} & & & \\ -\bar{v} & -\bar{a} & \bar{v} & -2\bar{a} & \bar{v} & 2\bar{a} & & \\ \bar{b} & \bar{v} & -\bar{b} & \bar{v} & -2\bar{b} & -\bar{v} & 2\bar{b} & \\ -\bar{v} & -2\bar{a} & \bar{v} & -\bar{a} & \bar{v} & \bar{a} & -\bar{v} & 2\bar{a} \end{bmatrix}$$

(7.2.17)

$$\bar{b} = \frac{8}{3}b^2$$

$$\bar{a} = \frac{8}{3}a^2$$

$$\bar{v} = 4vab \tag{7.2.18}$$

and

$$[K_2] = \frac{Eh}{8ab(1+v)} \begin{bmatrix} a^2 & & & & & & & \\ ab & b^2 & & & & \text{sym} & & \\ a^2 & ab & a^2 & & & & & \\ -ab & -b^2 & -ab & b^2 & & & & \\ -a^2 & -ab & -a^2 & ab & a^2 & & & \\ -ab & -b^2 & -ab & b^2 & ab & b^2 & & \\ -a^2 & -ab & -a^2 & ab & a^2 & ab & a^2 & \\ ab & b^2 & ab & -b^2 & -ab & -b^2 & -ab & b^2 \end{bmatrix}$$

(7.2.19)

Dynamics of Plates

Elemental Equation Following the derivation of section 7.1, we have

$$[\hat{K}]\{\hat{d}\} = \{\hat{f}\}$$

(7.2.20)

where the nodal force is given by

$$\{\hat{f}\} = \{\hat{P}\} + \iint_A [N]^T \{T\} dA$$

(7.1.26)

Consistent Mass Matrix
The mass matrix is given by

$$[\widehat{M}] = \iint_A \rho h [N]^T [N] dA$$

(7.1.46)

which can be written as

$$[\widehat{M}] = \rho h a b \int_{-1}^{1} \int_{-1}^{1} N_i N_j \, d\xi \, d\eta$$

(7.2.21)

Substituting from equation (7.2.5), the above equation (7.2.21) becomes

$$[\widehat{M}] = \frac{1}{16} \rho h a b \int_{-1}^{1} (1 + \xi_i \xi)(1 + \xi_j \xi) d\xi \int_{-1}^{1} (1 + \eta_i \eta)(1 + \eta_j \eta) d\eta$$

(7.2.22)

Upon evaluating the integrals,

$$[\widehat{M}] = \frac{1}{4} \rho h a b \left(1 + \frac{1}{3} \xi_i \xi_j\right)\left(1 + \frac{1}{3} \eta_i \eta_j\right)$$

(7.2.22)

which is

$$[\widehat{M}] = \frac{1}{9} \rho h a b \begin{bmatrix} 4 & & & & & & & \\ 0 & 4 & & & & & & \\ 2 & 0 & 4 & & & & & \\ 0 & 2 & 0 & 4 & & & & \\ 1 & 0 & 2 & 0 & 4 & & & \\ 0 & 1 & 0 & 2 & 0 & 4 & & \\ 2 & 0 & 1 & 0 & 2 & 0 & 4 & \\ 0 & 2 & 0 & 1 & 0 & 2 & 0 & 4 \end{bmatrix}$$
sym

(7.2.23)

Example

Consider the example of previous section, we will use here just one element to illustrate the process. Applying the boundary condition, the system equation can be written as follows:

$$\bar{b} = \frac{800}{3}, \bar{a} = \frac{3200}{3}, \bar{v} = \frac{720}{3}$$

$$\frac{Eh}{16ab(1-v^2)} = \frac{2\times 10^7 \times 1}{16\times 200(1-0.09)} = 6868.13$$

$$[K_1] = 228937.73 \begin{bmatrix} 16 & & & & & & & \\ 7.2 & 64 & & & & & \text{sym} & \\ -16 & -7.2 & 16 & & & & & \\ 7.2 & 32 & -7.2 & 64 & & & & \\ -8 & -7.2 & 8 & -7.2 & 16 & & & \\ -7.2 & -32 & 7.2 & -64 & 7.2 & 64 & & \\ 8 & 7.2 & -8 & 7.2 & -16 & -7.2 & 16 & \\ -7.2 & -64 & 7.2 & -32 & 7.2 & 32 & -7.2 & 64 \end{bmatrix}$$

$$\frac{Eh}{8ab(1+v)} = \frac{2\times 10^7 \times 1}{8\times 200 \times 1.3} = 9615.3846$$

$$[K_2] = 961538.46 \begin{bmatrix} 4 & & & & & & & \\ 2 & 1 & & & & & & \\ 4 & 2 & 4 & & & & & \\ -2 & -1 & -2 & 1 & & & & \\ -4 & -2 & -4 & 2 & 4 & & & \\ -2 & -1 & -2 & 1 & 2 & 1 & & \\ -4 & -2 & -4 & 2 & 4 & 2 & 4 & \\ 2 & 1 & 2 & -1 & -2 & -1 & -2 & 1 \end{bmatrix}_{\text{sym}}$$

$$[K_2] = 228937.73 \begin{bmatrix} 16.8 & & & & & & & \\ 8.4 & 4.2 & & & & & \text{sym} & \\ 16.8 & 8.4 & 16.8 & & & & & \\ -8.4 & -4.2 & -8.4 & 4.2 & & & & \\ -16.8 & -8.4 & -16.8 & 8.4 & 16.8 & & & \\ -8.4 & -4.2 & -8.4 & 4.2 & 8.4 & 4.2 & & \\ -16.8 & -8.4 & -16.8 & 8.4 & 16.8 & 8.4 & 16.8 & \\ 8.4 & 4.2 & 8.4 & -4.2 & -8.4 & -4.2 & -8.4 & 4.2 \end{bmatrix}$$

274 Dynamics of Plates

$$[K] = 228937.73 \begin{bmatrix} 32.8 & & & & & & & \\ 15.6 & 68.2 & & & & \text{sym} & & \\ 0.8 & 1.2 & 32.8 & & & & & \\ -1.2 & 27.8 & -15.6 & 68.2 & & & & \\ -24.8 & -15.6 & -8.8 & 1.2 & 32.8 & & & \\ -15.6 & -36.2 & -1.2 & -59.8 & 15.6 & 68.2 & & \\ -8.8 & -1.2 & -24.8 & 15.6 & 0.8 & 1.2 & 32.8 & \\ 1.2 & -59.8 & 15.6 & -36.2 & -1.2 & 27.8 & -15.6 & 68.2 \end{bmatrix}$$

$$228937.73 \begin{bmatrix} 32.8 & -15.6 & -8.8 & -1.2 \\ -15.6 & 68.2 & 1.2 & -59.8 \\ -8.8 & 1.2 & 32.8 & 15.6 \\ -1.2 & -59.8 & 15.6 & 68.2 \end{bmatrix} \begin{Bmatrix} u_2 \\ v_2 \\ u_3 \\ v_3 \end{Bmatrix} = \begin{Bmatrix} 5000 \\ 0 \\ 5000 \\ 0 \end{Bmatrix}$$

The solution of the above equation gives

$$u_2 = 0.9759 \times 10^{-3}$$
$$v_2 = 0.1098 \times 10^{-3}$$
$$u_3 = 0.9759 \times 10^{-3} \quad \text{cm}$$
$$v_3 = -0.1098 \times 10^{-3}$$

The mass matrix from (7.2.23) is

$$[\widehat{M}] = \frac{16}{90} \begin{bmatrix} 4 & 0 & 2 & 0 \\ 0 & 4 & 0 & 2 \\ 2 & 0 & 4 & 0 \\ 0 & 2 & 0 & 4 \end{bmatrix}$$

The eigen value problem is

$$\left| 228937.73 \begin{bmatrix} 32.8 & -15.6 & -8.8 & -1.2 \\ -15.6 & 68.2 & 1.2 & -59.8 \\ -8.8 & 1.2 & 32.8 & 15.6 \\ -1.2 & -59.8 & 15.6 & 68.2 \end{bmatrix} - \frac{16}{90} p^2 \begin{bmatrix} 4 & 0 & 2 & 0 \\ 0 & 4 & 0 & 2 \\ 2 & 0 & 4 & 0 \\ 0 & 2 & 0 & 4 \end{bmatrix} \right|$$

$$= 0$$

The eigen values of the above problem are

$$p^2 = 1 \times 10^7 \begin{Bmatrix} 1.5968 \\ 4.2110 \\ 1.5968 \\ 4.2110 \end{Bmatrix}$$

The two natural frequencies are

$$p_1 = 3996 \text{ rad/s}$$
$$p_2 = 6489$$

Considering pure extensional vibrations, the first mode natural frequency can be obtained from

$$p_1 = \frac{\pi}{2L}\sqrt{\frac{E}{\rho}}$$

$$= \frac{\pi}{2 \times 20}\sqrt{\frac{2 \times 10^7}{0.008}}$$

$$= 3927 \text{ rad/s}$$

7.3 TRIANGULAR PLATE ELEMENT WITH IN-PLANE STRESSES

Three dimensional thin structures can be analyzed using plate elements which provide for bending as well as in-plane forces. Here, we assume that the in-plane and bending stiffnesses remain uncoupled in an individual element. We further assume that rotation about $\hat{z}$ axis is not required in defining an individual element, however, provision is to be made for the rotation $\theta_{\hat{z}}$ as well as its counter part $M_{\hat{z}}$ by inserting zeros at appropriate locations in the stiffness matrix. This allows us to utilize superposition principle to combine the bending and in-plane effects from sections 5.1 and 7.1 for a triangular plate element. Thus, each node has six degrees of freedom and the element has 18 degrees of freedom as shown in Fig. 7.3.1. The in-plane stiffness matrix is given by equation (7.1.31), which can be written as

$$[\hat{K}]_{m(6 \times 6)} = \begin{bmatrix} [K_{11}]_{m(2\times2)} & [K_{12}]_{m(2\times2)} & [K_{13}]_{m(2\times2)} \\ [K_{21}]_{m(2\times2)} & [K_{22}]_{m(2\times2)} & [K_{23}]_{m(2\times2)} \\ [K_{31}]_{m(2\times2)} & [K_{32}]_{m(2\times2)} & [K_{33}]_{m(2\times2)} \end{bmatrix}$$

(7.3.1)

276 *Dynamics of Plates*

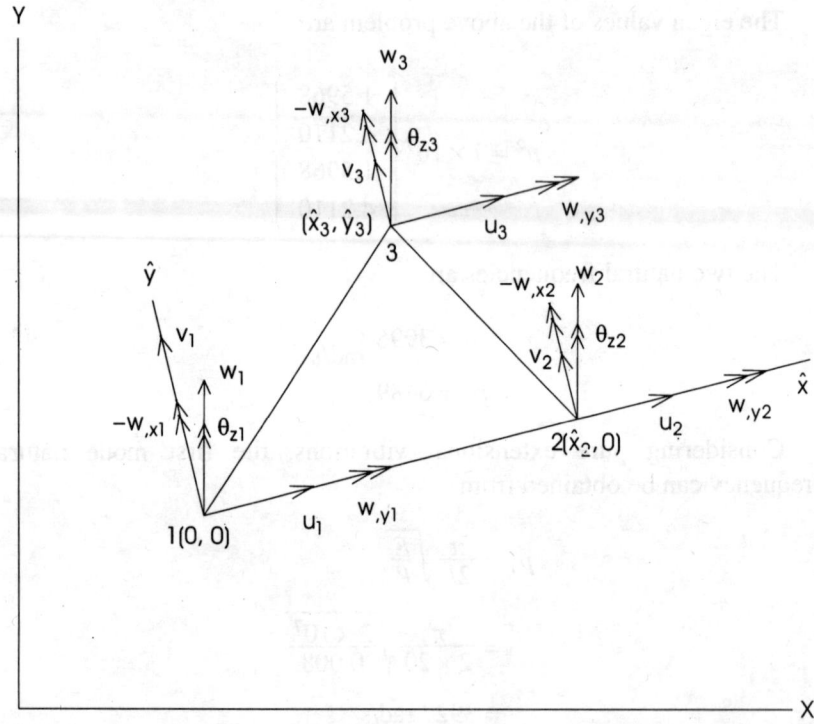

Fig. 7.3.1 Combined Bending and In-Plane Triangular Element

where the subscript m denotes membrane effect. Equation (7.1.33) defines the force displacement relationship through the above stiffness matrix, as

$$[\hat{K}]_{m(6\times6)}\{\hat{d}\}_{m(6\times1)} = \{\hat{f}\}_{m(6\times1)}$$

$$\begin{bmatrix} [K_{11}]_{m(2\times2)} & [K_{12}]_{m(2\times2)} & [K_{13}]_{m(2\times2)} \\ [K_{21}]_{m(2\times2)} & [K_{22}]_{m(2\times2)} & [K_{23}]_{m(2\times2)} \\ [K_{31}]_{m(2\times2)} & [K_{32}]_{m(2\times2)} & [K_{33}]_{m(2\times2)} \end{bmatrix} \begin{Bmatrix} \{\hat{d}_1\}_{m(2\times1)} \\ \{\hat{d}_2\}_{m(2\times1)} \\ \{\hat{d}_3\}_{m(2\times1)} \end{Bmatrix} = \begin{Bmatrix} \{\hat{f}_1\}_{m(2\times1)} \\ \{\hat{f}_2\}_{m(2\times1)} \\ \{\hat{f}_3\}_{m(2\times1)} \end{Bmatrix}$$

(7.3.2)

In a similar manner, we can write equation (5.1.22) to give the stiffness matrix for the triangular element in its own local coordinates

$$[\hat{K}]_{b(9\times9)} = \begin{bmatrix} [K_{11}]_{b(3\times3)} & [K_{12}]_{b(3\times3)} & [K_{13}]_{b(3\times3)} \\ [K_{21}]_{b(3\times3)} & [K_{22}]_{b(3\times3)} & [K_{23}]_{b(3\times3)} \\ [K_{31}]_{b(3\times3)} & [K_{32}]_{b(3\times3)} & [K_{33}]_{b(3\times3)} \end{bmatrix}$$

(7.3.3)

where subscript b denotes plate bending effect. Equation (5.1.21) defines the force displacement relationship through the above stiffness matrix, as

$$[\hat{K}]_{b(9\times 9)} \{\hat{d}\}_{b(9\times 1)} = \{\hat{f}\}_{b(9\times 1)}$$

$$\begin{bmatrix} [K_{11}]_{b(3\times 3)} & [K_{12}]_{b(3\times 3)} & [K_{13}]_{b(3\times 3)} \\ [K_{21}]_{b(3\times 3)} & [K_{22}]_{b(3\times 3)} & [K_{23}]_{b(3\times 3)} \\ [K_{31}]_{b(3\times 3)} & [K_{32}]_{b(3\times 3)} & [K_{33}]_{b(3\times 3)} \end{bmatrix} \begin{Bmatrix} \{\hat{d}_1\}_{b(3\times 1)} \\ \{\hat{d}_2\}_{b(3\times 1)} \\ \{\hat{d}_3\}_{b(3\times 1)} \end{Bmatrix} = \begin{Bmatrix} \{\hat{f}_1\}_{b(3\times 1)} \\ \{\hat{f}_2\}_{b(3\times 1)} \\ \{\hat{f}_3\}_{b(3\times 1)} \end{Bmatrix}$$

(7.3.4)

The combined stiffness matrix is now obtained by a superposition of the stiffness matrices in (7.3.1) and (7.3.3) to give

$$[\hat{K}]_{(18\times 18)}$$

$$= \begin{bmatrix} [K_{11}]_{m(2\times 2)} & 0 & 0 & [K_{12}]_{m(2\times 2)} & 0 & 0 & [K_{13}]_{m(2\times 2)} & 0 & 0 \\ 0 & [K_{11}]_{b(3\times 3)} & 0 & 0 & [K_{12}]_{b(3\times 3)} & 0 & 0 & [K_{13}]_{b(3\times 3)} & 0 \\ 0 & 0 & 0 & 0 & 0 & 0 & 0 & 0 & 0 \\ [K_{21}]_{m(2\times 2)} & 0 & 0 & [K_{22}]_{m(2\times 2)} & 0 & 0 & [K_{23}]_{m(2\times 2)} & 0 & 0 \\ 0 & [K_{21}]_{b(3\times 3)} & 0 & 0 & [K_{22}]_{b(3\times 3)} & 0 & 0 & [K_{23}]_{b(3\times 3)} & 0 \\ 0 & 0 & 0 & 0 & 0 & 0 & 0 & 0 & 0 \\ [K_{31}]_{m(2\times 2)} & 0 & 0 & [K_{32}]_{m(2\times 2)} & 0 & 0 & [K_{33}]_{m(2\times 2)} & 0 & 0 \\ 0 & [K_{31}]_{b(3\times 3)} & 0 & 0 & [K_{32}]_{b(3\times 3)} & 0 & 0 & [K_{33}]_{b(3\times 3)} & 0 \\ 0 & 0 & 0 & 0 & 0 & 0 & 0 & 0 & 0 \end{bmatrix}$$

(7.3.5)

The elemental degrees of freedom are now defined by

$$\{\hat{d}\}_{(18\times 1)} = \begin{Bmatrix} \{\hat{d}_1\}_{m(2\times 1)} \\ \{\hat{d}_1\}_{b(3\times 1)} \\ \theta_{z1} \\ \{\hat{d}_2\}_{m(2\times 1)} \\ \{\hat{d}_2\}_{b(3\times 1)} \\ \theta_{z2} \\ \{\hat{d}_3\}_{m(2\times 1)} \\ \{\hat{d}_3\}_{b(3\times 1)} \\ \theta_{z3} \end{Bmatrix}$$

(7.3.6)

i.e.,

$$\{\hat{d}\}_{(18\times 1)} = \begin{Bmatrix} u_1 & v_1 & w_1 & w_{,y1} & -w_{,x1} & \theta_{z1} & u_2 & v_2 & w_2 \\ -w_{,x2} & \theta_{z2} & u_3 & v_3 & w_3 & w_{,y3} & -w_{,x3} & \theta_{z3} \end{Bmatrix}^T$$

(7.3.7)

The elemental stiffness matrix in global coordinates is obtained from

$$[K] = [R]^T [\hat{K}] [R]$$

(7.3.8)

where $[R]$ is the transformation matrix. This matrix for three dimensional coordinate system is defined by

$$[R]_{(18\times 18)} = \begin{bmatrix} [L_3] & & \\ & [L_3] & \\ & & [L_3] \end{bmatrix}$$

(7.3.9)

where

$[L_3]_{(6\times 6)} =$

$$\begin{bmatrix} \begin{bmatrix} \cos(x,X) & \cos(x,Y) & \cos(x,Z) \\ \cos(y,X) & \cos(y,Y) & \cos(y,Z) \\ \cos(z,X) & \cos(z,Y) & \cos(z,Z) \end{bmatrix} & \begin{bmatrix} 0 & 0 & 0 \\ 0 & 0 & 0 \\ 0 & 0 & 0 \end{bmatrix} \\ \begin{bmatrix} 0 & 0 & 0 \\ 0 & 0 & 0 \\ 0 & 0 & 0 \end{bmatrix} & \begin{bmatrix} \cos(x,X) & \cos(x,Y) & \cos(x,Z) \\ \cos(y,X) & \cos(y,Y) & \cos(y,Z) \\ \cos(z,X) & \cos(z,Y) & \cos(z,Z) \end{bmatrix} \end{bmatrix}$$

(7.3.10)

The directional cosines in the above are obtained from the following equations.

$$\cos(x,X) = \frac{X_2 - X_1}{\sqrt{(X_2 - X_1)^2 + (Y_2 - Y_1)^2 + (Z_2 - Z_1)^2}}$$

$$\cos(x, Y) = \frac{Y_2 - Y_1}{\sqrt{(X_2 - X_1)^2 + (Y_2 - Y_1)^2 + (Z_2 - Z_1)^2}}$$

$$\cos(x, Z) = \frac{Z_2 - Z_1}{\sqrt{(X_2 - X_1)^2 + (Y_2 - Y_1)^2 + (Z_2 - Z_1)^2}}$$

(7.3.11)

$$\cos(z, X) = -\frac{Y_{21}Z_{31} - Y_{31}Z_{21}}{\sqrt{(X_{21}Y_{31} - X_{31}Y_{21})^2 + (Y_{21}Z_{31} - Y_{31}Z_{21})^2 + (Z_{21}X_{31} - Z_{31}X_{21})^2}}$$

$$\cos(z, Y) = -\frac{Z_{21}X_{31} - Z_{31}X_{21}}{\sqrt{(X_{21}Y_{31} - X_{31}Y_{21})^2 + (Y_{21}Z_{31} - Y_{31}Z_{21})^2 + (Z_{21}X_{31} - Z_{31}X_{21})^2}}$$

$$\cos(z, Z) = -\frac{X_{21}Y_{31} - X_{31}Y_{21}}{\sqrt{(X_{21}Y_{31} - X_{31}Y_{21})^2 + (Y_{21}Z_{31} - Y_{31}Z_{21})^2 + (Z_{21}X_{31} - Z_{31}X_{21})^2}}$$

(7.3.12)

where

$$X_{ij} = X_i - X_j$$
$$Y_{ij} = Y_i - Y_j$$
$$Z_{ij} = Z_i - Z_j$$

(7.3.13)

and the third set of directional cosines are determined from

$$\cos(y, X) = \cos(z, Y)\cos(x, Z) - \cos(z, Z)\cos(x, Y)$$
$$\cos(y, Y) = \cos(z, Z)\cos(x, X) - \cos(z, X)\cos(x, Z)$$
$$\cos(y, Z) = \cos(z, X)\cos(x, Y) - \cos(z, Y)\cos(x, X)$$

(7.3.14)

7.4 ISOPARAMETRIC FORMULATION OF QUADRILATERAL ELEMENT

We considered two different shapes for the elements, either triangular or rectangular in nature in the previous sections of this chapter as well in chapter 5. In reality, a structure has curved edges or surfaces and one requires a large number of elements to accurately model the system. In order to reduce the number of elements in this modeling, it is desirable to have elements with curved boundaries. This means, we have to express the boundary of the element by a suitable function of the plane or space coordinates and can be expressed by means of a polynomial function, in a similar manner to that of the displacement functions. A simple choice is to use the displacement functions directly to describe the element geometry, in which case, the element is called *Isoparametric Element*. If the displacement functions of two neighboring elements satisfy compatibility along the common boundary, then the isoparametric element will also satisfy continuity condition along the boundary.

For an isoparametric element, the geometric functions and the displacement functions are of the same order. We can have other choices as well, e.g., the geometric function chosen can be of a higher order polynomial, then the element is called *superparametric element*. If the geometric function used to describe the geometry is of lower order than than the displacement function, then we have a *subparametric element*.

While deriving the rectangular or triangular elements before, closed form integration has been adopted. This is a tedious process and becomes a major disadvantage in handling elements with curved edges or surfaces. Numerical integration is to be adopted for this purpose. To make the integration process simpler, *natural coordinates* are used by the nondimensionalization of Cartesian coordinates so that the integration limits for each coordinate variable over an element area are converted to become constants.

A square linear element with eight degrees of freedom is shown in Fig. 7.4.1 (see the membrane element of Fig. 7.2.1) in $\hat{x}, \hat{y}$ natural coordinate system attached to its center. The coordinates are so nondimensionalized such that its four corner nodes have the coordinate values ±1. The displacement functions are chosen as

$$u(\hat{x}, \hat{y}) = c_1 + c_2 \hat{x} + c_3 \hat{y} + c_4 \hat{x} \hat{y}$$

$$v(\hat{x}, \hat{y}) = c_1 + c_2 \hat{x} + c_3 \hat{y} + c_4 \hat{x} \hat{y}$$

(7.4.1)

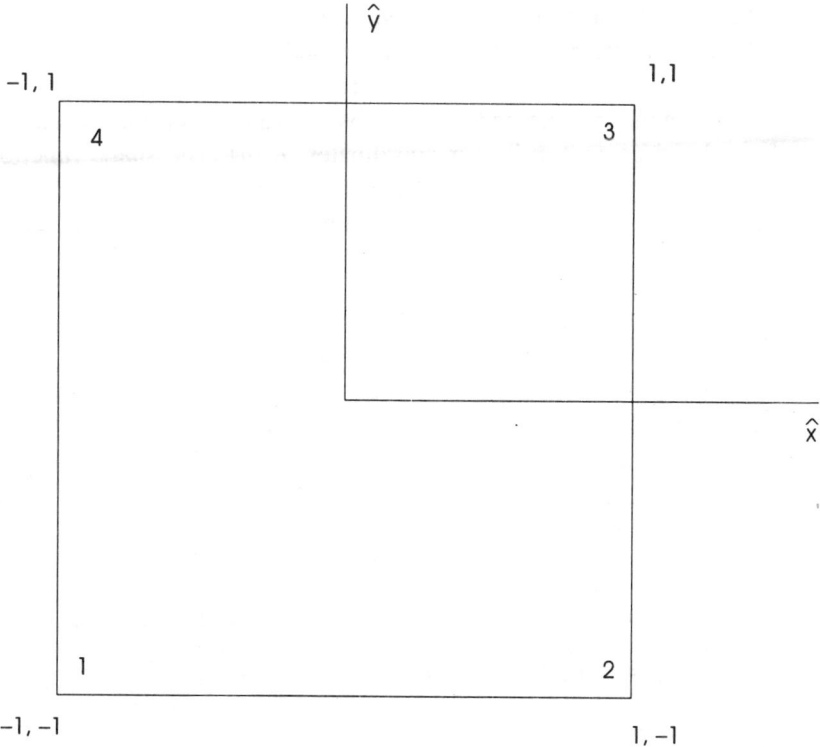

Fig. 7.4.1 Linear Square element in $\hat{x}, \hat{y}$ natural coordinates

The constants can be evaluated by substituting the nodal values in the above equation (7.4.1). After rearranging, we get

$$u(\hat{x},\hat{y}) = N_1(\hat{x},\hat{y})\hat{u}_1 + N_2(\hat{x},\hat{y})\hat{u}_2 + N_3(\hat{x},\hat{y})\hat{u}_3 + N_4(\hat{x},\hat{y})\hat{u}_4$$
$$v(\hat{x},\hat{y}) = N_1(\hat{x},\hat{y})\hat{v}_1 + N_2(\hat{x},\hat{y})\hat{v}_2 + N_3(\hat{x},\hat{y})\hat{v}_3 + N_4(\hat{x},\hat{y})\hat{v}_4$$

(7.4.2)

where

$$N_1(\hat{x},\hat{y}) = \tfrac{1}{4}(1-\hat{x})(1-\hat{y})$$

$$N_2(\hat{x},\hat{y}) = \tfrac{1}{4}(1+\hat{x})(1-\hat{y})$$

$$N_3(\hat{x},\hat{y}) = \tfrac{1}{4}(1+\hat{x})(1+\hat{y})$$

$$N_4(\hat{x},\hat{y}) = \tfrac{1}{4}(1-\hat{x})(1+\hat{y})$$

(7.4.3)

282 Dynamics of Plates

The above are non dimensional shape functions. N_i become unity at ith nodal point and are zero at the other three nodal points.

Consider now an actual quadrilateral element shown in Fig. 7.4.2 whose four nodes are defined in the global coordinates (x, y). The element is formulated using curvilinear coordinates (ξ, η). The shape functions in (7.4.3) are now used to map the square element of Fig. 7.4.1 in isoparametric coordinates (ξ, η) to the quadrilateral of Fig. 7.4.2 in the same form to define

$$x(\xi, \eta) = c_1 + c_2\xi + c_3\eta + c_4\xi\eta$$
$$y(\xi, \eta) = c_1 + c_2\xi + c_3\eta + c_4\xi\eta$$

(7.4.4)

and establish

$$x(\xi, \eta) = N_1(\xi, \eta)x_1 + N_2(\xi, \eta)x_2 + N_3(\xi, \eta)x_3 + N_4(\xi, \eta)x_4$$
$$y(\xi, \eta) = N_1(\xi, \eta)y_1 + N_2(\xi, \eta)y_2 + N_3(\xi, \eta)y_3 + N_4(\xi, \eta)y_4$$

(7.4.5)

where

$$N_1(\xi, \eta) = \frac{1}{4}(1-\xi)(1-\eta)$$
$$N_2(\xi, \eta) = \frac{1}{4}(1+\xi)(1-\eta)$$
$$N_3(\xi, \eta) = \frac{1}{4}(1+\xi)(1+\eta)$$
$$N_4(\xi, \eta) = \frac{1}{4}(1-\xi)(1+\eta)$$

(7.4.6)

The displacement functions for the quadrilateral element are

$$u(\xi, \eta) = N_1(\xi, \eta)\hat{u}_1 + N_2(\xi, \eta)\hat{u}_2 + N_3(\xi, \eta)\hat{u}_3 + N_4(\xi, \eta)\hat{u}_4$$
$$v(\xi, \eta) = N_1(\xi, \eta)\hat{v}_1 + N_2(\xi, \eta)\hat{v}_2 + N_3(\xi, \eta)\hat{v}_3 + N_4(\xi, \eta)\hat{v}_4$$

(7.4.7)

Following the procedure in section 7.1, the elemental stiffness matrix of the rectangular element in Fig. 7.4.1 is given by

$$[\hat{K}] = \iiint_V [B]^T [D][B] dV$$

(7.1.27)

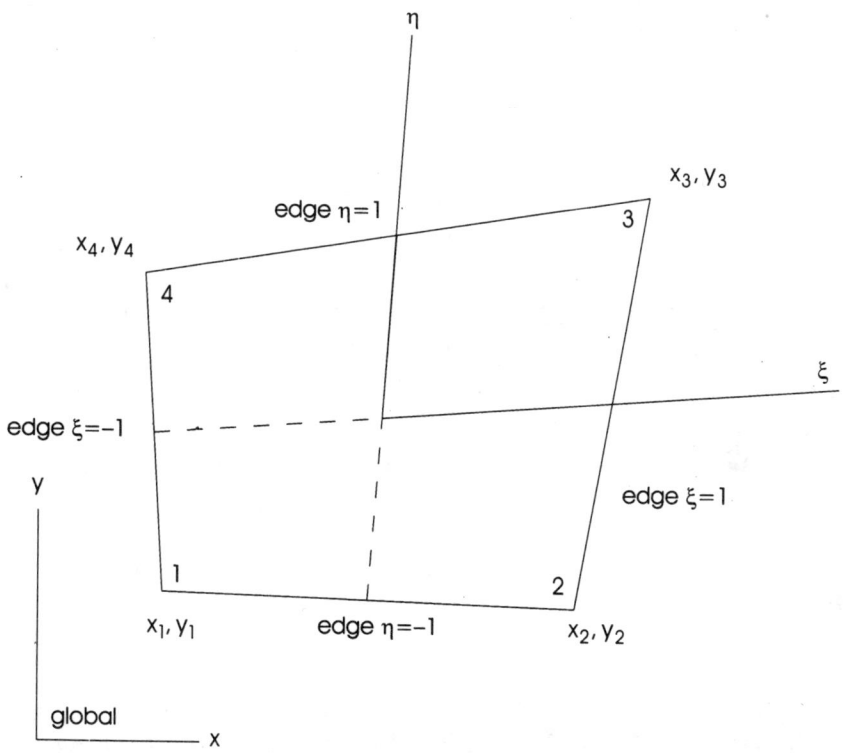

Fig. 7.4.2 Square Element mapped into quadrilateral in curvilinear coordinates

where

$$[D] = \frac{E}{1-v^2} \begin{bmatrix} 1 & v & 0 \\ v & 1 & 0 \\ 0 & 0 & \frac{1-v}{2} \end{bmatrix}$$

(5.1.16)

and $[B]$ relates strains to nodal displacements. The strain matrix given by equation (7.2.8) can be expressed in operator form

284 Dynamics of Plates

$$\left\{\begin{array}{c}\varepsilon_{xx}\\ \varepsilon_{yy}\\ \gamma_{xy}\end{array}\right\}=[B]\{\hat{d}\}$$

$$=\begin{bmatrix}\frac{\partial}{\partial x}(\ ) & 0\\ 0 & \frac{\partial}{\partial y}(\ )\\ \frac{\partial}{\partial y}(\ ) & \frac{\partial}{\partial x}(\ )\end{bmatrix}\left\{\begin{array}{c}\hat{u}\\ \hat{v}\end{array}\right\}$$

(7.4.8)

where

$$[B]=\begin{bmatrix}N_{1,\hat{x}} & N_{2,\hat{x}} & N_{3,\hat{x}} & N_{4,\hat{x}} & 0 & 0 & 0 & 0\\ 0 & 0 & 0 & 0 & N_{1,\hat{y}} & N_{2,\hat{y}} & N_{3,\hat{y}} & N_{4,\hat{y}}\\ N_{1,\hat{y}} & N_{2,\hat{y}} & N_{3,\hat{y}} & N_{4,\hat{y}} & N_{1,\hat{x}} & N_{2,\hat{x}} & N_{3,\hat{x}} & N_{4,\hat{y}}\end{bmatrix}$$

$$\{\hat{d}\}=\left\{\begin{array}{c}\hat{u}\\ \hat{v}\end{array}\right\}\qquad \{\hat{u}\}=\left\{\begin{array}{c}\hat{u}_1\\ \hat{u}_2\\ \hat{u}_3\\ \hat{u}_4\end{array}\right\}\text{ and }\{\hat{v}\}\left\{\begin{array}{c}\hat{v}_1\\ \hat{v}_2\\ \hat{v}_3\\ \hat{v}_4\end{array}\right\}$$

(7.4.9)

Here, N_i are defined in terms of ξ and η for the quadrilateral element in the curvilinear coordinates of Fig. 7.4.2, hence we have to change the derivatives from $\frac{\partial}{\partial x}$ and $\frac{\partial}{\partial y}$ to $\frac{\partial}{\partial \xi}$ and $\frac{\partial}{\partial \eta}$. This can be accomplished by using the chain rule of partial differentiation

$$\left\{\begin{array}{c}\frac{\partial N_i}{\partial \xi}\\ \frac{\partial N_i}{\partial \eta}\end{array}\right\}=\begin{bmatrix}\frac{\partial x}{\partial \xi} & \frac{\partial y}{\partial \xi}\\ \frac{\partial x}{\partial \eta} & \frac{\partial y}{\partial \eta}\end{bmatrix}\left\{\begin{array}{c}\frac{\partial N_i}{\partial x}\\ \frac{\partial N_i}{\partial y}\end{array}\right\}$$

$$=[J]\left\{\begin{array}{c}\frac{\partial N_i}{\partial x}\\ \frac{\partial N_i}{\partial y}\end{array}\right\}$$

(7.4.10)

where

$$[J] = \begin{bmatrix} \dfrac{\partial x}{\partial \xi} & \dfrac{\partial y}{\partial \xi} \\ \dfrac{\partial x}{\partial \eta} & \dfrac{\partial y}{\partial \eta} \end{bmatrix}$$

(7.4.11)

is the *Jacobian matrix*. The above relation can be used to transform the derivatives from ξ and η coordinates to x and y coordinates and vice versa. From (7.4.10), we can obtain

$$\left\{ \begin{array}{c} \dfrac{\partial N_i}{\partial x} \\ \dfrac{\partial N_i}{\partial y} \end{array} \right\} = \dfrac{1}{\det[J]} \left\{ \begin{array}{c} \dfrac{\partial y}{\partial \eta} \dfrac{\partial N_i}{\partial \xi} - \dfrac{\partial y}{\partial \xi} \dfrac{\partial N_i}{\partial \eta} \\ \dfrac{\partial x}{\partial \xi} \dfrac{\partial N_i}{\partial \eta} - \dfrac{\partial x}{\partial \eta} \dfrac{\partial N_i}{\partial \xi} \end{array} \right\}$$

(7.4.12)

Substituting (7.4.5) into (7.4.11), we get

$$[J] = \begin{bmatrix} \dfrac{\partial x}{\partial \xi} & \dfrac{\partial y}{\partial \xi} \\ \dfrac{\partial x}{\partial \eta} & \dfrac{\partial y}{\partial \eta} \end{bmatrix}$$

$$= \begin{bmatrix} \dfrac{\partial N_1}{\partial \xi} & \dfrac{\partial N_2}{\partial \xi} & \dfrac{\partial N_3}{\partial \xi} & \dfrac{\partial N_4}{\partial \xi} \\ \dfrac{\partial N_1}{\partial \eta} & \dfrac{\partial N_2}{\partial \eta} & \dfrac{\partial N_3}{\partial \eta} & \dfrac{\partial N_4}{\partial \eta} \end{bmatrix} \begin{bmatrix} x_1 & y_1 \\ x_2 & y_2 \\ x_3 & y_3 \\ x_4 & y_4 \end{bmatrix}$$

(7.4.13)

With the help of (7.4.6), the Jacobian in the above equation becomes

$$[J] = \dfrac{1}{4} \begin{bmatrix} -(1-\eta) & (1-\eta) & (1+\eta) & -(1+\eta) \\ -(1-\xi) & -(1+\xi) & (1+\xi) & (1-\xi) \end{bmatrix} \begin{bmatrix} x_1 & y_1 \\ x_2 & y_2 \\ x_3 & y_3 \\ x_4 & y_4 \end{bmatrix}$$

(7.4.14)

The determinant of the Jacobian in the above equation (7.4.14) can be obtained as

$$\det[J] = \frac{1}{8}\begin{bmatrix} x_1 & x_2 & x_3 & x_4 \end{bmatrix}$$

$$\times \begin{bmatrix} 0 & (1-\eta) & (\eta-\xi) & (\xi-1) \\ (\eta-1) & 0 & (\xi+1) & -(\xi+\eta) \\ (\xi-\eta) & -(\xi+1) & 0 & (\eta+1) \\ (1-\xi) & (\xi+\eta) & -(\eta+1) & 0 \end{bmatrix} \begin{Bmatrix} y_1 \\ y_2 \\ y_3 \\ y_4 \end{Bmatrix}$$

(7.4.15)

The Jacobian above is a function of ξ and η and known values of global coordinates $x_1, y_1, x_2, y_2, x_3, y_3, x_4$ and y_4. The element area in the curvilinear coordinates is given by

$$dA = \left| \vec{d\xi} \times \vec{d\eta} \right|$$

$$= \left(\frac{\partial x}{\partial \xi} \frac{\partial y}{\partial \eta} - \frac{\partial y}{\partial \xi} \frac{\partial x}{\partial \eta} \right) d\xi \, d\eta$$

$$= \det[J] d\xi \, d\eta$$

(7.4.16)

The stiffness matrix can now be written as

$$[\hat{K}] = h \int_{-1}^{1} \int_{-1}^{1} [B]^T [D][B] \det[J] d\xi \, d\eta$$

(7.4.17)

where h is the thickness of the element, matrix $[D]$ is given by equation (5.1.16), $\det[J]$ is given by equation (7.4.15) and matrix $[B]$ given by equation (7.4.8) should be converted into the natural coordinates by using (7.4.12).

$$\begin{Bmatrix} \varepsilon_{xx} \\ \varepsilon_{yy} \\ \gamma_{xy} \end{Bmatrix} = \frac{1}{\det[J]} \begin{bmatrix} \frac{\partial y}{\partial \eta}\frac{\partial}{\partial \xi}(\ ) - \frac{\partial y}{\partial \xi}\frac{\partial}{\partial \eta}(\ ) & 0 \\ 0 & \frac{\partial x}{\partial \xi}\frac{\partial}{\partial \eta}(\ ) - \frac{\partial x}{\partial \eta}\frac{\partial}{\partial \xi}(\ ) \\ \frac{\partial x}{\partial \xi}\frac{\partial}{\partial \eta}(\ ) - \frac{\partial x}{\partial \eta}\frac{\partial}{\partial \xi}(\ ) & \frac{\partial y}{\partial \eta}\frac{\partial}{\partial \xi}(\ ) - \frac{\partial y}{\partial \xi}\frac{\partial}{\partial \eta}(\ ) \end{bmatrix}$$

$$\times \begin{Bmatrix} \hat{u} \\ \hat{v} \end{Bmatrix}$$

$$= [D'][N]\{\hat{d}\}$$

$$= [\underline{B}]\{\hat{d}\} \quad (7.4.18)$$

where

$$[D'] = \frac{1}{\det[J]} \begin{bmatrix} \frac{\partial y}{\partial \eta}\frac{\partial}{\partial \xi}(\ ) - \frac{\partial y}{\partial \xi}\frac{\partial}{\partial \eta}(\ ) & 0 \\ 0 & \frac{\partial x}{\partial \xi}\frac{\partial}{\partial \eta}(\ ) - \frac{\partial x}{\partial \eta}\frac{\partial}{\partial \xi}(\ ) \\ \frac{\partial x}{\partial \xi}\frac{\partial}{\partial \eta}(\ ) - \frac{\partial x}{\partial \eta}\frac{\partial}{\partial \xi}(\ ) & \frac{\partial y}{\partial \eta}\frac{\partial}{\partial \xi}(\ ) - \frac{\partial y}{\partial \xi}\frac{\partial}{\partial \eta}(\ ) \end{bmatrix}$$

$$(7.4.19)$$

and

$$[N] = \begin{bmatrix} N_1 & 0 & N_2 & 0 & N_3 & 0 & N_4 & 0 \\ 0 & N_1 & 0 & N_2 & 0 & N_3 & 0 & N_4 \end{bmatrix}$$

$$(7.4.20)$$

$$[\underline{B}] = [D'][N] \quad (7.4.21)$$

The $[\underline{B}]$ matrix in the above equation (7.4.21) can be written in an explicit form after performing the required matrix multiplication

$$[\underline{B}(\xi,\eta)]_{3\times 8} = \frac{1}{\det[J]}\begin{bmatrix} [\underline{B}]_1 & [\underline{B}]_2 & [\underline{B}]_3 & [\underline{B}]_4 \end{bmatrix}$$

$$(7.4.22)$$

where

$$[\underline{B}]_i = \begin{bmatrix} a\frac{\partial}{\partial \xi}(N_i) - b\frac{\partial}{\partial \eta}(N_i) & 0 \\ 0 & c\frac{\partial}{\partial \eta}(N_i) - d\frac{\partial}{\partial \xi}(N_i) \\ c\frac{\partial}{\partial \eta}(N_i) - d\frac{\partial}{\partial \xi}(N_i) & a\frac{\partial}{\partial \xi}(N_i) - b\frac{\partial}{\partial \eta}(N_i) \end{bmatrix} \quad i = 1,2,3,4$$

$$(7.4.23)$$

$$a = \frac{1}{4}[(1-\xi)(y_4 - y_1) + (1+\xi)(y_3 - y_2)]$$

$$b = \frac{1}{4}[(1-\eta)(y_2 - y_1) + (1+\eta)(y_3 - y_4)]$$

$$c = \frac{1}{4}[(1-\eta)(x_2 - x_1) + (1+\eta)(x_3 - x_4)]$$

$$d = \frac{1}{4}[(1-\xi)(x_4 - x_1) + (1+\xi)(x_3 - x_2)]$$

(7.4.24)

and from (7.4.6)

$$\frac{\partial}{\partial \xi}(N_1) = \frac{1}{4}(\eta - 1)$$

$$\frac{\partial}{\partial \eta}(N_1) = \frac{1}{4}(\xi - 1)$$

$$\cdots$$

(7.4.25)

The stiffness matrix of the element is now given by

$$[\hat{K}] = h \int_{-1}^{1} \int_{-1}^{1} [\underline{B}]^T [D][\underline{B}] \det[J] d\xi\, d\eta$$

(7.4.26)

The integration in the above equation is best carried by numerical integration using *Guassian quadrature* formula.

$$\int_{-1}^{1} \int_{-1}^{1} f(\xi, \eta) d\xi\, d\eta = \sum_i \sum_j W_i W_j f(\xi_i, \eta_i)$$

(7.4.27)

where $i=1$ to M and $j=1$ to N and W are corresponding *weight factors* for the *Gaussian points* (ξ_i, η_i). For $M=N=2$, the Guassian points are given by $\xi_i, \eta_i = \pm\frac{1}{\sqrt{3}}$ with the weight factors equal to unity. With the help of the quadrature formula in (7.4.27), the stiffness matrix in (7.4.26) can be evaluated as

$$[\hat{K}] = h\left[\underline{B}\left(-\frac{1}{\sqrt{3}}, -\frac{1}{\sqrt{3}}\right)\right]^T [D]\left[\underline{B}\left(-\frac{1}{\sqrt{3}}, -\frac{1}{\sqrt{3}}\right)\right] \det\left[J\left(-\frac{1}{\sqrt{3}}, -\frac{1}{\sqrt{3}}\right)\right]$$

$$+ h\left[\underline{B}\left(-\frac{1}{\sqrt{3}}, \frac{1}{\sqrt{3}}\right)\right]^T [D]\left[\underline{B}\left(-\frac{1}{\sqrt{3}}, \frac{1}{\sqrt{3}}\right)\right] \det\left[J\left(-\frac{1}{\sqrt{3}}, \frac{1}{\sqrt{3}}\right)\right]$$

$$+ h\left[\underline{B}\left(\frac{1}{\sqrt{3}}, -\frac{1}{\sqrt{3}}\right)\right]^T [D]\left[\underline{B}\left(\frac{1}{\sqrt{3}}, -\frac{1}{\sqrt{3}}\right)\right] \det\left[J\left(\frac{1}{\sqrt{3}}, -\frac{1}{\sqrt{3}}\right)\right]$$

$$+ h\left[\underline{B}\left(\frac{1}{\sqrt{3}}, \frac{1}{\sqrt{3}}\right)\right]^T [D]\left[\underline{B}\left(\frac{1}{\sqrt{3}}, \frac{1}{\sqrt{3}}\right)\right] \det\left[J\left(\frac{1}{\sqrt{3}}, \frac{1}{\sqrt{3}}\right)\right]$$

(7.4.28)

The elemental mass matrix, see equation (7.1.46) is also evaluated in a similar manner by writing

$$[\widehat{M}] = \rho h \int_{-1}^{1} \int_{-1}^{1} [N]^T [N] \det[J] d\xi\, d\eta$$

(7.4.29)

The force matrix (7.1.25a) is also written in a similar way to give

$$\{\hat{f}\} = \{\hat{P}\} + \int_{-1}^{1} \int_{-1}^{1} [N]^T \{T\} \det[J] d\xi\, d\eta$$

(7.4.30)

7.5 EIGHT NODED ISOPARAMETRIC ELEMENT

In section 7.4, we have developed an isoparametric quadrilateral element, which enables the element edges to be straight lines not necessarily perpendicular to each other. To improve the accuracy of modeling, we need a curve to represent the edge rather than a straight line. In order to achieve this capability, we employ additional nodes, e.g., eight instead of four in the quadrilateral element. Increasing the number of nodes in this manner gives us *higher order elements*. We first consider an eight noded sixteen degrees of freedom element.

Figure 7.5.1 shows the square element with eight nodes, four corner nodes 1, 2, 3 and 4 and four mid side nodes 5, 6, 7 and 8 in $\hat{x}$, $\hat{y}$ natural coordinate system attached to its center. The coordinates are so non-dimensionalized that its four corner nodes have the coordinate values ± 1 and mid side nodes have the coordinate values $\pm\frac{1}{2}$. Since we have eight nodes, the displacement functions are chosen as

290 Dynamics of Plates

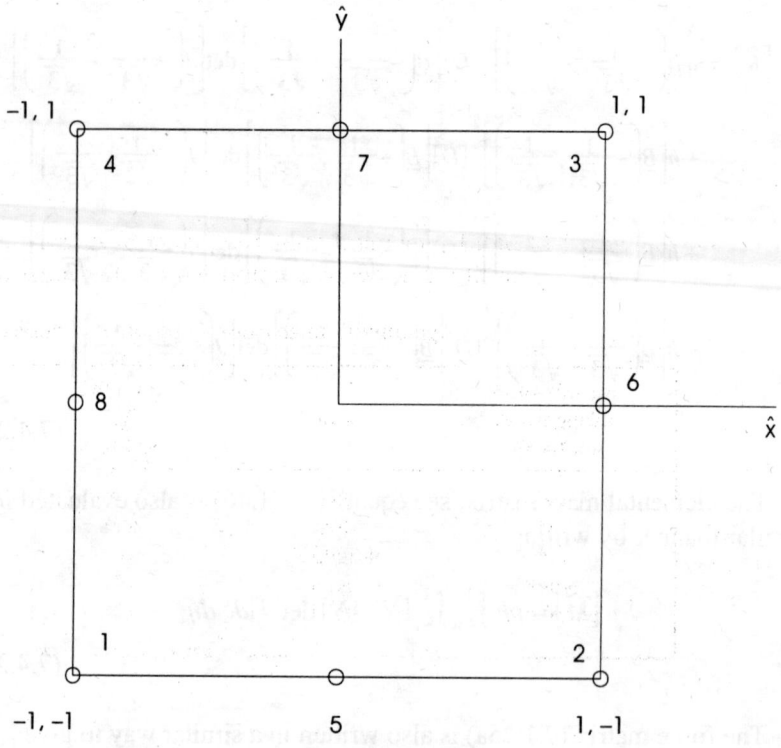

Fig. 7.5.1 Square Element in natural coordinates

$$u(\hat{x},\hat{y}) = c_1 + c_2\hat{x} + c_3\hat{y} + c_4\hat{x}^2 + c_5\hat{x}\hat{y} + c_6\hat{y}^2 + c_7\hat{x}^2\hat{y} + c_8\hat{x}\hat{y}^2$$
$$v(\hat{x},\hat{y}) = c_1 + c_2\hat{x} + c_3\hat{y} + c_4\hat{x}^2 + c_5\hat{x}\hat{y} + c_6\hat{y}^2 + c_7\hat{x}^2\hat{y} + c_8\hat{x}\hat{y}^2$$

(7.5.1)

The constants c_1 to c_8 can be evaluated by substituting the nodal values in the above equation (7.5.1). After rearranging, we get

$$u(\hat{x},\hat{y}) = \sum_{i=1}^{8} N_i(\hat{x},\hat{y})\hat{u}_i$$

$$v(\hat{x},\hat{y}) = \sum_{i=1}^{8} N_i(\hat{x},\hat{y})\hat{v}_i$$

(7.5.2)

$$N_1(\hat{x},\hat{y}) = -\frac{1}{4}(\hat{x}-1)(\hat{y}-1)(\hat{x}+\hat{y}+1)$$

$$N_2(\hat{x},\hat{y}) = -\frac{1}{4}(\hat{x}+1)(\hat{y}-1)(\hat{x}-\hat{y}-1)$$

$$N_3(\hat{x},\hat{y}) = \frac{1}{4}(\hat{x}+1)(\hat{y}+1)(\hat{x}+\hat{y}-1)$$

$$N_4(\hat{x},\hat{y}) = \frac{1}{4}(\hat{x}-1)(\hat{y}+1)(\hat{x}-\hat{y}+1)$$

$$N_5(\hat{x},\hat{y}) = \frac{1}{2}(\hat{x}-1)(\hat{x}+1)(\hat{y}-1)$$

$$N_6(\hat{x},\hat{y}) = -\frac{1}{2}(\hat{x}+1)(\hat{y}+1)(\hat{y}-1)$$

$$N_7(\hat{x},\hat{y}) = -\frac{1}{2}(\hat{x}-1)(\hat{x}+1)(\hat{y}+1)$$

$$N_8(\hat{x},\hat{y}) = \frac{1}{2}(\hat{x}-1)(\hat{y}-1)(\hat{y}+1)$$

(7.5.3)

Consider now the isoparametric element in Fig. 7.5.2 whose eight nodes are defined in the global coordinates (x, y). The element is formulated using curvilinear coordinates (ξ, η). The shape functions in (7.5.3) are now used to map the square element of Fig. 7.5.1 in isoparametric coordinates (ξ, η) to the curvilinear element of Fig. 7.5.2 in the same form to define

$$x(\xi,\eta) = c_1 + c_2\xi + c_3\eta + c_4\xi^2 + c_5\xi\eta + c_6\eta^2 + c_7^2\xi\eta + c_8\xi\eta^2$$

$$y(\xi,\eta) = c_1 + c_2\xi + c_3\eta + c_4\xi^2 + c_5\xi\eta + c_6\eta^2 + c_7^2\xi\eta + c_8\xi\eta^2$$

(7.5.4)

and establish the geometric functions

$$x(\xi,\eta) = \sum_{i=1}^{8} N_i(\xi,\eta) x_i$$

$$y(\xi,\eta) = \sum_{i=1}^{8} N_i(\xi,\eta) y_i$$

(7.5.5)

where

$$N_1(\xi,\eta) = -\frac{1}{4}(\xi-1)(\eta-1)(\xi+\eta+1)$$

$$N_2(\xi,\eta) = -\frac{1}{4}(\xi+1)(\eta-1)(\xi-\eta-1)$$

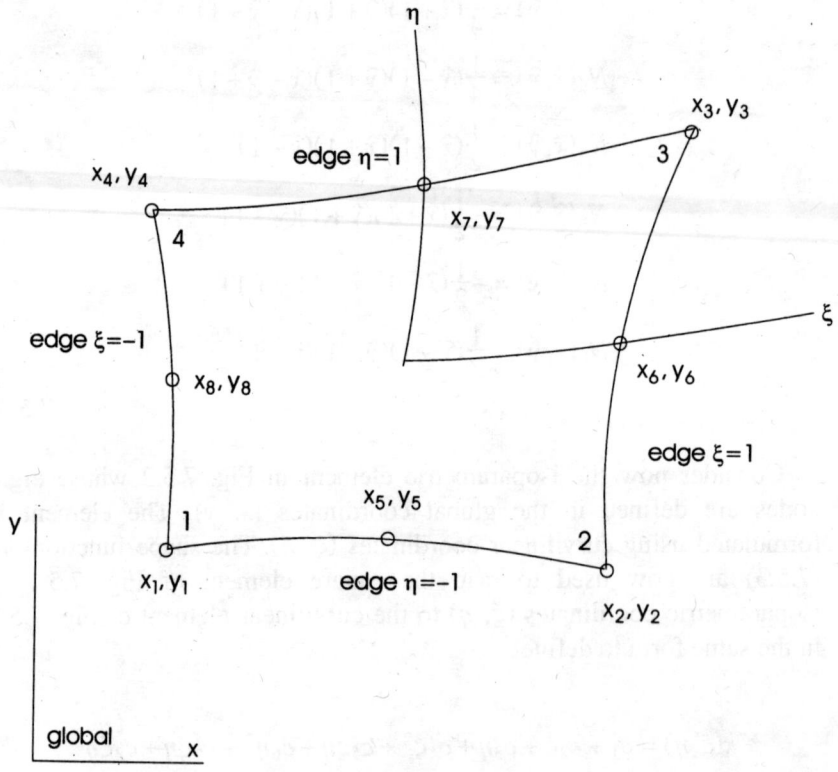

Fig. 7.5.2 Eight Noded Isoparametric Element

$$N_3(\xi,\eta) = \frac{1}{4}(\xi+1)(\eta+1)(\xi+\eta-1)$$

$$N_4(\xi,\eta) = \frac{1}{4}(\xi-1)(\eta+1)(\xi-\eta+1)$$

$$N_5(\xi,\eta) = \frac{1}{2}(\xi-1)(\xi+1)(\eta-1)$$

$$N_6(\xi,\eta) = -\frac{1}{2}(\xi+1)(\eta+1)(\eta-1)$$

$$N_7(\xi,\eta) = -\frac{1}{2}(\xi-1)(\xi+1)(\eta+1)$$

$$N_8(\xi,\eta) = \frac{1}{2}(\xi-1)(\eta-1)(\eta+1)$$

(7.5.6)

The displacement functions for the distorted element are

$$u(\xi,\eta) = \sum_{i=1}^{8} N_i(\xi,\eta) u_i$$

$$v(\xi,\eta) = \sum_{i=1}^{8} N_i(\xi,\eta) v_i$$

(7.5.7)

The strain displacement relations are expressed as

$$\left\{\begin{array}{c} \varepsilon_{xx} \\ \varepsilon_{yy} \\ \gamma_{xy} \end{array}\right\} = \left[\begin{array}{cc} \frac{\partial}{\partial x}(N_i) & 0 \\ 0 & \frac{\partial}{\partial y}(N_i) \\ \frac{\partial}{\partial y}(N_i) & \frac{\partial}{\partial x}(N_i) \end{array}\right]_{3\times 16} \left\{\begin{array}{c} \hat{u} \\ \hat{v} \end{array}\right\}_{16\times 1}$$

(7.5.8)

The transformation rule by Jacobian matrix remains same as

$$\left\{\begin{array}{c} \frac{\partial N_i}{\partial \xi} \\ \frac{\partial N_i}{\partial \eta} \end{array}\right\} = \left[\begin{array}{cc} \frac{\partial x}{\partial \xi} & \frac{\partial y}{\partial \xi} \\ \frac{\partial x}{\partial \eta} & \frac{\partial y}{\partial \eta} \end{array}\right] \left\{\begin{array}{c} \frac{\partial N_i}{\partial x} \\ \frac{\partial N_i}{\partial y} \end{array}\right\}$$

$$= [J] \left\{\begin{array}{c} \frac{\partial N_i}{\partial x} \\ \frac{\partial N_i}{\partial y} \end{array}\right\}$$

(7.4.10)

where the Jacobian now is given by

$$[J]_{2\times 2} = \left[\begin{array}{cc} \frac{\partial x}{\partial \xi} & \frac{\partial y}{\partial \xi} \\ \frac{\partial x}{\partial \eta} & \frac{\partial y}{\partial \eta} \end{array}\right]$$

$$= \left[\begin{array}{cccc} \frac{\partial N_1}{\partial \xi} & \frac{\partial N_2}{\partial \xi} & \frac{\partial N_3}{\partial \xi} & \cdots \\ \frac{\partial N_1}{\partial \eta} & \frac{\partial N_2}{\partial \eta} & \frac{\partial N_3}{\partial \eta} & \cdots \end{array}\right]_{2\times 8} \left[\begin{array}{cc} x_1 & y_1 \\ x_2 & y_2 \\ x_3 & y_3 \\ \vdots & \vdots \end{array}\right]_{8\times 2}$$

(7.5.9)

Equation (7.5.8) above is written as

$$\left\{ \begin{array}{c} \varepsilon_{xx} \\ \varepsilon_{yy} \\ \gamma_{xy} \end{array} \right\} = [B]_{3 \times 16} \{\hat{d}\}_{16 \times 1}$$

(7.5.10)

where

$$[B] = \begin{bmatrix} N_{i,\hat{x}} & 0 \\ 0 & N_{i,\hat{y}} \\ N_{i,\hat{y}} & N_{i,\hat{x}} \end{bmatrix}_{3 \times 16} \quad \{\hat{d}\}^T = \begin{bmatrix} \hat{u}_1 & \hat{u}_2 & \ldots & \hat{u}_8 & \hat{v}_1 & \hat{v}_2 & \ldots & \hat{v}_8 \end{bmatrix}$$

(7.5.10)

The stiffness matrix of the element is given by

$$[\hat{K}] = h \int_{-1}^{1} \int_{-1}^{1} [B]^T [D][B] \det[J] d\xi \, d\eta$$

(7.4.17)

where h is the thickness of the element, matrix $[D]$ is given by equation (5.1.16), Jacobian $[J]$ is given by equation (7.5.9) and matrix $[B]$ given by equation (7.5.10) should be converted into the natural coordinates by using (7.4.12).

$$\left\{ \begin{array}{c} \varepsilon_{xx} \\ \varepsilon_{yy} \\ \gamma_{xy} \end{array} \right\} = \frac{1}{\det[J]} \begin{bmatrix} \frac{\partial y}{\partial \eta}\frac{\partial}{\partial \xi}(\ ) - \frac{\partial y}{\partial \xi}\frac{\partial}{\partial \eta}(\ ) & 0 \\ 0 & \frac{\partial x}{\partial \xi}\frac{\partial}{\partial \eta}(\ ) - \frac{\partial x}{\partial \eta}\frac{\partial}{\partial \xi}(\ ) \\ \frac{\partial x}{\partial \xi}\frac{\partial}{\partial \eta}(\ ) - \frac{\partial x}{\partial \eta}\frac{\partial}{\partial \xi}(\ ) & \frac{\partial y}{\partial \eta}\frac{\partial}{\partial \xi}(\ ) - \frac{\partial y}{\partial \xi}\frac{\partial}{\partial \eta}(\ ) \end{bmatrix}$$

$$\times \left\{ \begin{array}{c} \hat{u} \\ \hat{v} \end{array} \right\}$$

$$= [D']_{3 \times 2} [N]_{2 \times 16} \{\hat{d}\}_{16 \times 1}$$

$$= [\underline{B}]_{3 \times 16} \{\hat{d}\}$$

(7.5.11)

where

$$[D'] = \frac{1}{\det[J]} \begin{bmatrix} \frac{\partial y}{\partial \eta}\frac{\partial}{\partial \xi}(\ ) - \frac{\partial y}{\partial \xi}\frac{\partial}{\partial \eta}(\ ) & 0 \\ 0 & \frac{\partial x}{\partial \xi}\frac{\partial}{\partial \eta}(\ ) - \frac{\partial x}{\partial \eta}\frac{\partial}{\partial \xi}(\ ) \\ \frac{\partial x}{\partial \xi}\frac{\partial}{\partial \eta}(\ ) - \frac{\partial x}{\partial \eta}\frac{\partial}{\partial \xi}(\ ) & \frac{\partial y}{\partial \eta}\frac{\partial}{\partial \xi}(\ ) - \frac{\partial y}{\partial \xi}\frac{\partial}{\partial \eta}(\ ) \end{bmatrix}$$

(7.5.12)

and

$$[N] = \begin{bmatrix} N_1 & 0 & N_2 & 0 & \cdots & N_8 & 0 \\ 0 & N_1 & 0 & N_2 & \cdots & 0 & N_8 \end{bmatrix}_{2 \times 16}$$

(7.5.13)

$$[\underline{B}]_{3 \times 16} = [D']_{3 \times 2}[N]_{2 \times 16}$$

(7.5.14)

Continuing further on similar lines as in section 7.4, we get the elemental stiffness matrix.

$$[\hat{k}]_{16 \times 16} = h \int_{-1}^{1} \int_{-1}^{1} [\underline{B}]^T [D][\underline{B}] \det[J] d\xi \, d\eta$$

(7.5.15)

With an increase in the number of nodes of the element, the geometry of the boundaries is significantly improved to fit any arbitrary curve. Further the element accuracy in general increases since the shape functions are of higher order. However, it is easy to observe that the integration effort also increases significantly with increase in the number of elements.

7.6 EIGHT NODED ISOPARAMETRIC MINDLIN PLATE ELEMENT

In the previous section, we considered an eight noded isoparametric membrane element. Here, this element is extended to a plate element taking into account rotary inertia terms in accordance to Mindlin plate theory. The geometric shape functions of the isoparametric element in Fig. 7.5.2 are now extended to

296 Dynamics of Plates

$$x(\xi,\eta) = \sum_{i=1}^{8} N_i(\xi,\eta)x_i$$

$$y(\xi,\eta) = \sum_{i=1}^{8} N_i(\xi,\eta)y_i$$

$$z(\xi,\eta) = \sum_{i=1}^{8} N_i(\xi,\eta)z_i$$

(7.6.1)

where x_i, y_i, z_i are the nodal coordinates and the shape functions are defined in equation (7.5.6). Equation (7.6.1) can be written as

$$\left\{\begin{array}{c} x(\xi,\eta) \\ y(\xi,\eta) \\ z(\xi,\eta) \end{array}\right\} = [\overline{N}_1] \left\{\begin{array}{c} x_1 \\ y_1 \\ z_1 \\ x_2 \\ y_2 \\ z_2 \\ \vdots \\ x_8 \\ y_8 \\ z_8 \end{array}\right\}$$

$$\{\bar{x}\} = [\overline{N}_1]\{\hat{x}\}$$

(7.6.2)

where

$$[\overline{N}_1] = \begin{bmatrix} N_1(\xi,\eta) & 0 & 0 & \cdots & N_8(\xi,\eta) & 0 & 0 \\ 0 & N_1(\xi,\eta) & 0 & \cdots & 0 & N_8(\xi,\eta) & 0 \\ 0 & 0 & N_1(\xi,\eta) & \cdots & 0 & 0 & N_8(\xi,\eta) \end{bmatrix}_{3\times 24}$$

(7.6.3)

and

$$\{\hat{x}\}^T = \begin{bmatrix} x_1 & y_1 & z_1 & x_2 & y_2 & z_2 & \cdots & x_8 & y_8 & z_8 \end{bmatrix}$$

(7.6.4)

The displacement field from Mindlin plate theory is given by equation (3.8.1), which is modified to include the middle plane displacements u_0, v_0, w_0 to give

$$u(x,y,t) = u_0(x,y,t) + z\psi_x(x,y,t)$$
$$v(x,y,t) = v_0(x,y,t) + z\psi_y(x,y,t)$$
$$w(x,y,t) = w_0(x,y,t)$$

(7.6.5)

where ψ_x, ψ_y are the slopes due to bending, taken positive in the counterclockwise directions i.e., in the positive directions of y and x axes. In accordance to isoparametric formulation, the displacement field above is expressed as

$$u(\xi,\eta) = \sum_{i=1}^{8} N_i(\xi,\eta) u_i + z \sum_{i=1}^{8} N_i(\xi,\eta) \psi_{xi}$$

$$v(\xi,\eta) = \sum_{i=1}^{8} N_i(\xi,\eta) v_i + z \sum_{i=1}^{8} N_i(\xi,\eta) \psi_{yi}$$

$$w(\xi,\eta) = \sum_{i=1}^{8} N_i(\xi,\eta) w_i$$

(7.6.6)

The above equation can be written as

$$\left\{ \begin{array}{c} u(\xi,\eta) \\ v(\xi,\eta) \\ w(\xi,\eta) \end{array} \right\} = \left[\overline{N}_2 \right]_{3\times 40} \{\hat{d}\}_{40\times 1}$$

$$\{\delta\} = \left[\overline{N}_2 \right]_{3\times 40} \{\hat{d}\}_{40\times 1}$$

(7.6.7)

where

$$\left[\overline{N}_2(\xi,\eta) \right] = \begin{bmatrix} N_1 & 0 & 0 & zN_1 & 0 & \cdots & N_8 & 0 & 0 & zN_8 & 0 \\ 0 & N_1 & 0 & 0 & zN_1 & \cdots & 0 & N_8 & 0 & 0 & zN_8 \\ 0 & 0 & N_1 & 0 & 0 & \cdots & 0 & 0 & N_8 & 0 & 0 \end{bmatrix}_{3\times 40}$$

(7.6.8)

$$\{\delta\}^T = \{ u \ v \ w \}$$

(7.6.9)

and the nodal degrees of freedom (40) for the element are

$$\{\hat{d}\}^T = \begin{bmatrix} u_1 & v_1 & w_1 & \psi_{x1} & \psi_{y1} & \cdots & u_8 & v_8 & w_8 & \psi_{x8} & \psi_{y8} \end{bmatrix}_{1 \times 40}$$

(7.6.10)

The strain displacement relations in equation (3.8.2) are modified taking into account of the middle plane displacements as well as the positive directions of slopes to give

$$\varepsilon_{xx} = \frac{\partial u_0}{\partial x} + z \frac{\partial \psi_x}{\partial x}$$

$$\varepsilon_{yy} = \frac{\partial v_0}{\partial y} + z \frac{\partial \psi_y}{\partial y}$$

$$\gamma_{xy} = \frac{\partial u_0}{\partial y} + \frac{\partial v_0}{\partial x} + z \left(\frac{\partial \psi_x}{\partial y} + \frac{\partial \psi_y}{\partial x} \right)$$

$$\gamma_{yz} = \frac{\partial w_0}{\partial y} + \psi_y$$

$$\gamma_{zx} = \frac{\partial w_0}{\partial x} + \psi_x$$

(7.6.11)

The above relations can be expressed in matrix form to give

$$\begin{Bmatrix} \varepsilon_{xx}(x,y) \\ \varepsilon_{yy}(x,y) \\ \gamma_{xy}(x,y) \\ \gamma_{yz}(x,y) \\ \gamma_{zx}(x,y) \end{Bmatrix} = \left[\overline{N}_3 \right]_{5 \times 40} \{\hat{d}\}_{40 \times 1}$$

(7.6.12)

where

$$[\overline{N}_3] = \begin{bmatrix} N_{1,x} & 0 & 0 & zN_{1,x} & 0 & \cdots & N_{8,x} & 0 & 0 & zN_{8,x} & 0 \\ 0 & N_{1,y} & 0 & 0 & zN_{1,y} & \cdots & 0 & N_{8,y} & 0 & 0 & zN_{8,y} \\ N_{1,y} & N_{1,x} & 0 & zN_{1,y} & zN_{1,x} & \cdots & N_{8,y} & N_{8,x} & 0 & zN_{8,y} & zN_{8,x} \\ 0 & 0 & N_{1,y} & 0 & N_1 & \cdots & 0 & 0 & N_{8,y} & 0 & N_8 \\ 0 & 0 & N_{1,x} & N_1 & 0 & \cdots & 0 & 0 & N_{8,x} & N_8 & 0 \end{bmatrix}_{5 \times 40}$$

(7.6.13)

We can use equation (7.4.12) to get $N_{i,x}$ and $N_{i,y}$ the partial derivatives of the shape functions with respect to the local coordinates

$$\begin{Bmatrix} N_{i,x} \\ N_{i,y} \end{Bmatrix} = [J]^{-1} \begin{Bmatrix} N_{i,\xi} \\ N_{i,\eta} \end{Bmatrix}$$

(7.6.14)

The stress strain relationship for the plate can be written as

$$\begin{Bmatrix} \tau_{xx} \\ \tau_{yy} \\ \tau_{xy} \\ \tau_{yz} \\ \tau_{zx} \end{Bmatrix} = \frac{E}{1-v^2} \begin{bmatrix} 1 & v & 0 & 0 & 0 \\ v & 1 & 0 & 0 & 0 \\ 0 & 0 & \frac{1-v}{2} & 0 & 0 \\ 0 & 0 & 0 & \frac{1-v}{2} & 0 \\ 0 & 0 & 0 & 0 & \frac{1-v}{2} \end{bmatrix} \begin{Bmatrix} \varepsilon_{xx} \\ \varepsilon_{yy} \\ \gamma_{xy} \\ \gamma_{yz} \\ \gamma_{zx} \end{Bmatrix}$$

$$\{\tau\} = [D]\{\varepsilon\}$$

(7.6.15)

The strain energy in the plate consists of two parts, one due to bending and the other due to in plane loads. The strain energy due to bending is

$$U_b = \frac{1}{2} \iiint_V \{\tau\}^T \{\varepsilon\} dV$$

$$= \frac{1}{2} \iiint_V \{\varepsilon\}^T [D] \{\varepsilon\} dV$$

(7.6.16)

300 Dynamics of Plates

Let the plate be subjected to an initial in plane stress field, τ_{xx}^0, τ_{yy}^0 and τ_{xy}^0, then the corresponding strain energy is

$$U_P = \iiint_V \{ \tau_{xx}^0 \quad \tau_{yy}^0 \quad \tau_{xy}^0 \} \begin{Bmatrix} \frac{1}{2} w_{,x}^2 \\ \frac{1}{2} w_{,y}^2 \\ w_{,x} w_{,y} \end{Bmatrix} dV$$

$$= \frac{1}{2} \iiint_V \{ w_{,x} \quad w_{,y} \} \begin{bmatrix} \tau_{xx}^0 & \tau_{xy}^0 \\ \tau_{xy}^0 & \tau_{yy}^0 \end{bmatrix} \begin{Bmatrix} w_{,x} \\ w_{,y} \end{Bmatrix} dV$$

$$= \frac{1}{2} \iiint_V \{\varepsilon_g\}^T [\tau^0] \{\varepsilon_g\} dV$$

(7.6.17)

where

$$\{\varepsilon_g\} = \begin{Bmatrix} w_{,x} \\ w_{,y} \end{Bmatrix} = [\overline{N}_4]\{\hat{a}\}$$

(7.6.18)

and

$$[\overline{N}_4] = \begin{bmatrix} 0 & 0 & N_{1,x} & 0 & 0 & \ldots & 0 & 0 & N_{8,x} & 0 & 0 \\ 0 & 0 & N_{1,y} & 0 & 0 & \ldots & 0 & 0 & N_{8,y} & 0 & 0 \end{bmatrix}_{2 \times 40}$$

(7.6.19)

Making use of (7.6.12) and (7.6.18), equations (7.6.16) and (7.6.17) become

$$U_b = \frac{1}{2} \{\hat{a}\}^T \iiint_V [\overline{N}_3]^T [D][\overline{N}_3] dV \{\hat{a}\}$$

(7.6.20)

$$U_P = \frac{1}{2} \{\hat{a}\}^T \iiint_V [\overline{N}_4]^T [\tau^0][\overline{N}_4] dV \{\hat{a}\}$$

(7.6.21)

Using Lagrangian approach, the elemental stiffness matrix is obtained as

$$[\hat{K}] = \iiint_V [\overline{N}_3]^T [D][\overline{N}_3] dV + \iiint_V [\overline{N}_4]^T [\tau^0][\overline{N}_4] dV$$

$$= [\hat{K}_e] + [\hat{K}_g]$$

(7.6.22)

where $[\hat{K}_e]$ is the *elastic stiffness matrix*

$$[\hat{K}_e] = \iiint_V [\overline{N}_3]^T [D][\overline{N}_3] dV \tag{7.6.23}$$

and $[\hat{K}_g]$ called *geometric stiffness matrix* is

$$[\hat{K}_g] = \iiint_V [\overline{N}_4]^T [\tau^0][\overline{N}_4] dV \tag{7.6.24}$$

With the help of equation (7.4.16), we can write the above as

$$[\hat{K}] = \iiint_V [\overline{N}_3]^T [D][\overline{N}_3] \det[J] d\xi\, d\eta\, dz$$
$$+ \iiint_V [\overline{N}_4]^T [\tau^0][\overline{N}_4] \det[J] d\xi\, d\eta\, dz \tag{7.6.25}$$

In the above equation, the partial derivatives $\frac{\partial}{\partial x}$ and $\frac{\partial}{\partial y}$ are to be converted to $\frac{\partial}{\partial \xi}$ and $\frac{\partial}{\partial \eta}$ respectively by making use of (7.6.14) before applying the Gaussian quadrature formula to evaluate the elemental stiffness matrix.

The mass matrix can be easily shown to be

$$[\hat{K}] = \iiint_V \varrho [\overline{N}_2]^T [\overline{N}_2] \det[J] d\xi\, d\eta\, dz \tag{7.6.26}$$

Let the external force be periodic with F_1, F_2 and F_3 as the force components per unit area, then elemental excitation force matrix is

$$\{\hat{F}\} = \left(\iint_A [\overline{N}_2]^T \begin{Bmatrix} F_1 \\ F_2 \\ F_3 \end{Bmatrix} dA \right) \cos(\omega t - \phi) \tag{7.6.27}$$

Example
Consider a freely vibrating square rectangular cantilever plate 0.05 by 0.05 m with 1 mm thickness. Use 16 elements with 65 nodes. Take $E =$

710 GPa, G = 280 GPa, Poisson's ratio = 0.25 and density = 2300 kg/m³. The plate is divided into 16 elements with 65 nodes.

The non dimensional frequencies obtained by a computer program using the above analysis are given below.

Non dimensional Frequency $pa^2\sqrt{\dfrac{\varrho h}{D}}$

Mode No.	Frequency
1. I Bending	3.353
2. I Torsion	8.889
3. II Bending	23.242
4. Plate Mode	28.600
5. II Torsion	33.510
6. Plate Mode	61.621

The corresponding mode shapes of the above frequencies are given in Figs. 7.6.1 to 7.6.6.

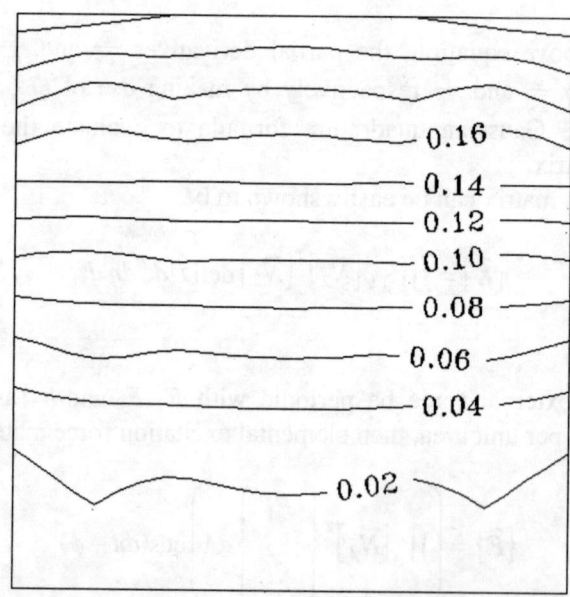

Fig. 7.6.1 The 1st Mode Shape of Cantilever Square Plate

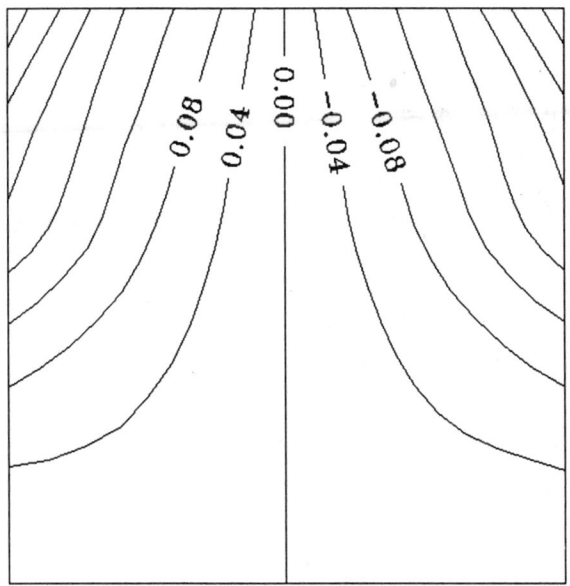

Fig. 7.6.2 The 2nd Mode Shape of Cantilever Square Plate

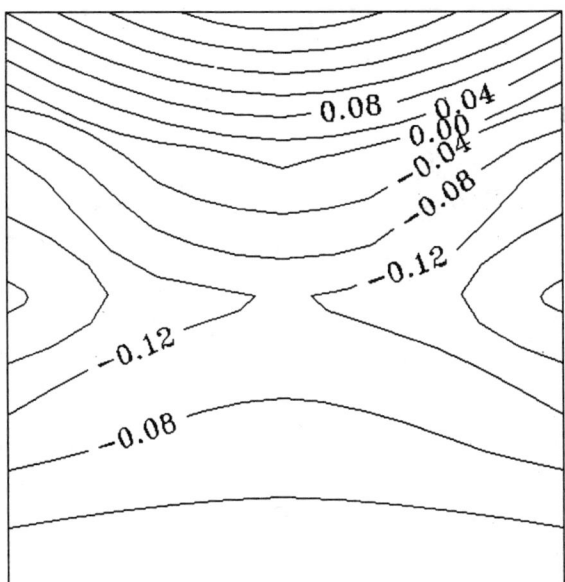

Fig. 7.6.3 The 3rd Mode Shape of Cantilever Square Plate

304 *Dynamics of Plates*

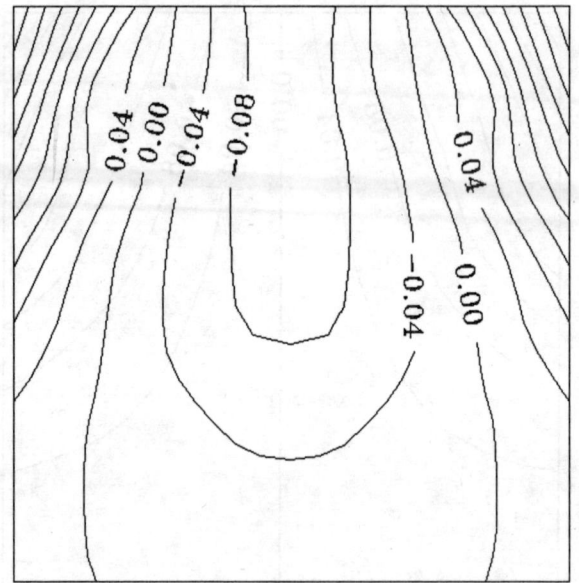

Fig. 7.6.4 The 4th Mode Shape of Cantilever Square Plate

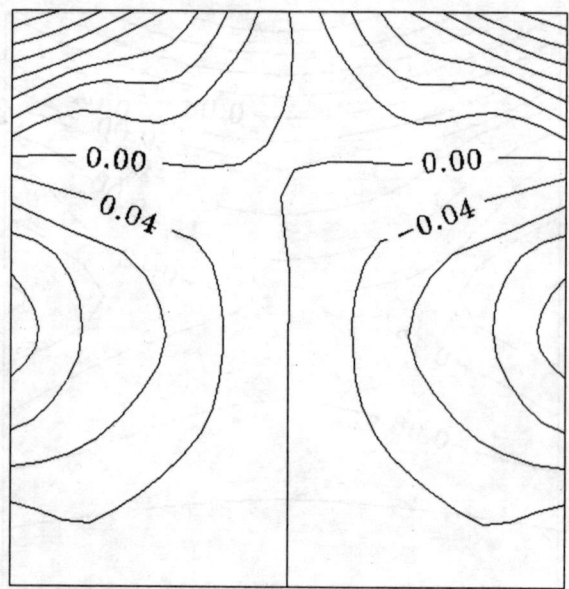

Fig. 7.6.5 The 5th Mode Shape of Cantilever Square Plate

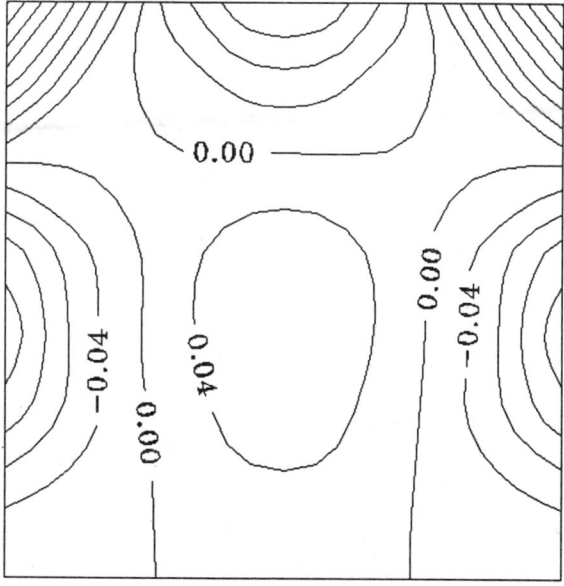

Fig. 7.6.6 The 6th Mode Shape of Cantilever Square Plate

7.7 EXAMPLE OF A ROTATING PLATE

Fig. 7.7.1 shows a plate of dimensions $a \times b$ attached to a disc of radius R rotating at an angular velocity ω rad/s. XYZ is the fixed axes system and xyz coordinate system is fixed to the rotating plate as shown. The y axis of the plate is located at a *setting angle* φ from the tangent to the disc periphery in the XY plane or at a *stagger angle* $\phi = 90^0 - \varphi$ from the cylinder axis Z.

Kinetic Energy

We will adopt the 8 noded isoparametric plate bending element with in plane displacements described in section 7.6, see Figs. 7.5.1 and 7.5.2. First consider the kinetic energy of the rotating plate. The position vector of point P after deformation in the xyz axis system can be written as

$$\overrightarrow{OP} = \begin{Bmatrix} x \\ y \\ z \end{Bmatrix} + \begin{Bmatrix} u \\ v \\ w \end{Bmatrix}$$

$$= \{\bar{x}\} + \{\delta\}$$

(7.7.1)

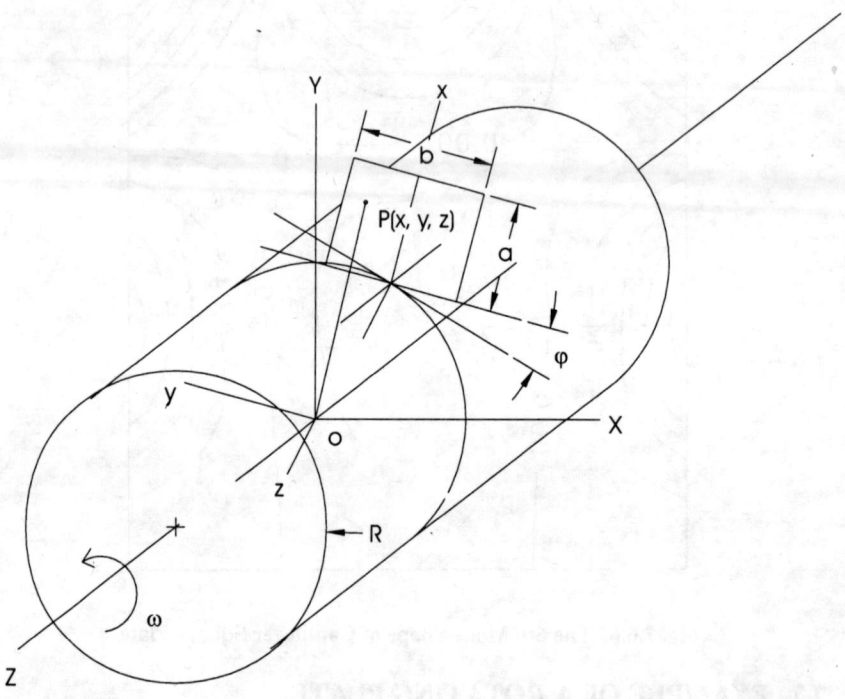

Fig. 7.7.1 A Rotating Plate

where $\{\bar{x}\}^T = \{ \; x \; y \; z \; \}$ repesents the vector of point P and its displacements are given by the vector $\{\delta\}^T = \{ \; u \; v \; w \; \}$. The angular velocity vector is written as

$$\vec{\omega} = \begin{Bmatrix} \omega_x \\ \omega_y \\ \omega_z \end{Bmatrix} = \begin{Bmatrix} 0 \\ \omega \sin\varphi \\ \omega \cos\varphi \end{Bmatrix} = \begin{Bmatrix} 0 \\ \omega \cos\phi \\ \omega \sin\phi \end{Bmatrix}$$

(7.7.2)

The velocity of point P is

$$\vec{v} = \overrightarrow{OP} + \vec{\omega} \times \overrightarrow{OP}$$

$$= \begin{Bmatrix} \dot{u} \\ \dot{v} \\ \dot{w} \end{Bmatrix} + \left(\omega_y \hat{j} + \omega_z \hat{k} \right) \times \left[(x+u)\hat{i} + (y+v)\hat{j} + (z+w)\hat{k} \right]$$

$$\begin{Bmatrix} \dot{u} \\ \dot{v} \\ \dot{w} \end{Bmatrix} + \begin{bmatrix} 0 & -\omega_z & \omega_y \\ \omega_z & 0 & 0 \\ -\omega_y & 0 & 0 \end{bmatrix} \begin{Bmatrix} x+u \\ y+v \\ z+w \end{Bmatrix}$$

$$= \{\dot{d}\} + [A](\{\bar{x}\} + \{\delta\})$$

(7.7.3)

where

$$[A] = \begin{bmatrix} 0 & -\omega_z & \omega_y \\ \omega_z & 0 & 0 \\ -\omega_y & 0 & 0 \end{bmatrix}$$

(7.7.4)

is the angular velocity matrix. The kinetic energy of the plate can now be written as

$$T = \frac{1}{2} \iiint_V \rho \vec{v} \cdot \vec{v} \, dV$$

$$= \frac{1}{2} \iiint_V \rho [\{\dot{d}\} + [A](\{\bar{x}\} + \{\delta\})] \cdot [\{\dot{d}\} + [A](\{\bar{x}\} + \{\delta\})] dV$$

$$= T_2 + T_1 + T_0$$

(7.7.5)

where

$$T_2 = \frac{1}{2} \iiint_V \rho \{\dot{\delta}\}^T \{\dot{\delta}\} dV$$

$$T_1 = \iiint_V \rho \{\dot{\delta}\}^T [A] \{\delta\} dV + \iiint_V \rho \{\dot{\delta}\}^T [A] \{\bar{x}\} dV$$

$$T_0 = \frac{1}{2} \iiint_V \rho \{\delta\}^T [A]^T [A] \{\delta\} dV + \frac{1}{2} \iiint_V \rho \{\bar{x}\}^T [A]^T [A] \{\bar{x}\} dV$$

$$+ \iiint_V \rho \{\delta\}^T [A]^T [A] \{\bar{x}\} dV$$

(7.7.6)

Elemental Equation

The position coordinates vector $\{\bar{x}\}$ and the displacement field $\{\delta\}$ for the element are given by equations (7.6.2) and (7.6.7) respectively in the curvilinear coordinates of the isoparametric element

$$\{\bar{x}\} = [\bar{N}_1]_{3\times 24}\{\hat{x}\}_{24\times 1}$$

(7.6.2)

$$\{\delta\} = [\bar{N}_2]_{3\times 40}\{\hat{d}\}_{40\times 1}$$

(7.6.7)

Substituting the above, the three components of the kinetic energy of the element from equation (7.7.6) can be obtained as

$$T_2 = \tfrac{1}{2}\iiint_V \rho\{\hat{d}\}^T[\bar{N}_2]^T[\bar{N}_2]\{\hat{d}\}dV$$

$$T_1 = \iiint_V \rho\{\hat{d}\}^T[\bar{N}_2]^T[A][\bar{N}_2]\{\hat{d}\}dV$$

$$+ \iiint_V \rho\{\hat{d}\}^T[\bar{N}_2]^T[A][\bar{N}_1]\{\hat{x}\}dV$$

$$T_0 = \tfrac{1}{2}\iiint_V \rho\{\hat{d}\}^T[\bar{N}_2]^T[A]^T[A][\bar{N}_2]\{\hat{d}\}dV$$

$$+ \tfrac{1}{2}\iiint_V \rho\{\hat{x}\}^T[\bar{N}_1]^T[A]^T[A][\bar{N}_1]\{\hat{x}\}dV$$

$$+ \iiint_V \rho\{\hat{d}\}^T[\bar{N}_2]^T[A]^T[A][\bar{N}_1]\{\hat{x}\}dV$$

(7.7.7)

The potential energy of the element is already obtained in section 7.6, equations (7.6.16) and (7.6.17). Forming the Lagrangian expression $L = T - U$, and applying Lagrangian equation

$$\frac{d}{dt}\left(\frac{\partial L}{\partial \dot{q}_i}\right) - \frac{\partial L}{\partial q_i} = Q_i$$

(7.7.8)

the elemental equation can be obtained as

$$[\hat{M}]\{\ddot{\hat{d}}\} + [\hat{G}]\{\dot{\hat{d}}\} + [\hat{K}]\{\hat{d}\} + \{\hat{f}_a\} + \{\hat{f}_c\} = \{\hat{F}_g\}$$

(7.7.9)

where the various elemental matrices in the above equation are

Mass Matrix:

$$[\widehat{M}] = \iiint_V \rho [\overline{N}_2]^T [\overline{N}_2] dV \tag{7.7.10}$$

Coriolis Matrix:

$$[\widehat{G}] = 2 \iiint_V \rho [\overline{N}_2]^T [A][\overline{N}_2] dV \tag{7.7.11}$$

Stiffness Matrix:

$$[\widehat{K}] = [\widehat{K}_r] + [\widehat{K}_a] + [\widehat{K}_e] + [\widehat{K}_g] \tag{7.7.12}$$

$$[\widehat{K}_r] = -\iiint_V \rho [\overline{N}_2]^T [A]^T [A][\overline{N}_2] dV \tag{7.7.13}$$

$$[\widehat{K}_a] = \iiint_V \rho [\overline{N}_2]^T [\dot{A}][\overline{N}_2] dV \tag{7.7.14}$$

$$[\widehat{K}_e] = \iiint_V [\overline{N}_3]^T [D][\overline{N}_3] dV \tag{7.6.23}$$

$$[\widehat{K}_g] = \iiint_V [\overline{N}_4]^T [\tau^0][\overline{N}_4] dV \tag{7.6.24}$$

$$\{\widehat{f}_a\} = -\iiint_V \rho [\overline{N}_2]^T [\dot{A}]^T [\overline{N}_1] dV \{\hat{x}\} \tag{7.7.15}$$

$$\{\widehat{f}_c\} = -\iiint_V \rho [\overline{N}_2]^T [A]^T [A][\overline{N}_1] dV \{\hat{x}\} \tag{7.7.16}$$

To obtain the initial stress vector in equation (7.6.24) under the centrifugal field, we first find the initial nodal displacements from

$$[\widehat{K}_e]\{\hat{d}_c\} = \{\hat{f}_c\} \tag{7.7.17}$$

310 Dynamics of Plates

From (7.6.12) and (7.6.15), the stress vector is given by

$$\{\tau^0\} = [D][\overline{N}_3]\{\hat{d}_c\}$$

(7.7.18)

Now, we are in a position to obtain all the elemental matrices in equations (7.7.9). Gaussian quadrature can be used to evaluate the integrals with Jacobian transformation wherever required. Assembling the elemental equations, we get the global equations of motion for the rotating plate

$$[M]\{\ddot{d}\} + [G]\{\dot{d}\} + [K]\{d\} + \{f_a\} = \{F_g\}$$

(7.7.19)

Let the excitation force be harmonic and have three components, then

$$\vec{F} = \begin{Bmatrix} F_x \\ F_y \\ F_z \end{Bmatrix} \cos(\Omega t - \Phi)$$

(7.7.20)

The elemental nodal force is then given by

$$\{\hat{F}_g\} = \left(\iint_A [\overline{N}_2]^T \begin{Bmatrix} F_x \\ F_y \\ F_z \end{Bmatrix} dA \right) \cos(\Omega t - \Phi)$$

(7.7.21)

Example
A plate of dimensions 0.05×0.05 m with 1 mm thickness mounted on a disc of radius 0.05 m at a setting angle $0°$ is chosen to calculate the non dimensional natural frequencies. The Young's modulus and shear modulus of the material are taken as $0.71 \times 10^{12}, 0.28 \times 10^{12}$ Pa. The density of the material is taken as 2300 kg/m³ and Poisson's ratio is assumed 0.25.

As in the previous section, the plate is divided into 16 elements with 65 nodes.

The first five non dimensional natural frequencies $pa^2 \sqrt{\frac{\rho h}{D}}$ of the stationary plate are given in the previous section 7.6.

For a speed of rotation equal to half the first natural frequency of the stationary plate, the first six non dimensional natural frequencies are:

4.601, 9.507, 24.186, 28.969, 34.275 and 62.109.

8

Anisotropic Plates

So far we considered the plates to be isotropic in nature, i.e., the material properties are same in any direction; and it does not matter how the coordinate axes are chosen to describe the plate static or dynamic behavior. Most of the metals are isotropic in nature and their behavior as a plate can be determined using the appropriate theories developed in the earlier chapters. In general, the materials can be considered to be *anisotropic* in nature exhibiting different material properties in different directions.

8.1 GENERALIZED HOOKE'S LAW

We discussed the generalized Hooke's law in Appendix A3.8 of *Advanced Theory of Vibration* book, which in tensor notation is

$$\tau_{ijkl} = C_{ijkl}\varepsilon_{kl}$$

(8.1.1)

In matrix form, using Voigt notation, this is

$$\begin{Bmatrix} \tau_{xx} \\ \tau_{yy} \\ \tau_{zz} \\ \tau_{yz} \\ \tau_{xz} \\ \tau_{xy} \end{Bmatrix} = \begin{bmatrix} C_{11} & C_{12} & C_{13} & C_{14} & C_{15} & C_{16} \\ & C_{22} & C_{23} & C_{24} & C_{25} & C_{26} \\ & & C_{33} & C_{34} & C_{35} & C_{36} \\ & & & C_{44} & C_{45} & C_{46} \\ & \text{Sym} & & & C_{55} & C_{56} \\ & & & & & C_{66} \end{bmatrix} \begin{Bmatrix} \varepsilon_{xx} \\ \varepsilon_{yy} \\ \varepsilon_{zz} \\ \gamma_{yz} \\ \gamma_{xz} \\ \gamma_{xy} \end{Bmatrix}$$

$$\{\tau\} = [C]\{\varepsilon\}$$

(8.1.2)

Anisotropic Plates

where $\{\tau\}, \{\varepsilon\}$ are the stress and strain vectors related through $[C]$ called *stiffness matrix*. The stiffness matrix is symmetrical as shown in the above equation and contains 21 independent elastic constants. Notice the order of the stress and strain components in the respective vectors, which is the convention followed for anisotropic plates. Here, each of the stress component is dependent on all the strain components. It is also the convention to use the engineering strains in the strain vector, i.e.,

$$\varepsilon_{xx} = u_{x,x}$$
$$\varepsilon_{yy} = u_{y,y}$$
$$\varepsilon_{zz} = u_{z,z}$$
$$\gamma_{yz} = u_{y,z} + u_{z,y}$$
$$\gamma_{xz} = u_{x,z} + u_{z,x}$$
$$\gamma_{xy} = u_{x,y} + u_{y,x}$$

(8.1.3)

where u_x, u_y and u_z are the displacements in the x, y, z directions respectively.

Inverting the stiffness matrix in equation (8.1.2), we can write the following relationship

$$\{\varepsilon\} = [S]\{\tau\}$$

(8.1.4)

where $[S] = [C]^{-1}$ is also a symmetric matrix called the *compliance* matrix.

8.2 TRANSFORMATION LAWS

For anisotropic plates, the material properties are dependent on the directions of the coordinate system in which they are determined. Therefore, it is convenient to write the constitutive laws for the stress and strain vectors. A special case of interest will be the rotation of coordinate system about z axis. Consider such a rotation as shown in Fig. 8.2.1, then it is easy to show the following transformation law for the strain vectors.

314 *Dynamics of Plates*

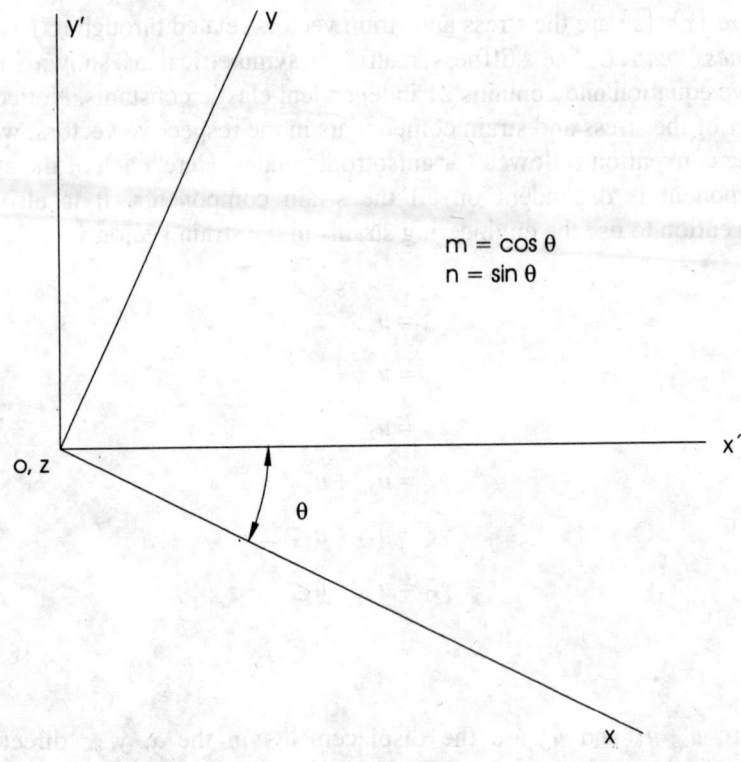

Fig. 8.2.1 Rotation of Axes

$$\{\varepsilon'\} = \begin{bmatrix} m^2 & n^2 & 0 & 0 & 0 & mn \\ n^2 & m^2 & 0 & 0 & 0 & -mn \\ 0 & 0 & 1 & 0 & 0 & 0 \\ 0 & 0 & 0 & m & -n & 0 \\ 0 & 0 & 0 & n & m & 0 \\ -2mn & 2mn & 0 & 0 & 0 & m^2 - n^2 \end{bmatrix} \{\varepsilon\}$$

$$\{\varepsilon'\} = [T_z^*]\{\varepsilon\}$$

(8.2.1)

where $[T_z^*]$ is the transformation matrix for rotation about z axis for the strain vector and

$$m = \cos\theta$$
$$n = \sin\theta$$

(8.2.2)

Similarly, for the stress vector,

$$\{\tau'\} = \begin{bmatrix} m^2 & n^2 & 0 & 0 & 0 & 2mn \\ n^2 & m^2 & 0 & 0 & 0 & -2mn \\ 0 & 0 & 1 & 0 & 0 & 0 \\ 0 & 0 & 0 & m & -n & 0 \\ 0 & 0 & 0 & n & m & 0 \\ -mn & mn & 0 & 0 & 0 & m^2-n^2 \end{bmatrix} \{\tau\}$$

$$\{\tau'\} = [T_z]\{\tau\}$$

(8.2.3)

where $[T_z]$ is the transformation matrix for rotation about z axis for the stress vector.

It can be shown that

$$[T_z]^{-1} = [T_z^*]^T$$
$$[T_z^*]^{-1} = [T_z]^T$$

(8.2.4)

The stress vector in the transformed coordinates is

$$\begin{aligned}\{\tau'\} &= [T_z]\{\tau\} \\ &= [T_z][C]\{\varepsilon\} \\ &= [T_z][C][T_z^*]^{-1}\{\varepsilon'\} \\ &= [T_z][C][T_z]^T\{\varepsilon'\} \\ &= [C']\{\varepsilon'\}\end{aligned}$$

(8.2.5)

where

$$[C'] = [T_z][C][T_z]^T$$

(8.2.6)

With the help of equation (8.2.3), the above equation (8.2.6) gives the elements of $[C']$ matrix which is symmetric, as

$$C'_{11} = C_{11}m^4 + 2m^2n^2(C_{12} + 2C_{66}) + 4mn(C_{16}m^2 + C_{26}n^2) + C_{22}n^4$$

$$C'_{12} = m^2n^2(C_{11} + C_{22} - 4C_{66}) - 2mn(C_{16} - C_{26})(m^2 - n^2) + C_{12}(m^4 + n^4)$$

$$C'_{13} = C_{13}m^2 + C_{23}n^2 + 2C_{36}mn$$

$$C'_{14} = C_{14}m^3 + mn[(2C_{46} - C_{15})m + (C_{24} - 2C_{56})n] - C_{25}n^3$$

$$C'_{15} = C_{15}m^3 + mn[(C_{14} + 2C_{56})m + (C_{25} + 2C_{46})n] + C_{24}n^3$$

$$C'_{16} = C_{16}m^2(m^2 - 3n^2) - mn[C_{11}m^2 - C_{22}n^2 - (C_{12} + 2C_{66})(m^2 - n^2)]$$
$$+ C_{26}n^2(3m^2 - n^2)$$

$$C'_{22} = C_{11}n^4 + 2m^2n^2(C_{12} + 2C_{66}) - 4mn(C_{26}m^2 + C_{16}n^2) + C_{22}m^4$$

$$C'_{23} = C_{13}n^2 + C_{23}m^2 - 2C_{36}mn$$

$$C'_{24} = C_{24}m^3 - mn[(C_{25} + 2C_{46})m - (C_{14} + 2C_{56})n] - C_{15}n^3$$

$$C'_{25} = C_{25}m^3 + mn[(C_{24} - 2C_{56})m + (C_{15} - 2C_{46})n] + C_{14}n^3$$

$$C'_{26} = C_{26}m^2(m^2 - 3n^2) - mn[C_{11}n^2 - C_{22}m^2 + (C_{12} + 2C_{66})(m^2 - n^2)]$$
$$+ C_{16}n^2(3m^2 - n^2)$$

$$C'_{33} = C_{33}$$

$$C'_{34} = C_{34}m - C_{35}n$$

$$C'_{35} = C_{34}n + C_{35}m$$

$$C'_{36} = C_{36}(m^2 - n^2) + (C_{23} - C_{13})mn$$

$$C'_{44} = C_{44}m^2 + C_{55}n^2 - 2C_{45}mn$$

$$C'_{45} = C_{45}(m^2 - n^2) + (C_{44} - C_{55})mn$$

$$C'_{46} = (C_{46}m - C_{56}n)(m^2 - n^2) + (C_{24} - C_{14})m^2n + (C_{15} - C_{25})mn^2$$

$$C'_{55} = C_{55}m^2 + C_{44}n^2 + 2C_{45}mn$$

$$C'_{56} = (C_{56}m + C_{46}n)(m^2 - n^2) + (C_{25} - C_{15})m^2n + (C_{24} - C_{14})mn^2$$

$$C'_{66} = (C_{11} + C_{22} - 2C_{12})m^2n^2 + 2(C_{26} - C_{16})(m^2 - n^2)mn$$
$$+ C_{66}(m^2 - n^2)^2$$

(8.2.7)

In a similar manner, we can show that

$$\{\varepsilon'\} = [S']\{\tau'\}$$

(8.2.8)

where

$$[S'] = [T_z^*][S][T_z^*]^T$$

(8.2.9)

8.3 MATERIAL SYMMETRIES

We have seen that there are 21 independent elastic constants for a general anisotropic material. Such a very general anisotropic material is rare and most of the materials exhibit symmetries which can be taken into account to reduce the number of elastic constants that describe the material behavior. This will also reduce the number of tests to be conducted in characterizing the material.

Two fold or Monoclinic Symmetry

In Fig. 8.3.1, xy axis system is in the plane of the paper with the z axis coming out of it. Consider a rotation of the xy axis system from the given position (designated 0 position) through an angle π giving the $x'y'$ axis system as shown. If the material properties are same in the quadrants A and B as marked, then the material is said to posses *two fold symmetry*. Another rotation through an angle π from the $x'y'$ axis system about z axis gives the original coordinate system. Therefore, $0, \pi, 2\pi$ rotations about z axis all give the same physical form for the two fold symmetry. In general, an n fold symmetry system is one which exhibits same physical form through rotations $\frac{2\pi i}{n}$ with i ranging from 0, 1, 2, ... n. The specific case of 2 fold symmetry is of the lowest order of symmetry and is termed *monoclinic* symmetry. Therefore from equation (8.2.6), we have

$$[C'] = [T_z(\pi)][C][T_z(\pi)]^T$$

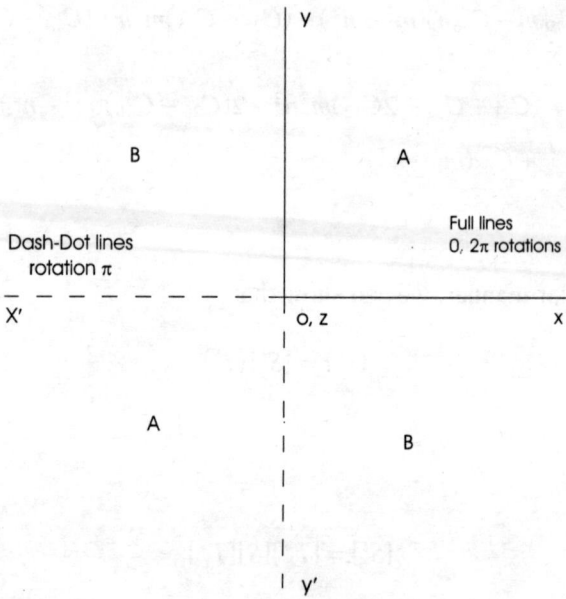

Fig. 8.3.1 Twofold Symmetry about z axis

$$[C'] = [C] \tag{8.3.1}$$

With the help of (8.2.7), the above equation gives the stiffness matrix of a monoclinically symmetric anisotropic material as

$$[C] = \begin{bmatrix} C_{11} & C_{12} & C_{13} & 0 & 0 & C_{16} \\ & C_{22} & C_{23} & 0 & 0 & C_{26} \\ & & C_{33} & 0 & 0 & C_{36} \\ & & & C_{44} & C_{45} & 0 \\ & \text{Sym} & & & C_{55} & 0 \\ & & & & & C_{66} \end{bmatrix} \tag{8.3.2}$$

In the above, the coefficients $C_{14}, C_{15}, C_{24}, C_{25}, C_{34}, C_{35}, C_{46}$ and C_{56} are negative and therefore set to zero. When compared to a general anisotropic material, the elastic constants of a two fold symmetric anisotropic material are reduced to 13 from 21.

Orthotropic Materials
In addition to a two fold symmetry about the z axis, if a material is monoclinic for the rotation about another axis, say, x axis, i.e., the

material is monoclinic for rotations about two mutually perpendicular axes, the material is said to be *orthotropic*. In this case, the stiffness matrix simplifies to

$$[C] = \begin{bmatrix} C_{11} & C_{12} & C_{13} & 0 & 0 & 0 \\ & C_{22} & C_{23} & 0 & 0 & 0 \\ & & C_{33} & 0 & 0 & 0 \\ & & & C_{44} & 0 & 0 \\ & \text{Sym} & & & C_{55} & 0 \\ & & & & & C_{66} \end{bmatrix}$$

(8.3.3)

As can be seen from the above, there are only nine elastic constants for orthotropic materials. The compliance matrix for orthotropic materials is

$$[S] = \begin{bmatrix} S_{11} & S_{12} & S_{13} & 0 & 0 & 0 \\ & S_{22} & S_{23} & 0 & 0 & 0 \\ & & S_{33} & 0 & 0 & 0 \\ & & & S_{44} & 0 & 0 \\ & \text{Sym} & & & S_{55} & 0 \\ & & & & & S_{66} \end{bmatrix}$$

(8.3.4)

Aragonite crystal is a typical example of an orthotropic material which has three mutually orthogonal planes of elastic symmetry. Boron and graphite fibers have a very high strength and modulus compared to conventional metals, they can be bonded together in plastic or metal binders to form orthotropic layers. In practice, several layers of such fiber composites are laminated together to form a *layered plate*. Each layer may have the fibers oriented in different directions to achieve the desired property of a *laminate*. An added advantage of this type of construction is light weight for the structural element compared to the conventional metals. The *composite plates* have therefore found excellent application in aircraft industry and also becoming common as turbomachine blades.

Transversely Isotropic Materials
Let yz be a plane of isotropy and x axis be perpendicular to it, then, the elastic constants reduce to only five in number with $C_{33} = C_{22}$, $C_{13} = C_{12}$, $C_{55} = C_{66}$ and $C_{44} = \frac{1}{2}(C_{22} - C_{23})$. The corresponding stiffness matrix for this case is

$$[C] = \begin{bmatrix} C_{11} & C_{12} & C_{12} & 0 & 0 & 0 \\ & C_{22} & C_{23} & 0 & 0 & 0 \\ & & C_{22} & 0 & 0 & 0 \\ & & & \frac{1}{2}(C_{22}-C_{23}) & 0 & 0 \\ & \text{Sym} & & & C_{66} & 0 \\ & & & & & C_{66} \end{bmatrix}$$

(8.3.5)

Completely Isotropic Materials

For complete isotropy, $C_{11} = C_{22}$, $C_{23} = C_{12}$, and $C_{66} = \frac{1}{2}(C_{11} - C_{12})$. In this case, we have only two elastic constants and the stiffness matrix reduces to

$$[C] = \begin{bmatrix} C_{11} & C_{12} & C_{12} & 0 & 0 & 0 \\ & C_{11} & C_{12} & 0 & 0 & 0 \\ & & C_{11} & 0 & 0 & 0 \\ & & & \frac{1}{2}(C_{11}-C_{12}) & 0 & 0 \\ & \text{Sym} & & & \frac{1}{2}(C_{11}-C_{12}) & 0 \\ & & & & & \frac{1}{2}(C_{11}-C_{12}) \end{bmatrix}$$

(8.3.6)

8.4 ENGINEERING CONSTANTS (ORTHOTROPIC MATERIALS)

For orthotropic materials, the strain vector is given by, from (8.3.4)

$$[\varepsilon] = \begin{bmatrix} S_{11} & S_{12} & S_{13} & 0 & 0 & 0 \\ & S_{22} & S_{23} & 0 & 0 & 0 \\ & & S_{33} & 0 & 0 & 0 \\ & & & S_{44} & 0 & 0 \\ & \text{Sym} & & & S_{55} & 0 \\ & & & & & S_{66} \end{bmatrix} \{\tau\}$$

(8.4.1)

Tests can be conducted in the three different directions x, y and z and the compliances in the above equation can be expressed as

$$S_{11} = \frac{1}{E_{xx}} \quad S_{12} = \frac{-\nu_{xy}}{E_{xx}} \quad S_{13} = \frac{-\nu_{xz}}{E_{xx}}$$

$$S_{22} = \frac{1}{E_{yy}} \quad S_{23} = \frac{-\nu_{yz}}{E_{yy}}$$

$$S_{33} = \frac{1}{E_{zz}}$$

$$S_{44} = \frac{1}{G_{yz}} \quad S_{55} = \frac{1}{G_{xz}} \quad S_{66} = \frac{1}{G_{xy}}$$

(8.4.2)

where E are the Young's modulii in tension or compression along the respective directions and v_{ij} are the Poisson's ratios as determined from the contraction in the j (y) direction during a tensile test in i (x) direction and G_{ij} are shear modulii in the ij (xy) planes etc. The symmetry of compliances in equation (8.4.1) gives the following relationships

$$E_{xx}v_{yx} = E_{yy}v_{xy}$$

$$E_{yy}v_{zy} = E_{zz}v_{yz}$$

$$E_{zz}v_{xz} = E_{xx}v_{zx}$$

(8.4.3)

The elements of the stiffness matrix can be obtained by the inverse of compliance matrix elements given above in (8.4.2).

$$C_{11} = \frac{E_{xx}}{DE_{yy}}(E_{yy} - E_{zz}v_{yz}^2)$$

$$C_{12} = C_{21} = \frac{1}{D}(E_{yy}v_{xy} + E_{zz}v_{xz}v_{yz})$$

$$C_{13} = C_{31} = \frac{E_{zz}}{D}(v_{xz} + v_{xy}v_{yz})$$

$$C_{22} = \frac{E_{yy}}{DE_{xx}}(E_{xx} - E_{zz}v_{xz}^2)$$

$$C_{23} = C_{32} = \frac{1}{D}\left(E_{zz}v_{yz} + \frac{E_{yy}E_{zz}}{E_{xx}}v_{xy}v_{xz}\right)$$

$$C_{22} = \frac{E_{zz}}{DE_{xx}}(E_{xx} - E_{yy}v_{xy}^2)$$

$$C_{44} = G_{yz} \quad C_{55} = G_{xz} \quad C_{66} = G_{xy}$$

(8.4.4)

where
$$D = 1 - 2\frac{E_{zz}}{E_{xx}}v_{xy}v_{yz}v_{xz} - v_{xz}^2\frac{E_{zz}}{E_{xx}} - v_{yz}^2\frac{E_{zz}}{E_{yy}} - v_{xy}^2\frac{E_{yy}}{E_{xx}}$$

(8.4.5)

8.5 STIFFNESS TRANSFORMATIONS AND REDUCED STIFFNESS MATRIX

In plate theory we are interested in the plane stress situation, equations (3.2.18) and (5.1.15), which for orthotropic plates is given by

$$\begin{Bmatrix} \tau_{xx} \\ \tau_{yy} \\ \tau_{xy} \end{Bmatrix} = \begin{bmatrix} \frac{E_{xx}}{(1-v_{xy}v_{yx})} & \frac{v_{xy}E_{yy}}{(1-v_{xy}v_{yx})} & 0 \\ & \frac{E_{yy}}{(1-v_{xy}v_{yx})} & 0 \\ \text{Sym} & & G_{xy} \end{bmatrix} \begin{Bmatrix} \varepsilon_{xx} \\ \varepsilon_{yy} \\ v_{xy} \end{Bmatrix}$$

(8.5.1)

In the above, xyz are the axes of symmetry of the plate. The corresponding compliance equation is

$$\begin{Bmatrix} \varepsilon_{xx} \\ \varepsilon_{yy} \\ v_{xy} \end{Bmatrix} = \begin{bmatrix} \frac{1}{E_{xx}} & -\frac{v_{xy}}{E_{xx}} & 0 \\ & \frac{1}{E_{yy}} & 0 \\ \text{Sym} & & \frac{1}{G_{xy}} \end{bmatrix} \begin{Bmatrix} \tau_{xx} \\ \tau_{yy} \\ \tau_{xy} \end{Bmatrix}$$

(8.5.2)

The global or geometric plate axes of convenience $x'y'z'$ are not necessarily aligned with the material symmetry directions. Therefore, it is necessary to perform a rotation about the z axis. For a rotation of coordinates as given in Fig. 8.2.1, using the following transformations, see equations (8.2.1) and (8.2.3)

$$\begin{Bmatrix} \varepsilon_{x'x'} \\ \varepsilon_{y'y'} \\ v_{x'y'} \end{Bmatrix} = \begin{bmatrix} m^2 & n^2 & mn \\ n^2 & m^2 & -mn \\ -2mn & 2mn & (m^2-n^2) \end{bmatrix} \begin{Bmatrix} \varepsilon_{xx} \\ \varepsilon_{yy} \\ v_{xy} \end{Bmatrix}$$

$$\begin{Bmatrix} \varepsilon_{y'z'} \\ \varepsilon_{x'z'} \end{Bmatrix} = \begin{bmatrix} m & -n \\ m & n \end{bmatrix} \begin{Bmatrix} \varepsilon_{yz} \\ \varepsilon_{xz} \end{Bmatrix}$$

$$\begin{Bmatrix} \tau_{x'x'} \\ \tau_{y'y'} \\ \tau_{x'y'} \end{Bmatrix} = \begin{bmatrix} m^2 & n^2 & 2mn \\ n^2 & m^2 & -2mn \\ -mn & mn & (m^2-n^2) \end{bmatrix} \begin{Bmatrix} \tau_{xx} \\ \tau_{yy} \\ \tau_{xy} \end{Bmatrix}$$

$$\begin{Bmatrix} \tau_{y'z'} \\ \tau_{x'z'} \end{Bmatrix} = \begin{bmatrix} m & -n \\ m & n \end{bmatrix} \begin{Bmatrix} \tau_{yz} \\ \tau_{xz} \end{Bmatrix}$$

(8.5.3)

the above strain-stress relation (8.5.2) can be shown to be

$$\begin{Bmatrix} \varepsilon_{x'x'} \\ \varepsilon_{y'y'} \\ v_{x'y'} \end{Bmatrix} = \begin{bmatrix} \frac{1}{E_{x'x'}} & -\frac{v_{x'y'}}{E_{x'x'}} & -\frac{\eta_{x'y'}}{E_{x'x'}} \\ & \frac{1}{E_{y'y'}} & -\frac{\eta_{x'y'}}{E_{y'y'}} \\ \text{Sym} & & \frac{1}{G_{x'y'}} \end{bmatrix} \begin{Bmatrix} \tau_{x'x'} \\ \tau_{y'y'} \\ \tau_{x'y'} \end{Bmatrix}$$

(8.5.4)

where

$$\frac{1}{E_{x'x'}} = \frac{m^4}{E_{xx}} + \left(\frac{1}{G_{xy}} - \frac{2v_{xy}}{E_{xx}}\right)m^2n^2 + \frac{n^4}{E_{yy}}$$

$$\frac{1}{E_{y'y'}} = \frac{n^4}{E_{xx}} + \left(\frac{1}{G_{xy}} - \frac{2v_{xy}}{E_{xx}}\right)m^2n^2 + \frac{m^4}{E_{yy}}$$

$$\frac{1}{G_{x'y'}} = \frac{1}{G_{xy}} + 4\left(\frac{1+2v_{xy}}{E_{xx}} + \frac{1}{E_{yy}} - \frac{1}{G_{xy}}\right)m^2n^2$$

$$v_{x'y'} = E_{x'x'}\left[\frac{v_{xy}}{E_{xx}} - \left(\frac{1+2v_{xy}}{E_{xx}} + \frac{1}{E_{yy}} - \frac{1}{G_{xy}}\right)m^2n^2\right]$$

(8.5.5a)

and the shear coupling coefficients

$$\eta_{x'y'} = E_{x'x'}\left[-\frac{2m^3n}{E_{xx}} + \frac{2mn^3}{E_{yy}} + \left(\frac{1}{G_{xy}} - \frac{2v_{xy}}{E_{xx}}\right)(m^2-n^2)mn\right]$$

$$\eta_{y'x'} = E_{y'y'}\left[-\frac{2mn^3}{E_{xx}} + \frac{2m^3n}{E_{yy}} - \left(\frac{1}{G_{xy}} - \frac{2v_{xy}}{E_{xx}}\right)(m^2-n^2)mn\right]$$

(8.5.5b)

A general case of plate problem includes all the shear components and the compliance equation (8.4.1) can be written as

$$\begin{Bmatrix} \varepsilon_{xx} \\ \varepsilon_{yy} \\ \gamma_{yz} \\ \gamma_{xz} \\ \gamma_{xy} \end{Bmatrix} = \begin{bmatrix} S_{11} & S_{12} & 0 & 0 & 0 \\ & S_{22} & 0 & 0 & 0 \\ & & S_{44} & 0 & 0 \\ & \text{Sym} & & S_{55} & 0 \\ & & & & S_{66} \end{bmatrix} \begin{Bmatrix} \tau_{xx} \\ \tau_{yy} \\ \tau_{yz} \\ \tau_{xz} \\ \tau_{xy} \end{Bmatrix}$$

(8.5.6)

The stiffness relationship is obtained by inverting the above as

$$\begin{Bmatrix} \tau_{xx} \\ \tau_{yy} \\ \tau_{yz} \\ \tau_{xz} \\ \tau_{xy} \end{Bmatrix} = \begin{bmatrix} Q_{11} & Q_{12} & 0 & 0 & 0 \\ & Q_{22} & 0 & 0 & 0 \\ & & Q_{44} & 0 & 0 \\ & \text{Sym} & & Q_{55} & 0 \\ & & & & Q_{66} \end{bmatrix} \begin{Bmatrix} \varepsilon_{xx} \\ \varepsilon_{yy} \\ \gamma_{yz} \\ \gamma_{xz} \\ \gamma_{xy} \end{Bmatrix}$$

(8.5.7)

where $[Q]$ is called *reduced stiffness matrix*. By inverting the compliance matrix in (8.5.6), we get the reduced stiffness matrix elements as

$$Q_{11} = \frac{S_{22}}{S_{11}S_{22} - S_{12}^2} = \frac{E_{xx}}{1 - v_{xy}v_{yx}}$$

$$Q_{12} = -\frac{S_{12}}{S_{11}S_{22} - S_{12}^2} = \frac{v_{xy}E_{yy}}{1 - v_{xy}v_{yx}}$$

$$Q_{22} = \frac{S_{11}}{S_{11}S_{22} - S_{12}^2} = \frac{E_{yy}}{1 - v_{xy}v_{yx}}$$

$$Q_{44} = C_{44} = \frac{1}{S_{44}} = G_{yz}$$

$$Q_{55} = C_{55} = \frac{1}{S_{55}} = G_{xz}$$

$$Q_{66} = C_{66} = \frac{1}{S_{66}} = G_{xy}$$

(8.5.8)

It may be noted that the normal stress τ_{zz} is set to zero in the above as per plane stress problem and that the normal strain $\varepsilon_{zz} = S_{13}\tau_{xx} + S_{23}\tau_{yy}$ is dropped in the above equations. Therefore, the elastic constants E_{zz}, v_{xz} and v_{yz} are not used in the constitutive relations above. The reduced stiffness matrix coefficients (8.5.8) can also be shown to be given by the following compact relation

$$Q_{ij} = C_{ij} - \frac{C_{i3}C_{j3}}{C_{33}} \quad i,j = 1,2,4,5,6$$

(8.5.9)

The transformed stiffness relation for application in the global axes system $x'y'z'$ is then

$$\begin{Bmatrix} \tau_{x'x'} \\ \tau_{y'y'} \\ \tau_{y'z'} \\ \tau_{x'z'} \\ \tau_{x'y'} \end{Bmatrix} = \begin{bmatrix} Q'_{11} & Q'_{12} & 0 & 0 & Q'_{16} \\ & Q'_{22} & 0 & 0 & Q'_{26} \\ & & Q'_{44} & Q'_{45} & 0 \\ & \text{Sym} & & Q'_{55} & 0 \\ & & & & Q'_{66} \end{bmatrix} \begin{Bmatrix} \varepsilon_{x'x'} \\ \varepsilon_{y'y'} \\ \gamma_{y'z'} \\ \gamma_{x'z'} \\ \gamma_{x'y'} \end{Bmatrix}$$

(8.5.10)

where the transformed reduced stiffness matrix elements are

$$Q'_{ij} = C'_{ij} - \frac{C'_{i3}C'_{j3}}{C'_{33}} \text{ for } \tau_{zz} = 0$$

$$Q'_{ij} = C'_{ij} \text{ for } \tau_{zz} \neq 0$$

(8.5.11)

i.e.,

$$Q'_{11} = Q_{11}m^4 + 2(Q_{12} + 2Q_{66})m^2n^2 + Q_{22}n^4$$

$$Q'_{12} = (Q_{11} + Q_{22} - 4Q_{66})m^2n^2 + Q_{12}(m^4 + n^4)$$

$$Q'_{16} = (Q_{11} - Q_{12} - 2Q_{66})m^3n + (Q_{12} - Q_{22} + 2Q_{66})mn^3$$

$$Q'_{22} = Q_{11}n^4 + 2(Q_{12} + 2Q_{66})m^2n^2 + Q_{22}m^4$$

$$Q'_{26} = (Q_{11} - Q_{12} - 2Q_{66})mn^3 + (Q_{12} - Q_{22} + 2Q_{66})m^3n$$

$$Q'_{44} = Q_{44}m^2 + Q_{55}n^2$$

$$Q'_{45} = (Q_{55} - Q_{44})mn$$

$$Q'_{55} = Q_{55}m^2 + Q_{44}n^2$$

$$Q'_{66} = (Q_{11} + Q_{22} - 2Q_{12} - 2Q_{66})m^2n^2 + Q_{66}(m^4 + n^4)$$

(8.5.12)

Specially Orthotropic Plates

For plane stress case, equation (8.5.4) shows the presence of shear coupling when the geometric plate axes and the material symmetry directions are different from each other. In the special case, when these two axes systems coincide, i.e., the material symmetry directions and the global axes are coincident, the shear coupling term disappears. Such plates are called *Specially orthotropic* plates.

8.6 SPECIALLY ORTHOTROPIC PLATE EQUATIONS

Let us consider the plate element as shown in Fig. 3.2.4 with the xyz axes system coinciding with the axes of symmetry. Proceeding as in section 3.2, we can determine the expressions for the bending and twisting moments with the displacement and strain fields as defined in equations (3.2.15) and (3.2.16).

$$u_x = -zw_{,x}$$

$$u_y = -zw_{,y}$$

$$u_z = w$$

(3.2.15)

$$\varepsilon_{xx} = -zw_{,xx}$$

$$\varepsilon_{yy} = -zw_{,yy}$$

$$\gamma_{xy} = -2zw_{,xy}$$

(3.2.16)

The stresses are defined in equation (8.5.1) which can be written as

$$\tau_{xx} = \frac{E_{xx}}{1 - v_{xy}v_{yx}}\varepsilon_{xx} + \frac{v_{xy}E_{yy}}{1 - v_{xy}v_{yx}}\varepsilon_{yy}$$

$$\tau_{yy} = \frac{v_{xy}E_{yy}}{1 - v_{xy}v_{yx}}\varepsilon_{xx} + \frac{E_{yy}}{1 - v_{xy}v_{yx}}\varepsilon_{yy}$$

$$\tau_{xy} = G_{xy}\gamma_{xy}$$

(8.6.1)

Substituting for the strains in equation (3.2.16) above, (8.6.1) gives

$$\tau_{xx} = -z\left(\frac{E_{xx}}{1-v_{xy}v_{yx}}w_{,xx} + \frac{v_{xy}E_{yy}}{1-v_{xy}v_{yx}}w_{,yy}\right)$$

$$\tau_{yy} = -z\left(\frac{v_{xy}E_{yy}}{1-v_{xy}v_{yx}}w_{,xx} + \frac{E_{yy}}{1-v_{xy}v_{yx}}w_{,yy}\right)$$

$$\tau_{xy} = -2G_{xy}zw_{,xy}$$

(8.6.2)

The bending moments (per unit length) M_x and M_y and the twisting moment M_{xy} are then determined as

$$M_x = \int_{-h/2}^{h/2} \tau_{xx} z \, dz$$

$$= -\int_{-h/2}^{h/2} z^2 \left(\frac{E_{xx}}{1-v_{xy}v_{yx}}w_{,xx} + \frac{v_{xy}E_{yy}}{1-v_{xy}v_{yx}}w_{,yy}\right) dz$$

$$= -(D_{11}w_{,xx} + D_{12}w_{,yy})$$

$$M_y = \int_{-h/2}^{h/2} \tau_{yy} z \, dz$$

$$= -\int_{-h/2}^{h/2} z^2 \left(\frac{v_{xy}E_{yy}}{1-v_{xy}v_{yx}}w_{,xx} + \frac{E_{yy}}{1-v_{xy}v_{yx}}w_{,yy}\right) dz$$

$$= -(D_{22}w_{,yy} + D_{12}w_{,xx})$$

$$M_{xy} = -\int_{-h/2}^{h/2} \tau_{xy} z \, dz$$

$$= \int_{-h/2}^{h/2} 2G_{xy} z^2 w_{,xy} \, dz$$

$$= 2D_{66}w_{,xy}$$

(8.6.3)

where

$$D_{11} = \frac{E_{xx}h^3}{12(1-v_{xy}v_{yx})}$$

$$D_{22} = \frac{E_{yy}h^3}{12(1-v_{xy}v_{yx})}$$

$$D_{12} = \frac{v_{xy}E_{yy}h^3}{12(1-v_{xy}v_{yx})}$$

$$D_{66} = \frac{G_{xy}h^3}{12}$$

(8.6.4)

In the particular case of isotropy in xy plane, the above plate parameters are

$$D_{11} = D_{22} = \frac{Eh^3}{12(1-v^2)}$$

$$D_{12} = \frac{vEh^3}{12(1-v^2)}$$

$$D_{66} = \frac{Gh^3}{12} = \frac{Eh^3}{24(1+v)}$$

$$D_{12} + 2D_{66} = \frac{Eh^3}{12(1-v^2)}$$

(8.6.4a)

Equation (3.2.44) gives the differential equation for the plate

$$M_{x,xx} - 2M_{xy,xy} + M_{y,yy} = -q$$

(3.2.44)

Substituting for the bending and twisting moments from equation (8.6.3), the above equation gives

$$-(D_{11}w_{,xx} + D_{12}w_{,yy})_{,xx} - 4(D_{66}w_{,xy})_{,xy} - (D_{22}w_{,yy} + D_{12}w_{,xx})_{,yy} = -q$$

$$D_{11}w_{,xxxx} + 2(D_{12} + 2D_{66})w_{,xxyy} + D_{22}w_{,yyyy} = q$$

(8.6.5)

The shear forces of equations (3.2.47) and (3.2.48) for the present case are

$$V_x = M_{yx,y} + M_{x,x}$$

$$= -(2D_{66}w_{,xy})_{,y} - (D_{11}w_{,xx} + D_{12}w_{,yy})_{,x}$$

$$= -\left[D_{11}w_{,xx} + (D_{12} + 2D_{66})w_{,yy}\right]_{,x}$$

(8.6.6)

$$V_y = -M_{xy,x} + M_{y,y}$$
$$= -(2D_{66}w_{,xy})_{,x} - (D_{22}w_{,yy} + D_{12}w_{,xx})_{,y}$$
$$= -\left[D_{22}w_{,yy} + (D_{12} + 2D_{66})w_{,xx}\right]_{,y}$$

(8.6.7)

Strain Energy

The strain energy of a plate element is given by the work done by the bending and twisting moments, see equations (3.2.28) and (3.2.31),

$$dU = -\frac{1}{2}(M_x w_{,xx} + M_y w_{,yy} - 2M_{xy}w_{,xy})dx\,dy$$

(8.6.8)

Substituting for the bending and twisting moment expressions from equations (8.6.3), and integrating over the plate area, we get

$$U = \frac{1}{2}\iint_A \left[D_{11}w_{,xx}^2 + 2D_{12}w_{,yy}w_{,xx} + D_{22}w_{,yy}^2 + 4D_{66}w_{,xy}^2\right]dx\,dy$$

(8.6.9)

For a vibrating plate, the differential equation (8.6.5) is modified as

$$D_{11}w_{,xxxx} + 2(D_{12} + 2D_{66})w_{,xxyy} + D_{22}w_{,yyyy} + \rho h\,\ddot{w} = q(x,y,t)$$

(8.6.10)

The kinetic energy expression remains same as (3.2.52) and Hamilton's principle for the present problem is given by, see equation (3.2.53)

$$\delta \int_{t_1}^{t_2} \left[\begin{array}{c} \frac{1}{2}\iint_R \rho h\,\dot{w}^2\,dx\,dy \\ -\frac{1}{2}\iint_A \left[\begin{array}{c} D_{11}w_{,xx}^2 + 2D_{12}w_{,yy}w_{,xx} \\ +D_{22}w_{,yy}^2 + 4D_{66}w_{,xy}^2 \end{array} \right]dx\,dy \\ +\iint_R wq\,dx\,dy \end{array} \right] dt = 0$$

(8.6.11)

8.7 SIMPLY SUPPORTED ORTHOTROPIC RECTANGULAR PLATES

Navier's Solution

This is a simple case and we can obtain Navier's solution of section 3.3 as follows. Let the plate be of width a in x direction and length b in y direction with thickness h and acted on by a static force $q(x,y)$. The differential equation for this problem is given by equation (8.6.5).

$$D_{11} w_{,xxxx} + 2(D_{12} + 2D_{66}) w_{,xxyy} + D_{22} w_{,yyyy} = q(x,y)$$

(8.6.5)

The boundary conditions are

$$M_x = -(D_{11} w_{,xx} + D_{12} w_{,yy}) = 0 \text{ on the edge } x = 0 \text{ and } x = a$$

$$w = 0 \text{ on the edge } x = 0 \text{ and } x = a$$

and

$$M_y = -(D_{22} w_{,yy} + D_{12} w_{,xx}) = 0 \text{ on the edge } y = 0 \text{ and } y = b$$

$$w = 0 \text{ on the edge } y = 0 \text{ and } y = b$$

(8.7.1)

As in section 3.3, the external force is expressed in the form of a double trigonometric series

$$q(x,y) = \sum_{m=1}^{\infty} \sum_{n=1}^{\infty} q_{mn} \sin \frac{m\pi x}{a} \sin \frac{n\pi y}{b}$$

(3.3.1)

where

$$q_{mn} = \frac{4}{ab} \int_0^a \int_0^b q(x,y) \sin \frac{m\pi x}{a} \sin \frac{n\pi y}{b} dx\, dy$$

(3.3.4)

Therefore the problem to be solved is reduced to the following differential equation

Anisotropic Plates

$$D_{11}w_{,xxxx} + 2(D_{12}+2D_{66})w_{,xxyy} + D_{22}w_{,yyyy}$$
$$= \sum_{m=1}^{\infty}\sum_{n=1}^{\infty} q_{mn} \sin\frac{m\pi x}{a} \sin\frac{n\pi y}{b}$$

(8.7.2)

where q_{mn} are first obtained from the integral in (3.3.4). Since the problem is linear, we can find the solution for each term in (3.3.4) and sum up the solutions to find the final answer. Therefore, it is sufficient for us to obtain the solution of

$$D_{11}w_{,xxxx} + 2(D_{12}+2D_{66})w_{,xxyy} + D_{22}w_{,yyyy}$$
$$= q_{mn} \sin\frac{m\pi x}{a} \sin\frac{n\pi y}{b}$$

(8.7.3)

The solution can be assumed once again in the form

$$w = C \sin\frac{m\pi x}{a} \sin\frac{n\pi y}{b}$$

(3.3.7)

which satisfies all the boundary conditions in (8.7.1). Substituting the above and its required differentials in (8.7.3), we get

$$w = \frac{q_{mn}}{\pi^4 \left[D_{11}\left(\frac{m}{a}\right)^4 + 2(D_{12}+2D_{66})\left(\frac{mn}{ab}\right)^2 + D_{22}\left(\frac{n}{b}\right)^4 \right]} \sin\frac{m\pi x}{a} \sin\frac{n\pi y}{b}$$

(8.7.4)

The complete solution is

$$w = \frac{1}{\pi^4}\sum_{m=1}^{\infty}\sum_{n=1}^{\infty} \frac{q_{mn}}{\left[D_{11}\left(\frac{m}{a}\right)^4 + 2(D_{12}+2D_{66})\left(\frac{mn}{ab}\right)^2 + D_{22}\left(\frac{n}{b}\right)^4 \right]} \sin\frac{m\pi x}{a} \sin\frac{n\pi y}{b}$$

(8.7.5)

For uniformly distributed force of magnitude q, the above relation gives, see equation (3.3.4a)

$$w = \frac{16q}{\pi^6} \sum_{m=1,3,5,...}^{\infty} \sum_{n=1,3,5,...}^{\infty} \frac{1}{mn \left[D_{11}\left(\frac{m}{a}\right)^4 + 2(D_{12}+2D_{66})\left(\frac{mn}{ab}\right)^2 + D_{22}\left(\frac{n}{b}\right)^4 \right]}$$

$$\times \sin\frac{m\pi x}{a} \sin\frac{n\pi y}{b}$$

(8.7.6)

The maximum deflection of the plate is

$$w_{max} = \frac{16q}{\pi^6} \sum_{m=1,3,5,...}^{\infty} \sum_{n=1,3,5,...}^{\infty} \frac{(-1)^{\frac{m+n}{2}-1}}{mn \left[D_{11}\left(\frac{m}{a}\right)^4 + 2(D_{12}+2D_{66})\left(\frac{mn}{ab}\right)^2 + D_{22}\left(\frac{n}{b}\right)^4 \right]}$$

(8.7.7)

For a square plate, the above further simplifies to

$$w_{max} = \frac{16qa^4}{\pi^6} \sum_{m=1,3,5,...}^{\infty} \sum_{n=1,3,5,...}^{\infty} \frac{(-1)^{\frac{m+n}{2}-1}}{mn[D_{11}m^4 + 2(D_{12}+2D_{66})m^2n^2 + D_{22}n^4]}$$

(8.7.8)

Levy Solution
This is a case most suitable for application similar to Nadai's approach as in section 3.6. Let the origin of the xy coordinate system be located at the middle of the edge $y=0$, see Fig. 8.7.1. The boundary conditions for the edges $y=0$ and $y=b$. are

$$M_y = -(D_{22}w_{,yy} + D_{12}w_{,xx}) = 0 \quad \text{i.e., } w_{,yy} = 0$$

and $w = 0$

(8.7.9)

The solution for the deflected surface is taken in two parts. The first part is taken to represent the deflection of a uniformly loaded strip parallel to y axis

$$w_1 = \frac{q}{24D_{22}}(y^4 - 2by^3 + b^3y)$$

(8.7.10)

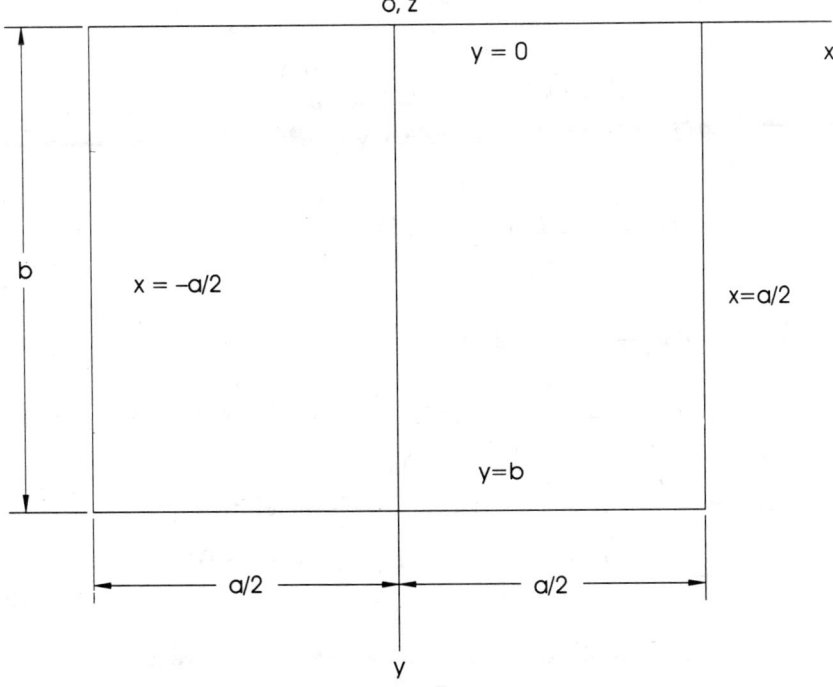

Fig. 8.7.1 Simply Supported Rectangular Plate

which satisfies the differential equation and also the boundary conditions at y=0 and y=b. The second part of the solution should satisfy

$$D_{11}w_{2,xxxx} + 2(D_{12} + 2D_{66})w_{2,xxyy} + D_{22}w_{2,yyyy} = 0$$
(8.7.11)

Let us consider the solution as in section 2.7 for a symmetrical deflection about y axis

$$w_2(x,y) = \sum_{m=1,3,5\cdots}^{\infty} X_m(x) \sin \frac{m\pi y}{b}$$
(8.7.12)

which satisfies the boundary conditions of (8.7.9). X_m is determined to satisfy the boundary conditions on the other two edges $x = \pm\frac{a}{2}$.

$$M_x = -(D_{11}w_{,xx} + D_{12}w_{,yy}) = 0 \text{ on the edge } x = -\frac{a}{2} \text{ and } x = \frac{a}{2}$$

$$w = 0 \text{ on the edge } x = -\frac{a}{2} \text{ and } x = \frac{a}{2}$$
(8.7.13)

Substituting (8.7.13) in (8.7.11), we get

$$\sum_{m=1,3,5\cdots}^{\infty} \left[\begin{array}{c} D_{11} X_m'''' - 2(D_{12} + 2D_{66})\left(\frac{m\pi}{b}\right)^2 X_m'' \\ + D_{22}\left(\frac{m\pi}{b}\right)^4 X_m \end{array} \right] \sin\frac{m\pi x}{b} = 0$$

i.e., $\quad D_{11} X_m'''' - 2(D_{12} + 2D_{66})\left(\frac{m\pi}{b}\right)^2 X_m'' + D_{22}\left(\frac{m\pi}{b}\right)^4 X_m = 0$

(8.7.14)

A particular solution of the above equation (8.7.14) is

$$X_m = \exp\left(\frac{m\pi}{b} sx\right)$$

(8.7.15)

The characteristic equation of (8.7.14) is therefore

$$D_{11} s^4 - 2(D_{12} + 2D_{66}) s^2 + D_{22} = 0$$

(8.7.16)

The two roots of the quadratic equation (8.7.16) above are

$$s_{1,2}^2 = \frac{D_{12} + 2D_{66}}{D_{11}} \pm \sqrt{\left(\frac{D_{12} + 2D_{66}}{D_{11}}\right)^2 - \frac{D_{22}}{D_{11}}}$$

(8.7.17)

Let the two roots above be real and unequal, then

$$X_m = A_m \cosh\frac{m\pi s_1 x}{b} + B_m \sinh\frac{m\pi s_1 x}{b} + C_m \cosh\frac{m\pi s_2 x}{b}$$
$$+ D_m \sinh\frac{m\pi s_2 x}{b}$$

(8.7.18)

The total solution is given by

$$w = w_1 + w_2$$
$$= \frac{q}{24 D_{22}}(y^4 - 2by^3 + b^3 y) +$$
$$+ \left\{ \begin{array}{c} A_m \cosh\frac{m\pi s_1 x}{b} + B_m \sinh\frac{m\pi s_1 x}{b} + C_m \cosh\frac{m\pi s_2 x}{b} \\ + D_m \sinh\frac{m\pi s_2 x}{b} \end{array} \right\} \sin\frac{m\pi y}{b}$$

(8.7.19)

The first term in the above equation is expressed in series form

$$\frac{q}{24D_{22}}(y^4 - 2by^3 + b^3y) = \frac{4qb^4}{\pi^5 D_{22}} \sum_{m=1,3,5\cdots}^{\infty} \frac{1}{m^5} \sin \frac{m\pi y}{b}$$

(8.7.20)

so that

$$w = \sum_{m=1,3,5\cdots}^{\infty} \left(\begin{array}{c} \frac{4qb^4}{\pi^5 D_{22}} \frac{1}{m^5} + A_m \cosh \frac{m\pi s_1 x}{b} + B_m \sinh \frac{m\pi s_1 x}{b} \\ + C_m \cosh \frac{m\pi s_2 x}{b} + D_m \sinh \frac{m\pi s_2 x}{b} \end{array} \right) \sin \frac{m\pi y}{b}$$

(8.7.21)

The boundary conditions at $x = \pm \frac{a}{2}$

$$M_x = -(D_{12} w_{,yy} + D_{11} w_{,xx}) = 0 \quad \text{i.e., } w_{,xx} = 0$$

and $w = 0$

(8.7.22)

Since the deflection surface is symmetrical with respect to y axis, we simplify (8.7.21) by setting $B_m = D_m = 0$ to give

$$w = \sum_{m=1,3,5\cdots}^{\infty} \left(\begin{array}{c} \frac{4qb^4}{\pi^5 D_{22}} \frac{1}{m^5} + A_m \cosh \frac{m\pi s_1 x}{b} \\ + C_m \cosh \frac{m\pi s_2 x}{b} \end{array} \right) \sin \frac{m\pi y}{b}$$

$$w_{,xx} = \sum_{m=1,3,5\cdots}^{\infty} \left[\begin{array}{c} A_m \left(\frac{m\pi s_1}{b}\right)^2 \cosh \frac{m\pi s_1 x}{b} \\ + C_m \left(\frac{m\pi s_2}{b}\right)^2 \cosh \frac{m\pi s_2 x}{b} \end{array} \right] \sin \frac{m\pi y}{b}$$

(8.7.23)

Applying (8.7.22) to the above equation (8.7.23) gives

$$\frac{4qb^4}{\pi^5 D_{22}} \frac{1}{m^5} + A_m \cosh \frac{m\pi s_1 a}{2b} + C_m \cosh \frac{m\pi s_2 a}{2b} = 0$$

$$A_m s_1^2 \cosh \frac{m\pi s_1 a}{2b} + C_m s_2^2 \cosh \frac{m\pi s_2 a}{2b} = 0$$

(8.7.24)

Therefore,

$$A_m = \frac{4qb^4}{\pi^5 D_{22}}\left(\frac{s_2^2}{s_1^2-s_2^2}\right)\left[\frac{1}{m^5 \cosh\frac{m\pi s_1 a}{2b}}\right]$$

$$C_m = -\frac{4qb^4}{\pi^5 D_{22}}\left(\frac{s_1^2}{s_1^2-s_2^2}\right)\frac{1}{m^5 \cosh\frac{m\pi s_2 a}{2b}}$$

(8.7.25)

The final solution is then

$$w = \frac{q}{24 D_{22}}(y^4 - 2by^3 + b^3 y)$$

$$+ \frac{4qb^4}{\pi^5 D_{22}} \frac{1}{s_1^2 - s_2^2} \sum_{m=1,3,5\cdots}^{\infty} \frac{1}{m^5}\left[\frac{s_2^2 \cosh\frac{m\pi s_1 x}{b}}{\cosh\frac{m\pi s_1 a}{2b}} - \frac{s_1^2 \cosh\frac{m\pi s_2 x}{b}}{\cosh\frac{m\pi s_2 a}{2b}}\right]$$

$$\times \sin\frac{m\pi y}{b}$$

(8.7.26)

Natural Frequencies
For free vibrations, equation (8.6.10) is

$$D_{11} w_{,xxxx} + 2(D_{12} + 2D_{66}) w_{,xxyy} + D_{22} w_{,yyyy} + \rho h \ddot{w} = 0$$

(8.6.10)

The solution to the above satisfying the boundary conditions in (8.7.1) can be written as

$$w = A \sin\frac{m\pi x}{a} \sin\frac{n\pi y}{b} \sin pt$$

(8.7.27)

where m and n are integers. Substituting the above and its derivatives in (8.6.10), we get

$$D_{11}\left(\frac{m\pi}{a}\right)^4 + 2(D_{12} + 2D_{66}) m^2 n^2 \frac{\pi^4}{a^2 b^2} + D_{22}\left(\frac{n\pi}{b}\right)^4 = \rho h p_{mn}^2$$

(8.7.28)

The natural frequency equation is therefore

$$p_{mn}^2 = \frac{\pi^4}{\rho h a^4}[D_{11}m^4 + 2(D_{12} + 2D_{66})m^2n^2k^2 + D_{22}n^4k^4]$$

(8.7.29)

where $k = a/b$. Corresponding to the isotropic case in (3.3.18), let us define

$$\lambda_{mn}^2 = a^4 \frac{\rho h}{D_{11}} p_{mn}^2$$

$$= \pi^4\left[m^4 + 2\frac{(D_{12} + 2D_{66})}{D_{11}}m^2n^2k^2 + \frac{D_{22}}{D_{11}}n^4k^4\right]$$

(8.7.30)

Let us now consider the case of a square plate, $k = 1$, and $\frac{D_{22}}{D_{11}} = \frac{(D_{12} + 2D_{66})}{D_{11}} = 0.1$ then

$$\lambda_{mn}^2 = \pi^4[m^4 + 0.2m^2n^2 + 0.1n^4]$$

(8.7.31)

The values of the nondimensional frequency above are given in Table 8.7.1.

Table 8.7.1 Frequency Parameters λ_{mn}^2 for a Square Plate

m	n	λ_{mn}^2	Mode Shape
1	1	126.63	$\sin\frac{\pi x}{a} \sin\frac{\pi y}{b}$
1	2	331.19	$\sin\frac{\pi x}{a} \sin\frac{2\pi y}{b}$
2	1	1,646.21	$\sin\frac{2\pi x}{a} \sin\frac{\pi y}{b}$
2	2	2,026.11	$\sin\frac{2\pi x}{a} \sin\frac{2\pi y}{b}$
1	3	1,061.76	$\sin\frac{\pi x}{a} \sin\frac{3\pi y}{b}$
3	1	8,075.21	$\sin\frac{3\pi x}{a} \sin\frac{\pi y}{b}$
2	3	3,048.9	$\sin\frac{2\pi x}{a} \sin\frac{3\pi y}{b}$
3	2	8,747.34	$\sin\frac{3\pi x}{a} \sin\frac{2\pi y}{b}$

338 Dynamics of Plates

Comparing with Table 3.3.1, we find that
1. $\lambda_{mn} \neq \lambda_{nm}$
2. The natural frequencies do not increase with the order m and n as in the case of isotropic plates. As an example, the natural frequency with two nodal lines parallel to x axis ($m = 1$ and $n = 3$) is lower in magnitude when compared with the frequency having only one nodal line parallel to y axis ($m = 2$ and $n = 1$). The frequencies, however, increase with n for a given m or with m for a given n.

8.8 PLATES SIMPLY SUPPORTED AT $y = 0, b$ AND CLAMPED AT $x = \pm a/2$ OR $x = 0, a$

The solution for this problem follows the one given in the previous section until equation (8.7.21). The boundary conditions at $x = \pm \frac{a}{2}$

$$w_{,x} = 0 \text{ and } w = 0 \tag{8.8.1}$$

Once again, since the deflection surface is symmetrical with respect to y axis, we simplify (8.7.21) by setting $B_m = D_m = 0$ to give

$$w = \sum_{m=1,3,5\ldots}^{\infty} \left(\frac{4qb^4}{\pi^5 D_{22}} \frac{1}{m^5} + A_m \cosh \frac{m\pi s_1 x}{b} + C_m \cosh \frac{m\pi s_2 x}{b} \right) \sin \frac{m\pi y}{b}$$

$$w_{,x} = \sum_{m=1,3,5\ldots}^{\infty} \left(A_m \frac{m\pi s_1}{b} \sinh \frac{m\pi s_1 x}{b} + C_m \frac{m\pi s_2}{b} \cosh \frac{m\pi s_2 x}{b} \right) \sin \frac{m\pi y}{b}$$

$$\tag{8.8.2}$$

Applying (8.7.22) to the above equation (8.7.23) gives

$$\frac{4qb^4}{\pi^5 D_{22}} \frac{1}{m^5} + A_m \cosh \frac{m\pi s_1 a}{2b} + C_m \cosh \frac{m\pi s_2 a}{2b} = 0$$

$$A_m s_1 \sinh \frac{m\pi s_1 a}{2b} + C_m s_2 \sinh \frac{m\pi s_2 a}{2b} = 0$$

$$\tag{8.8.3}$$

Therefore,

$$A_m = \frac{\dfrac{4qb^4}{\pi^5 m^5 D_{22}}}{\left[s_1 \sinh \dfrac{m\pi s_1 a}{2b} \cosh \dfrac{m\pi s_2 a}{2b} - s_2 \sinh \dfrac{m\pi s_2 a}{2b} \cosh \dfrac{m\pi s_1 a}{2b} \right]} s_2 \sinh \dfrac{m\pi s_2 a}{2b}$$

$$C_m = \frac{4qb^4}{\pi^5 m^5 D_{22}}$$

$$\left[\frac{s_1 \sinh\frac{m\pi s_1 a}{2b}}{s_1 \sinh\frac{m\pi s_1 a}{2b}\cosh\frac{m\pi s_2 a}{2b} - s_2 \sinh\frac{m\pi s_2 a}{2b}\cosh\frac{m\pi s_1 a}{2b}}\right]$$

(8.8.4)

The final solution is then

$$w = \frac{q}{24 D_{22}}(y^4 - 2by^3 + b^3 y) +$$

$$+ \frac{4qb^4}{\pi^5 D_{22}} \frac{1}{s_1^2 - s_2^2} \sum_{m=1,3,5\cdots}^{\infty} \frac{1}{m^5} \left[\frac{s_1 \sinh\frac{m\pi s_2 a}{2b}\cosh\frac{m\pi s_1 x}{b} - s_1 \sinh\frac{m\pi s_1 a}{2b}\cosh\frac{m\pi s_2 x}{b}}{s_1 \sinh\frac{m\pi s_1 a}{2b}\cosh\frac{m\pi s_2 a}{2b} - s_2 \sinh\frac{m\pi s_2 a}{2b}\cosh\frac{m\pi s_1 a}{2b}}\right]$$

$$\times \sin\frac{m\pi y}{b}$$

(8.8.5)

Natural Frequencies

We take the clamped edges to be $x = 0$ and $x = a$. The method of Ritz is convenient to determine the natural frequencies. The mode shapes are assumed from beam functions. These functions are

For simply supported edges $y = 0$ and $y = b$,

$$g_j(y) = \sin\frac{j\pi y}{b}$$

(8.8.6)

For clamped edges $x = 0$ and $x = a$,

$$f_i(x) = (\cos r_i a - \cosh r_i a)(\cos r_i x - \cosh r_i x)$$
$$+ (\sin r_i a + \sinh r_i a)(\sin r_i x - \sinh r_i x)$$

(8.8.7)

where $r_i a$ are the roots of the frequency equation

$$1 - \cos ra \cosh ra = 0 \qquad (8.8.8)$$

The first five roots of $r_i a$ are

$$r_1 a = 4.73$$
$$r_2 a = 7.853$$
$$r_3 a = 10.996$$
$$r_4 a = 14.137$$
$$r_5 a = 17.279 \qquad (8.8.9)$$

For free vibration, the solution is now assumed

$$w = \sum_{i=1}^{m} \sum_{j=1}^{n} A_{ij} f_i(x) g_j(y) \sin pt \qquad (8.8.10)$$

The Lagrangian of the plate from equation (8.6.11) can be written as

$$L = \frac{1}{2} \iint_A \left[D_{11} w_{,xx}^2 + 2D_{12} w_{,yy} w_{,xx} + D_{22} w_{,yy}^2 + 4D_{66} w_{,xy}^2 \right] dx\, dy$$

$$- \frac{1}{2} \iint_R \rho h \, \dot{w}^2 \, dx\, dy$$

$$(8.8.11)$$

Substituting equation (8.8.9) in the above equation (8.8.10) and minimizing the resulting Lagrangian with respect to A_{ij}, we get

$$\sum_{k=1}^{m} \sum_{l=1}^{n} \begin{bmatrix} D_{11} \int_0^a f_{k,xx} f_{i,xx}\, dx \int_0^b g_l g_j\, dy \\ + D_{12} \int_0^a f_{k,xx} f_i\, dx \int_0^b g_l g_{j,yy}\, dy \\ + D_{12} \int_0^a f_{i,xx} f_k\, dx \int_0^b g_j g_{l,yy}\, dy \\ + D_{22} \int_0^a f_k f_i\, dx \int_0^b g_{l,yy} g_{j,yy}\, dy \\ + 4 D_{66} \int_0^a f_{k,x} f_{i,x}\, dx \int_0^b g_{l,y} g_{j,y}\, dy \end{bmatrix} A_{kl}$$

$$= \sum_{k=1}^{m} \sum_{l=1}^{n} \left[\rho h p^2 \int_0^a f_i f_k\, dx \int_0^b g_l g_j\, dy \right] A_{kl} \qquad i = 1, 2, \ldots m \quad j = 1, 2, \ldots n$$

$$(8.8.12)$$

Hearmon restricted to the first approximation for each of the natural frequencies and simplified the above to

$$D_{11}\frac{a_1^4}{a^4} + 2(D_{12} + 2D_{66})\frac{a_2}{a^2 b^2} + D_{22}\frac{a_3^4}{b^4}$$
$$= \rho h p^2$$

(8.8.13)

where a are obtained from the respective integrals in equation (8.8.12)

$$a_1 = 4.73$$
$$a_3 = n\pi$$
$$a_2 = 12.30\, n^2\pi^2 \text{ for } m = 1, n = 1, 2, 3 \ldots$$
$$a_1 = \left(m + \frac{1}{2}\right)\pi$$
$$a_3 = n\pi$$
$$a_2 = a_1(a_1 - 2)n^2\pi^2 \text{ for } m = 2, 3, 4, \ldots n = 1, 2, 3 \ldots$$

(8.8.14)

The first mode natural frequency for a square plate can be obtained from the above as

$$\rho h a^4 p_{11}^2 = 500.547 D_{11} + 2 \times 121.396(D_{12} + 2D_{66}) + 97.409 D_{22}$$

(8.8.15)

For an isotropic plate, $D_{11} = D_{22} = D_{12} + 2D_{66}$, the above equation further reduces to

$$\frac{\rho h a^4}{D} p_{11}^2 = 840.748$$

i.e.,

$$p_{11} = 28.9957 \sqrt{\frac{D}{\rho h a^4}}$$

(8.8.16)

8.9 CLAMPED PLATES

For this case, it is not possible to get an exact solution as in the previous two sections. We can choose an appropriate approximate method to solve this problem. Let us consider Ritz method for this purpose. The strain energy of the plate is given by equation (8.6.9)

$$U = \frac{1}{2} \iint_A \left[D_{11} w_{,xx}^2 + 2D_{12} w_{,yy} w_{,xx} + D_{22} w_{,yy}^2 + 4D_{66} w_{,xy}^2 \right] dx\, dy \tag{8.6.9}$$

The work done by a uniformly distributed external force q is

$$W = -V = \iint_A q w\, dx\, dy \tag{8.9.1}$$

Let the coordinate system be located at the left hand top corner with the x axis lying on the top edge of width a and the y axis lying on the left side edge of width b. The boundary conditions for the clamped plate are

$$w = w_{,x} = 0 \text{ for } x = 0, a$$
$$w = w_{,y} = 0 \text{ for } y = 0, b \tag{8.9.2}$$

Let us assume the following series solution for w

$$w(x, y) = \sum_{i=1}^{\infty} \sum_{j=1}^{\infty} a_{ij} f_i(x) g_i(y) \tag{8.9.3}$$

where f and g satisfy the boundary conditions on the edges x constant and y constant respectively. The Lagrangian of the plate is given by

$$L = \frac{1}{2} \iint_A \left[D_{11} w_{,xx}^2 + 2D_{12} w_{,yy} w_{,xx} + D_{22} w_{,yy}^2 + 4D_{66} w_{,xy}^2 \right] dx\, dy$$
$$- \iint_A q w\, dx\, dy \tag{8.9.4}$$

Substituting (8.9.3) in the above equation (8.9.4), we get

$$L = \frac{1}{2}\iint_A \left[\begin{array}{l} D_{11}\left(\sum_{i=1}^{\infty}\sum_{j=1}^{\infty} a_{ij}f_{i,xx}g_i\right)^2 \\ +2D_{12}\left(\sum_{i=1}^{\infty}\sum_{j=1}^{\infty} a_{ij}f_i g_{i,yy}\right)\left(\sum_{i=1}^{\infty}\sum_{j=1}^{\infty} a_{ij}f_{i,xx}g_i\right) \\ +D_{22}\left(\sum_{i=1}^{\infty}\sum_{j=1}^{\infty} a_{ij}f_i g_{i,yy}\right)^2 \\ +4D_{66}\left(\sum_{i=1}^{\infty}\sum_{j=1}^{\infty} a_{ij}f_{i,x}g_{i,y}\right)^2 \end{array} \right] dx\,dy$$

$$-\iint_A q\left(\sum_{i=1}^{\infty}\sum_{j=1}^{\infty} a_{ij}f_i g_i\right) dx\,dy$$

(8.9.5)

For Ritz approximation with m number of terms in f and n terms in g, we have

$$\frac{\partial L}{\partial a_{ij}} = 0 \text{ for } i = 1, 2, \ldots m \text{ and } j = 1, 2, \ldots n$$

(8.9.6)

$$\sum_{k=1}^{m}\sum_{l=1}^{n} \left[\begin{array}{l} D_{11}\int_0^a f_{k,xx}f_{i,xx}\,dx \int_0^b g_l g_j\,dy \\ +D_{12}\int_0^a f_{k,xx}f_i\,dx \int_0^b g_l g_{j,yy}\,dy \\ +D_{12}\int_0^a f_{i,xx}f_k\,dx \int_0^b g_j g_{l,yy}\,dy \\ +D_{22}\int_0^a f_k f_i\,dx \int_0^b g_{l,yy}g_{j,yy}\,dy \\ +4D_{66}\int_0^a f_{k,x}f_{i,x}\,dx \int_0^b g_{l,y}g_{j,y}\,dy \end{array} \right] a_{kl}$$

$$= q\int_0^a f_i\,dx \int_0^b g_j\,dy \quad i = 1, 2, \ldots m \quad j = 1, 2, \ldots n$$

(8.9.7)

For a one term approximation, the assumed solution that satisfies the boundary conditions can be taken as

$$f_1(x) = (x^2 - ax)^2$$
$$g_1(y) = (y^2 - by)^2$$

(8.9.8)

Equation (8.9.7) then becomes

$$a_{11}D_{11} \int_0^a (12x^2 + 2a^2 - 12ax)^2 \, dx \int_0^b (y^2 - by)^4 \, dy$$
$$+ 2a_{11}D_{12} \int_0^a (12x^2 + 2a^2 - 12ax)(x^2 - ax)^2 \, dx$$
$$\times \int_0^b (y^2 - by)^2 (12y^2 + 2b^2 - 12by) \, dy$$
$$+ a_{11}D_{22} \int_0^a (x^2 - ax)^4 \, dx \int_0^b (12y^2 + 2b^2 - 12by)^2 \, dy$$
$$+ 4a_{11}D_{66} \int_0^a (4x^3 + 2a^2 x - 6ax^2)^2 \, dx \int_0^b (4y^3 + 2b^2 y - 6by^2)^2 \, dy$$
$$= q \int_0^a (x^2 - ax)^2 \, dx \int_0^b (y^2 - by)^2 \, dy$$

(8.9.9)

Upon evaluation of the integrals in the above equation and rearranging, we get

$$a_{11} = \frac{\frac{49}{8} q}{7D_{11}b^4 + 4(D_{12} + 2D_{66})a^2 b^2 + 7D_{22}a^4}$$

(8.9.10)

Therefore, the deflection w is given by

$$w = \frac{\frac{49}{8} q (x^2 - ax)^2 (y^2 - by)^2}{7D_{11}b^4 + 4(D_{12} + 2D_{66})a^2 b^2 + 7D_{22}a^4}$$

(8.9.11)

The maximum deflection occurs at the center of the plate $x = a/2$ and $y = b/2$,

$$w_{max} = \frac{49 q a^4 b^4}{2048[7D_{11}b^4 + 4(D_{12} + 2D_{66})a^2 b^2 + 7D_{22}a^4]}$$

(8.9.12)

Natural Frequencies
Here, we follow the procedure outlined in section 8.8. Since all edges are clamped, we assume the solution to be

$$f_i(x) = (\cos r_i a - \cosh r_i a)(\cos r_i x - \cosh r_i x)$$
$$+ (\sin r_i a + \sinh r_i a)(\sin r_i x - \sinh r_i x)$$

$$g_i(y) = (\cos r_i b - \cosh r_i b)(\cos r_i y - \cosh r_i y)$$
$$+ (\sin r_i b + \sinh r_i b)(\sin r_i y - \sinh r_i y)$$

(8.9.13)

where the roots of $r_i a$ and $r_i b$ are given in equation (8.8.9). Following the procedure closely with the above shape functions in equation (8.8.12), we get

$$D_{11}\frac{a_1^4}{a^4} + 2(D_{12} + 2D_{66})\frac{a_2}{a^2 b^2} + D_{22}\frac{a_3^4}{b^4} = \rho h p^2$$

(8.8.13a)

where a for the present case are

$a_1 = 4.73$

$a_3 = 4.73$

$a_2 = 151.3$ for $m = 1, n = 1$

$a_1 = 4.73$

$a_3 = \left(n + \frac{1}{2}\right)\pi$

$a_2 = 12.3 a_3 (a_3 - 2)$ for $m = 1, n = 2, 3, 4, \ldots$

$a_1 = \left(m + \frac{1}{2}\right)\pi$

$a_3 = 4.73$

$a_2 = 12.3 a_1 (a_1 - 2)$ for $m = 2, 3, 4, \ldots n = 1$

$a_1 = \left(m + \frac{1}{2}\right)\pi$

$a_3 = \left(n + \frac{1}{2}\right)\pi$

$a_2 = a_1(a_1 - 2)a_3(a_3 - 2)$ for $m = 2, 3, 4, \ldots n = 2, 3, 4, \ldots$

(8.9.14)

The first mode natural frequency for a square plate can be obtained from the above as

$$\rho h a^4 p_{11}^2 = 500.547 D_{11} + 2 \times 151.3(D_{12} + 2D_{66}) + 500.547 D_{22}$$

(8.9.15)

For an isotropic plate, $D_{11} = D_{22} = D_{12} + 2D_{66}$, the above equation further reduces to

$$\frac{\rho h a^4}{D} p_{11}^2 = 1303.694$$

i.e.,

$$p_{11} = 36.1 \sqrt{\frac{D}{\rho h a^4}}$$

(8.9.16)

8.10 FIBER REINFORCED LAMINAE

A *lamina* usually refers to a fiber reinforced composite. The *composite* consists of high strength and modulus fibers embedded in a *matrix* or bonded to a matrix. Both the fibers and the matrix retain their distinct properties and together they produce properties which cannot be achieved individually. The fibers are usually the principal load carrying elements in the composite and the purpose of the matrix is essentially to keep the fibers in the desired location. The matrix also serves the purpose of protecting the fiber from heat, corrosion and other environmental damages.

The fibers may be glass, carbon, boron, silicon carbide, aluminum oxide etc. One of the popular commercial fiber is Kevlar 49. These fibers may be embedded in the matrix either in a continuous form or in discontinuous form (chopped pieces of different lengths). The matrix usually consists of a polymer, metal or a ceramic.

A *laminate* or *laminated plate* consists of several laminae each with a fiber oriented at a particular angle. This is generally made by stacking several thin layers of fibers at the desired locations and angles in a matrix and consolidating them to give the required thickness. The fiber orientation in each thin layer can be arranged in a specific manner so as to achieve the required properties of the structural member. Since the fibers are of high strength in their axial direction and at the same time very light compared to conventional metals, laminated plates, they find many applications in aeronautical industry, automobile industry etc.

Unidirectional Continuous Fiber 0 Degree Lamina

Fig. 8.10.1 shows a simple lamina with a unidirectional continuous fiber at 0°. The fibers are assumed to be uniformly distributed throughout the matrix and a perfect bonding exists between the fibers and the matrix. Also, the matrix is assumed to be free of any voids. The fibers as well as matrix are further assumed to behave like elastic materials.

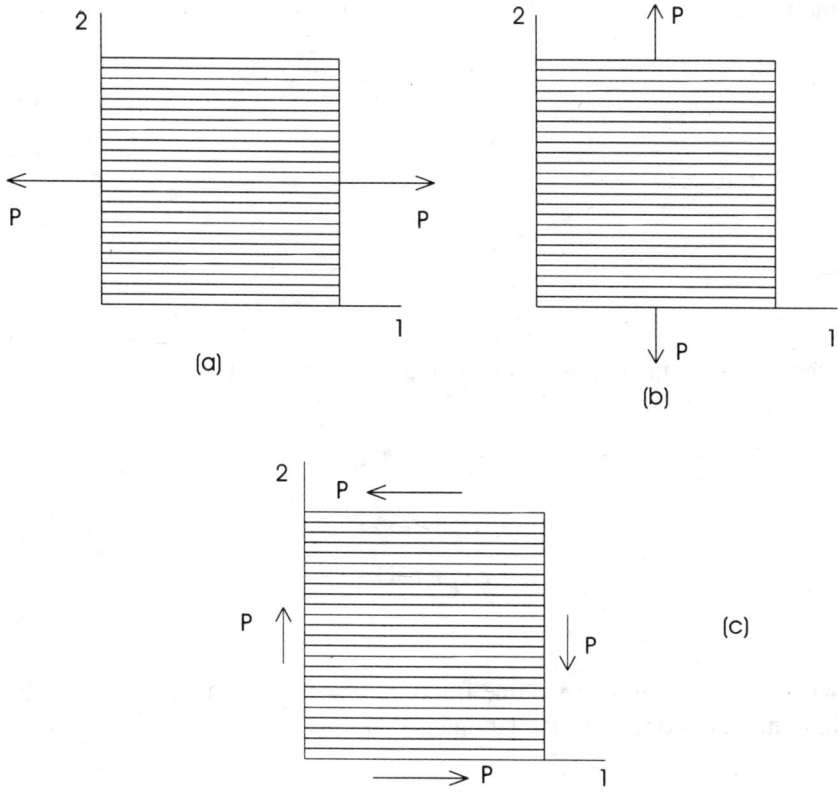

Fig. 8.10.1 Unidirectional Continuous Fiber 0 Deg Lamina

1-2 axes system denotes the direction of the fiber. For an applied load P parallel to the direction of fibers, the strains in the fiber, matrix as well as the composite are all equal, i.e.,

$$\varepsilon_f = \varepsilon_m = \varepsilon_c$$

(8.10.1)

where suffix f refers to the fiber, m refers to the matrix and c refers to the composite. The stresses are given by

$$\tau_f = E_f \varepsilon_f = E_f \varepsilon_c$$
$$\tau_m = E_m \varepsilon_m = E_m \varepsilon_c$$

(8.10.2)

The total load applied on the lamina is shared between the fiber and the matrix

$$P = P_f + P_m$$

(8.10.3)

Therefore,

$$\tau_c A_c = \tau_f A_f + \tau_m A_m$$

(8.10.4)

where A is the net cross-sectional area. Equation (8.10.4) gives

$$\tau_c = \frac{1}{A_c}(\tau_f A_f + \tau_m A_m)$$
$$= \tau_f v_f + \tau_m v_m$$
$$= \tau_f v_f + \tau_m (1 - v_f)$$

(8.10.5)

where v represents the volume fraction, $v_f = A_f/A_c$ and $v_m = A_m/A_c$. The longitudinal modulus of the lamina is then given by

$$E_{11} = \frac{\tau_c}{\varepsilon_c}$$
$$= E_f v_f + E_m (1 - v_f)$$

(8.10.6)

The corresponding Poisson's ratio (major) is

$$\nu_{12} = \nu_f v_f + \nu_m v_m$$

(8.10.7)

The transverse modulus and the minor Poisson's ratio for the loading transverse to the fiber direction as shown in Fig. 8.10.1b are

$$E_{22} = \frac{E_f E_m}{E_f v_m + E_m v_f}$$

$$v_{21} = \frac{E_{22}}{E_{11}} v_{12}$$

(8.10.8)

The longitudinal modulus, E_{11} is greater than the transverse modulus E_{22}. The ratio $\frac{E_{11}}{E_{22}}$ is a measure of orthotropy of the lamina. Also, the major Poisson's ratio v_{12} is greater than the minor one v_{21}. Note that they are related through the longitudinal and transverse modulii and therefore, only one of them is considered as an independent quantity. For a shear force loading as shown in Fig. 8.10.1c, we have

$$G_{12} = \frac{G_f G_m}{G_f v_m + G_m v_f}$$

(8.10.9)

There are in all four independent quantities, E_{11}, E_{22}, v_{12} and G_{12} to describe the elastic behavior of the lamina, as given by equations (8.4.1) and (8.4.2) for the case of plane stress.

Unidirectional Continuous Fiber Angle-Ply Lamina

Fig. 8.10.2 shows a lamina with a unidirectional continuous fiber inclined at an angle θ with the positive direction of x axis. In this case, we can write from equations (8.5.4)

$$\frac{1}{E_{xx}} = \frac{m^4}{E_{11}} + \left(\frac{1}{G_{12}} - \frac{2v_{12}}{E_{11}}\right)m^2 n^2 + \frac{n^4}{E_{22}}$$

$$\frac{1}{E_{yy}} = \frac{n^4}{E_{11}} + \left(\frac{1}{G_{12}} - \frac{2v_{12}}{E_{11}}\right)m^2 n^2 + \frac{m^4}{E_{22}}$$

$$\frac{1}{G_{xy}} = \frac{1}{G_{12}} + 4\left(\frac{1+2v_{12}}{E_{11}} + \frac{1}{E_{22}} - \frac{1}{G_{12}}\right)m^2 n^2$$

$$v_{xy} = E_{xx}\left[\frac{v_{12}}{E_{11}} - \left(\frac{1+2v_{12}}{E_{11}} + \frac{1}{E_{22}} - \frac{1}{G_{12}}\right)m^2 n^2\right]$$

$$v_{yx} = \frac{E_{yy}}{E_{xx}} v_{xy}$$

(8.10.10)

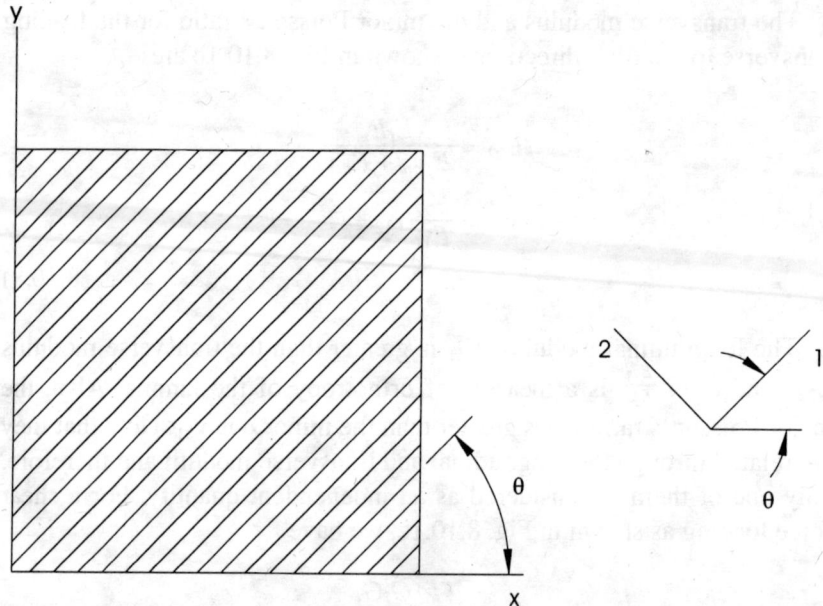

Fig. 8.10.2 Unidirectional Continuous Fiber Angle-Ply Lamina

Unidirectional Discontinuous Fiber 0 Degree Lamina

For the unidirectional discontinuous fiber lamina shown in Fig. 8.10.3, the following Halpin-Tsai relations are used to determine the elastic properties.

$$E_{11} = \frac{1 + 2a_f \eta_l v_f}{1 - \eta_l v_f} E_m$$

$$E_{22} = \frac{1 + 2\eta_T v_f}{1 - \eta_T v_f} E_m$$

$$G_{12} = G_{21} = \frac{1 + \eta_G v_f}{1 - \eta_G v_f} G_m$$

$$v_{12} = v_f v_f + v_m v_m$$

(8.10.11)

where

$$\eta_l = \frac{\dfrac{E_f}{E_m} - 1}{\dfrac{E_f}{E_m} + 2a_f}$$

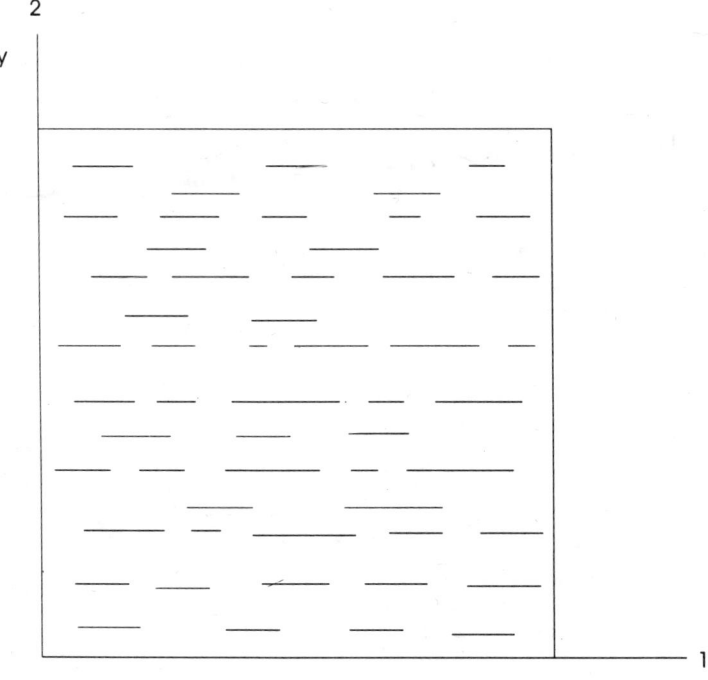

Fig. 8.10.3 Unidirectional Discontinuous Lamina

$$\eta_T = \frac{\frac{E_f}{E_m} - 1}{\frac{E_f}{E_m} + 2}$$

$$\eta_G = \frac{\frac{G_f}{G_m} - 1}{\frac{G_f}{G_m} + 1}$$

(8.10.12)

and $a_f = l_f/d_f$ is the ratio of average fiber length to fiber diameter.

Randomly Oriented Discontinuous Fiber Lamina

Let E_{11} and E_{22} be the longitudinal and transverse moduli defined by (8.5.11) for a unidirectional discontinuous fiber 0° lamina of the same fiber aspect ratio and fiber volume fraction as the randomly oriented discontinuous fiber lamina shown in Fig. 8.10.4. Since the fiber is randomly oriented, the lamina exhibits isotropic behavior. The Young's modulus and shear modulus of such a plate are given by

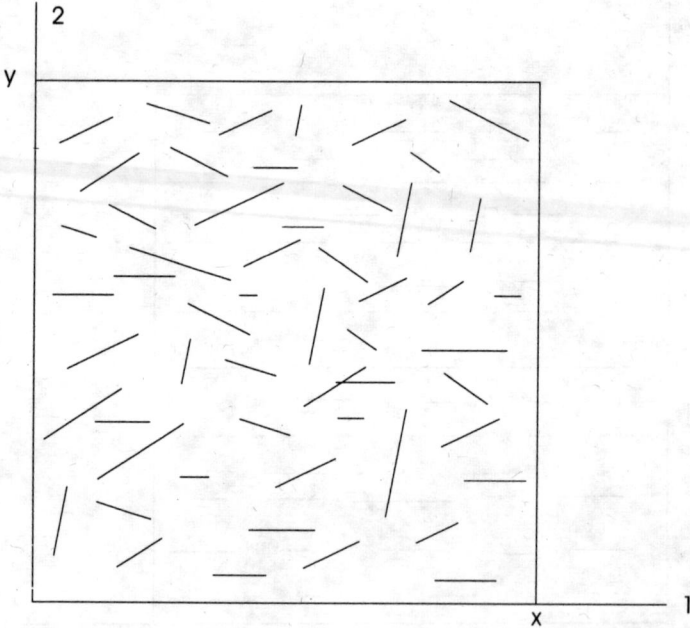

Fig. 8.10.4   Randomly Oriented Discontinuous Fiber Lamina

$$E_{random} = \frac{3}{8}E_{11} + \frac{5}{8}E_{22}$$

$$G_{random} = \frac{1}{8}E_{11} + \frac{1}{4}E_{22}$$

(8.10.13)

Stress Strain Relations for a Thin Lamina

For plane stress, the stress strain relations for an orthotropic lamina shown in Fig. 8.10.5 are given by, see equations (8.5.4),

$$\left\{\begin{array}{c}\varepsilon_{xx}\\ \varepsilon_{yy}\\ \nu_{xy}\end{array}\right\} = \begin{bmatrix}\frac{1}{E_{xx}} & -\frac{\nu_{xy}}{E_{xx}} & -\frac{\eta_{xy}}{E_{xx}}\\ & \frac{1}{E_{yy}} & -\frac{\eta_{xy}}{E_{yy}}\\ \text{Sym} & & \frac{1}{G_{xy}}\end{bmatrix}\left\{\begin{array}{c}\tau_{xx}\\ \tau_{yy}\\ \tau_{xy}\end{array}\right\}$$

(8.10.14)

with the shear coupling coefficients (also called *coefficients of mutual influence*)

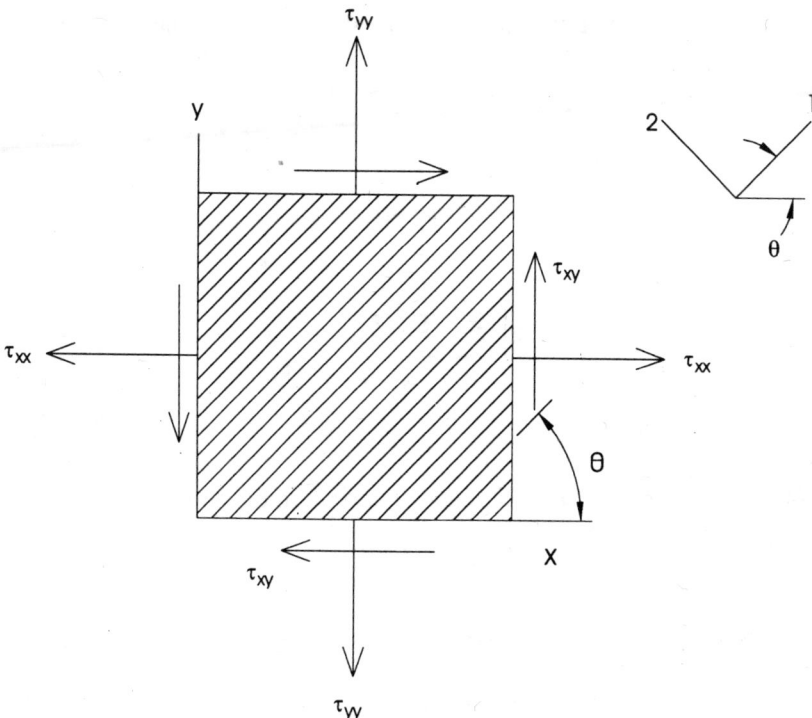

Fig. 8.10.5 Stresses in a General Orthotropic Lamina

$$\eta_{xy} = E_{xx}\left[-\frac{2m^3n}{E_{xx}} + \frac{2mn^3}{E_{yy}} + \left(\frac{1}{G_{xy}} - \frac{2v_{xy}}{E_{xx}}\right)(m^2 - n^2)mn\right]$$

$$\eta_{yx} = E_{yy}\left[-\frac{2mn^3}{E_{xx}} + \frac{2m^3n}{E_{yy}} - \left(\frac{1}{G_{xy}} - \frac{2v_{xy}}{E_{xx}}\right)(m^2 - n^2)mn\right]$$

(8.10.15)

As discussed before, for $\theta = 0$, the shear coupling coefficients are zero and the orthotropic lamina becomes *specially orthotropic* lamina.

$$\left\{\begin{array}{c}\varepsilon_{xx} = \varepsilon_{11}\\ \varepsilon_{yy} = \varepsilon_{22}\\ v_{xy} = \gamma_{12}\end{array}\right\} = \left[\begin{array}{ccc}\frac{1}{E_{11}} & -\frac{v_{12}}{E_{22}} & 0\\ & \frac{1}{E_{22}} & 0\\ \text{Sym} & & \frac{1}{G_{12}}\end{array}\right]\left\{\begin{array}{c}\tau_{xx}\\ \tau_{yy}\\ \tau_{xy}\end{array}\right\}$$

(8.10.16)

Stiffness and Compliance Matrices
The stress strain relations for a general orthotropic lamina are, see (8.5.4)

$$\left\{\begin{array}{c}\varepsilon_{xx}\\ \varepsilon_{yy}\\ v_{xy}\end{array}\right\} = [\overline{S}]\left\{\begin{array}{c}\tau_{xx}\\ \tau_{yy}\\ \tau_{xy}\end{array}\right\}$$

$$= \left[\begin{array}{ccc}\overline{S}_{11} & \overline{S}_{12} & \overline{S}_{16}\\ \overline{S}_{12} & \overline{S}_{22} & \overline{S}_{26}\\ \overline{S}_{16} & \overline{S}_{26} & \overline{S}_{66}\end{array}\right]\left\{\begin{array}{c}\tau_{xx}\\ \tau_{yy}\\ \tau_{xy}\end{array}\right\}$$

(8.10.17)

where the elements of the compliance matrix $\overline{S}$ in the above equation are, see equations (8.4.2) and (8.5.5)

$$\overline{S}_{11} = \frac{1}{E_{xx}} = S_{11}m^4 + (2S_{12} + S_{66})m^2n^2 + S_{22}n^4$$

$$\overline{S}_{12} = \frac{-v_{xy}}{E_{xx}} = (S_{11} + S_{22} - S_{66})m^2n^2 + S_{12}(m^4 + n^4)$$

$$\overline{S}_{16} = -\frac{\eta_{xy}}{E_{xx}} = (2S_{11} - 2S_{12} - S_{66})m^3n + (2S_{12} - 2S_{22} + S_{66})mn^3$$

$$\overline{S}_{22} = \frac{1}{E_{yy}} = S_{11}n^4 + (2S_{12} + S_{66})m^2n^2 + S_{22}m^4$$

$$\overline{S}_{26} = -\frac{\eta_{xy}}{E_{yy}} = (2S_{11} - 2S_{12} - S_{66})mn^3 + (2S_{12} - 2S_{22} + S_{66})m^3n$$

$$\overline{S}_{66} = \frac{1}{G_{xy}} = 2(2S_{11} + 2S_{22} - 4S_{12} - S_{66})m^2n^2 + S_{66}(m^4 + n^4)$$

(8.10.18)

A useful form of the above equations (8.10.18) can be written as

$$\overline{S}_{11} = V_1 + V_2 \cos 2\theta + V_3 \cos 4\theta$$

$$\overline{S}_{12} = V_4 - V_3 \cos 4\theta$$

$$\overline{S}_{16} = V_2 \sin 2\theta + 2V_3 \sin 4\theta$$

$$\overline{S_{22}} = V_1 - V_2 \cos 2\theta + V_3 \cos 4\theta$$

$$\overline{S_{26}} = V_2 \sin 2\theta - 2V_3 \sin 4\theta$$

$$\overline{S_{66}} = V_5 - 4V_3 \cos 4\theta$$

(8.10.18a)

where

$$V_1 = \tfrac{1}{8}(3S_{11} + 3S_{22} + 2S_{12} + S_{66})$$

$$V_2 = \tfrac{1}{2}(S_{11} - S_{22})$$

$$V_3 = \tfrac{1}{8}(S_{11} + S_{22} - 2S_{12} - S_{66})$$

$$V_4 = \tfrac{1}{8}(S_{11} + S_{22} + 6S_{12} - S_{66})$$

$$V_5 = 2(V_1 - V_4)$$

(8.10.18b)

The corresponding stiffness matrix is obtained by inverting the compliance matrix given above, also see equation (8.5.10)

$$\left\{\begin{array}{c} \tau_{xx} \\ \tau_{yy} \\ \tau_{xy} \end{array}\right\} = \left[\begin{array}{ccc} \overline{Q_{11}} & \overline{Q_{12}} & \overline{Q_{16}} \\ \overline{Q_{12}} & \overline{Q_{22}} & \overline{Q_{26}} \\ \overline{Q_{16}} & \overline{Q_{26}} & \overline{Q_{66}} \end{array}\right] \left\{\begin{array}{c} \varepsilon_{xx} \\ \varepsilon_{yy} \\ \gamma_{xy} \end{array}\right\}$$

(8.10.19)

where in terms of equation (8.5.8)

$$\overline{Q_{11}} = Q_{11}m^4 + 2(Q_{12} + 2Q_{66})m^2n^2 + Q_{22}n^4$$

$$\overline{Q_{12}} = (Q_{11} + Q_{22} - 4Q_{66})m^2n^2 + Q_{12}(m^4 + n^4)$$

$$\overline{Q_{16}} = (Q_{11} - Q_{12} - 2Q_{66})m^3n + (Q_{12} - Q_{22} + 2Q_{66})mn^3$$

$$\overline{Q_{22}} = Q_{11}n^4 + 2(Q_{12} + 2Q_{66})m^2n^2 + Q_{22}m^4$$

$$\overline{Q_{26}} = (Q_{11} - Q_{12} - 2Q_{66})mn^3 + (Q_{12} - Q_{22} + 2Q_{66})m^3n$$

$$\overline{Q_{66}} = (Q_{11} + Q_{22} - 2Q_{12} - 2Q_{66})m^2n^2 + Q_{66}(m^4 + n^4)$$

(8.10.20)

As in the case of compliance matrix, the stiffness matrix elements can be written down in a convenient form as given below.

$$\overline{Q_{11}} = U_1 + U_2 \cos 2\theta + U_3 \cos 4\theta$$

$$\overline{Q_{12}} = U_4 - U_3 \cos 4\theta$$

$$\overline{Q_{16}} = \frac{1}{2}U_2 \sin 2\theta + U_3 \sin 4\theta$$

$$\overline{Q_{22}} = U_1 - U_2 \cos 2\theta + U_3 \cos 4\theta$$

$$\overline{Q_{26}} = \frac{1}{2}U_2 \sin 2\theta - U_3 \sin 4\theta$$

$$\overline{Q_{66}} = U_5 - U_3 \cos 4\theta$$

(8.10.20a)

where

$$U_1 = \frac{1}{8}(3Q_{11} + 3Q_{22} + 2Q_{12} + 4Q_{66})$$

$$U_2 = \frac{1}{2}(Q_{11} - Q_{22})$$

$$U_3 = \frac{1}{8}(Q_{11} + Q_{22} - 2Q_{12} - 4Q_{66})$$

$$U_4 = \frac{1}{8}(Q_{11} + Q_{22} + 6Q_{12} - 4Q_{66})$$

$$U_5 = \frac{1}{2}(U_1 - U_4)$$

(8.10.20b)

Example

Let us consider a 45 degree angle ply lamina containing 60% carbon fibers in an epoxy resin. The Young's moduli of the fiber and the resin are 220 and 4 GPa, Shear moduli are 90 and 1 GPa and the Poisson's ratios are 0.2 and 0.3 respectively.

$$v_f = 0.6$$
$$v_m = 0.4$$

From (8.10.6) to (8.10.9)

$$E_{11} = E_f v_f + E_m(1 - v_f)$$
$$= 220 \times 0.6 + 4 \times 0.4$$
$$= 133.6 \text{ GPa}$$

$$v_{12} = v_f v_f + v_m v_m$$
$$= 0.2 \times 0.6 + 0.3 \times 0.4$$
$$= 0.24$$

$$E_{22} = \frac{E_f E_m}{E_f v_m + E_m v_f}$$
$$= \frac{220 \times 4}{220 \times 0.4 + 4 \times 0.6}$$
$$= 9.7345 \text{ GPa}$$

$$v_{21} = \frac{E_{22}}{E_{11}} v_{12} = 0.0175$$

$$G_{12} = \frac{G_f G_m}{G_f v_m + G_m v_f}$$
$$= \frac{90 \times 1}{90 \times 0.4 + 1 \times 0.6}$$
$$= 2.459 \text{ GPa}$$

From equation (8.5.8)

$$Q_{11} = \frac{E_{11}}{1 - v_{12} v_{21}} = \frac{133.6}{1 - 0.24 \times 0.0175} = 134.1635 \text{ GPa}$$

$$Q_{12} = \frac{v_{12} E_{22}}{1 - v_{12} v_{21}} = \frac{0.24 \times 9.7345}{1 - 0.24 \times 0.0175} = 2.3461 \text{ GPa}$$

$$Q_{22} = \frac{E_{22}}{1 - v_{12} v_{21}} = \frac{9.7345}{1 - 0.24 \times 0.0175} = 9.7756 \text{ GPa}$$

$$Q_{66} = G_{12} = 2.459 \text{ GPa}$$

358 *Dynamics of Plates*

From equation (8.10.20b)

$$U_1 = \frac{1}{8}(3 \times 134.1635 + 3 \times 9.7756 + 2 \times 2.3461 + 4 \times 2.459) = 55.79 \text{ GPa}$$

$$U_2 = \frac{1}{2}(134.1635 - 9.7756) = 62.19 \text{ GPa}$$

$$U_3 = \frac{1}{8}(134.1635 + 9.7756 - 2 \times 2.3461 - 4 \times 2.459) = 16.18 \text{ GPa}$$

$$U_4 = \frac{1}{8}(134.1635 + 9.7756 + 6 \times 2.3461 - 4 \times 2.459) = 18.52 \text{ GPa}$$

$$U_5 = \frac{1}{2}(55.79 - 18.52) = 18.64 \text{ GPa}$$

From equation (8.10.20a), the stiffness matrix elements are

$$\overline{Q}_{11} = 55.79 + 62.19 \cos 90° + 16.18 \cos 180° = 39.61 \text{ GPa}$$

$$\overline{Q}_{12} = 18.52 - 16.18 \cos 180° = 34.7 \text{ GPa}$$

$$\overline{Q}_{16} = \frac{1}{2} \times 62.19 \sin 90° + 16.18 \sin 180° = 31.1 \text{ GPa}$$

$$\overline{Q}_{22} = 55.79 - 62.19 \cos 90° + 16.18 \cos 180° = 39.61 \text{ GPa}$$

$$\overline{Q}_{26} = \frac{1}{2} \times 62.19 \sin 90° - 16.18 \sin 180° = 31.1 \text{ GPa}$$

$$\overline{Q}_{66} = 18.64 - 16.18 \cos 180° = 34.82 \text{ GPa}$$

8.11 LAMINATED PLATES

A *laminated plate* consists of several laminae each with a fiber oriented at a specified angle. This is generally made by stacking several thin layers of fibers at the desired locations and angles in a matrix and consolidating them to give the required thickness. The fiber orientation in each thin layer can be arranged in a specific manner so as to achieve the required properties of the structural member. Most commonly used forms are:

1. *Unidirectional laminates*, where the fiber orientation angle is same in all laminae.
2. *Mid plane symmetric laminates*, where the ply (lamina) orientation is symmetrical about the center line of the laminated plate; that is, for each lamina above the mid plane, there is an

identical lamina in all respects (material, thickness, fiber orientation) at an equal distance below the mid plane. Examples are:

a. $[0/45/90]_S$ which contains six layers [0/45/90/90/45/0]

b. $[0/45/\overline{90}]_S$ which contains five layers [0/45/90/45/0], where the bar over 90 indicates that the plane of symmetry passes midway through this lamina

c. $[0/\pm 45/\overline{90}]_S$ which contains seven layers [0/+45/-45/90/-45/+45/0]

3. *Angle-ply laminates*, where the orientation angles in alternate layers are $\cdots/\theta/-\theta/\theta/-\theta/\cdots$ when $\theta \neq 0$ or $90°$, for example, $[\pm\theta/\theta/-\overline{\theta}]_S$ which contains seven layers $[\theta/-\theta/\theta/-\theta/\theta/-\theta/\theta]$

4. *Cross-ply laminates*, where the orientation angles in alternate layers are $\cdots/0/90/0/90°/\cdots$, for example, $[(0/90)_2/\overline{0}]_S$ which contains nine layers [0/90/0/90/0/90/0/90/0]

5. *Balanced laminates* where for every lamina of $+\theta°$ orientation, there is an identical $-\theta°$ orientation lamina, somewhere within the laminate.

6. *Anti symmetric laminates*, where for every lamina of $+\theta°$ orientation at distance z, there is an identical $-\theta°$ orientation lamina at distance $-z$.

Laminate Geometry
The geometry of an N layer laminate is shown in Fig. 8.11.1. The z coordinate is measured positive downward from the mid plane of the laminate. The total thickness of the laminate is h. The first lamina, of thickness t_1 is located at the top of the stacking. The jth lamina of thickness t_j is at a distance h_{j-1}. The laminate itself is considered thin as in the case of thin plate theory and a perfect inter laminar bond exists between all the laminae. Each lamina is also assumed to be macroscopically homogeneous and behaves in a linear elastic manner.

Strain Displacement Relations
In accordance to thin plate theory with in-plane forces, the displacement field is assumed to be

360 Dynamics of Plates

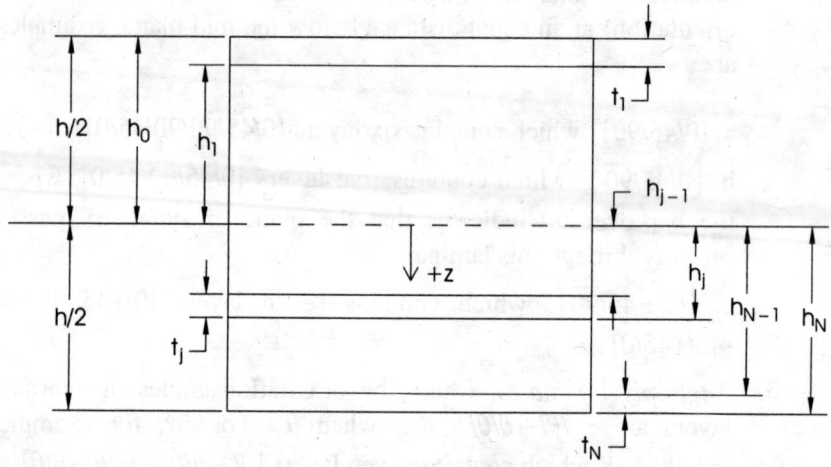

Fig. 8.11.1 Laminate Geometry

$$u_x = u^0(x,y) - zw_{,x}$$
$$u_y = v^0(x,y) - zw_{,y}$$
$$u_z = w(x,y)$$

(8.11.1)

where u^0 and v^0 are the tangential displacements of the mid plane of the laminate. The strains are therefore given by

$$\varepsilon_{xx} = \varepsilon_{xx}^0(x,y) + z\chi_x$$
$$\varepsilon_{yy} = \varepsilon_{yy}^0(x,y) + z\chi_y$$
$$\gamma_{xy} = \gamma_{xy}^0 + z\chi_{xy}$$

(8.11.2)

where the mid plane normal and shear strains are

$$\varepsilon_{xx}^0 = u_{,x}^0$$
$$\varepsilon_{yy}^0 = v_{,y}^0$$
$$\gamma_{xy}^0 = u_{,y}^0 + v_{,x}^0$$

(8.11.3)

and the bending and twisting curvatures

$$\chi_x = -w_{,xx}$$
$$\chi_y = -w_{,yy}$$
$$\chi_{xy} = -2w_{,xy}$$

(8.11.4)

Stress Strain Relation
Depending on the individual lamina properties, the stresses are determined from equation (8.10.19)

$$\left\{\begin{array}{c} \tau_{xx} \\ \tau_{yy} \\ \tau_{xy} \end{array}\right\} = \left[\begin{array}{ccc} \overline{Q}_{11} & \overline{Q}_{12} & \overline{Q}_{16} \\ \overline{Q}_{12} & \overline{Q}_{22} & \overline{Q}_{26} \\ \overline{Q}_{16} & \overline{Q}_{26} & \overline{Q}_{66} \end{array}\right] \left\{\begin{array}{c} \varepsilon_{xx} \\ \varepsilon_{yy} \\ \gamma_{xy} \end{array}\right\}$$

$$= \left[\begin{array}{ccc} \overline{Q}_{11} & \overline{Q}_{12} & \overline{Q}_{16} \\ \overline{Q}_{12} & \overline{Q}_{22} & \overline{Q}_{26} \\ \overline{Q}_{16} & \overline{Q}_{26} & \overline{Q}_{66} \end{array}\right] \left\{\begin{array}{c} \varepsilon_{xx}^0 + z\chi_x \\ \varepsilon_{yy}^0 + z\chi_y \\ \gamma_{xy}^0 + z\chi_{xy} \end{array}\right\}$$

(8.11.5)

Axial Stress Resultants
The axial stress resultants per unit width are obtained from equations (6.1.6).

$$N_x = \int_{-h/2}^{h/2} \tau_{xx}\, dz = \sum_{j=1}^{N} \int_{h_{j-1}}^{h_j} \tau_{xx}^j\, dz$$

$$N_y = \int_{-h/2}^{h/2} \tau_{yy}\, dz = \sum_{j=1}^{N} \int_{h_{j-1}}^{h_j} \tau_{yy}^j\, dz$$

$$N_{xy} = \int_{-h/2}^{h/2} \tau_{xy}\, dz = \sum_{j=1}^{N} \int_{h_{j-1}}^{h_j} \tau_{xy}^j\, dz$$

(8.11.6)

With the help of (8.11.5), the above equation (8.11.6) can be written as

$$\left\{\begin{array}{c} N_x \\ N_y \\ N_{xy} \end{array}\right\} = \left[\begin{array}{ccc} A_{11} & A_{12} & A_{16} \\ A_{12} & A_{22} & A_{26} \\ A_{16} & A_{26} & A_{66} \end{array}\right] \left\{\begin{array}{c} \varepsilon^0_{xx} \\ \varepsilon^0_{yy} \\ \gamma^0_{xy} \end{array}\right\} + \left[\begin{array}{ccc} B_{11} & B_{12} & B_{16} \\ B_{12} & B_{22} & B_{26} \\ B_{16} & B_{26} & B_{66} \end{array}\right] \left\{\begin{array}{c} \chi_x \\ \chi_y \\ \chi_{xy} \end{array}\right\}$$

(8.11.7)

where

$$A_{mn} = \sum_{j=1}^{N} \overline{(Q_{mn})}_j (h_j - h_{j-1})$$

$$B_{mn} = \frac{1}{2} \sum_{j=1}^{N} \overline{(Q_{mn})}_j (h_j^2 - h_{j-1}^2)$$

(8.11.8)

Bending and Twisting Moment Resultants

The moment resultants per unit width are obtained from equations (6.1.5).

$$M_x = \int_{-h/2}^{h/2} \tau_{xx} z \, dz = \sum_{j=1}^{N} \int_{h_{j-1}}^{h_j} \tau^j_{xx} z \, dz$$

$$M_y = \int_{-h/2}^{h/2} \tau_{yy} z \, dz = \sum_{j=1}^{N} \int_{h_{j-1}}^{h_j} \tau^j_{yy} z \, dz$$

$$M_{xy} = -\int_{-h/2}^{h/2} \tau_{xy} z \, dz = -\sum_{j=1}^{N} \int_{h_{j-1}}^{h_j} \tau^j_{xy} z \, dz$$

(8.11.9)

With the help of (8.11.5), the above equation (8.11.9) can be written as

$$\left\{\begin{array}{c} M_x \\ M_y \\ -M_{xy} \end{array}\right\} = \left[\begin{array}{ccc} B_{11} & B_{12} & B_{16} \\ B_{12} & B_{22} & B_{26} \\ B_{16} & B_{26} & B_{66} \end{array}\right] \left\{\begin{array}{c} \varepsilon^0_{xx} \\ \varepsilon^0_{yy} \\ \gamma^0_{xy} \end{array}\right\} + \left[\begin{array}{ccc} D_{11} & D_{12} & D_{16} \\ D_{12} & D_{22} & D_{26} \\ D_{16} & D_{26} & D_{66} \end{array}\right] \left\{\begin{array}{c} \chi_x \\ \chi_y \\ \chi_{xy} \end{array}\right\}$$

(8.11.10)

where B_{mn} are given in equation (8.11.8) and

$$D_{mn} = \frac{1}{3} \sum_{j=1}^{N} \overline{(Q_{mn})}_j (h_j^3 - h_{j-1}^3)$$

(8.11.11)

Laminate Constitutive Equations

Equations (8.11.7) and (8.11.11) can be combined to give the constitutive equations of the laminate as

$$\left\{ \begin{array}{c} \{N\} \\ \{M\} \end{array} \right\} = \left[\begin{array}{cc} [A] & [B] \\ [B] & [D] \end{array} \right] \left\{ \begin{array}{c} \{\varepsilon^0\} \\ \{\chi\} \end{array} \right\}$$

(8.11.12)

where

$$\{N\} = \left\{ \begin{array}{c} N_x \\ N_y \\ N_{xy} \end{array} \right\} \quad \{M\} = \left\{ \begin{array}{c} M_x \\ M_y \\ -M_{xy} \end{array} \right\}$$

$$\{\varepsilon^0\} = \left\{ \begin{array}{c} \varepsilon^0_{xx} \\ \varepsilon^0_{yy} \\ \gamma^0_{xy} \end{array} \right\} \quad \{\chi\} = \left\{ \begin{array}{c} \chi_x \\ \chi_y \\ \chi_{xy} \end{array} \right\}$$

(8.11.13)

and the stiffness matrices

$$[A] = \left[\begin{array}{ccc} A_{11} & A_{12} & A_{16} \\ A_{12} & A_{22} & A_{26} \\ A_{16} & A_{26} & A_{66} \end{array} \right]$$

$$[B] = \left[\begin{array}{ccc} B_{11} & B_{12} & B_{16} \\ B_{12} & B_{22} & B_{26} \\ B_{16} & B_{26} & B_{66} \end{array} \right]$$

$$[D] = \left[\begin{array}{ccc} D_{11} & D_{12} & D_{16} \\ D_{12} & D_{22} & D_{26} \\ D_{16} & D_{26} & D_{66} \end{array} \right]$$

(8.11.14)

We note the following from equation (8.11.12):

1. [B] is the coupling matrix; it will give rise to extension and shear deformations as well as bending and twisting curvatures. Such an extension-bending coupling may be present even if the individual laminae are isotropic or orthotropic.

2. For a symmetric laminate, we find that the extension-bending coupling matrix is zero. Therefore, the extension and bending are decoupled in the case of symmetric angle-ply or cross-ply laminates.
3. For balanced laminates, A_{16} and A_{26} terms in the [A] matrix are zero. This indicates that the normal stress and shear strain are decoupled for balanced laminates.
4. For anti symmetric laminates, D_{16} and D_{26} terms in the [D] matrix are zero. This indicates that the bending and twisting curvatures are decoupled for such laminates. Such decoupling also occurs in the case of symmetric laminates with $\theta = 0$ or $90°$.

Example
Consider an unsymmetrical angle-ply laminate [45/0/-45] made with 6 mm thick lamina of the example of Section 8.10

The laminate arrangement is shown in Fig. 8.11.2. The reduced stiffness matrices for the three laminae are

$$[\bar{Q}]_{+45} = \begin{bmatrix} 39.61 & 34.7 & 31.1 \\ 31.1 & 39.61 & 31.1 \\ 31.1 & 31.1 & 34.82 \end{bmatrix}$$

$$[\bar{Q}]_{0} = \begin{bmatrix} 134.16 & 2.34 & 0 \\ 2.34 & 9.78 & 0 \\ 0 & 0 & 2.46 \end{bmatrix}$$

$$[\bar{Q}]_{-45} = \begin{bmatrix} 39.61 & 34.7 & -31.1 \\ 34.7 & 39.61 & -31.1 \\ -31.1 & -31.1 & 34.82 \end{bmatrix}$$

From (8.11.8)

$$A_{mn} = \overline{(Q_{mn})}_{+45}(h_1 - h_0) + \overline{(Q_{mn})}_{0}(h_2 - h_1)$$
$$+ \overline{(Q_{mn})}_{-45}(h_3 - h_2)$$
$$= 6 \times 10^{-3} \left[\overline{(Q_{mn})}_{+45} + \overline{(Q_{mn})}_{0} + \overline{(Q_{mn})}_{-45} \right]$$

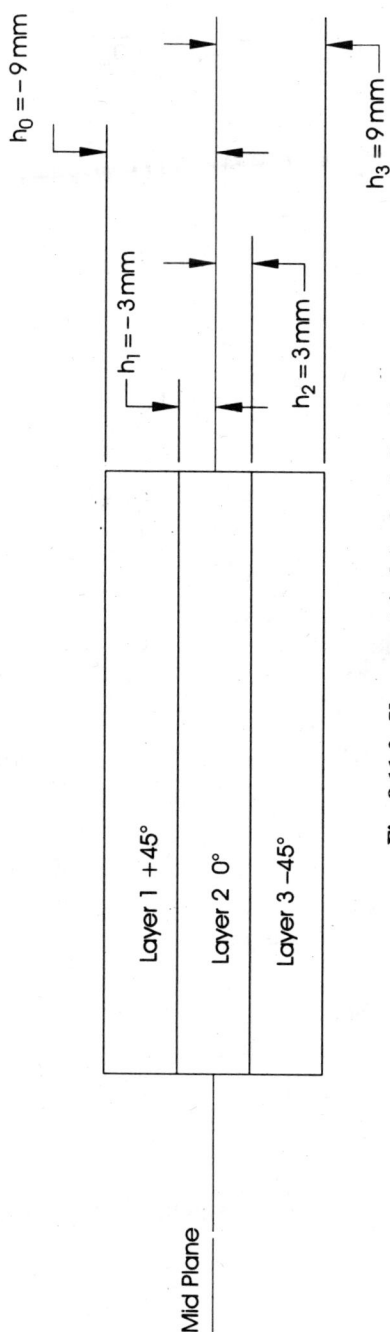

Fig. 8.11.2 Unsymmetrical Angle-ply Laminate

$$B_{mn} = \frac{1}{2}\left[\begin{array}{c}(\overline{Q_{mn}})_{+45}(h_1^2 - h_0^2) + (\overline{Q_{mn}})_0 (h_2^2 - h_1^2) \\ + (\overline{Q_{mn}})_{-45}(h_3^2 - h_2^2)\end{array}\right]$$

$$= 36 \times 10^{-6}\left[-(\overline{Q_{mn}})_{+45} + (\overline{Q_{mn}})_{-45}\right]$$

From (8.11.11)

$$D_{mn} = \frac{1}{3}\left[\begin{array}{c}(\overline{Q_{mn}})_{+45}(h_1^3 - h_0^3) + (\overline{Q_{mn}})_0 (h_2^3 - h_1^3) \\ + (\overline{Q_{mn}})_{-45}(h_3^3 - h_2^3)\end{array}\right]$$

$$= 18 \times 10^{-9}\left[13(\overline{Q_{mn}})_{+45} + (\overline{Q_{mn}})_0 + 13(\overline{Q_{mn}})_{-45}\right]$$

The stiffness matrices are

$$[A] = 10^6 \begin{bmatrix} 1280.28 & 430.44 & 0 \\ 408.84 & 534 & 0 \\ 0 & 0 & 72.1 \end{bmatrix} \text{N/m}$$

$$[B] = 10^3 \begin{bmatrix} 0 & 0 & -2239.2 \\ 0 & 0 & -2239.2 \\ -2239.2 & -2239.2 & 0 \end{bmatrix} \text{N}$$

$$[D] = \begin{bmatrix} 20952.36 & 16281.72 & 0 \\ 15439.32 & 18713.52 & 0 \\ 0 & 0 & 16340.04 \end{bmatrix} \text{Nm}$$

Invariants of Stiffness Matrices

The following are invariants of terms in the stiffness matrices.

Matrix $\overline{Q}$:

$$L_1 = \overline{Q_{11}} + \overline{Q_{22}} + 2\overline{Q_{12}}$$
$$L_2 = \overline{Q_{66}} - \overline{Q_{12}}$$

(8.11.15)

Matrix [A]:

$$P_1 = A_{11} + A_{22} + 2A_{12} = hL_1$$
$$P_2 = A_{66} - A_{12} = hL_2$$

(8.11.16a)

Matrix [B]:

$$P_3 = B_{11} + B_{22} + 2B_{12} = 0$$
$$P_4 = B_{66} - B_{12} = 0$$

(8.11.16b)

Matrix [D]:

$$P_5 = D_{11} + D_{22} + 2D_{12} = \frac{h^3}{12}L_1$$
$$P_6 = D_{66} - D_{12} = \frac{h^3}{12}L_2$$

(8.11.16c)

Laminate Equations

The equilibrium relations for the laminate remain same as given by equations (6.1.15) for a homogeneous plate. Here, we assume that the in-plane forces have negligible effect on bending of the laminate and write the equations (6.1.15) as

$$N_{x,x} + N_{xy,y} = 0$$
$$N_{y,y} + N_{xy,x} = 0$$
$$M_{x,xx} + M_{y,yy} - 2M_{xy,xy} + q = 0$$

(8.11.17)

Substituting the constitutive relations for the laminate given by equation (8.11.12) in the above equation (8.11.17) we get

$$A_{11}\varepsilon^0_{xx,x} + A_{12}\varepsilon^0_{yy,x} + A_{16}\gamma^0_{xy,x} + B_{11}\chi_{x,x} + B_{12}\chi_{y,x} + B_{16}\chi_{xy,x} +$$
$$A_{16}\varepsilon^0_{xx,y} + A_{26}\varepsilon^0_{yy,y} + A_{66}\gamma^0_{xy,y} + B_{16}\chi_{x,y} + B_{26}\chi_{y,y} + B_{66}\chi_{xy,y}$$
$$= 0$$

(8.11.18)

$$A_{12}\varepsilon_{xx,y}^0 + A_{22}\varepsilon_{yy,y}^0 + A_{26}\gamma_{xy,y}^0 + B_{12}\chi_{x,y} + B_{22}\chi_{y,y} + B_{26}\chi_{xy,y}$$
$$+ A_{16}\varepsilon_{xx,x}^0 + A_{26}\varepsilon_{yy,x}^0 + A_{66}\gamma_{xy,x}^0 + B_{16}\chi_{x,x} + B_{26}\chi_{y,x} + B_{66}\chi_{xy,x}$$
$$= 0$$
(8.11.19)

$$B_{11}\varepsilon_{xx,xx}^0 + B_{12}\varepsilon_{yy,xx}^0 + B_{16}\gamma_{xy,xx}^0 + D_{11}\chi_{x,xx} + D_{12}\chi_{y,xx} + D_{16}\chi_{xy,xx}$$
$$+ B_{12}\varepsilon_{xx,yy}^0 + B_{22}\varepsilon_{yy,yy}^0 + B_{26}\gamma_{xy,yy}^0 + D_{12}\chi_{x,yy} + D_{22}\chi_{y,yy} + D_{26}\chi_{xy,yy}$$
$$+ 2B_{16}\varepsilon_{xx,xy}^0 + 2B_{26}\varepsilon_{yy,xy}^0 + 2B_{66}\gamma_{xy,xy}^0 + 2D_{16}\chi_{x,xy} + 2D_{26}\chi_{y,xy}$$
$$+ 2D_{66}\chi_{xy,xy} + q$$
$$= 0$$
(8.11.20)

Now, using equations (8.11.3) and (8.11.4), the above equations (8.11.18) to (8.11.20) can be expressed in terms of laminate mid plane displacements as

$$A_{11}u_{,xx}^0 + 2A_{16}u_{,xy}^0 + A_{66}u_{,yy}^0$$
$$+ A_{16}v_{,xx}^0 + (A_{12}+A_{66})v_{,xy}^0 + A_{26}v_{,yy}^0$$
$$= B_{11}w_{,xxx} + (B_{12}+2B_{66})w_{,xyy} + 3B_{16}w_{,xxy} + B_{26}w_{,yyy}$$
(8.11.21)

$$A_{16}u_{,xx}^0 + (A_{12}+A_{66})u_{,xy}^0 + A_{26}u_{,yy}^0$$
$$+ A_{66}v_{,xx}^0 + 2A_{26}v_{,xy}^0 + A_{22}v_{,yy}^0$$
$$= B_{16}w_{,xxx} + 3B_{26}w_{,xyy} + (B_{12}+2B_{66})w_{,xxy} + B_{22}w_{,yyy}$$
(8.11.22)

$$B_{11}u_{,xxx}^0 + (B_{12}+2B_{66})u_{,xyy}^0 + 3B_{16}u_{,xxy}^0 + B_{26}u_{,yyy}^0$$
$$+ B_{16}v_{,xxx}^0 + 3B_{26}v_{,xyy}^0 + (B_{12}+2B_{66})v_{,xxy}^0 + B_{22}v_{,yyy}^0 + q$$
$$= D_{11}w_{,xxxx} + 4D_{16}w_{,xxxy} + 2(D_{12}+2D_{66})w_{,xxyy} + 4D_{26}w_{,xyyy} + D_{22}w_{,yyyy}$$
(8.11.23)

Symmetric Laminates:
For symmetric laminated plates, the extension-bending coupling matrix [B] is zero and the above equations simplify to

$$A_{11}u^0_{,xx} + 2A_{16}u^0_{,xy} + A_{66}u^0_{,yy} + A_{16}v^0_{,xx} + (A_{12}+A_{66})v^0_{,xy} + A_{26}v^0_{,yy} = 0$$

(8.11.24)

$$A_{16}u^0_{,xx} + (A_{12}+A_{66})u^0_{,xy} + A_{26}u^0_{,yy} + A_{66}v^0_{,xx} + 2A_{26}v^0_{,xy} + A_{22}v^0_{,yy} = 0$$

(8.11.25)

$$D_{11}w_{,xxxx} + 4D_{16}w_{,xxxy} + 2(D_{12}+2D_{66})w_{,xxyy} + 4D_{26}w_{,xyyy} + D_{22}w_{,yyyy} = q$$

(8.11.26)

Specially Orthotropic Plates:
For specially orthotropic plates, $A_{16} = A_{26} = D_{16} = D_{26} = 0$ and the above equations (8.11.24) to (8.11.26) further reduce to

$$A_{11}u^0_{,xx} + A_{66}u^0_{,yy} + (A_{12}+A_{66})v^0_{,xy} = 0$$

(8.11.27)

$$(A_{12}+A_{66})u^0_{,xy} + A_{66}v^0_{,xx} + A_{22}v^0_{,yy} = 0$$

(8.11.28)

$$D_{11}w_{,xxxx} + 2(D_{12}+2D_{66})w_{,xxyy} + D_{22}w_{,yyyy} = q$$

(8.11.29)

Equation (8.11.29) was the subject of sections 8.6 to 8.9.

In Plane Forces

The in plane forces in terms of displacements are obtained by substituting equations (8.11.3) and (8.11.4) in equation (8.11.7) to give

$$N_x = A_{11}u^0_{,x} + A_{16}\left(u^0_{,y} + v^0_{,x}\right) + A_{12}v^0_{,y}$$
$$\quad - B_{11}w_{,xx} - 2B_{16}w_{,xy} - B_{12}w_{,yy}$$
$$N_y = A_{12}u^0_{,x} + A_{26}\left(u^0_{,y} + v^0_{,x}\right) + A_{22}v^0_{,y}$$
$$\quad - B_{12}w_{,xx} - 2B_{26}w_{,xy} - B_{22}w_{,yy}$$
$$N_{xy} = A_{16}u^0_{,x} + A_{66}\left(u^0_{,y} + v^0_{,x}\right) + A_{26}v^0_{,y}$$
$$\quad - B_{16}w_{,xx} - 2B_{66}w_{,xy} - B_{26}w_{,yy}$$

(8.11.30)

Bending and Twisting Moments

The bending and twisting moments in terms of displacements are obtained by substituting equations (8.11.3) and (8.11.4) in equation (8.11.10) to give

$$M_x = B_{11}u^0_{,x} + B_{12}v^0_{,y} + B_{16}\left(u^0_{,y} + v^0_{,x}\right)$$
$$- D_{11}w_{,xx} - 2D_{16}w_{,xy} - D_{12}w_{,yy}$$

$$M_y = B_{12}u^0_{,x} + B_{26}\left(u^0_{,y} + v^0_{,x}\right) + B_{22}v^0_{,y}$$
$$- D_{12}w_{,xx} - 2D_{26}w_{,xy} - D_{22}w_{,yy}$$

$$-M_{xy} = B_{16}u^0_{,x} + B_{66}\left(u^0_{,y} + v^0_{,x}\right) + B_{26}v^0_{,y}$$
$$- D_{16}w_{,xx} - 2D_{66}w_{,xy} - D_{26}w_{,yy}$$

(8.11.31)

Shear Resultants

The shear stress resultants per unit width are obtained from equations (3.2.40)

$$V_x = \int_{-h/2}^{h/2} \tau_{xz}\,dz = \sum_{j=1}^{N} \int_{h_{j-1}}^{h_j} \tau^j_{xz}\,dz$$

$$V_y = \int_{-h/2}^{h/2} \tau_{yz}\,dz = \sum_{j=1}^{N} \int_{h_{j-1}}^{h_j} \tau^j_{yz}\,dz$$

(8.11.32)

The above can be evaluated in a similar manner to that of in-plane forces and bending and twisting moments to give

$$V_x = B_{11}u^0_{,xx} + 2B_{16}u^0_{,xy} + B_{66}u^0_{,yy} + B_{16}v^0_{,xx} + (B_{12}+B_{66})v^0_{,xy} + B_{26}v^0_{,yy}$$
$$- D_{11}w_{,xxx} - 3D_{16}w_{,xxy} - (D_{12}+2D_{66})w_{,xyy} - D_{26}w_{,yyy}$$

$$V_y = B_{16}u^0_{,xx} + (B_{12}+B_{66})u^0_{,xy} + B_{26}u^0_{,yy} + B_{66}v^0_{,xx} + 2B_{26}v^0_{,xy} + B_{22}v^0_{,yy}$$
$$- D_{16}w_{,xxx} - (D_{12}+2D_{66})w_{,xxy} - 3D_{26}w_{,xyy} - D_{22}w_{,yyy}$$

(8.11.33)

Lamina Stresses - *j*th Layer

From (8.11.5), we can obtain

$$\tau_{xx}^j = \overline{Q_{11}^j} u_{,x}^0 + \overline{Q_{16}^j}(u_{,y}^0 + v_{,x}^0) + \overline{Q_{12}^j} v_{,y}^0$$
$$- z\left(\overline{Q_{11}^j} w_{,xx} + 2\overline{Q_{16}^j} w_{,xy} + \overline{Q_{12}^j} w_{,yy}\right)$$

$$\tau_{yy}^j = \overline{Q_{12}^j} u_{,x}^0 + \overline{Q_{26}^j}(u_{,y}^0 + v_{,x}^0) + \overline{Q_{22}^j} v_{,y}^0$$
$$- z\left(\overline{Q_{12}^j} w_{,xx} + 2\overline{Q_{26}^j} w_{,xy} + \overline{Q_{22}^j} w_{,yy}\right)$$

$$\tau_{xy}^j = \overline{Q_{16}^j} u_{,x}^0 + \overline{Q_{66}^j}(u_{,y}^0 + v_{,x}^0) + \overline{Q_{26}^j} v_{,y}^0$$
$$- z\left(\overline{Q_{16}^j} w_{,xx} + 2\overline{Q_{66}^j} w_{,xy} + \overline{Q_{26}^j} w_{,yy}\right)$$

(8.11.34)

Inter laminar Shear Stresses

The inter laminar shear stresses are determined from the first two equations of equilibrium

$$\tau_{xx,x} + \tau_{xy,y} + \tau_{xz,z} = 0$$
$$\tau_{xy,x} + \tau_{yy,y} + \tau_{yz,z} = 0$$

(8.11.35)

Integrating the above equations (8.11.35) with respect to z and using equations (8.11.34), we get

$$\tau_{xz}^j = \frac{1}{2} z^2 \left[\overline{Q_{11}^j} w_{,xxx} + 3\overline{Q_{16}^j} w_{,xxy} + \left(\overline{Q_{12}^j} + 2\overline{Q_{66}^j}\right) w_{,xyy} + \overline{Q_{26}^j} w_{,yyy} \right]$$
$$- z \left[\begin{array}{c} \overline{Q_{11}^j} u_{,xx}^0 + 2\overline{Q_{16}^j} u_{,xy}^0 + \overline{Q_{66}^j} u_{,yy}^0 + \overline{Q_{16}^j} v_{,xx}^0 \\ + \left(\overline{Q_{12}^j} + \overline{Q_{66}^j}\right) v_{,xy}^0 + \overline{Q_{26}^j} v_{,yy}^0 \end{array} \right] + f^j(x,y)$$

$$\tau_{yz}^j = \frac{1}{2}z^2\left[\overline{Q_{16}}^j w_{,xxx} + \left(\overline{Q_{12}}^j + 2\overline{Q_{66}}^j\right)w_{,xxy} + 3\overline{Q_{26}}^j w_{,xyy} + \overline{Q_{22}}^j w_{,yyy}\right]$$

$$-z\left[\begin{array}{c}\overline{Q_{16}}^j u_{,xx}^0 + \left(\overline{Q_{12}}^j + \overline{Q_{66}}^j\right)u_{,xy}^0 + \overline{Q_{26}}^j u_{,yy}^0 \\ +\overline{Q_{66}}^j v_{,xx}^0 + 2\overline{Q_{26}}^j v_{,xy}^0 + \overline{Q_{22}}^j v_{,yy}^0\end{array}\right] + g^j(x,y)$$

(8.11.36)

where f and g are the functions of integration to be determined from the inter laminar continuity conditions for the intermediate layers and zero shear traction condition on the top and bottom surfaces.

Stress Function Equations

We have three equations for the laminated plate expressed in terms of the transverse and in plane displacements. We can reduce these to two equations by defining a stress function F as in equation (6.1.29). From (8.11.12), we can write

$$\{N\} = [A]\{\varepsilon^0\} + [B]\{\chi\}$$
$$\{M\} = [B]\{\varepsilon^0\} + [D]\{\chi\}$$

(8.11.37)

Solving for $\{\varepsilon^0\}$ from the first equation and substituting the same in second equation of (8.11.37) above, we get

$$\left\{\begin{array}{c}\{\varepsilon^0\} \\ \{M\}\end{array}\right\} = \left[\begin{array}{cc}[\overline{A}] & [\overline{B}] \\ [-\overline{B}]^T & [\overline{D}]\end{array}\right]\left\{\begin{array}{c}\{N\} \\ \{\chi\}\end{array}\right\}$$

(8.11.38)

where

$$[\overline{A}] = [A]^{-1}$$
$$[\overline{B}] = -[A]^{-1}[B]$$
$$[\overline{D}] = [D] - [B][A]^{-1}[B]$$

(8.11.39)

It may be noted that $\overline{A}, \overline{D}$ matrices above are symmetrical while $\overline{B}$ is not. With the help of equation (8.11.38), the third equation of (8.11.17) gives

$$\overline{D}_{11}w_{,xxxx} + 4\overline{D}_{16}w_{,xxxy} + 2(\overline{D}_{12} + \overline{D}_{66})w_{,xxyy} + 4\overline{D}_{26}w_{,xyyy} + \overline{D}_{22}w_{,yyyy}$$
$$+ \overline{B}_{21}F_{,xxxx} + (2\overline{B}_{26} - \overline{B}_{61})F_{,xxxy} + (\overline{B}_{11} + \overline{B}_{22} - 2\overline{B}_{66})F_{,xxyy}$$
$$+ (2\overline{B}_{16} - \overline{B}_{62})F_{,xyyy} + \overline{B}_{12}F_{,yyyy}$$
$$= q$$

(8.11.40)

To obtain the second equation, we use the following strain compatibility relation

$$(\varepsilon_{xx}^0)_{,yy} + (\varepsilon_{yy}^0)_{,xx} = (\gamma_{xy}^0)_{,xy}$$

(8.11.41)

Using (8.11.38), (6.1.29), (8.11.3) and (8.11.4) in the above equation (8.11.41), we get

$$\overline{A}_{22}F_{,xxxx} - 2\overline{A}_{26}F_{,xxxy} + 2(\overline{A}_{12} + \overline{A}_{66})F_{,xxyy}$$
$$- 2\overline{A}_{16}F_{,xyyy} + \overline{A}_{11}F_{,yyyy} - \overline{B}_{21}w_{,xxxx}$$
$$- (2\overline{B}_{26} - \overline{B}_{61})w_{,xxxy} - (\overline{B}_{11} + \overline{B}_{22} - 2\overline{B}_{66})w_{,xxyy}$$
$$- (2\overline{B}_{16} - \overline{B}_{62})w_{,xyyy} - \overline{B}_{12}w_{,yyyy}$$
$$= 0$$

(8.11.42)

For symmetric laminates, $[\overline{B}] = 0$ and $[\overline{D}] = [D]$, equation (8.11.40) reduces to (8.11.26) and (8.11.42) reduces to (8.11.43) given below.

$$D_{11}w_{,xxxx} + 4D_{16}w_{,xxxy} + 2(D_{12} + D_{66})w_{,xxyy} + 4D_{26}w_{,xyyy} + D_{22}w_{,yyyy} = q$$

(8.11.26)

$$\overline{A_{22}}F_{,xxxx} - 2\overline{A_{26}}F_{,xxxy} + 2(\overline{A_{12}} + \overline{A_{66}})F_{,xxyy} - 2\overline{A_{16}}F_{,xyyy} + \overline{A_{11}}F_{,yyyy} = 0$$

(8.11.43)

The moment resultants in terms of w and F are obtained from equation (8.11.38) as

$$\begin{Bmatrix} M_x \\ M_y \\ -M_{xy} \end{Bmatrix} = -\begin{Bmatrix} (\overline{B_{21}}F_{,xx} - \overline{B_{61}}F_{,xy} + \overline{B_{11}}F_{,yy} + \overline{D_{11}}w_{,xx} + 2\overline{D_{16}}w_{,xy} + \overline{D_{12}}w_{,yy}) \\ (\overline{B_{22}}F_{,xx} - \overline{B_{62}}F_{,xy} + \overline{B_{12}}F_{,yy} + \overline{D_{12}}w_{,xx} + 2\overline{D_{26}}w_{,xy} + \overline{D_{22}}w_{,yy}) \\ (\overline{B_{26}}F_{,xx} - \overline{B_{66}}F_{,xy} + \overline{B_{16}}F_{,yy} + \overline{D_{16}}w_{,xx} + 2\overline{D_{66}}w_{,xy} + \overline{D_{26}}w_{,yy}) \end{Bmatrix}$$

(8.11.44)

The shear resultants in terms of w and F are obtained from equations (3.2.42) and (3.2.43) and the above equation (8.11.44). The mid plane displacements are obtained by integrating the first two equations in (8.11.38).

Equations of Motion for Laminated Plates

For a vibrating plate, we modify the equations of motion in (8.11.21) to (8.11.23) by including the inertia forces to give

$$A_{11}u^0_{,xx} + 2A_{16}u^0_{,xy} + A_{66}u^0_{,yy} + A_{16}v^0_{,xx} + (A_{12} + A_{66})v^0_{,xy} + A_{26}v^0_{,yy}$$
$$= B_{11}w_{,xxx} + (B_{12} + 2B_{66})w_{,xyy} + 3B_{16}w_{,xxy} + B_{26}w_{,yyy} + \varrho h u^0_{,tt}$$

(8.11.21a)

$$A_{16}u^0_{,xx} + (A_{12} + A_{66})u^0_{,xy} + A_{26}u^0_{,yy} + A_{66}v^0_{,xx} + 2A_{26}v^0_{,xy} + A_{22}v^0_{,yy}$$
$$= B_{16}w_{,xxx} + 3B_{26}w_{,xyy} + (B_{12} + 2B_{66})w_{,xxy} + B_{22}w_{,yyy} + \varrho h v^0_{,tt}$$

(8.11.22a)

$$B_{11}u^0_{,xxx} + (B_{12} + 2B_{66})u^0_{,xyy} + 3B_{16}u^0_{,xxy} + B_{26}u^0_{,yyy}$$
$$+ B_{16}v^0_{,xxx} + 3B_{26}v^0_{,xyy} + (B_{12} + 2B_{66})v^0_{,xxy} + B_{22}v^0_{,yyy} + q(x,y,t)$$
$$= D_{11}w_{,xxxx} + 4D_{16}w_{,xxxy} + 2(D_{12} + 2D_{66})w_{,xxyy}$$
$$+ 4D_{26}w_{,xyyy} + D_{22}w_{,yyyy} + \rho h w_{,tt}$$

(8.11.23a)

If we adopt, the stress function approach, then equation (8.11.40) is modified to

$$\overline{D_{11}}w_{,xxxx} + 4\overline{D_{16}}w_{,xxxy} + 2(\overline{D_{12}} + \overline{D_{66}})w_{,xxyy} + 4\overline{D_{26}}w_{,xyyy} + \overline{D_{22}}w_{,yyyy}$$
$$+ \overline{B_{21}}F_{,xxxx} + (2\overline{B_{26}} - \overline{B_{61}})F_{,xxxy} + (\overline{B_{11}} + \overline{B_{22}} - 2\overline{B_{66}})F_{,xxyy}$$
$$+ (2\overline{B_{16}} - \overline{B_{62}})F_{,xyyy} + \overline{B_{12}}F_{,yyyy} + \rho h \ddot{w}$$
$$= q$$

(8.11.40a)

The second equation (8.11.42) remains unaltered.

Energy Formulation

It is often found that energy methods are easy to apply for complicated problems like laminated plates. The strain energy in bending is given by

$$U = \frac{1}{2} \iiint_V (\tau_{xx}\varepsilon_{xx} + \tau_{yy}\varepsilon_{yy} + \tau_{xy}\gamma_{xy})dV$$

(8.11.45)

The above expression for the laminated plate can be written with the help of equation (8.11.5) as

$$U = \frac{1}{2} \iint_A \sum_{j=1}^{N} \int_{h_{j-1}}^{h_j} \left[\begin{array}{c} (\overline{Q_{11}}^j \varepsilon_{xx} + \overline{Q_{12}}^j \varepsilon_{yy} + \overline{Q_{16}}^j \gamma_{xy})\varepsilon_{xx} + \\ (\overline{Q_{12}}^j \varepsilon_{xx} + \overline{Q_{22}}^j \varepsilon_{yy} + \overline{Q_{26}}^j \gamma_{xy})\varepsilon_{yy} + \\ (\overline{Q_{16}}^j \varepsilon_{xx} + \overline{Q_{26}}^j \varepsilon_{yy} + \overline{Q_{66}}^j \gamma_{xy})\gamma_{xy} \end{array} \right] dz \, dA$$

(8.11.46)

Using (8.11.2) to (8.114), the above energy expression is

$$U = \frac{1}{2} \iint_A \left[\sum_{j=1}^{N} \int_{h_{j-1}}^{h_j} \left\{ \begin{array}{l} \overline{Q_{11}}^j (u^0_{,x})^2 + 2\overline{Q_{12}}^j u^0_{,x} v^0_{,y} + \overline{Q_{22}}^j (v^0_{,y})^2 \\ + 2(u^0_{,y} + v^0_{,x})\left(\overline{Q_{16}}^j u^0_{,x} + \overline{Q_{26}}^j v^0_{,y}\right) \\ + \overline{Q_{66}}^j (u^0_{,y} + v^0_{,x})^2 \end{array} \right\} dz \right.$$

$$- \sum_{j=1}^{N} \int_{h_{j-1}}^{h_j} \left\{ \begin{array}{l} \overline{Q_{11}}^j u^0_{,x} w_{,xx} + 2\overline{Q_{12}}^j (u^0_{,x} w_{,yy} + v^0_{,y} w_{,xx}) \\ + \overline{Q_{22}}^j v^0_{,y} w_{,yy} + \\ 2\overline{Q_{16}}^j (u^0_{,y} w_{,xx} + v^0_{,x} w_{,xx} + 2u^0_{,x} w_{,xy}) + \\ 2\overline{Q_{26}}^j (u^0_{,y} w_{,yy} + v^0_{,x} w_{,yy} + 2v^0_{,y} w_{,xy}) + \\ 4\overline{Q_{66}}^j w_{,xy}(u^0_{,y} + v^0_{,x}) \end{array} \right\} z\, dz$$

$$+ \sum_{j=1}^{N} \int_{h_{j-1}}^{h_j} \left\{ \begin{array}{l} \overline{Q_{11}}^j (w_{,xx})^2 + 2\overline{Q_{12}}^j w_{,xx} w_{,yy} \\ + \overline{Q_{22}}^j (w_{,yy})^2 + \\ 4\left(\overline{Q_{16}}^j w_{,xx} + \overline{Q_{26}}^j w_{,yy}\right) w_{,xy} \\ + 4\overline{Q_{66}}^j (w_{,xy})^2 \end{array} \right\} z^2 dz \left. \right] dA$$

(8.11.47)

Since all the displacements are Independent of thickness within each layer, we can carry out the integration with respect to z to give

$$U = \frac{1}{2} \iint_A \left[\begin{array}{l} A_{11}(u^0_{,x})^2 + 2A_{12} u^0_{,x} v^0_{,y} + Q_{22}(v^0_{,y})^2 \\ + 2(u^0_{,y} + v^0_{,x})(A_{16} u^0_{,x} + A_{26} v^0_{,y}) + A_{66}(u^0_{,y} + v^0_{,x})^2 \\ - B_{11} u^0_{,x} w_{,xx} - 2B_{12}(u^0_{,x} w_{,yy} + v^0_{,y} w_{,xx}) \\ - B_{22} v^0_{,y} w_{,yy} - 2B_{16}(u^0_{,y} w_{,xx} + v^0_{,x} w_{,xx} + 2u^0_{,x} w_{,xy}) \\ - 2B_{26}(u^0_{,y} w_{,yy} + v^0_{,x} w_{,yy} + 2v^0_{,y} w_{,xy}) - 4B_{66} w_{,xy}(u^0_{,y} + v^0_{,x}) \\ + D_{11}(w_{,xx})^2 + 2D_{12} w_{,xx} w_{,yy} + D_{22}(w_{,yy})^2 \\ + 4(D_{16} w_{,xx} + D_{26} w_{,yy}) w_{,xy} + 4D_{66}(w_{,xy})^2 \end{array} \right] dA$$

(8.11.48)

where A, B and D are given in equations (8.11.8) and (8.11.11). For symmetric plates, the above equation reduces to

$$U_{sym} = \frac{1}{2} \iint_A \begin{bmatrix} A_{11}(u^0_{,x})^2 + 2A_{12} u^0_{,x} v^0_{,y} + Q_{22}(v^0_{,y})^2 \\ +2(u^0_{,y}+v^0_{,x})(A_{16}u^0_{,x}+A_{26}v^0_{,y})+A_{66}(u^0_{,y}+v^0_{,x})^2 \\ +D_{11}(w_{,xx})^2 + 2D_{12} w_{,xx} w_{,yy} + D_{22}(w_{,yy})^2 \\ +4(D_{16}w_{,xx}+D_{26}w_{,yy})w_{,xy} + 4D_{66}(w_{,xy})^2 \end{bmatrix} dA$$

(8.11.49)

In the above expression, the strain energy due to in plane displacements and the strain energy due to transverse displacement are uncoupled. Therefore, the strain energy due to in plane displacements can be considered as a constant with respect to the transverse displacement and we can simplify the above as

$$U_{sym} = \frac{1}{2} \iint_A \begin{bmatrix} D_{11}(w_{,xx})^2 + 2D_{12} w_{,xx} w_{,yy} + D_{22}(w_{,yy})^2 \\ +4(D_{16}w_{,xx}+D_{26}w_{,yy})w_{,xy} + 4D_{66}(w_{,xy})^2 \end{bmatrix} dA + C$$

(8.11.50)

Potential Energy due to Applied Loads

Let q be the applied transverse force, then the corresponding strain energy is

$$V_{transverse} = -\iint_A qw \, dA$$

(8.11.51)

Let the total in plane stress resultants be N_x, N_y and N_{xy} and consider the strains due to transverse displacement as derived in Figs. 6.1.1a and 6.1.1b, then the corresponding potential energy is

$$U_{in\,plane} = \iint_A [N_x \varepsilon_{xx} + N_y \varepsilon_{yy} + N_{xy} \gamma_{xy}] dA$$

$$= \frac{1}{2} \iint_A [N_x(w_{,x})^2 + N_y(w_{,y})^2 + 2N_{xy} w_{,x} w_{,y}] dA$$

(8.11.52)

378 Dynamics of Plates

The total potential energy is given by the sum of the two expressions in (8.11.51) and (8.11.52) together with the expression (8.11.48).

Kinetic Energy
The kinetic energy is given by

$$T = \frac{1}{2} \iint_A \rho h \left[(\dot{u}^0)^2 + (\dot{v}^0)^2 + (\dot{w})^2 \right] dA$$

(8.11.53)

where ρ is the average density of the laminate.

Ritz Method
Equations (8.11.48), (8.11.51) and (8.11.52) can be used in the Ritz formulation for a laminated plate subjected to transverse and in plane forces. For a vibrating plate, the kinetic energy expression in (8.11.53) should be included in the Lagrangian expression. For a symmetric laminated plate, equation (8.11.49) is used instead of (8.11.48).

8.12 SIMPLY SUPPORTED CROSS PLY LAMINATES

For cross ply laminates, the material directions are oriented at 0 or 90 degrees, therefore from equation (8.10.20), we have

$$(\overline{Q_{16}})_0 = (\overline{Q_{16}})_{90} = 0$$
$$(\overline{Q_{26}})_0 = (\overline{Q_{26}})_{90} = 0$$
$$(\overline{Q_{11}})_0 = (\overline{Q_{22}})_{90}$$
$$(\overline{Q_{22}})_0 = (\overline{Q_{11}})_{90}$$
$$(\overline{Q_{12}})_0 = (\overline{Q_{12}})_{90}$$
$$(\overline{Q_{66}})_0 = (\overline{Q_{66}})_{90}$$

(8.12.1)

Then, from (8.11.8) and (8.11.11), we find the following,

$$A_{16} = A_{26} = B_{12} = B_{16} = B_{26} = B_{66} = D_{16} = D_{26} = 0$$

(8.12.2)

$$A_{22} = A_{11}$$
$$D_{22} = D_{11}$$
$$B_{22} = -B_{11}$$

(8.12.3)

Let the transverse force applied be q uniformly distributed over the plate surface, then the static case of equations (8.11.21) to (8.11.23) for the present case become

$$A_{11}u^0_{,xx} + A_{66}u^0_{,yy} + (A_{12}+A_{66})v^0_{,xy} - B_{11}w_{,xxx} = 0$$

$$(A_{12}+A_{66})u^0_{,xy} + A_{66}v^0_{,xx} + A_{11}v^0_{,yy} + B_{11}w_{,yyy} = 0$$

$$D_{11}(w_{,xxxx} + w_{,yyyy}) + 2(D_{12}+2D_{66})w_{,xxyy} - B_{11}\left(u^0_{,xxx} - v^0_{,yyy}\right) = q$$

(8.12.4)

Let the plate of dimensions $a \times b \times h$ be hinged on all the edges and free in the normal direction, then the boundary conditions are, see (8.11.30) and (8.11.31)

For $x = 0, a$

$$w = 0, \quad M_x = B_{11}u^0_{,x} - D_{11}w_{,xx} - D_{12}w_{,yy} = 0$$
$$v^0 = 0, \quad N_x = A_{11}u^0_{,x} + A_{12}v^0_{,y} - B_{11}w_{,xx} = 0$$

For $y = 0, b$

$$w = 0, \quad M_y = -B_{11}v^0_{,y} - D_{12}w_{,xx} - D_{11}w_{,yy} = 0$$
$$u^0 = 0, \quad N_y = A_{12}u^0_{,x} + A_{11}v^0_{,y} + B_{11}w_{,yy} = 0$$

(8.12.5)

We can obtain a Navier type solution by writing the transverse force in the form of double Fourier series, see equation (3.3.1)

$$q(x,y) = \sum_{m=1}^{\infty}\sum_{n=1}^{\infty} q_{mn} \sin\frac{m\pi x}{a} \sin\frac{n\pi y}{b}$$

(3.3.1)

380 Dynamics of Plates

The solution to equations (8.12.4) is assumed as

$$u^0(x,y) = \sum_{m=1}^{\infty}\sum_{n=1}^{\infty} A_{mn} \cos\frac{m\pi x}{a} \sin\frac{n\pi y}{b}$$

$$v^0(x,y) = \sum_{m=1}^{\infty}\sum_{n=1}^{\infty} B_{mn} \sin\frac{m\pi x}{a} \cos\frac{n\pi y}{b}$$

$$w(x,y) = \sum_{m=1}^{\infty}\sum_{n=1}^{\infty} C_{mn} \sin\frac{m\pi x}{a} \sin\frac{n\pi y}{b}$$

(8.12.6)

which satisfies the boundary conditions in (8.12.5). Substituting the above solutions (8.12.6) in (8.12.5), the coefficients A, B and C are obtained as

$$A_{mn} = q_{mn}\frac{k^3 b^3 B_{11} m}{\pi^3 D_{mn}}\{A_{66}m^4 + A_{11}m^2n^2k^2 + (A_{12}+A_{66})n^4k^4\}$$

$$B_{mn} = -q_{mn}\frac{k^4 b^3 B_{11} n}{\pi^3 D_{mn}}\{(A_{12}+A_{66})m^4 + A_{11}m^2n^2k^2 + A_{66}n^4k^4\}$$

$$C_{mn} = q_{mn}\frac{k^4 b^4}{\pi^4 D_{mn}}\left\{\begin{array}{c}(A_{11}m^2+A_{66}n^2k^2)(A_{66}m^2+A_{11}n^2k^2)\\-(A_{12}+A_{66})^2 m^2 n^2 k^2\end{array}\right\}$$

(8.12.7)

where k is the aspect ratio equal to a/b and

$$D_{mn} = \left[\begin{array}{c}\left\{\begin{array}{c}(A_{11}m^2+A_{66}n^2k^2)(A_{66}m^2+A_{11}n^2k^2)-\\-(A_{12}+A_{66})^2 m^2 n^2 k^2\end{array}\right\}\times\\ \times\left\{\begin{array}{c}D_{11}(m^4+n^4k^4)+\\+2(D_{12}+2D_{66})m^2n^2k^2\end{array}\right\}-\\-B_{11}^2\left\{\begin{array}{c}A_{11}m^2n^2k^2(m^4+n^4k^4)+\\+2(A_{12}+A_{66})m^4n^4k^4+A_{66}(m^8+n^8k^8)\end{array}\right\}\end{array}\right]$$

(8.12.8)

Free Vibration
The equation of motion is given by, see equation (8.11.21) to (8.11.23) and (8.12.4)

$$A_{11}u^0_{,xx} + A_{66}u^0_{,yy} + (A_{12}+A_{66})v^0_{,xy} - B_{11}w_{,xxx} = 0$$

$$(A_{12} + A_{66})u^0_{,xy} + A_{66}v^0_{,xx} + A_{11}v^0_{,yy} + B_{11}w_{,yyy} = 0$$

$$D_{11}(w_{,xxxx} + w_{,yyyy}) + 2(D_{12} + 2D_{66})w_{,xxyy} - B_{11}(u^0_{,xxx} - v^0_{,yyy}) + ph\ddot{w} = 0$$

(8.12.9)

In the above, the transverse inertia is only included and the in plane inertia forces are neglected. The solution to the above system (8.12.9) is written as

$$u^0(x, y, t) = A \cos\frac{m\pi x}{a} \sin\frac{n\pi y}{b} \sin pt$$

$$v^0(x, y, t) = B \sin\frac{m\pi x}{a} \cos\frac{n\pi y}{b} \sin pt$$

$$w(x, y, t) = C \sin\frac{m\pi x}{a} \sin\frac{n\pi y}{b} \sin pt$$

(8.12.10)

which satisfy the boundary conditions for a simply supported rectangular plate given in (8.12.5). Substituting (8.12.10) in (8.12.9), we get

$$\begin{bmatrix} A_{mn} & B_{mn} & C_{mn} \\ & D_{mn} & E_{mn} \\ \text{sym} & & F_{mn} \end{bmatrix} \begin{Bmatrix} A \\ B \\ C \end{Bmatrix} - \begin{bmatrix} 0 & 0 & 0 \\ 0 & 0 & 0 \\ 0 & 0 & \rho h p^2 k^2 b^2/\pi^2 \end{bmatrix} \begin{Bmatrix} A \\ B \\ C \end{Bmatrix} = 0$$

(8.12.11)

where

$$A_{mn} = A_{11}m^2 + A_{66}n^2k^2$$

$$B_{mn} = (A_{11} + A_{66})mnk$$

$$C_{mn} = -B_{11}n^3k\pi$$

$$D_{mn} = A_{66}m^2 + A_{11}n^2k^2$$

$$E_{mn} = B_{11}n^3k\pi$$

$$F_{mn} = \frac{\pi^2 k^2}{b^2}[D_{11}(m^4 + n^4k^4) + 2(D_{12} + 2D_{66})m^2n^2k^2]$$

(8.12.12)

382 Dynamics of Plates

The frequency determinant is therefore given by

$$\begin{vmatrix} A_{mn} & B_{mn} & C_{mn} \\ & D_{mn} & E_{mn} \\ \text{sym} & F_{mn} & -\dfrac{\rho h p^2 k^2 b^2}{\pi^2} \end{vmatrix} = 0$$

(8.12.13)

Expanding the determinant in the above equation (8.12.13), the frequency equation is obtained.

$$\frac{\rho h a^4}{\pi^4} p_{mn}^2 = D_{11}(m^4 + n^4 k^4) + 2(D_{12} + 2D_{66})m^2 n^2 k^2$$
$$- B_{11}^2 \frac{J_1 m^4 + J_2 n^4 k^4}{J_3}$$

(8.12.14)

where

$$J_1 = A_{66} m^4 + A_{11} m^2 n^2 k^2 + (A_{12} + A_{66}) n^4 k^4$$
$$J_2 = (A_{12} + A_{66}) m^4 + A_{11} m^2 n^2 k^2 + A_{66} n^4 k^4$$
$$J_3 = (A_{11} m^2 + A_{66} n^2 k^2)(A_{66} m^2 + A_{11} n^2 k^2) - (A_{12} + A_{66})^2 m^2 n^2 k^2$$

(8.12.15)

Example
Consider a symmetric graphite/epoxy cross ply /0/90/90/0/ of total thickness h equal to 0.00052 m with a equal to 0.1143 m and aspect ratio k equal to 3. The material properties are given by

$E_{11} = 128 \times 10^9, E_{22} = 11 \times 10^9, G_{12} = 4.48 \times 10^9 \text{Pa}, v_{12} = 0.25,$

$v_{21} = 0.0215, \rho = 1500 \text{ kg/m}^3$

From (8.5.8)

$$Q_{11} = \frac{128 \times 10^9}{1 - 0.25 \times 0.0215} = 128.692 \times 10^9 \text{Pa}$$

$$Q_{12} = \frac{0.25 \times 11 \times 10^9}{1 - 0.25 \times 0.0215} = 2.765 \times 10^9 \text{Pa}$$

$$Q_{22} = \frac{11 \times 10^9}{1 - 0.25 \times 0.0215} = 11.059 \times 10^9 \text{Pa}$$
$$Q_{66} = G_{12} = 4.48 \times 10^9 \text{Pa}$$

From (8.10.20)

$$(\overline{Q_{11}})_0 = (\overline{Q_{22}})_{90} = 128.692 \times 10^9 \text{Pa}$$
$$(\overline{Q_{12}})_0 = (\overline{Q_{12}})_{90} = 2.765 \times 10^9 \text{Pa}$$
$$(\overline{Q_{22}})_0 = (\overline{Q_{11}})_{90} = 11.059 \times 10^9 \text{Pa}$$
$$(\overline{Q_{66}})_0 = (\overline{Q_{66}})_{90} = 4.48 \times 10^9 \text{Pa}$$
$$(\overline{Q_{16}})_0 = (\overline{Q_{26}})_0 = (\overline{Q_{16}})_{90} = (\overline{Q_{26}})_{90} = 0$$

For the laminate under consideration, we have from (8.11.8)

$$A_{ij} = (\overline{Q_{ij}})_0 \left[-\frac{h}{4} - \left(-\frac{h}{2} \right) \right] + (\overline{Q_{ij}})_{90} \left[0 - \left(-\frac{h}{4} \right) \right]$$
$$+ (\overline{Q_{ij}})_{90} \left[\frac{h}{4} - 0 \right] + (\overline{Q_{ij}})_0 \left[\frac{h}{2} - \frac{h}{4} \right]$$
$$= \frac{h}{2} \left[(\overline{Q_{ij}})_0 + (\overline{Q_{ij}})_{90} \right]$$

$$B_{ij} = 0$$

$$\therefore \quad [A] = \frac{h}{2} \begin{bmatrix} 139.751 & 5.53 & 0 \\ & 139.751 & 0 \\ & & 8.96 \end{bmatrix} 10^9 \text{ N/m}$$

$$[B] = 0$$

From equation (8.11.11), we have

$$D_{ij} = \frac{1}{3}\left\{\begin{array}{l}(\overline{Q_{ij}})_0\left[\left(-\frac{h}{4}\right)^3 - \left(-\frac{h}{2}\right)^3\right] + (\overline{Q_{ij}})_{90}\left[0 - \left(-\frac{h}{4}\right)^3\right] \\ + (\overline{Q_{ij}})_{90}\left[\left(\frac{h}{4}\right)^3 - 0\right] + (\overline{Q_{ij}})_0\left[\left(\frac{h}{2}\right)^3 - \left(\frac{h}{4}\right)^3\right]\end{array}\right\}$$

$$= \frac{h^3}{96}\left[7(\overline{Q_{ij}})_0 + (\overline{Q_{ij}})_{90}\right]$$

$$\therefore \quad [D] = \frac{h^3}{96}\begin{bmatrix} 911.903 & 22.12 & 0 \\ & 2206.105 & 0 \\ & & 35.84 \end{bmatrix} 10^9 \text{ Nm}$$

Equation (8.12.14) gives the natural frequencies. Since $[B]$ is zero, no coupling exists and the influence of $[A]$ stiffness terms is also not present.

$$p_{mn}^2 = \frac{0.00052^2 \times \pi^4}{96 \times 1500 \times 0.1143^4} \times 10^9[911.903(m^4 + 81n^4) + 18 \times 93.8 m^2 n^2]$$

$$= 1071.66[911.903(m^4 + 81n^4) + 1688.4 m^2 n^2]$$

$$p_{11} = \sqrt{1071.66[911.903 \times 82 + 1688.4]}$$

$$= 9052.3 \text{ rad/s}$$

8.13 SIMPLY SUPPORTED ANGLE PLY LAMINATES

For angle ply laminates, the material directions are oriented at $+\theta$ or $-\theta$ degrees, therefore from equation (8.10.20), we have

$$(\overline{Q_{16}})_\theta = -(\overline{Q_{16}})_{-\theta}$$
$$(\overline{Q_{26}})_\theta = -(\overline{Q_{26}})_{-\theta}$$
$$(\overline{Q_{11}})_\theta = (\overline{Q_{11}})_{-\theta}$$
$$(\overline{Q_{22}})_\theta = (\overline{Q_{22}})_{-\theta}$$
$$(\overline{Q_{12}})_\theta = (\overline{Q_{12}})_{-\theta}$$

$$(\overline{Q_{66}})_\theta = (\overline{Q_{66}})_{-\theta}$$

(8.13.1)

Then, from (8.11.8) and (8.11.11), we find the following,

$$A_{16} = A_{26} = B_{11} = B_{12} = B_{22} = B_{66} = D_{16} = D_{26} = 0$$

(8.13.2)

For a transverse force q uniformly distributed over the plate surface, the static case of equations (8.11.21) to (8.11.23) for angle ply laminates is

$$A_{11} u^0_{,xx} + A_{66} u^0_{,yy} + (A_{12} + A_{66}) v^0_{,xy} - 3B_{16} w_{,xxy} - B_{26} w_{,yyy} = 0$$

$$(A_{12} + A_{66}) u^0_{,xy} + A_{66} v^0_{,xx} + A_{22} v^0_{,yy} - 3B_{26} w_{,xyy} - B_{16} w_{,xxx} = 0$$

$$D_{11} w_{,xxxx} + 2(D_{12} + 2D_{66}) w_{,xxyy} + D_{22} w_{,yyyy}$$

$$- B_{16} \left(3 u^0_{,xxy} + v^0_{,xxx} \right) - B_{26} \left(u^0_{,yyy} + 3 v^0_{,xyy} \right)$$

$$= q$$

(8.13.3)

Let the plate be hinged on all the edges and free in the tangential direction, then

For x = 0, a

$$w = 0, \quad M_x = B_{16} \left(u^0_{,y} + v^0_{,x} \right) - D_{11} w_{,xx} - D_{12} w_{,yy} = 0$$

$$u^0 = 0, \quad N_{xy} = A_{66} \left(u^0_{,y} + v^0_{,x} \right) - B_{16} w_{,xx} - B_{26} w_{,yy} = 0$$

For y = 0, b

$$w = 0, \quad M_y = B_{26} \left(u^0_{,y} + v^0_{,x} \right) - D_{12} w_{,xx} - D_{22} w_{,yy} = 0$$

$$v^0 = 0, \quad N_{xy} = A_{66} \left(u^0_{,y} + v^0_{,x} \right) - B_{16} w_{,xx} - B_{26} w_{,yy} = 0$$

(8.13.4)

The transverse force is expressed again in the form of double Fourier series,

$$q(x,y) = \sum_{m=1}^{\infty} \sum_{n=1}^{\infty} q_{mn} \sin \frac{m\pi x}{a} \sin \frac{n\pi y}{b}$$

(3.3.1)

The solution to equations (8.13.3) is now assumed as

$$u^0(x,y) = \sum_{m=1}^{\infty} \sum_{n=1}^{\infty} A_{mn} \sin \frac{m\pi x}{a} \cos \frac{n\pi y}{b}$$

$$v^0(x,y) = \sum_{m=1}^{\infty} \sum_{n=1}^{\infty} B_{mn} \cos \frac{m\pi x}{a} \sin \frac{n\pi y}{b}$$

$$w(x,y) = \sum_{m=1}^{\infty} \sum_{n=1}^{\infty} C_{mn} \sin \frac{m\pi x}{a} \sin \frac{n\pi y}{b}$$

(8.13.5)

which satisfies the boundary conditions in (8.13.4). Substituting the above solutions (8.13.5) in (8.13.3), the coefficients A, B and C are obtained as

$$A_{mn} = q_{mn} \frac{k^4 b^3 n}{\pi^3 D_{mn}} \left\{ \begin{array}{l} (A_{66}m^2 + A_{22}n^2 k^2)(3B_{16}m^2 + B_{26}n^2 k^2) \\ -m^2(A_{12} + A_{66})(B_{16}m^2 + 3B_{26}n^2 k^2) \end{array} \right\}$$

$$B_{mn} = q_{mn} \frac{k^3 b^3 m}{\pi^3 D_{mn}} \left\{ \begin{array}{l} (A_{11}m^2 + A_{66}n^2 k^2)(B_{16}m^2 + 3B_{26}n^2 k^2) \\ -n^2 k^2 (A_{12} + A_{66})(3B_{16}m^2 + B_{26}n^2 k^2) \end{array} \right\}$$

$$C_{mn} = q_{mn} \frac{k^4 b^4}{\pi^4 D_{mn}} \left\{ \begin{array}{l} (A_{11}m^2 + A_{66}n^2 k^2)(A_{66}m^2 + A_{22}n^2 k^2) \\ -(A_{12} + A_{66})^2 m^2 n^2 k^2 \end{array} \right\}$$

(8.13.6)

where

$$D_{mn} = \begin{array}{l} \{(A_{11}m^2 + A_{66}n^2 k^2)(A_{66}m^2 + A_{22}n^2 k^2) - (A_{12} + A_{66})^2 m^2 n^2 k^2\} \\ \times \{D_{11}m^4 + 2(D_{12} + 2D_{66})m^2 n^2 k^2 + D_{22}n^4 k^4\} \\ + 2m^2 n^2 k^2 (A_{12} + A_{66})(3B_{16}m^2 + B_{26}n^2 k^2)(B_{16}m^2 + 3B_{26}n^2 k^2) \\ - n^2 k^2 (A_{66}m^2 + A_{22}n^2 k^2)(3B_{16}m^2 + B_{26}n^2 k^2)^2 \\ - m^2 (A_{11}m^2 + A_{66}n^2 k^2)(B_{16}m^2 + 3B_{26}n^2 k^2)^2 \end{array}$$

(8.13.7)

Free Vibration
The equation of motion is given by, see equation (8.11.21) to (8.11.23) and (8.13.2)

$$A_{11}u^0_{,xx} + A_{66}u^0_{,yy} + (A_{12}+A_{66})v^0_{,xy} - 3B_{16}w_{,xxy} - B_{26}w_{,yyy} = 0$$

$$(A_{12}+A_{66})u^0_{,xy} + A_{66}v^0_{,xx} + A_{22}v^0_{,yy} - B_{16}w_{,xxx} - 3B_{26}w_{,xyy} = 0$$

$$D_{11}w_{,xxxx} + 2(D_{12}+2D_{66})w_{,xxyy} + D_{22}w_{,yyyy}$$
$$- B_{16}\left(3u^0_{,xxy}+v^0_{,xxx}\right) - B_{26}\left(u^0_{,yyy}+3v^0_{,xyy}\right) + ph\ddot{w}$$
$$= 0$$

(8.13.8)

In the above, the transverse inertia is only included and the in plane inertia forces are neglected. The solution to the above system (8.13.8) is written as

$$u^0(x,y,t) = A \sin\frac{m\pi x}{a} \cos\frac{n\pi y}{b} \sin pt$$

$$v^0(x,y,t) = B \cos\frac{m\pi x}{a} \sin\frac{n\pi y}{b} \sin pt$$

$$w(x,y,t) = C \sin\frac{m\pi x}{a} \sin\frac{n\pi y}{b} \sin pt$$

(8.13.9)

which satisfy the boundary conditions given in (8.13.4). Substituting (8.13.9) in (8.13.8), we get

$$\begin{bmatrix} A_{mn} & B_{mn} & C_{mn} \\ & D_{mn} & E_{mn} \\ \text{sym} & & F_{mn} \end{bmatrix}\begin{Bmatrix} A \\ B \\ C \end{Bmatrix} - \begin{bmatrix} 0 & 0 & 0 \\ 0 & 0 & 0 \\ 0 & 0 & \rho h p^2 k^2 b^2/\pi^2 \end{bmatrix}\begin{Bmatrix} A \\ B \\ C \end{Bmatrix} = 0$$

(8.13.10)

where

$$A_{mn} = A_{11}m^2 + A_{66}n^2k^2$$

$$B_{mn} = (A_{11}+A_{66})mnk$$

$$C_{mn} = -\frac{n\pi}{b}(3B_{16}m^2 + B_{26}n^2k^2)$$

$$D_{mn} = A_{66}m^2 + A_{11}n^2k^2$$

$$E_{mn} = -\frac{m\pi}{kb}(B_{16}m^2 + 3B_{26}n^2k^2)$$

$$F_{mn} = \frac{\pi^2 k^2}{b^2}[D_{11}(m^4 + n^4k^4) + 2(D_{12} + 2D_{66})m^2n^2k^2]$$

(8.13.11)

The frequency determinant is written down similar to that in the previous section 8.12 and expanding, we get the frequency equation for cross ply plates as

$$\frac{\rho h a^4}{\pi^4}p_{mn}^2 = D_{11}m^4 + 2(D_{12} + 2D_{66})m^2n^2k^2 + D_{22}n^4k^4$$
$$- \frac{J_1 m(B_{16}m^2 + 3B_{26}n^2k^2) + J_2 nk(3B_{16}m^2 + B_{26}n^2k^2)}{J_3}$$

(8.13.12)

where

$$J_1 = (A_{11}m^2 + A_{66}n^2k^2)(B_{16}m^2 + 3B_{26}n^2k^2)$$
$$- n^2k^2(A_{12} + A_{66})(3B_{16}m^2 + B_{26}n^2k^2)$$
$$J_2 = (A_{66}m^2 + A_{22}n^2k^2)(3B_{16}m^2 + B_{26}n^2k^2)$$
$$- n^2k^2(A_{12} + A_{66})(B_{16}m^2 + 3B_{26}n^2k^2)$$
$$J_3 = (A_{11}m^2 + A_{66}n^2k^2)(A_{66}m^2 + A_{22}n^2k^2)$$
$$- m^2n^2k^2(A_{12} + A_{66})^2$$

(8.13.13)

Example

Consider a two lamina square plate \45\-45\ with properties given in the example of section 8.10 and each 0.2 m by 0.2 m with 6 mm thickness. The density of the plate is 1500 kg/m^3

The stiffness matrices are

$$[\overline{Q}]_{45} = \begin{bmatrix} 39.61 & 34.7 & 31.1 \\ & 39.61 & 31.1 \\ & & 34.82 \end{bmatrix} \text{Gpa}$$

$$[\overline{Q}]_{-45} = \begin{bmatrix} 39.61 & 34.7 & -31.1 \\ & 39.61 & -31.1 \\ & & 34.82 \end{bmatrix} \text{GPa}$$

For the laminate under consideration, we have the stiffness matrix [A] from (8.11.8)

$$A_{ij} = \left(\overline{Q_{ij}}\right)_{45}[0-(-0.006)] + \left(\overline{Q_{ij}}\right)_{-45}[0.006-0]$$

$$= 0.006\left[\left(\overline{Q_{ij}}\right)_{45} + \left(\overline{Q_{ij}}\right)_{-45}\right] \text{N/m}$$

$$\therefore [A] = \begin{bmatrix} 475.32 & 416.4 & 0 \\ 416.4 & 475.32 & 0 \\ 0 & 0 & 417.84 \end{bmatrix} 10^6 \text{ N/m}$$

Here, [B] is not zero, since the laminate is not symmetric about the mid plane, it is given by equation (8.11.8)

$$B_{ij} = \frac{1}{2}\left[\left(\overline{Q_{ij}}\right)_{45}\{0-(-0.006)^2\} + \left(\overline{Q_{ij}}\right)_{-45}\{0.006^2 - 0\}\right]$$

$$= 18 \times 10^{-6}\left[-\left(\overline{Q_{ij}}\right)_{45} + \left(\overline{Q_{ij}}\right)_{-45}\right] \text{N}$$

$$\therefore [B] = \begin{bmatrix} 0 & 0 & -1119.6 \\ 0 & 0 & -1119.6 \\ -1119.6 & -1119.6 & 0 \end{bmatrix} 10^6 \text{ N/m}$$

From equation (8.11.11), we have the [D] stiffness matrix

$$D_{ij} = \frac{1}{3}\left\{\left(\overline{Q_{ij}}\right)_{45}[0-(-0.006)^3] + \left(\overline{Q_{ij}}\right)_{-45}[0.006^3 - 0]\right\}$$

$$= 72 \times 10^{-9}\left[\left(\overline{Q_{ij}}\right)_0 + \left(\overline{Q_{ij}}\right)_{90}\right]$$

$$\therefore [D] = \begin{bmatrix} 5703 & 4996 & 0 \\ & 5703 & 0 \\ & & 5014 \end{bmatrix} \text{Nm}$$

From (8.13.13)

$$J_1 = -10^9(475.32m^2 + 417.84n^2)(1119.6m^2 + 3358.8n^2)$$
$$+ 10^9 n^2(834.24)(3358.8m^2 + 1119.6n^2)$$

390 Dynamics of Plates

$$J_2 = -10^9(417.84m^2 + 475.32n^2)(3358.8m^2 + 1119.6n^2)$$
$$+ 10^9 n^2(834.24)(1119.6m^2 + 3358.8n^2)$$
$$J_3 = 10^9(475.32m^2 + 417.84n^2)(417.84m^2 + 475.32n^2)$$
$$- 10^9 m^2 n^2(834.24)^2$$

Let us consider the first mode $m = 1$ and $n = 1$, then

$$J_1 = -263867.328 \times 10^9$$
$$J_2 = -263867.328 \times 10^9$$
$$J_3 = 101778.408 \times 10^9$$

The frequency equation in (8.13.12) then becomes

$$\frac{1500 \times 0.012 \times 0.2^4}{\pi^4} p_{11}^2 = 5703 + 2(4996 + 2 \times 5014) + 5703$$

$$- \frac{263867.328}{101778.408} \times 2 \times 4 \times 1119.6$$

$$= 18232.897$$

$$p_{11} = 7852.9 \text{ rad/s}$$

8.14 BERGER'S APPROXIMATION FOR ORTHOTROPIC PLATES

Here, we consider large deformation of a specially rectangular orthotropic plate resting on a viscoelastic foundation. The nonlinear strain displacement relations of the plate are given by equation (6.1.4)

$$\varepsilon_{xx} = u_{,x} + \frac{1}{2}w_{,x}^2 - zw_{,xx}$$
$$\varepsilon_{yy} = v_{,y} + \frac{1}{2}w_{,y}^2 - zw_{,yy}$$
$$\gamma_{xy} = u_{,y} + v_{,x} - 2zw_{,xy} + w_{,x}w_{,y}$$

(6.6.1)

The stress strain relations for a specially orthotropic plate are given by (8.6.1)

$$\tau_{xx} = \frac{E_{xx}}{1 - \nu_{xy}\nu_{yx}}(\varepsilon_{xx} + \nu_{yx}\varepsilon_{yy})$$

$$\tau_{yy} = \frac{E_{yy}}{1-v_{xy}v_{yx}}(\varepsilon_{yy} + v_{xy}\varepsilon_{xx})$$

$$\tau_{xy} = G_{xy}\gamma_{xy}$$

(8.6.1)

The strain energy due to bending and stretching of the middle surface of the plate is given by

$$V = \frac{1}{2}\iiint(\tau_{xx}\varepsilon_{xx} + \tau_{yy}\varepsilon_{yy} + \tau_{xy}\gamma_{xy})dx\,dy\,dz$$

(6.6.3)

Substituting equations (6.6.1) and (6.6.2) in the above equation (6.6.3) and integrating with respect to z, we get

$$V = \frac{1}{2}\frac{hE_{xx}}{1-v_{xy}v_{yx}}\iint\left\{\begin{array}{c}\bar{\varepsilon}_{xx}^2 + \frac{E_{yy}}{E_{xx}}\bar{\varepsilon}_{yy}^2 + \left(v_{yx} + \frac{E_{yy}}{E_{xx}}v_{xy}\right)\bar{\varepsilon}_{yy}\bar{\varepsilon}_{xx} \\ + \frac{1-v_{xy}v_{yx}}{E_{xx}}\frac{E_{xx}}{2(1-v_{xy})}\bar{\gamma}_{xy}^2\end{array}\right\}dx\,dy$$

$$+ \frac{1}{2}D_{11}\iint\left\{\begin{array}{c}w_{,xx}^2 + v_{yx}w_{,xx}w_{,yy} + \frac{E_{yy}}{E_{xx}}(w_{,yy}^2 + v_{xy}w_{,xx}w_{,yy}) \\ + \frac{2(1-v_{xy}v_{yx})}{1+v_{xy}}w_{,xy}^2\end{array}\right\}dx\,dy$$

(8.14.1)

where the middle surface strains are given by equation (6.6.6) and

$$D_{11} = \frac{E_{xx}h^3}{12(1-v_{xy}v_{yx})}$$

(8.14.2)

The strain invariants for the orthotropic plate are

$$\bar{e}_1 = \bar{\varepsilon}_{xx} + m\bar{\varepsilon}_{yy}$$

$$\bar{e}_2 = \bar{\varepsilon}_{xx}\bar{\varepsilon}_{yy} - \frac{1-v_{xy}v_{yx}}{4(1-v_{xy})(m-v_{yx})}\bar{\gamma}_{xy}^2$$

(8.14.3)

where

$$m = \sqrt{E_{yy}/E_{xx}}$$

(8.14.4)

Now, equation (8.14.1) can be written as

$$V = \frac{1}{2}\frac{hE_{xx}}{1-v_{xy}v_{yx}} \iint \{\bar{e}_1^2 - 2(m-v_{yx})\bar{e}_2\}dx\,dy$$

$$+ \frac{1}{2}D_{11} \iint \left\{ w_{,xx}^2 + 2v_{yx}w_{,xx}w_{,yy} + m^2 w_{,yy}^2 + \frac{2(1-v_{xy}v_{yx})}{1+v_{xy}}w_{,xy}^2 \right\}dx\,dy$$

(8.14.5)

On lines of Berger's approximation, the second strain invariant is neglected to simplify the above strain energy approximation

$$V = \frac{1}{2}D_{11} \iint \left[\begin{array}{l} \frac{12}{h^2}\bar{e}_1^2 + w_{,xx}^2 + m^2 w_{,yy}^2 + \\ 2v_{yx}w_{,xx}w_{,yy} + \frac{2(1-v_{xy}v_{yx})}{1+v_{xy}}w_{,xy}^2 \end{array} \right] dx\,dy$$

(8.14.6)

Following Hamilton's principle, we can write the following variational equation for the plate on a Kelvin foundation.

$$\delta \int_{t_0}^{t_1} \iint \left[\frac{1}{2}D_{11} \left\{ \begin{array}{l} \frac{12}{h^2}\bar{e}_1^2 + w_{,xx}^2 + m^2 w_{,yy}^2 + 2v_{yx}w_{,xx}w_{,yy} \\ + \frac{2(1-v_{xy}v_{yx})}{1+v_{xy}}w_{,xy}^2 \\ -\frac{1}{2}\rho h(\dot{u}^2 + \dot{v}^2 + \dot{w}^2) + \frac{1}{2}kw^2 \end{array} \right\} \right] dx\,dy\,dt$$

$$= \int_{t_0}^{t_1} \iint [p(x,y,t) - c\,\dot{w}]\delta w\,dx\,dy\,dt$$

(8.14.7)

Taking the variation, we get

$$\bar{e}_{1,x} = \frac{\rho h^3}{12 D_{11}} \ddot{u}$$

$$\bar{e}_{1,y} = \frac{\rho h^3}{12 D_{11}} \ddot{v}$$

(8.14.8)

and

$$D_{11}\left[w_{,xxxx} + 2\gamma^2 w_{,xxyy} + m^2 w_{,yyyy}\right]$$
$$- \frac{12 D_{11}}{h^2}\left[(\bar{e}_1 w_{,x})_{,x} + (\bar{e}_1 w_{,y})_{,y}\right] + \rho h \ddot{w} + c \dot{w} + k w$$
$$= p(x, y, t)$$

(8.14.9)

where

$$\gamma^2 = v_{yx} + \frac{1 - v_{xy} v_{yx}}{1 + v_{xy}}$$

(8.14.10)

Neglecting the in-plane inertia, we have equations (8.14.8)

$$\bar{e}_{1,x} = 0$$
$$\bar{e}_{1,y} = 0$$

(8.14.11)

From the above, we deduce that $\bar{e}_1$ is independent of x and y, and a solution of these equations leads to

$$\bar{e}_1 = \frac{1}{12} a_1^2 h^2$$

(8.14.12)

Dynamics of Plates

Equation (8.14.9) can now be written as

$$D_{11}[w_{,xxxx} + 2\gamma^2 w_{,xxyy} + m^2 w_{,yyyy}]$$
$$- D_{11} a_1^2 \nabla^2 w + \rho h \ddot{w} + c \dot{w} + k w$$
$$= p(x, y, t)$$

(8.14.13)

From equations (8.14.12), (8.14.3) and (6.6.1), we can write

$$\frac{1}{12} a_1^2 h^2 = u_{,x} + m v_{,y} + \frac{1}{2}(w_{,x}^2 + m w_{,y}^2)$$

(8.14.14)

Integrating the above equation over the area of the plate, we get

$$\iint \frac{a_1^2 h^2}{12} dx\, dy = \iint (u_{,x} + m v_{,y}) dx\, dy + \frac{1}{2} \iint (w_{,x}^2 + m w_{,y}^2) dx\, dy$$

(8.14.15)

The boundary conditions for the in-plane displacements are

$$u = 0 \quad \text{at } x = 0, a \text{ and } 0 \leq y \leq b$$
$$v = 0 \quad \text{at } y = 0, b \text{ and } 0 \leq x \leq a$$

(8.14.16)

In view of the above conditions, equation (8.14.15) reduces to

$$\frac{a_1^2 h^2}{12} ab = \frac{1}{2} \iint (w_{,x}^2 + m w_{,y}^2) dx\, dy$$

(8.14.17)

Integrating by parts

$$\frac{a_1^2 h^2}{12} ab = \frac{1}{2} \int [w_{,x} w]_{x=0}^{x=a} dy + \frac{1}{2} m \int [w_{,y} w]_{y=0}^{y=b} dx$$
$$- \frac{1}{2} \iint w(w_{,xx} + m w_{,yy}) dx\, dy$$

(8.14.18)

The boundary conditions for the transverse deflection of the plate are

$$w = w_{,xx} = 0 \quad \text{at } x = 0, a \text{ and } 0 \leq y \leq b$$
$$w = w_{,yy} = 0 \quad \text{at } y = 0, b \text{ and } 0 \leq x \leq a$$

(8.14.19)

With the help of the above boundary conditions, equation (8.14.18) simplifies to

$$a_1^2 = -\frac{6}{abh^2} \iint w(w_{,xx} + mw_{,yy}) dx\, dy$$

(8.14.20)

Substituting the above result in equation (8.14.13), we get

$$D_{11}\left[w_{,xxxx} + 2\gamma^2 w_{,xxyy} + m^2 w_{,yyyy}\right] +$$
$$D_{11}\left[\frac{6}{abh^2} \iint w(w_{,xx} + mw_{,yy}) dx\, dy\right]\nabla^2 w + ph\, \ddot{w} + c\, \dot{w} + kw$$
$$= p(x,y,t)$$

(8.14.21)

To solve the above nonlinear equation, we can employ Galerkin's method with a one term approximation

$$w(x,y,t) = \phi(x,y) f(t)$$
$$= \sin\frac{\pi x}{a} \sin\frac{\pi y}{b} f(t)$$

(8.14.22)

which satisfies the boundary conditions and the linear differential equation

$$\left[w_{,xxxx} + 2\gamma^2 w_{,xxyy} + m^2 w_{,yyyy}\right] - \lambda^4 w = 0$$

(8.14.21a)

where the eigen values are

$$\lambda^4 = \frac{ph}{D_{11}} p^2$$

(8.14.23)

396 Dynamics of Plates

Substituting the above (8.14.22), and making use of (8.14.21a) we get the error in the differential equation (8.14.21) as

$$\epsilon = \lambda^4 \phi(x,y) f(t)$$
$$+ \left[\frac{6}{abh^2} \iint \phi(x,y)(\phi_{,xx} + m\phi_{,yy}) dx\,dy \right] \nabla^2 \phi(x,y) f^2(t)$$
$$+ \frac{\rho h}{D_{11}} \phi(x,y) \ddot{f} + \frac{c}{D_{11}} \phi(x,y) \dot{f} + \frac{k}{D_{11}} \phi(x,y) f(t) - \frac{p(x,y,t)}{D_{11}}$$

(8.14.24)

According to Galerkin's method,

$$\lambda^4 f(t) \iint \phi^2(x,y) dx\,dy$$
$$+ f^3(t) \iint \left[\frac{6}{abh^2} \iint \phi(x,y)(\phi_{,xx} + m\phi_{,yy}) dx\,dy \right] \phi(x,y) \nabla^2 \phi(x,y) dx\,dy$$
$$+ \frac{\rho h}{D_{11}} \ddot{f} \iint \phi^2(x,y) dx\,dy + \frac{c}{D_{11}} \dot{f} \iint \phi^2(x,y) dx\,dy$$
$$+ \frac{k}{D_{11}} f(t) \iint \phi^2(x,y) dx\,dy$$
$$= \frac{1}{D_{11}} \iint p(x,y,t) \phi(x,y) dx\,dy$$

(8.14.25)

Noting that

$$\int_0^a \int_0^b \phi^2 dx\,dy = \frac{ab}{4}$$
$$\int_0^a \int_0^b \phi \nabla^2 \phi\, dx\,dy = -\frac{a^2+b^2}{4ab} \pi^2$$
$$\int_0^a \int_0^b \phi(\phi_{,xx} + m\phi_{,yy}) dx\,dy = -\frac{ma^2+b^2}{4ab} \pi^2$$

(8.14.26)

Equation (8.14.25) becomes

$$\lambda^4 f(t)\frac{ab}{4} + f^3(t)\frac{3\pi^4}{8a^3b^3h^2}(ma^2+b^2)(a^2+b^2)$$

$$+ \frac{\rho h}{D_{11}}\ddot{f}\frac{ab}{4} + \frac{c}{D_{11}}\dot{f}\frac{ab}{4} + \frac{k}{D_{11}}f(t)\frac{ab}{4}$$

$$= \frac{1}{D_{11}}\iint p(x,y,t)\phi(x,y)\,dx\,dy$$

(8.14.27)

Expressing the excitation force in the form

$$p(x,y,t) = p_1 \sin\omega t + p_2 \cos\omega t$$

(6.6.33)

and making use of the result in (3.3.3), the above equation (8.14.27) further reduces to

$$\lambda^4 f(t) + f^3(t)\frac{3\pi^4}{2a^4b^4h^2}(ma^2+b^2)(a^2+b^2)$$

$$+ \frac{\rho h}{D_{11}}\ddot{f} + \frac{c}{D_{11}}\dot{f} + \frac{k}{D_{11}}f(t)$$

$$= \frac{1}{D_{11}}(p_1 \sin\omega t + p_2 \cos\omega t)$$

(8.14.28)

The above equation can be rearranged as

$$\ddot{f} + \frac{c}{\rho h}\dot{f} + \frac{(k+D_{11}\lambda^4)}{\rho h}f + \frac{3D_{11}\pi^4(ma^2+b^2)(a^2+b^2)}{2\rho h^3 a^4 b^4}f^3$$

$$= \frac{p_1}{\rho h}\sin\omega t + \frac{p_2}{\rho h}\cos\omega t$$

(8.14.29)

Let us introduce the non dimensional time parameter

$$T = \frac{\pi^2}{ab}\sqrt{\frac{D_{11}}{\rho h}}\,t$$

(8.14.30)

then, equation (8.14.29) becomes

$$\ddot{f} + 2a\dot{f} + \omega_1^2 f + \gamma^2 f^3 = P_1 \sin\Omega T + P_2 \cos\Omega T$$

(8.14.31)

where

$$2a = \frac{cab}{\pi^2 \sqrt{D_{11}\rho h}}$$

$$\omega_1^2 = \frac{(k + D_{11}\lambda^4)a^2 b^2}{\pi^4 D_{11}}$$

$$\gamma^2 = \frac{3}{2} \frac{(ma^2 + b^2)(a^2 + b^2)}{h^2 a^2 b^2}$$

$$P_1 = \frac{p_1 a^2 b^2}{\pi^4 D_{11}}$$

$$P_2 = \frac{p_2 a^2 b^2}{\pi^4 D_{11}}$$

$$\Omega = \frac{\omega ab}{\pi^2} \sqrt{\frac{\rho h}{D_{11}}}$$

(8.14.32)

Equation (8.14.31) is the familiar Duffing's equation, the solution of which can be obtained by using Harmonic Balancing method, see section 6.6.

8.15 VON KARMAN ORTHOTROPIC PLATES

Here, we consider large deformation of a specially orthotropic plate resting on Kelvin viscoelastic foundation in accordance to Von Karman's theory. The middle surface strain components for an orthotropic plate can be expressed using equations (6.1.28) as

$$\bar{\varepsilon}_{xx} = \frac{1}{hE_{xx}}(N_x - v_{xy}N_y)$$

$$\bar{\varepsilon}_{yy} = \frac{1}{hE_{yy}}(N_y - v_{yx}N_x)$$

$$\bar{\gamma}_{xy} = \frac{1}{hG_{xy}}N_{xy}$$

(8.15.1)

The strains of the middle surface are given by equation (6.6.6)

$$\bar{\varepsilon}_{xx} = u_{,x} + \frac{1}{2}w_{,x}^2$$

$$\bar{\varepsilon}_{yy} = v_{,y} + \frac{1}{2}w_{,y}^2$$

$$\bar{\gamma}_{xy} = u_{,y} + v_{,x} + w_{,x}w_{,y}$$

(6.6.6)

From the above, we can obtain the following strain compatibility relation

$$(\bar{\varepsilon}_{xx})_{,yy} + (\bar{\varepsilon}_{yy})_{,xx} - (\bar{\gamma}_{xy})_{,xy} = w_{,xy}^2 - w_{,xx}w_{,yy}$$

(8.15.2)

The stress resultants in terms of Airy stress function are given by equation (6.1.29)

$$N_x = F_{,yy}$$

$$N_y = F_{,xx}$$

$$N_{xy} = -F_{,xy}$$

(6.1.29)

Therefore, equations (8.15.1) become

$$\bar{\varepsilon}_{xx} = \frac{1}{hE_{xx}}(F_{,yy} - v_{xy}F_{,xx})$$

$$\bar{\varepsilon}_{yy} = \frac{1}{hE_{yy}}(F_{,xx} - v_{yx}F_{,yy})$$

$$\bar{\gamma}_{xy} = -\frac{1}{hG_{xy}}F_{,xy}$$

(8.15.3)

Substituting the above strain values in equation (8.15.2), we get

$$F_{,xxxx} + p^2 F_{,xyxy} + m^2 F_{,yyyy} = hE_{yy}\left(w_{,xy}^2 - w_{,xx}w_{,yy}\right)$$

(8.15.4)

where

$$p^2 = E_{yy}\left(\frac{1}{G_{xy}} - \frac{2v_{yx}}{E_{yy}}\right)$$

$$m^2 = \frac{E_{yy}}{E_{xx}}$$

(8.15.5)

Equation (6.1.16) for an orthotropic plate with the foundation resistance force $q(x,y,t)$ can be modified to give

$$M_{x,xx} + M_{y,yy} - 2M_{xy,xy} = -(N_x w_{,xx} + N_y w_{,yy} + 2N_{xy} w_{,xy} + q)$$

(8.15.6)

The bending moment and twisting moment relations are given by

$$M_x = -(D_{11} w_{,xx} + D_{12} w_{,yy})$$
$$M_y = -(D_{22} w_{,yy} + D_{12} w_{,xx})$$
$$M_{xy} = 2D_{66} w_{,xy}$$

(8.6.3)

and

$$D_{11} = \frac{E_{xx} h^3}{12(1 - v_{xy} v_{yx})}$$

$$D_{22} = \frac{E_{yy} h^3}{12(1 - v_{xy} v_{yx})}$$

$$D_{12} = \frac{v_{xy} E_{yy} h^3}{12(1 - v_{xy} v_{yx})}$$

$$D_{66} = \frac{G_{xy} h^3}{12}$$

(8.6.4)

Now, substituting equations (8.6.3) above in (8.15.6) and using (6.1.29), we get

$$(D_{11}w_{,xx} + D_{12}w_{,yy})_{,xx} + (D_{22}w_{,yy} + D_{12}w_{,xx})_{,yy} + 4D_{66}w_{,xyxy}$$
$$= (F_{,yy}w_{,xx} + F_{,xx}w_{,yy} - 2F_{,xy}w_{,xy} + q)$$

(8.15.7)

The above equation can be simplified as

$$D_{11}(w_{,xxxx} + 2\gamma^2 w_{,yyxx} + m^2 w_{,yyyy})$$
$$= (F_{,yy}w_{,xx} + F_{,xx}w_{,yy} - 2F_{,xy}w_{,xy} + q)$$

(8.15.8)

where

$$\gamma^2 = \left(\frac{D_{12}}{D_{11}} + 2\frac{D_{66}}{D_{11}}\right)$$
$$= v_{yx} + \frac{1 - v_{xy}v_{yx}}{1 + v_{xy}}$$

(8.15.9)

Equations (8.15.4) and (8.15.8) are the modified versions of equations (6.1.31) and (6.1.32) for the orthotropic plate. From Fig. 6.6.1, we have

$$q(x,y,t) = p(x,y,t) - kw(x,y,t) - c\,\dot{w}(x,y,t) - ph\,\ddot{w}(x,y,t)$$

(8.15.10)

Substituting the above equation (8.15.10) in (8.15.8), the following is obtained.

$$D_{11}(w_{,xxxx} + 2\gamma^2 w_{,yyxx} + m^2 w_{,yyyy}) + ph\,\ddot{w} + c\,\dot{w} + kw$$
$$= F_{,yy}w_{,xx} + F_{,xx}w_{,yy} - 2F_{,xy}w_{,xy} + p(x,y,t)$$

(8.15.11)

The boundary conditions for the transverse deflections of the plate are

$$w = w_{,xx} = 0 \quad \text{at } x = 0, a \text{ and } 0 \le y \le b$$
$$w = w_{,yy} = 0 \quad \text{at } y = 0, b \text{ and } 0 \le x \le a$$

(8.15.12)

402 Dynamics of Plates

The boundary conditions for the in-plane displacements are

$$u = 0 \text{ and } F_{,xy} = 0 \quad \text{at } x = 0, a \text{ and } 0 \le y \le b$$
$$v = 0 \text{ and } F_{,xy} = 0 \quad \text{at } y = 0, b \text{ and } 0 \le x \le a$$

(8.15.13)

where u and v are written from equations (6.8.4) and (6.8.5) for the orthotropic plate as

$$u = \int_0^x \left[\frac{1}{hE_{xx}}(F_{,yy} - v_{xy}F_{,xx}) - \frac{1}{2}w_{,x}^2 \right] dx$$

$$v = \int_0^y \left[\frac{1}{hE_{yy}}(F_{,xx} - v_{yx}F_{,yy}) - \frac{1}{2}w_{,y}^2 \right] dy$$

(8.15.14)

A one term approximation is assumed for the transverse deflection w as a solution of equations (8.15.4) and (8.15.11)

$$w(x, y, t) = \sin\frac{\pi x}{a} \sin\frac{\pi y}{b} f(t)$$

(8.15.15)

which satisfies the boundary conditions.

Substituting the above approximate solution in equation (8.15.4), we get

$$F_{,xxxx} + p^2 F_{,xyxy} + m^2 F_{,yyyy} = \frac{1}{2}hE_{yy}f^2 \frac{\pi^4}{a^2 b^2}\left(\cos\frac{2\pi x}{a} + \cos\frac{2\pi y}{b}\right)$$

(8.15.16)

Following the steps of solution of equation (6.8.9), the solution of the above equation is

$$F = \frac{1}{2}C_2(t)x^2 + \frac{1}{2}C_1(t)y^2 + \frac{1}{32}E_{yy}hf^2(t)\left(\frac{a^2}{b^2}\cos\frac{2\pi x}{a} + \frac{b^2}{a^2}\cos\frac{2\pi y}{b}\right)$$

(8.15.17)

The above equation satisfies the boundary conditions for the in-plane displacements in (8.15.13) for $x = 0$ and $y = 0$. For the conditions at $x = a$

and $y = b$, following the steps in section 6.8, we get the two constants of integration as

$$C_1(t) = \frac{hE_{xx}}{8(1-v_{xy}v_{yx})} f^2(t)\pi^2 \left(\frac{1}{a^2} + \frac{v_{xy}}{b^2}\right)$$

$$C_2(t) = \frac{hE_{yy}}{8(1-v_{xy}v_{yx})} f^2(t)\pi^2 \left(\frac{1}{b^2} + \frac{v_{yx}}{a^2}\right)$$

(8.15.18)

The stress function F is now obtained as

$$F = \frac{h}{16} f^2(t)\pi^2 \left[\frac{1}{(1-v_{xy}v_{yx})} \left\{E_{yy}\left(\frac{1}{b^2} + \frac{v_{yx}}{a^2}\right)x^2 + E_{xx}\left(\frac{1}{a^2} + \frac{v_{xy}}{b^2}\right)y^2\right\}\right]$$

$$+ \frac{h}{32} f^2(t) E_{yy}\left(\frac{a^2}{b^2}\cos\frac{2\pi x}{a} + \frac{b^2}{m^2 a^2}\cos\frac{2\pi y}{b}\right)$$

(8.15.19)

We can now get the error in the differential equation as

$$\epsilon = \left[\begin{array}{l} ph\ddot{f}(t) + c\dot{f}(t) \\ + \left\{D_{xx}\pi^4\left(\frac{1}{a^4} + \frac{2y^2}{a^2b^2} + \frac{m^2}{b^4}\right) + k\right\}f(t) \end{array}\right] \sin\frac{\pi x}{a} \sin\frac{\pi y}{b}$$

$$- \frac{h\pi^4}{8} f^3(t)\left[\frac{E_{xx}}{a^4}\cos\frac{2\pi y}{b} + \frac{E_{yy}}{b^4}\cos\frac{2\pi x}{a}\right]\sin\frac{\pi x}{a}\sin\frac{\pi y}{b}$$

$$+ \frac{h\pi^4}{8(1-v_{xy}v_{yx})} f^3(t)\left[\begin{array}{l} E_{xx}\left(\frac{1}{a^4} + \frac{v_{xy}}{a^2b^2}\right) \\ + E_{yy}\left(\frac{1}{b^4} + \frac{v_{yx}}{a^2b^2}\right) \end{array}\right]\sin\frac{\pi x}{a}\sin\frac{\pi y}{b}$$

$$- p(x, y, t)$$

(8.15.20)

Dynamics of Plates

Applying Galerkin's method,

$$\left[\rho h \ddot{f}(t) + c\dot{f}(t) + \left\{D_{xx}\pi^4\left(\frac{1}{a^4} + \frac{2\gamma^2}{a^2b^2} + \frac{m^2}{b^4}\right) + k\right\}f(t)\right]\iint \sin^2\frac{\pi x}{a}\sin^2\frac{\pi y}{b}dx\,dy$$

$$-\frac{h\pi^4}{8}f^3(t)\frac{E_{xx}}{a^4}\iint \cos\frac{2\pi y}{b}\sin^2\frac{\pi x}{a}\sin^2\frac{\pi y}{b}dx\,dy$$

$$-\frac{h\pi^4}{8}f^3(t)\frac{E_{yy}}{b^4}\iint \cos\frac{2\pi x}{a}\sin^2\frac{\pi x}{a}\sin^2\frac{\pi y}{b}dx\,dy$$

$$+\frac{h\pi^4}{8(1-v_{xy}v_{yx})}f^3(t)E_{xx}\left(\frac{1}{a^4} + \frac{v_{xy}}{a^2b^2}\right)\iint \sin^2\frac{\pi x}{a}\sin^2\frac{\pi y}{b}dx\,dy$$

$$+\frac{h\pi^4}{8(1-v_{xy}v_{yx})}f^3(t)E_{yy}\left(\frac{1}{b^4} + \frac{v_{yx}}{a^2b^2}\right)\iint \sin^2\frac{\pi x}{a}\sin^2\frac{\pi y}{b}dx\,dy$$

$$=\iint p(x,y,t)\sin\frac{\pi x}{a}\sin\frac{\pi y}{b}dx\,dy$$

(8.15.21)

Making use of the integrals in equations (6.8.23), the above equation simplifies to

$$\left[\ddot{f}(t) + \frac{c}{\rho h}\dot{f}(t) + \left\{\frac{D_{xx}}{\rho h}\pi^4\left(\frac{1}{a^4} + \frac{2\gamma^2}{a^2b^2} + \frac{m^2}{b^4} + \frac{k}{D_{xx}}\right)\right\}f(t)\right]$$

$$+\frac{\pi^4}{16\rho}f^3(t)\left(\frac{E_{xx}}{a^4} + \frac{E_{yy}}{b^4}\right)$$

$$+\frac{\pi^4}{8\rho(1-v_{xy}v_{yx})}f^3(t)\left\{E_{xx}\left(\frac{1}{a^4} + \frac{v_{xy}}{a^2b^2}\right) + E_{yy}\left(\frac{1}{b^4} + \frac{v_{yx}}{a^2b^2}\right)\right\}$$

$$=\frac{4Q_1(t)}{\rho hab}$$

(8.15.22)

where

$$Q_1(t) = \iint p(x,y,t)\sin\frac{\pi x}{a}\sin\frac{\pi y}{b}dx\,dy$$

(8.15.23)

Using the non dimensional time parameter

$$T = \frac{\pi^2}{ab}\sqrt{\frac{D_{xx}}{\varrho h}}\, t$$

(8.15.24)

we get

$$\ddot{f}(T) + 2a\dot{f} + \omega_1^2 f(T) + \beta_1^2 f^3(T)$$
$$= \frac{4ab}{\pi^4 D} Q_1(T)$$

(8.15.25)

the familiar Duffing's equation for the dynamic behavior of the plate. In the above

$$2a = \frac{cab}{\pi^2 \sqrt{D_{xx}\varrho h}}$$

$$\omega_1^2 = \frac{ka^2 b^2}{\pi^4 D_{xx}} + \left(\frac{b}{a}\right)^2 \left[1 + 2\gamma^2\left(\frac{a}{b}\right)^2 + m^2\left(\frac{a}{b}\right)^4\right]$$

$$\beta_1^2 = \frac{3b^2}{2a^2 h^2} \left[\begin{array}{l} \frac{1}{2}\left\{1 + m^2\left(\frac{a}{b}\right)^4\right\}(1 - v_{xy}v_{yx}) \\ + 1 + v_{xy}\left(\frac{a}{b}\right)^2 + m^2\left(\frac{a}{b}\right)^2 \left\{\left(\frac{a}{b}\right)^2 + v_{yx}\right\} \end{array} \right]$$

(8.15.26)

Let

$$p(x, y, T) = P \sin\frac{\pi x}{a} \sin\frac{\pi y}{b} F(T)$$

(8.15.27)

then

$$Q_1(T) = PF(t) \iint \sin^2\frac{\pi x}{a} \sin^2\frac{\pi y}{b} dx\, dy$$
$$= PF(t)\frac{ab}{4}$$

(8.15.28)

The plate equation can then be written as

$$\ddot{f}(T) + 2a\,\dot{f} + \omega_L^2 f(T) + \beta_L^2 f^3(T) = \frac{a^2 b^2}{\pi^4 D_{xx}} PF(T)$$

(8.15.29)

For a solution of the above equation, see section 6.6.

8.16 EIGHT NODED ISOPARAMETRIC ORTHOTROPIC PLATE ELEMENT

We will consider the eight noded thick plate element (Mindlin) of section 7.6 and extend it to the case of an orthotropic laminated plate. The element geometry is given in Figs. 7.5.1 and 7.5.2. The geometric shape functions are given by

$$\{\bar{x}\} = \left\{ \begin{array}{c} x(\xi,\eta) \\ y(\xi,\eta) \\ z(\xi,\eta) \end{array} \right\} = [\bar{N}_1]\{\hat{x}\}$$

(7.6.2)

where

$$[\bar{N}_1] = \begin{bmatrix} N_1(\xi,\eta) & 0 & 0 & \cdots & N_8(\xi,\eta) & 0 & 0 \\ 0 & N_1(\xi,\eta) & 0 & \cdots & 0 & N_8(\xi,\eta) & 0 \\ 0 & 0 & N_1(\xi,\eta) & \cdots & 0 & 0 & N_8(\xi,\eta) \end{bmatrix}_{3\times 24}$$

(7.6.3)

$$N_1(\xi,\eta) = -\frac{1}{4}(\xi-1)(\eta-1)(\xi+\eta+1)$$

$$N_2(\xi,\eta) = -\frac{1}{4}(\xi+1)(\eta-1)(\xi-\eta-1)$$

$$N_3(\xi,\eta) = \frac{1}{4}(\xi+1)(\eta+1)(\xi+\eta-1)$$

$$N_4(\xi,\eta) = \frac{1}{4}(\xi-1)(\eta+1)(\xi-\eta+1)$$

$$N_5(\xi,\eta) = \frac{1}{2}(\xi-1)(\xi+1)(\eta-1)$$

$$N_6(\xi,\eta) = -\frac{1}{2}(\xi+1)(\eta+1)(\eta-1)$$

$$N_7(\xi,\eta) = -\frac{1}{2}(\xi-1)(\xi+1)(\eta+1)$$

$$N_8(\xi,\eta) = \frac{1}{2}(\xi-1)(\eta-1)(\eta+1)$$

(7.5.6)

and

$$\{\hat{x}\}^T = \begin{bmatrix} x_1 & y_1 & z_1 & x_2 & y_2 & z_2 & \cdots & x_8 & y_8 & z_8 \end{bmatrix}$$

(7.6.4)

The displacement field is given by

$$\{\delta\} = \begin{Bmatrix} u(\xi,\eta) \\ v(\xi,\eta) \\ w(\xi,\eta) \end{Bmatrix} = [\overline{N}_2]_{3\times 40} \{\hat{d}\}_{40\times 1}$$

(7.6.7)

where

$$[\overline{N}_2(\xi,\eta)] = \begin{bmatrix} N_1 & 0 & 0 & zN_1 & 0 & \cdots & N_8 & 0 & 0 & zN_8 & 0 \\ 0 & N_1 & 0 & 0 & zN_1 & \cdots & 0 & N_8 & 0 & 0 & zN_8 \\ 0 & 0 & N_1 & 0 & 0 & \cdots & 0 & 0 & N_8 & 0 & 0 \end{bmatrix}_{3\times 40}$$

(7.6.8)

$$\{\delta\}^T = \begin{Bmatrix} u & v & w \end{Bmatrix}$$

(7.6.9)

and the nodal degrees of freedom (40) for the element are

$$\{\hat{d}\}^T = \begin{bmatrix} u_1 & v_1 & w_1 & \psi_{x1} & \psi_{y1} & \cdots & u_8 & v_8 & w_8 & \psi_{x8} & \psi_{y8} \end{bmatrix}_{1\times 40}$$

(7.6.10)

408 Dynamics of Plates

The strain displacement relations are given by

$$\begin{Bmatrix} \varepsilon_{xx}(x,y) \\ \varepsilon_{yy}(x,y) \\ \gamma_{xy}(x,y) \\ \gamma_{yz}(x,y) \\ \gamma_{zx}(x,y) \end{Bmatrix} = [\overline{N}_3]_{5\times 40} \{\hat{d}\}_{40\times 1}$$

(7.6.12)

where

$$[\overline{N}_3] =$$

$$\begin{bmatrix} N_{1,x} & 0 & 0 & zN_{1,x} & 0 & \cdots & N_{8,x} & 0 & 0 & zN_{8,x} & 0 \\ 0 & N_{1,y} & 0 & 0 & zN_{1,y} & \cdots & 0 & N_{8,y} & 0 & 0 & zN_{8,y} \\ N_{1,y} & N_{1,x} & 0 & zN_{1,y} & zN_{1,x} & \cdots & N_{8,y} & N_{8,x} & 0 & zN_{8,y} & zN_{8,x} \\ 0 & 0 & N_{1,y} & 0 & N_1 & \cdots & 0 & 0 & N_{8,y} & 0 & N_8 \\ 0 & 0 & N_{1,x} & N_1 & 0 & \cdots & 0 & 0 & N_{8,x} & N_8 & 0 \end{bmatrix}_{5\times 40}$$

(7.6.13)

We can use equation (7.4.12) to get $N_{i,x}$ and $N_{i,y}$ the partial derivatives of the shape functions with respect to the local coordinates

$$\begin{Bmatrix} N_{i,x} \\ N_{i,y} \end{Bmatrix} = [J]^{-1} \begin{Bmatrix} N_{i,\xi} \\ N_{i,\eta} \end{Bmatrix}$$

(7.6.14)

The stress strain relationship for an orthotropic material is adopted from (8.5.10), extended from equation (8.10.19) to include the in-plane shear stresses

$$\begin{Bmatrix} \tau_{xx} \\ \tau_{yy} \\ \tau_{xy} \\ \tau_{yz} \\ \tau_{xz} \end{Bmatrix} = \begin{bmatrix} \overline{Q}_{11} & \overline{Q}_{12} & \overline{Q}_{16} & 0 & 0 \\ & \overline{Q}_{22} & \overline{Q}_{26} & 0 & 0 \\ & & \overline{Q}_{66} & 0 & 0 \\ & \text{Sym} & & \overline{Q}_{44} & \overline{Q}_{45} \\ & & & & \overline{Q}_{55} \end{bmatrix} \begin{Bmatrix} \varepsilon_{xx} \\ \varepsilon_{yy} \\ \gamma_{xy} \\ \gamma_{yz} \\ \gamma_{xz} \end{Bmatrix}$$

$$\{\tau\} = [D]\{\varepsilon\}$$

(8.16.1)

where the transformed reduced stiffness matrix elements are

$$\overline{Q}_{11} = Q_{11}m^4 + 2(Q_{12} + 2Q_{66})m^2n^2 + Q_{22}n^4$$

$$\overline{Q}_{12} = (Q_{11} + Q_{22} - 4Q_{66})m^2n^2 + Q_{12}(m^4 + n^4)$$

$$\overline{Q}_{16} = (Q_{11} - Q_{12} - 2Q_{66})m^3n + (Q_{12} - Q_{22} + 2Q_{66})mn^3$$

$$\overline{Q}_{22} = Q_{11}n^4 + 2(Q_{12} + 2Q_{66})m^2n^2 + Q_{22}m^4$$

$$\overline{Q}_{26} = (Q_{11} - Q_{12} - 2Q_{66})mn^3 + (Q_{12} - Q_{22} + 2Q_{66})m^3n$$

$$\overline{Q}_{44} = Q_{44}m^2 + Q_{55}n^2$$

$$\overline{Q}_{45} = (Q_{55} - Q_{44})mn$$

$$\overline{Q}_{55} = Q_{55}m^2 + Q_{44}n^2$$

$$\overline{Q}_{66} = (Q_{11} + Q_{22} - 2Q_{12} - 2Q_{66})m^2n^2 + Q_{66}(m^4 + n^4)$$

(8.16.2)

and

$$m = \cos\theta$$
$$n = \sin\theta$$

(8.2.2)

Essentially, there is only one difference in the analysis from section 7.6 of the isotropic material, which is the stress-strain relation given above. The material property matrix [D] is to be determined from the above equation and the principles of laminated plates in the section 8.11.

Following the steps in section 7.6, we have the strain energy of the plate due to bending as

$$U_b = \frac{1}{2}\{\bar{a}\}^T \iiint_v [\overline{N}_3]^T [D][\overline{N}_3] dV\{\bar{a}\}$$

(8.16.3)

Let the plate be subjected to an initial in plane stress field, τ_{xx}^0, τ_{yy}^0 and τ_{xy}^0, then the corresponding strain energy is

$$U_P = \frac{1}{2}\{\hat{a}\}^T \iiint_V [\overline{N}_4]^T [\tau^0][\overline{N}_4] dV\{\hat{a}\} \tag{8.16.4}$$

where

$$[\overline{N}_4] = \begin{bmatrix} 0 & 0 & N_{1,x} & 0 & 0 & \ldots & 0 & 0 & N_{8,x} & 0 & 0 \\ 0 & 0 & N_{1,y} & 0 & 0 & \ldots & 0 & 0 & N_{8,y} & 0 & 0 \end{bmatrix}_{2\times 40} \tag{7.6.19}$$

Using Lagrangian approach, the elemental stiffness matrix is obtained as

$$[\hat{K}] = [\hat{K}_e] + [\hat{K}_g] \tag{7.6.22}$$

where

$$[\hat{K}_e] = \iiint_V [\overline{N}_3]^T [D][\overline{N}_3] dV \tag{7.6.23}$$

$$[\hat{K}_g] = \iiint_V [\overline{N}_4]^T [\tau^0][\overline{N}_4] dV \tag{7.6.24}$$

The mass and the forcing matrices are obtained in a similar manner to that of section 7.6. Once again, it may be noted that for a laminated plate, all that one needs is to obtain the transformed reduced stiffness matrix for a given configuration and use this in place of the material property matrix $[D]$.

Example
Consider a $[0/\pm 45/90]_{sym}$ composite mid plane symmetric laminated plate made of 0.13 mm thick graphite/epoxy layers. The material properties of the layer are

$$E_{xx} = 128 \text{ GPa}$$

$$E_{yy} = 11 \text{ GPa}$$

$$G_{xy} = G_{xz} = 4.48 \text{ GPa}$$

$G_{yz} = 1.53$ GPa

$v_{xy} = 0.25$

$\rho = 1500$ kg/m³

The nondimensional natural frequencies $pa^2\sqrt{\frac{\rho h}{D_{11}}}$ for bending and $pab\sqrt{\frac{\rho h}{48D_{66}}}$ for torsion obtained with a 16 element (65 nodes) 280 degrees of freedom model are:

	Nondimensional Frequency	
a/b.	$pa^2\sqrt{\frac{\rho h}{D_{11}}}$	$pab\sqrt{\frac{\rho h}{48D_{66}}}$
	I Bending	I Torsion
2	3.4698	1.8649
3	3.4628	1.7499
4	3.4561	1.6902
5	3.4503	1.5742

8.17 ROTATING ORTHOTROPIC LAMINATED PLATE

Consider the rotating plate in Fig. 7.7.1 to be a laminated plate. The three components of kinetic energy $T = T_2 + T_1 + T_0$ are given by equation (7.7.7)

$$T_2 = \frac{1}{2} \iiint_V \rho \{\dot{a}\}^T [\overline{N}_2]^T [\overline{N}_2] \{\dot{a}\} dV$$

$$T_1 = \iiint_V \rho \{\dot{a}\}^T [\overline{N}_2]^T [A][\overline{N}_2] \{\hat{a}\} dV$$

$$+ \iiint_V \rho \{\dot{a}\}^T [\overline{N}_2]^T [A][\overline{N}_1] \{\hat{x}\} dV$$

$$T_0 = \frac{1}{2} \iiint_V \rho \{\hat{a}\}^T [\overline{N}_2]^T [A]^T [A][\overline{N}_2] \{\hat{a}\} dV$$

$$+ \frac{1}{2} \iiint_V \rho \{\hat{x}\}^T [\overline{N}_1]^T [A]^T [A][\overline{N}_1] \{\hat{x}\} dV$$

$$+ \iiint_V \rho \{\hat{a}\}^T [\overline{N}_2]^T [A]^T [A][\overline{N}_1] \{\hat{x}\} dV$$

(7.7.7)

The two components of potential energy $U = U_b + U_P$ are given by equation (8.16.3) and (8.16.4)

$$U_b = \frac{1}{2}\{\hat{a}\}^T \iiint_V [\overline{N}_3]^T [D][\overline{N}_3] dV \{\hat{a}\} \tag{8.16.3}$$

$$U_P = \frac{1}{2}\{\hat{a}\}^T \iiint_V [\overline{N}_4]^T [\tau^0][\overline{N}_4] dV \{\hat{a}\} \tag{8.16.4}$$

The material property matrix in the above equation is given by the reduced stiffness matrix

$$[D] = \begin{bmatrix} \overline{Q}_{11} & \overline{Q}_{12} & \overline{Q}_{16} & 0 & 0 \\ & \overline{Q}_{22} & \overline{Q}_{26} & 0 & 0 \\ & & \overline{Q}_{66} & 0 & 0 \\ & \text{Sym} & & \overline{Q}_{44} & \overline{Q}_{45} \\ & & & & \overline{Q}_{55} \end{bmatrix} \tag{8.17.1}$$

with the elements defined in equation (8.16.2). Making use of Lagrangian equations, we get the elemental equation as

$$[\widehat{M}]\{\ddot{\hat{a}}\} + [\widehat{G}]\{\dot{\hat{a}}\} + [\widehat{K}]\{\hat{a}\} + \{\hat{f}_a\} + \{\hat{f}_c\} = \{\widehat{F}_g\} \tag{8.17.2}$$

where

$$[\widehat{M}] = \iiint_V \rho [\overline{N}_2]^T [\overline{N}_2] dV \tag{7.7.10}$$

$$[\widehat{G}] = 2 \iiint_V \rho [\overline{N}_2]^T [A][\overline{N}_2] dV \tag{7.7.11}$$

and

$$[\widehat{K}] = [\widehat{K}_r] + [\widehat{K}_a] + [\widehat{K}_e] + [\widehat{K}_g] \tag{7.7.12}$$

$$[\hat{K}_r] = -\iiint_V \rho [\bar{N}_2]^T [A]^T [A][\bar{N}_2] dV$$

(7.7.13)

$$[\hat{K}_a] = \iiint_V \rho [\bar{N}_2]^T [\dot{A}][\bar{N}_2] dV$$

(7.7.14)

$$[\hat{K}_e] = \iiint_V [\bar{N}_3]^T [D][\bar{N}_3] dV$$

(7.6.23)

$$[\hat{K}_g] = \iiint_V [\bar{N}_4]^T [\tau^0][\bar{N}_4] dV$$

(7.6.24)

$$\{\hat{f}_a\} = -\iiint_V \rho [\bar{N}_2]^T [\dot{A}]^T [\bar{N}_1] dV\{\hat{x}\}$$

(7.7.15)

$$\{\hat{f}_c\} = -\iiint_V \rho [\bar{N}_2]^T [A]^T [A][\bar{N}_1] dV\{\hat{x}\}$$

(7.7.16)

In order to determine the geometric stiffness matrix in equation (7.6.24) above, we require the initial stress field $[\tau^0]$ which is determined from

$$[\hat{K}_e]\{\hat{d}_c\} = \{\hat{f}_c\}$$

$$\{\tau^0\} = [D][\bar{N}_3]\{\hat{d}_c\}$$

(8.17.4)

Gaussian quadrature can be used to evaluate the integrals with Jacobian transformation wherever required. Assembling the elemental equations, we get the global equations of motion for the rotating orthotropic laminated plate

$$[M]\{\ddot{d}\} + [G]\{\dot{d}\} + [K]\{d\} + \{f_a\} = \{F_g\}$$

(8.17.5)

where

$$\{\hat{F}_g\} = \left(\iint_A [\overline{N}_2]^T \begin{Bmatrix} F_x \\ F_y \\ F_z \end{Bmatrix} dA \right) \cos(\Omega t - \Phi)$$

(7.7.21)

Example
Consider now the plate of section 8.16 mounted on a disc with $R/a = 1$ and at a stagger angle $30°$. The nondimnesional natural frequencies with the disc rotational speed equal to half of the first stationary bending natural frequency are:

Nondimensional Frequency

a/b	$pa^2\sqrt{\dfrac{\varrho h}{D_{11}}}$ I Bending	$pab\sqrt{\dfrac{\varrho h}{48 D_{66}}}$ I Torsion
2	4.5917	2.0504
3	4.5903	1.9316
4	4.5839	1.8801
5	4.5769	1.6513

9

Pre-Twisted Plates

So far, we considered the case of flat plates. The middle surface of the plate is in the x-y plane and the plate deformation is transverse to this plane in the z direction. In general, the middle surface of the plate is a surface rather than a plane as in the case of flat plates, such of those members are called *shells*. In this book, we will restrict ourselves to the case of plates which are pre-twisted. Imagine one of the edges of a flat plate to be held and twist the opposite side of the flat plate through a specified angle, called *pre-twist angle*, we have a *pre-twisted plate*. Pre-twisted plates are essentially shells, and we use shell theory to determine the static and dynamic characteristics. Here, we follow Gol'denveizer to define the shell surface, strain displacement relations and stress strain relations.

9.1 THEORY OF A SURFACE

Consider a surface shown in Fig. 9.1 1 which can be defined in three dimensions by its radius vector

$$\vec{M} = \vec{M}(a, \beta)$$

(9.1.1)

where a, β are arbitrary parameters. Given the values of the parameters a, β, we can obtain the corresponding point or points on the surface and equation (9.1.1) determines the geometric properties of the surface.

We can choose a value for the parameter a, say a_0, and let β vary, then we get a space curve on the given surface $\vec{M}(a, \beta)$. These curves are β- curves and each such curve in a family have different values of a_0. In a similar way, we can define a family of a- curves with different values of β_0. If we specify a_0, β_0 simultaneously, then the intersection points of these two curves are obtained. This is similar to specifying the two coordinates of a point on a plane, therefore, α- and β- curves form a system of curvilinear coordinates.

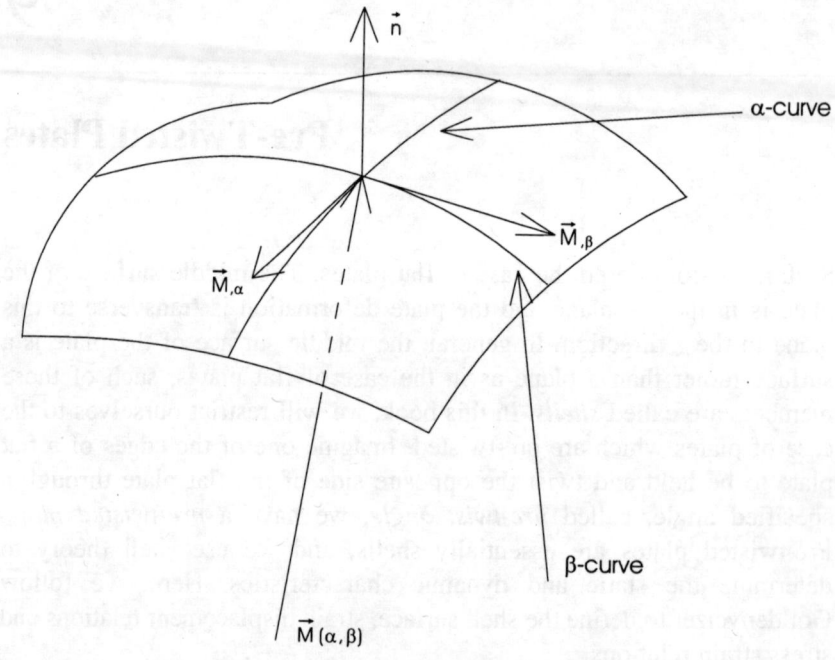

Fig. 9.1.1 A Surface with Basic Trihedra

In defining the surface in equation (9.1.1), we can choose a different set of independent parameters

$$\alpha' = \alpha'(\alpha, \beta)$$
$$\beta' = \beta'(\alpha, \beta)$$

(9.1.2)

then we obtain equation (9.1.1) in the form

$$\vec{M'} = \vec{M'}(\alpha', \beta')$$

(9.1.3)

For the special case when equation (9.1.2) takes the form

$$a' = a'(a)$$
$$\beta' = \beta'(\beta)$$

(9.1.2a)

the geometric nature of the coordinate lines remains same. Taking the derivatives of the radius vector in equation (9.1.1), we have

$$\vec{M},_a = \frac{\partial \vec{M}}{\partial a}$$

$$\vec{M},_\beta = \frac{\partial \vec{M}}{\partial \beta}$$

(9.1.4)

giving the directions of tangents to the a- and β- curves respectively. The lengths of the vectors in (9.1.4) are denoted by A and B, i.e.,

$$\left|\vec{M},_a\right| = A$$

$$\left|\vec{M},_\beta\right| = B$$

(9.1.5)

so that the unit vectors tangent to the coordinate curves are $\frac{1}{A}\vec{M},_a$ and $\frac{1}{B}\vec{M},_\beta$. If the angle between the two coordinate curves is χ, then

$$\frac{\vec{M},_a}{A} \cdot \frac{\vec{M},_\beta}{B} = \cos\chi$$

(9.1.6)

Since a- and β- curves are never tangent to each other $\sin\chi \neq 0$.

First Quadratic Form

Consider now, two points (a,β) and $(a+da, \beta+d\beta)$ on the surface which are close to each other in an arbitrary manner. Then, the increment in the vector $\vec{M}$ going from (a,β) to $(a+da, \beta+d\beta)$ is given by

$$d\vec{M} = \vec{M},_\alpha d\alpha + \vec{M},_\beta d\beta$$

(9.1.7)

Let $|d\vec{M}| = ds$, then

$$d\vec{M} \cdot d\vec{M} = (ds)^2 = A^2(d\alpha)^2 + 2AB\cos\chi\, d\alpha\, d\beta + B^2(d\beta)^2$$

(9.1.8)

The right hand side of the above equation is termed the *first quadratic form* of the surface.

$$I = A^2(d\alpha)^2 + 2AB\cos\chi\, d\alpha\, d\beta + B^2(d\beta)^2$$

(9.1.8a)

A, B and χ are called the *coefficients of the first quadratic form*. Once they are known, the first quadratic form of the surface is completely determined. The geometry of the surface is completely determined with the first quadratic form.

Basic Trihedra

Let the normal to the surface at any point be denoted by $\vec{n}$. This is orthogonal to the vectors $\vec{M},_\alpha$ and $\vec{M},_\beta$ and is connected by the following relation

$$\vec{n} = \frac{1}{AB\sin\chi}\left(\vec{M},_\alpha \times \vec{M},_\beta\right)$$

(9.1.9)

The three vectors $\vec{n}, \frac{1}{A}\vec{M},_\alpha$ and $\frac{1}{B}\vec{M},_\beta$ are called *basic vectors* or *principal vectors* of the surface and together they form a moving trihedron called *basic trihedron*, see Fig. 9.1.1. This trihedron forms a right handed coordinate system in which any arbitrary vector $\vec{S}$ can be expressed as

$$\vec{S} = s_\alpha \frac{\vec{M},_\alpha}{A} + s_\beta \frac{\vec{M},_\beta}{B} + s_n \vec{n}$$

(9.1.10)

where s_α, s_β, s_n are basic components of the vector $\vec{S}$. These basic components are also expressed as

$$s_a = \left|\vec{S}\right|_a, \quad s_\beta = \left|\vec{S}\right|_\beta, \quad s_n = \left|\vec{S}\right|_n$$
(9.1.11)

It may be noted that χ is not necessarily $\pi/2$ and therefore the basic components of the vector $\vec{S}$ are equal to its projections upon the axes of basic trihedron only when the surface is given by orthogonal coordinate system, i.e., when $\chi = \pi/2$.

Auxiliary Trihedra

Consider now two auxiliary vectors $\vec{N}_{(a)}, \vec{N}_{(\beta)}$ which lie in the tangent plane of the surface as shown in Fig. 9.1.2. The two auxiliary vectors are so chosen to satisfy the following.

1. $\vec{N}_{(a)}$ is orthogonal to the vector $\vec{M},_\beta$
2. $\vec{N}_{(\beta)}$ is orthogonal to the vector $\vec{M},_a$
3. The lengths of vectors $\vec{N}_{(a)}, \vec{N}_{(\beta)}$ are A and B respectively
4. $\vec{N}_{(a)}, \vec{N}_{(\beta)}$ make acute angles with $\vec{M},_a, \vec{M},_\beta$

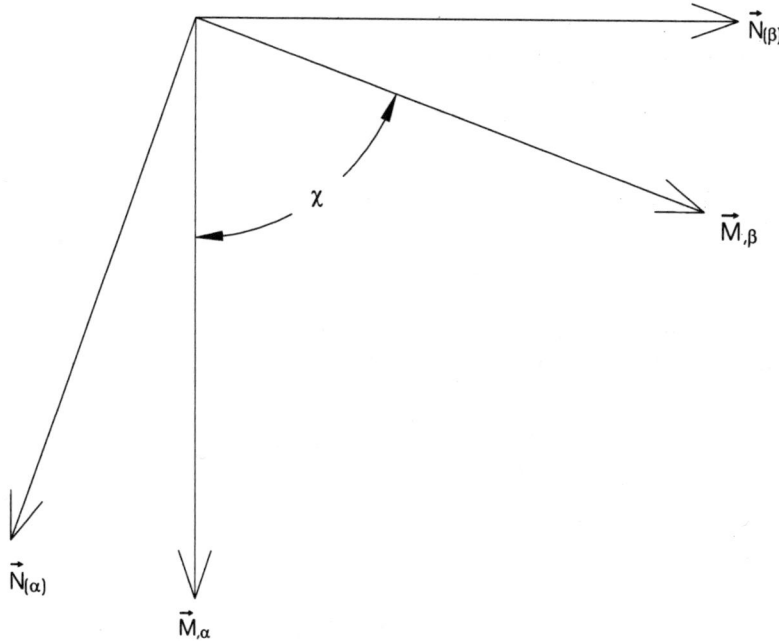

Fig. 9.1.2 Auxiliary Vectors

420 Dynamics of Plates

The above conditions imply that

$$\vec{N}_{(a)} \cdot \vec{n} = 0 \qquad \vec{N}_{(\beta)} \cdot \vec{n} = 0$$

$$\vec{N}_{(a)} \cdot \vec{M}_{,\beta} = 0 \qquad \vec{N}_{(\beta)} \cdot \vec{M}_{,a} = 0$$

$$\left|\vec{N}_{(a)}\right| = A \qquad \left|\vec{N}_{(\beta)}\right| = B$$

$$\frac{\vec{M}_{,a}}{A} \cdot \frac{\vec{N}_{(a)}}{A} = \sin\chi \qquad \frac{\vec{M}_{,\beta}}{B} \cdot \frac{\vec{N}_{(\beta)}}{B} = \sin\chi$$

(9.1.12)

The auxiliary vectors can be obtained from the basic vectors as

$$\frac{\vec{N}_{(a)}}{A} = \frac{1}{\sin\chi} \frac{\vec{M}_{,a}}{A} - \cot\chi \frac{\vec{M}_{,\beta}}{B}$$

$$\frac{\vec{N}_{(\beta)}}{B} = \frac{1}{\sin\chi} \frac{\vec{M}_{,\beta}}{B} - \cot\chi \frac{\vec{M}_{,a}}{A}$$

(9.1.13)

For the case when $\chi = \pi/2$

$$\vec{N}_{(a)} = \vec{M}_{,a}$$

$$\vec{N}_{(\beta)} = \vec{M}_{,\beta}$$

(9.1.13a)

Taking the scalar product of the vector $\vec{S}$ in (9.1.10) with $\frac{\vec{N}_{(a)}}{A}$ and using the results in equation (9.1.13), we get

$$\vec{S} \cdot \frac{\vec{N}_{(a)}}{A} = s_a \frac{\vec{M}_{,a}}{A} \cdot \frac{\vec{N}_{(a)}}{A} = s_a \sin\chi$$

(9.1.14)

Therefore

$$s_a = \frac{1}{\sin\chi} \vec{S} \cdot \frac{\vec{N}_{(a)}}{A}$$

(9.1.15a)

In a similar manner taking the scalar products of equation (9.1.10) with $\dfrac{\vec{N}_{(\beta)}}{B}$ and $\vec{n}$ we obtain

$$s_\beta = \dfrac{1}{\sin \chi} \vec{S} \cdot \dfrac{\vec{N}_{(\beta)}}{B}$$

(9.1.15b)

$$s_n = \vec{S} \cdot \vec{n}$$

(9.1.15c)

For a given vector, $\vec{T}$, given its components $\tau_\alpha, \tau_\beta, \tau_n$, i.e.,

$$\tau_\alpha = \vec{T} \cdot \dfrac{\vec{N}_{(\alpha)}}{A}$$

$$\tau_\beta = \vec{T} \cdot \dfrac{\vec{N}_{(\beta)}}{B}$$

$$\tau_n = \vec{T} \cdot \vec{n}$$

(9.1.16)

it can be expressed in the basic trihedron axes by

$$\vec{T} = \dfrac{\tau_\alpha}{\sin \chi} \dfrac{\vec{M}_{,\alpha}}{A} + \dfrac{\tau_\beta}{\sin \chi} \dfrac{\vec{M}_{,\beta}}{B} + \tau_n \vec{n}$$

(9.1.17)

The three unit vectors $\dfrac{\vec{N}_{(\alpha)}}{A}, \dfrac{\vec{N}_{(\beta)}}{B}, \vec{n}$ are called *auxiliary vectors* of the surface and together the system of these vectors is called *moving auxiliary trihedron*. The vector $\vec{S}$ in (9.1.10) can be expressed in an alternate way in terms of auxiliary vectors as

$$\vec{S} = s'_\alpha \dfrac{\vec{N}_{(\alpha)}}{A} + s'_\beta \dfrac{\vec{N}_{(\beta)}}{B} + s'_n \vec{n}$$

(9.1.18)

422 Dynamics of Plates

where s'_a, s'_β, s'_n are auxiliary components of the vector $\vec{S}$. Thus, every vector can be represented in two ways, one in basic components and the second in auxiliary components. Such a representation helps in carrying out vector operations in a convenient way. As an example, consider two vectors, $\vec{S}$ expressed in basic components and the other $\vec{T}$ in auxiliary components. Taking scalar product of these two vectors we have

$$\vec{S} \cdot \vec{T} = \left(s_a \frac{\vec{M}_{,a}}{A} + s_\beta \frac{\vec{M}_{,\beta}}{B} + s_n \vec{n} \right) \cdot \left(\tau'_a \frac{\vec{N}_{(a)}}{A} + \tau'_\beta \frac{\vec{N}_{(\beta)}}{B} + \tau'_n \vec{n} \right)$$

(9.1.19)

Making use of the relations in (9.1.12), the above gives

$$\vec{S} \cdot \vec{T} = \sin \chi \, s_a \tau'_a + \sin \chi \, s_\beta \tau'_\beta + s_n \tau'_n$$

(9.1.20)

We can also obtain relations similar to those in (9.1.15)

$$s'_a = \frac{1}{\sin \chi} \vec{S} \cdot \frac{\vec{M}_{,a}}{A}$$

$$s'_\beta = \frac{1}{\sin \chi} \vec{S} \cdot \frac{\vec{M}_{,\beta}}{B}$$

$$s'_n = s_n = \vec{S} \cdot \vec{n}$$

(9.1.21)

$\vec{S}$ is then given by equation (9.1.18). We can also establish the following relations

$$s_a = \frac{1}{\sin \chi} (s'_a - s'_\beta \cos \chi)$$

$$s_\beta = \frac{1}{\sin \chi} (s'_\beta - s'_a \cos \chi)$$

(9.1.22)

$$s'_a = \frac{1}{\sin \chi} (s_a + s_\beta \cos \chi)$$

$$s'_\beta = \frac{1}{\sin \chi} (s_\beta + s_a \cos \chi)$$

(9.1.23)

Finally, the vector products of the basic vectors with the auxiliary vectors are given by

$$\frac{\vec{M},_\alpha}{A} \times \frac{\vec{M},_\beta}{B} = \vec{n} \sin \chi$$

$$\frac{\vec{M},_\alpha}{A} \times \vec{n} = -\frac{\vec{N}_{(\beta)}}{B} \qquad \frac{\vec{M},_\beta}{B} \times \vec{n} = \frac{\vec{N}_{(\alpha)}}{A}$$

$$\frac{\vec{M},_\alpha}{A} \times \frac{\vec{N}_{(\alpha)}}{A} = -\vec{n} \cos \chi \qquad \frac{\vec{M},_\beta}{B} \times \frac{\vec{N}_{(\beta)}}{B} = \vec{n} \cos \chi$$

$$\frac{\vec{M},_\alpha}{A} \times \frac{\vec{N}_{(\beta)}}{B} = \vec{n} \qquad \frac{\vec{M},_\beta}{B} \times \frac{\vec{N}_{(\alpha)}}{A} = -\vec{n}$$

(9.1.24)

Coefficients of Second Quadratic Form

The second derivatives of the radius vector can be obtained as

$$\vec{M},_{\alpha\alpha} = \begin{Bmatrix} 11 \\ 1 \end{Bmatrix} \vec{M},_\alpha + \begin{Bmatrix} 11 \\ 2 \end{Bmatrix} \vec{M},_\beta + L\vec{n}$$

$$\vec{M},_{\alpha\beta} = \begin{Bmatrix} 12 \\ 1 \end{Bmatrix} \vec{M},_\alpha + \begin{Bmatrix} 12 \\ 2 \end{Bmatrix} \vec{M},_\beta + M\vec{n}$$

$$\vec{M},_{\beta\beta} = \begin{Bmatrix} 22 \\ 1 \end{Bmatrix} \vec{M},_\alpha + \begin{Bmatrix} 22 \\ 2 \end{Bmatrix} \vec{M},_\beta + N\vec{n}$$

(9.1.25)

where the coefficients in the above equation are denoted by Christoffel symbols, the top two numbers in the curved brackets representing $\alpha = 1$ and $\beta = 2$ or the left hand side derivative and the lower number representing the coefficient of $\vec{M},_\alpha = 1$ and $\vec{M},_\beta = 2$ and L, M, N are the coefficients of the second fundamental quadratic form.

In the above, the coefficients of the second quadratic form are

$$L = \vec{n} \cdot \vec{M},_{\alpha\alpha}$$

$$M = \vec{n} \cdot \vec{M},_{\alpha\beta}$$

$$N = \vec{n} \cdot \vec{M},_{\beta\beta}$$

(9.1.26)

424 *Dynamics of Plates*

and the coefficients represented by Christoffel symbols are

$$\left\{\begin{matrix}11\\1\end{matrix}\right\} = \frac{AB^2\frac{\partial A}{\partial \alpha} + A^2 B \cos\chi \frac{\partial A}{\partial \beta} - AB\cos\chi \frac{\partial}{\partial \alpha}(AB\cos\chi)}{A^2 B^2 \sin^2\chi}$$

$$\left\{\begin{matrix}11\\2\end{matrix}\right\} = \frac{-A^2 B \cos\chi \frac{\partial A}{\partial \alpha} - A^3 \frac{\partial A}{\partial \beta} + A^2 \frac{\partial}{\partial \alpha}(AB\cos\chi)}{A^2 B^2 \sin^2\chi}$$

$$\left\{\begin{matrix}12\\1\end{matrix}\right\} = \frac{AB^2 \frac{\partial A}{\partial \beta} - AB^2 \cos\chi \frac{\partial B}{\partial \alpha}}{A^2 B^2 \sin^2\chi}$$

$$\left\{\begin{matrix}12\\2\end{matrix}\right\} = \frac{A^2 B \frac{\partial B}{\partial \alpha} - A^2 B \cos\chi \frac{\partial A}{\partial \beta}}{A^2 B^2 \sin^2\chi}$$

$$\left\{\begin{matrix}22\\1\end{matrix}\right\} = \frac{-AB^2 \cos\chi \frac{\partial B}{\partial \beta} + B^2 \frac{\partial}{\partial \beta}(AB\cos\chi) - B^3 \frac{\partial B}{\partial \alpha}}{A^2 B^2 \sin^2\chi}$$

$$\left\{\begin{matrix}22\\2\end{matrix}\right\} = \frac{AB^2 \cos\chi \frac{\partial B}{\partial \alpha} + A^2 B \frac{\partial B}{\partial \beta} - AB\cos\chi \frac{\partial}{\partial \beta}(AB\cos\chi)}{A^2 B^2 \sin^2\chi}$$

(9.1.27)

The coefficients of the second quadratic form in (9.1.26) can be shown to be

$$L = -\vec{n}_{,\alpha} \cdot \vec{M}_{,\alpha}$$
$$M = -\vec{n}_{,\alpha} \cdot \vec{M}_{,\beta} = -\vec{n}_{,\beta} \cdot \vec{M}_{,\alpha}$$
$$N = -\vec{n}_{,\beta} \cdot \vec{M}_{,\beta}$$

(9.1.28)

where

$$\left\{\begin{matrix}\vec{n}_{,\alpha}\\ \vec{n}_{,\beta}\end{matrix}\right\} = \begin{bmatrix} -\frac{L}{A\sin^2\chi} + \frac{M\cos\chi}{B\sin^2\chi} & \frac{L\cos\chi}{A\sin^2\chi} - \frac{M}{B\sin^2\chi} \\ \frac{N\cos\chi}{B\sin^2\chi} - \frac{M}{A\sin^2\chi} & -\frac{N}{B\sin^2\chi} + \frac{M\cos\chi}{A\sin^2\chi} \end{bmatrix} \left\{\begin{matrix}\frac{\vec{M}_{,\alpha}}{A}\\ \frac{\vec{M}_{,\beta}}{B}\end{matrix}\right\}$$

(9.1.29)

or

$$\left\{ \begin{array}{c} \vec{n},_a \\ \vec{n},_\beta \end{array} \right\} = \left[\begin{array}{cc} -\dfrac{L}{A\sin\chi} & -\dfrac{M}{B\sin\chi} \\ -\dfrac{M}{A\sin\chi} & -\dfrac{N}{B\sin\chi} \end{array} \right] \left\{ \begin{array}{c} \dfrac{\vec{N}_{(a)}}{A} \\ \dfrac{\vec{N}_{(\beta)}}{B} \end{array} \right\}$$

(9.1.30)

The derivatives of basic vectors along the axes of auxiliary trihedron are given by

$$\frac{\partial}{\partial a}\left(\frac{\vec{M},_a}{A}\right) = \frac{B}{A}\sin\chi \left\{ \begin{array}{c} 11 \\ 2 \end{array} \right\} \frac{\vec{N}_{(\beta)}}{B} + \frac{L}{A}\vec{n}$$

$$\frac{\partial}{\partial \beta}\left(\frac{\vec{M},_a}{A}\right) = \frac{B}{A}\sin\chi \left\{ \begin{array}{c} 12 \\ 2 \end{array} \right\} \frac{\vec{N}_{(\beta)}}{B} + \frac{M}{A}\vec{n}$$

$$\frac{\partial}{\partial a}\left(\frac{\vec{M},_\beta}{B}\right) = \frac{A}{B}\sin\chi \left\{ \begin{array}{c} 12 \\ 1 \end{array} \right\} \frac{\vec{N}_{(a)}}{A} + \frac{M}{B}\vec{n}$$

$$\frac{\partial}{\partial \beta}\left(\frac{\vec{M},_\beta}{B}\right) = \frac{A}{B}\sin\chi \left\{ \begin{array}{c} 22 \\ 1 \end{array} \right\} \frac{\vec{N}_{(a)}}{A} + \frac{N}{B}\vec{n}$$

(9.1.31)

Consider now, the radius vector $\vec{M}$ at a point $a=a_0, \beta=\beta_0$ and expand this in power series about the point $a=a_0, \beta=\beta_0$,

$$\vec{M}(a,\beta) = \vec{M}(a_0,\beta_0) + \frac{a-a_0}{1!}\vec{M},_a(a_0,\beta_0) + \frac{\beta-\beta_0}{1!}\vec{M},_\beta(a_0,\beta_0) +$$

$$+ \frac{(a-a_0)^2}{2!}\vec{M},_{aa}(a_0,\beta_0)$$

$$+ \frac{(a-a_0)(\beta-\beta_0)}{2!}\vec{M},_{a\beta}(a_0,\beta_0)$$

$$+ \frac{(\beta-\beta_0)^2}{2!}\vec{M},_{\beta\beta}(a_0,\beta_0) + \cdots$$

(9.1.32)

Since the derivatives of the radius vector contain only the coefficients of first and second quadratic form, the right hand side of equation (9.1.32) is completely determined by the quantities $\vec{M}, \vec{M}_{,\alpha}, \vec{M}_{,\beta}, A, B, \chi, L, M, N, \partial A/\partial \alpha, \cdots$ at (α_0, β_0). This means that a surface is given, except for the position, by the coefficients of the first and second quadratic forms, viz., the six quantities A, B, χ, L, M and N. These six quantities should satisfy the following equations, first two Codazzi equations and the third one Gauss equation.

$$\frac{\partial L}{\partial \beta} - \frac{\partial M}{\partial \alpha} - \left\{\begin{array}{c} 12 \\ 1 \end{array}\right\} L + \left(\left\{\begin{array}{c} 11 \\ 1 \end{array}\right\} - \left\{\begin{array}{c} 12 \\ 2 \end{array}\right\}\right) M + \left\{\begin{array}{c} 11 \\ 2 \end{array}\right\} N = 0$$

$$\frac{\partial N}{\partial \alpha} - \frac{\partial M}{\partial \beta} - \left\{\begin{array}{c} 22 \\ 1 \end{array}\right\} L + \left(\left\{\begin{array}{c} 22 \\ 2 \end{array}\right\} - \left\{\begin{array}{c} 12 \\ 1 \end{array}\right\}\right) M + \left\{\begin{array}{c} 12 \\ 2 \end{array}\right\} N = 0$$

$$\frac{LN - M^2}{AB \sin \chi} + \frac{\partial^2 \chi}{\partial \alpha \partial \beta} + \frac{\partial}{\partial \alpha}\left[\frac{\frac{\partial B}{\partial \alpha} - \frac{\partial A}{\partial \beta} \cos \chi}{A \sin \chi}\right] + \frac{\partial}{\partial \beta}\left[\frac{\frac{\partial A}{\partial \beta} - \frac{\partial B}{\partial \alpha} \cos \chi}{B \sin \chi}\right] = 0$$

(9.1.33)

Derivatives of Radius Vectors

Consider the vector $\vec{S}$ in equation (9.1.10) in terms of basic components or (9.1.18) in terms of auxiliary components

$$\vec{S} = s_\alpha \frac{\vec{M}_{,\alpha}}{A} + s_\beta \frac{\vec{M}_{,\beta}}{B} + s_n \vec{n}$$

(9.1.10)

$$\vec{S} = s'_\alpha \frac{\vec{N}_{(\alpha)}}{A} + s'_\beta \frac{\vec{N}_{(\beta)}}{B} + s_n \vec{n}$$

(9.1.18)

From (9.1.21), we can write

$$\vec{S} \cdot \frac{\vec{M}_{,\alpha}}{A} = s'_\alpha \sin \chi$$

$$\vec{S} \cdot \frac{\vec{M}_{,\beta}}{B} = s'_\beta \sin \chi$$

$$\vec{S} \cdot \vec{n} = s_n$$

(9.1.21a)

Differentiating the above with respect to a, we have

$$\frac{\partial \vec{S}}{\partial a} \cdot \frac{\vec{M},_a}{A} = \frac{\partial}{\partial a}(s'_a \sin\chi) - \vec{S} \cdot \frac{\partial}{\partial a}\left(\frac{\vec{M},_a}{A}\right)$$

$$\frac{\partial \vec{S}}{\partial a} \cdot \frac{\vec{M},_\beta}{B} = \frac{\partial}{\partial a}(s'_\beta \sin\chi) - \vec{S} \cdot \frac{\partial}{\partial a}\left(\frac{\vec{M},_\beta}{B}\right)$$

$$\frac{\partial \vec{S}}{\partial a} \cdot \vec{n} = \frac{\partial s_n}{\partial a} - \vec{S} \cdot \frac{\partial \vec{n}}{\partial a}$$

(9.1.34)

Making use of (9.1.31) and (9.1.30) for the derivatives of basic vectors in the above equation (9.1.34), and also equation (9.1.20), we get

$$\frac{\partial \vec{S}}{\partial a} \cdot \frac{\vec{M},_a}{A} = \frac{\partial}{\partial a}(s'_a \sin\chi) - \frac{B}{A}\sin^2\chi \left\{ \begin{array}{c} 11 \\ 2 \end{array} \right\} s_\beta - \frac{L}{A} s_n$$

$$\frac{\partial \vec{S}}{\partial a} \cdot \frac{\vec{M},_\beta}{B} = \frac{\partial}{\partial a}(s'_\beta \sin\chi) - \frac{A}{B}\sin^2\chi \left\{ \begin{array}{c} 12 \\ 1 \end{array} \right\} s_a - \frac{M}{B} s_n$$

$$\frac{\partial \vec{S}}{\partial a} \cdot \vec{n} = \frac{\partial s_n}{\partial a} + \frac{L}{A} s_a + \frac{M}{B} s_\beta$$

(9.1.35)

Now, writing $\frac{\partial \vec{S}}{\partial a}$ in the form of (9.1.18)

$$\frac{\partial \vec{S}}{\partial a} = \left|\frac{\partial \vec{S}}{\partial a}\right|'_a \frac{\vec{N}_{(a)}}{A} + \left|\frac{\partial \vec{S}}{\partial a}\right|'_\beta \frac{\vec{N}_{(\beta)}}{B} + \left|\frac{\partial \vec{S}}{\partial a}\right|_n \vec{n}$$

(9.1.36)

Also, from (9.1.21), we can write

$$\left|\frac{\partial \vec{S}}{\partial a}\right|'_a = \frac{1}{\sin\chi} \frac{\partial \vec{S}}{\partial a} \cdot \frac{\vec{M},_a}{A}$$

$$\left|\frac{\partial \vec{S}}{\partial a}\right|'_\beta = \frac{1}{\sin\chi} \frac{\partial \vec{S}}{\partial a} \cdot \frac{\vec{M},_\beta}{B}$$

$$\left|\frac{\partial \vec{S}}{\partial a}\right|'_n = \left|\frac{\partial \vec{S}}{\partial a}\right|_n = \frac{\partial \vec{S}}{\partial a} \cdot \vec{n}$$

(9.1.37)

With the help of (9.1.35), the above equations become

$$\left|\frac{\partial \vec{S}}{\partial a}\right|'_a = \frac{1}{\sin\chi}\frac{\partial}{\partial a}(s'_a \sin\chi) - \frac{B}{A}\sin\chi\begin{Bmatrix}11\\2\end{Bmatrix}s_\beta - \frac{L}{A\sin\chi}s_n$$

$$\left|\frac{\partial \vec{S}}{\partial a}\right|'_\beta = \frac{1}{\sin\chi}\frac{\partial}{\partial a}(s'_\beta \sin\chi) - \frac{A}{B}\sin\chi\begin{Bmatrix}12\\1\end{Bmatrix}s_a - \frac{M}{B\sin\chi}s_n$$

$$\left|\frac{\partial \vec{S}}{\partial a}\right|'_n = \left|\frac{\partial \vec{S}}{\partial a}\right|_n = \frac{\partial s_n}{\partial a} + \frac{L}{A}s_a + \frac{M}{B}s_\beta$$

(9.1.37a)

In a similar manner, we can obtain

$$\left|\frac{\partial \vec{S}}{\partial \beta}\right|'_a = \frac{1}{\sin\chi}\frac{\partial}{\partial \beta}(s'_a \sin\chi) - \frac{B}{A}\sin\chi\begin{Bmatrix}12\\2\end{Bmatrix}s_\beta - \frac{M}{A\sin\chi}s_n$$

$$\left|\frac{\partial \vec{S}}{\partial \beta}\right|'_\beta = \frac{1}{\sin\chi}\frac{\partial}{\partial \beta}(s'_\beta \sin\chi) - \frac{A}{B}\sin\chi\begin{Bmatrix}22\\1\end{Bmatrix}s_a - \frac{N}{B\sin\chi}s_n$$

$$\left|\frac{\partial \vec{S}}{\partial \beta}\right|'_n = \left|\frac{\partial \vec{S}}{\partial \beta}\right|_n = \frac{\partial s_n}{\partial \beta} + \frac{N}{B}s_a + \frac{M}{A}s_\beta$$

(9.1.38)

Radius of Curvature : Second Quadratic Form

Let $\vec{M}$ be the radius vector of a surface, consider any curve on this surface with s representing the arc length on this given curve. Also, let $\vec{\tau}$ be the unit vector tangential to the given curve. Then,

$$\vec{\tau} = \frac{d\vec{M}}{ds} = \vec{M},_a a,_s + \vec{M},_\beta \beta,_s$$

(9.1.39)

From the theory of surfaces, Frenet's formula gives

$$\frac{d\vec{\tau}}{ds} = \frac{\vec{v}}{\rho}$$

(9.1.40)

where ρ is the radius of curvature and $\vec{v}$ is the unit vector along the principal normal of the curve. Substituting (9.1.39) in (9.1.40), we have

$$\frac{\vec{v}}{\rho} = \frac{d^2\vec{M}}{ds^2}$$

$$= \vec{M},_{\alpha\alpha}(\alpha,_s)^2 + 2\vec{M},_{\alpha\beta}\alpha,_s\beta,_s + \vec{M},_{\beta\beta}(\beta,_s)^2 + \vec{M},_\alpha\alpha,_{ss} + \vec{M},_\beta\beta,_{ss}$$

(9.1.41)

Let φ be the angle between the normal to the surface and the principal normal to the given curve, then

$$\vec{n} \cdot \vec{v} = \cos\varphi$$

(9.1.42)

Now, taking the scalar product of equation (9.1.41) with $\vec{n}$, we get

$$\frac{\cos\varphi}{\rho} = \vec{n} \cdot \vec{M},_{\alpha\alpha}(\alpha,_s)^2 + 2\vec{n} \cdot \vec{M},_{\alpha\beta}\alpha,_s\beta,_s + \vec{n} \cdot \vec{M},_{\beta\beta}(\beta,_s)^2$$

(9.1.43)

Using (9.1.26) and (9.1.8)

$$\frac{\cos\varphi}{\rho} = \frac{L(d\alpha)^2 + 2M\,d\alpha\,d\beta + N(d\beta)^2}{A^2(d\alpha)^2 + 2AB\cos\chi\,d\alpha\,d\beta + B^2(d\beta)^2}$$

(9.1.44)

The numerator of the right hand side in the above equation (9.1.44) is the *second quadratic form* of the surface, denoted by

$$\text{II} = L(d\alpha)^2 + 2M\,d\alpha\,d\beta + N(d\beta)^2$$

(9.1.45)

where L,M and N are the coefficients of the second quadratic form as noted in equation (9.1.25).

Let R be the radius of curvature of the surface, $(\rho = R, \varphi = 0)$, then equation (9.1.44) gives

$$-\frac{1}{R} = \frac{L(da)^2 + 2M\,da\,d\beta + N(d\beta)^2}{A^2(da)^2 + 2AB\cos\chi\,da\,d\beta + B^2(d\beta)^2}$$

(9.1.46)

In the above the negative sign implies that the vector $\vec{n}$ should be directed toward the convexity of the normal sections whose curvatures are positive.

The first quadratic form given in equation (9.1.8a)

$$I = A^2(da)^2 + 2AB\cos\chi\,da\,d\beta + B^2(d\beta)^2$$

(9.1.8a)

determines the lengths of the curves on the surface and the angles at which they intersect and the second quadratic form in equation (9.1.45) is closely connected with the curvatures of the curves on the surface. The curvature of the surface is defined in accordance to the following:

$LN - M^2 > 0$ positive Gaussian curvature

$LN - M^2 < 0$ negative Gaussian curvature

$LN - M^2 = 0$ zero Gaussian curvature

(9.1.47)

Dupin's Indicatrix and Gaussian Curvature

Consider a point P on a given surface and pass through it a plane E which is tangent to the surface. Next, cut the surface with a plane parallel to E arbitrarily close to it. The intersection of this plane with the surface gives a curve, which when projected on to plane E gives a curve S. Let us place a Cartesian coordinate system ξ, η at point P on the plane E, with their axes parallel to $\vec{M}_{,a}, \vec{M}_{,\beta}$, then the equation of the curve S can be written as

$$\frac{L}{A^2}\xi^2 + 2\frac{M}{AB}\xi\eta + \frac{N}{B^2}\eta^2 = \text{constant}$$

(9.1.48)

Dupin's Indicatrix is defined by

$$\frac{L}{A^2}\xi^2 + 2\frac{M}{AB}\xi\eta + \frac{N}{B^2}\eta^2 = \pm 1$$

(9.1.49)

and can be constructed for any surface where a unique tangent plane exists.

Equation (9.1.49) is a second degree curve. The directions of conjugate diameters, principal directions and the directions of the asymptotes of the Indicatrix are called *conjugate directions, principal directions* and *asymptotic directions* on the surface respectively. Two families of curves whose tangents are conjugate to each other at every point are *conjugate net of curves* of the surface. The surface curves whose tangents coincide with the principal directions on the surface are called *lines of curvature*. Likewise, surface curves whose tangents at every point coincide with the asymptotic directions are called *asymptotic lines* of the surface. When the Gaussian curvature is negative, two asymptotic lines pass through the point on the surface, they coincide when the Gaussian curvature is zero and become imaginary when the Gaussian curvature is positive. For a surface referred in curvilinear coordinates with $a-$ and $\beta-$ curves forming a conjugate net, $\chi = \pi/2$; $M = 0$, Dupin's Indicatrix given in equation (9.1.49) simplifies to

$$\frac{L}{A^2}\xi^2 + \frac{N}{B^2}\eta^2 = \pm 1$$

(9.1.50)

Every surface can be referred to lines of curvature, i.e., the lines of curvature are used as curvilinear coordinates. R_1 and R_2 are *principal radii of curvature* for the normal section through the lines of curvature of the surface. One of them is a maximum and the other is a minimum of the values of R. The two principal radii of curvature can be determined by first setting β constant and then a constant in equation (9.1.46).

$$\frac{1}{R_1} = -\frac{L}{A^2}$$
$$\frac{1}{R_2} = -\frac{N}{B^2}$$

(9.1.51)

The radii of curvature of the normal sections cut along the coordinate lines are given by

432 Dynamics of Plates

$$\frac{1}{R_1'} = -\frac{L}{A^2}$$

$$\frac{1}{R_2'} = -\frac{N}{B^2}$$

$$\frac{1}{R_{12}} = \frac{M}{AB}$$

(9.1.52)

where R_{12} is called *radius of torsion*.

It can be shown that

$$\frac{1}{R_{12}} = \sin\chi \cos\chi \left(\frac{1}{R_1} - \frac{1}{R_2}\right)$$

(9.1.53)

The Gaussian curvature K of a surface is defined by the product of the principal curvatures

$$K = \frac{1}{R_1} \frac{1}{R_2}$$

(9.1.54)

which in general can be shown to be

$$K = \frac{LN - M^2}{A^2 B^2 \sin^2\chi}$$

(9.1.55)

9.2 STRAIN-DISPLACEMENT RELATIONS

Elastic Displacement

Fig. 9.2.1 shows the middle surface of a shell. The *elastic displacement* $\vec{U} = \vec{U}(\alpha, \beta)$ is resolved along the axes of basic trihedron to represent it as

$$\vec{U} = u \frac{\vec{M}_{,\alpha}}{A} + v \frac{\vec{M}_{,\beta}}{B} - w \vec{n}$$

(9.2.1)

where u, v and w are the displacement components. If (α, β) are the curvilinear coordinates of an orthogonal system, then u, v and w are the projections of the total displacement along the positive directions of (α, β) and the negative direction of the nomal to the surfae.

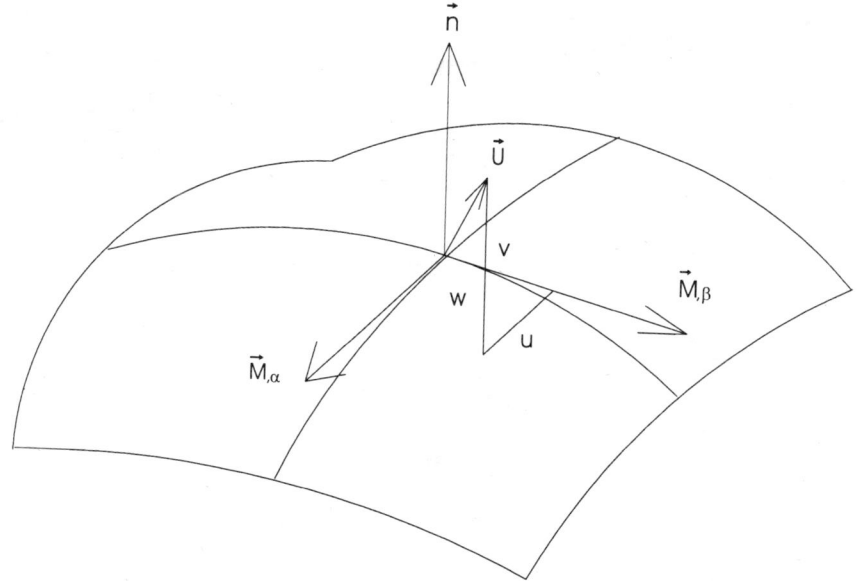

Fig. 9.2.1 Elastic Displacement

Elastic Rotations

Figure 9.2.2 shows the three planes $\vec{M},_\alpha \vec{n}$, $\vec{M},_\beta \vec{n}$ and $\vec{M},_\alpha \vec{M},_\beta$ separately. Let

1. γ_1 be the angle through which the element directed along the vector $\vec{M},_\alpha$ rotates towards the normal $\vec{n}$ to the surface in the plane $\vec{M},_\alpha \vec{n}$

2. γ_2 be the angle through which the element directed along the vector $\vec{M},_\beta$ rotates towards the normal $\vec{n}$ to the surface in the plane $\vec{M},_\beta \vec{n}$

3. ω_1 be the angle through which the element directed along the vector $\vec{M},_\alpha$ rotates towards $\vec{M},_\beta$ in the tangent plane $\vec{M},_\alpha \vec{M},_\beta$

4. ω_2 be the angle through which the element directed along the vector $\vec{M},_\beta$ rotates towards $\vec{M},_\alpha$ in the tangent plane $\vec{M},_\alpha \vec{M},_\beta$.

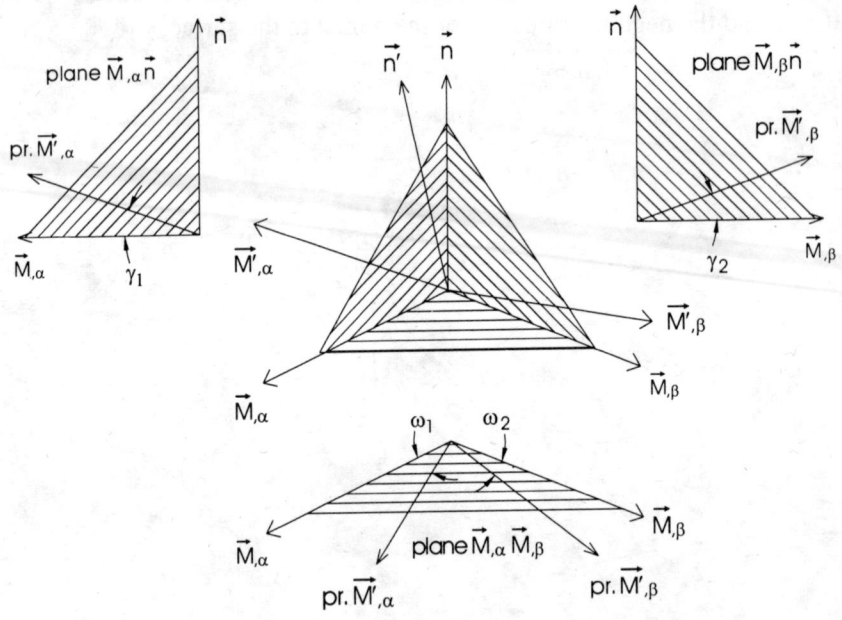

Fig. 9.2.2 Angles of Rotation

It can be seen that the change in the angle χ is given by $\omega = \omega_1 + \omega_2$ and $\delta = \frac{1}{2}(\omega_2 - \omega_1)$ gives a measure of the angle of rotation of an element of the middle surface around the normal $\vec{n}$ from $\vec{M}_{,\beta}$ to $\vec{M}_{,\alpha}$. A vector whose projection on $\vec{N}_{(\beta)}$ is equal to γ_1, on $\vec{N}_{(a)}$ is equal to $-\gamma_2$ and on vector $\vec{n}$ is equal to δ, is denoted by $\vec{\Omega}$ which is called the vector of *elastic rotation*, i.e.,

$$\vec{\Omega} \cdot \frac{\vec{N}_{(\beta)}}{B} = \gamma_1$$

$$\vec{\Omega} \cdot \frac{\vec{N}_{(a)}}{A} = -\gamma_2$$

$$\vec{\Omega} \cdot \vec{n} = \delta$$

(9.2.2)

The vector of elastic rotation can be expressed in the basic trihedron axes by using (9.1.17) as

$$\vec{\Omega} = -\frac{\gamma_2}{\sin\chi}\frac{\vec{M},_a}{A} + \frac{\gamma}{\sin 1\chi}\frac{\vec{M},_\beta}{B} + \delta\vec{n}$$

(9.2.3)

Let the *deformed surface* be denoted as $\vec{M'}$, so that

$$\vec{M'} = \vec{M} + \vec{U}$$

(9.2.4)

Then,

$$\vec{M'},_a = \vec{M},_a + \vec{U},_a$$
$$\vec{M'},_\beta = \vec{M},_\beta + \vec{U},_\beta$$

(9.2.5)

$$\vec{M'},_a \cdot \vec{n} = \vec{M},_a \cdot \vec{n} + \vec{U},_a \cdot \vec{n} = \vec{U},_a \cdot \vec{n}$$
$$\vec{M'},_\beta \cdot \vec{n} = \vec{M},_\beta \cdot \vec{n} + \vec{U},_\beta \cdot \vec{n} = \vec{U},_\beta \cdot \vec{n}$$

(9.2.6)

From Fig. 9.2.2, it can be seen that the angle between $\vec{M'},_a$ and $\vec{n}$ is $\pi/2 - \gamma_1$, and for small deformations, therefore

$$\vec{M'},_a \cdot \vec{n} = A' \sin\gamma_1$$
$$\simeq A\gamma_1$$

(9.2.7)

Similarly,

$$\vec{M'},_\beta \cdot \vec{n} = B' \sin\gamma_2$$
$$\simeq B\gamma_2$$

(9.2.7a)

436 Dynamics of Plates

Hence, equations (9.2.6) give

$$\frac{\vec{U}_{,\alpha}}{A} \cdot \vec{n} = \gamma_1$$

$$\frac{\vec{U}_{,\beta}}{B} \cdot \vec{n} = \gamma_2 \qquad (9.2.8)$$

Making use of the above result, we have from (9.2.2),

$$\frac{\vec{U}_{,\alpha}}{A} \cdot \vec{n} - \vec{\Omega} \cdot \frac{\vec{N}_{(\beta)}}{B} = 0$$

$$\frac{\vec{U}_{,\beta}}{B} \cdot \vec{n} + \vec{\Omega} \cdot \frac{\vec{N}_{(\alpha)}}{A} = 0 \qquad (9.2.9)$$

Proceeding in a similar manner,

$$\frac{\vec{M}'_{,\alpha}}{A'} \cdot \frac{\vec{N}_{(\beta)}}{B} = \frac{\vec{M}_{,\alpha}}{A'} \cdot \frac{\vec{N}_{(\beta)}}{B} + \frac{\vec{U}_{,\alpha}}{A'} \cdot \frac{\vec{N}_{(\beta)}}{B} = \frac{\vec{U}_{,\alpha}}{A'} \cdot \frac{\vec{N}_{(\beta)}}{B}$$

$$\frac{\vec{M}'_{,\beta}}{B'} \cdot \frac{\vec{N}_{(\alpha)}}{A} = \frac{\vec{M}_{,\beta}}{B'} \cdot \frac{\vec{N}_{(\alpha)}}{A} + \frac{\vec{U}_{,\beta}}{B'} \cdot \frac{\vec{N}_{(\alpha)}}{A} = \frac{\vec{U}_{,\beta}}{B'} \cdot \frac{\vec{N}_{(\alpha)}}{A} \qquad (9.2.10)$$

Noting

$$\frac{\vec{M}'_{,\alpha}}{A'} \cdot \frac{\vec{N}_{(\beta)}}{B} = \cos\left(\frac{\pi}{2} - \omega_1\right) \simeq \omega_1$$

$$\frac{\vec{M}'_{,\beta}}{B'} \cdot \frac{\vec{N}_{(\alpha)}}{A} = \cos\left(\frac{\pi}{2} - \omega_2\right) \simeq \omega_2 \qquad (9.2.11)$$

equations (9.2.10) give

$$\frac{\vec{U}_{,\alpha}}{A} \cdot \frac{\vec{N}_{(\beta)}}{B} = \omega_1$$

$$\frac{\vec{U},_{\beta}}{B} \cdot \frac{\vec{N}_{(a)}}{A} = \omega_2$$

(9.2.12)

Therefore

$$\frac{\vec{U},_{\beta}}{B} \cdot \frac{\vec{N}_{(a)}}{A} - \frac{\vec{U},_{a}}{A} \cdot \frac{\vec{N}_{(\beta)}}{B} = \omega_2 - \omega_1$$

$$= 2\delta$$

(9.2.13)

Making use of the result in (9.2.2), the above quation (9.2.13) gives

$$\frac{\vec{U},_{\beta}}{B} \cdot \frac{\vec{N}_{(a)}}{A} - \frac{\vec{U},_{a}}{A} \cdot \frac{\vec{N}_{(\beta)}}{B} - 2\vec{\Omega} \cdot \vec{n} = 0$$

(9.2.14)

Now, let $\vec{n}'$ be the unit vector normal to the deformed surface and let

$$\vec{m} = \vec{n}' - \vec{n}$$

(9.2.15)

Since the angle between the normals of the deformed and undeformed surfaces is small, we can write

$$\vec{n} \cdot \vec{n}' = 1$$

(9.2.16)

From Fig. 9.2.2, we note that

$$\frac{\vec{M},_{a}}{A} \cdot \vec{n}' = \cos\left(\frac{\pi}{2} + \gamma_1\right) \simeq -\gamma_1$$

$$\frac{\vec{M},_{\beta}}{B} \cdot \vec{n}' = \cos\left(\frac{\pi}{2} + \gamma_2\right) \simeq -\gamma_2$$

(9.2.17)

Making use of (9.1.18) and (9.1.21), and the above equations (9.2.16) and (9.2.17), we can write $\vec{n}'$ as

$$\vec{n}' = \vec{n}' \cdot \frac{\vec{M}_{,a}}{A} \frac{1}{\sin\chi} \frac{\vec{N}_{(a)}}{A} + \vec{n}' \cdot \frac{\vec{M}_{,\beta}}{B} \frac{1}{\sin\chi} \frac{\vec{N}_{(\beta)}}{B} + (\vec{n}' \cdot \vec{n})\vec{n}$$

$$= -\gamma_1 \frac{1}{\sin\chi} \frac{\vec{N}_{(a)}}{A} - \gamma_2 \frac{1}{\sin\chi} \frac{\vec{N}_{(\beta)}}{B} + \vec{n}$$

(9.2.18)

We can now obtain

$$\vec{m} = \vec{n}' - \vec{n}$$

$$= -\frac{\gamma_1}{\sin\chi} \frac{\vec{N}_{(a)}}{A} - \frac{\gamma_2}{\sin\chi} \frac{\vec{N}_{(\beta)}}{B}$$

(9.2.19)

Now, taking the vector product, see (9.2.3),

$$\vec{n} \times \vec{\Omega} = -\frac{\gamma_2}{\sin\chi}\left(\vec{n} \times \frac{\vec{M}_{,a}}{A}\right) + \frac{\gamma_1}{\sin\chi}\left(\vec{n} \times \frac{\vec{M}_{,\beta}}{B}\right) + \delta(\vec{n} \times \vec{n})$$

$$= -\frac{\gamma_2}{\sin\chi} \frac{\vec{N}_{(\beta)}}{B} - \frac{\gamma_1}{\sin\chi} \frac{\vec{N}_{(a)}}{A}$$

$$= \vec{m}$$

(9.2.20)

In-plane Strains

Let the undeformed surface $\vec{M}$ be referred to an orthogonal system of curvilinear coordinates, then an element of arc length is given by

$$(ds)^2 = A^2(da)^2 + B^2(d\beta)^2$$

(9.2.21)

For the deformed surface $\vec{M}'$, the corresponding arc length is

$$(ds')^2 = A'^2(da)^2 + 2A'B'\cos\chi'\, da\, d\beta + B'^2(d\beta)^2$$

(9.2.22)

where

$$A'^2 = \left(\vec{M'}_{,\alpha}\right)^2$$

$$B'^2 = \left(\vec{M'}_{,\beta}\right)^2$$

$$A'B'\cos\chi' = \vec{M'}_{,\alpha} \cdot \vec{M'}_{,\beta}$$

(9.2.23)

The *in-plane strain components* referred to the orthogonal system of coordinates are defined by

$$\varepsilon_{\alpha\alpha} = \frac{ds'_\alpha - ds_\alpha}{ds_\alpha}$$

$$\varepsilon_{\beta\beta} = \frac{ds'_\beta - ds_\beta}{ds_\beta}$$

$$\varepsilon_{\alpha\beta} = \omega = \frac{\pi}{2} - \chi' = -\delta\chi$$

(9.2.24)

where ds_α, ds_β are the differentials of the arc lengths of the $\alpha-$ and $\beta-$ lines on the undeformed surface and ds'_α, ds'_β are the differentials of the arc lengths of the $\alpha-$ and $\beta-$ lines on the deformed surface respectively.

In (9.2.21) and (9.2.22) let β be set constant first and then α be set constant, then

$$ds_\alpha = A\,d\alpha$$
$$ds_\beta = B\,d\beta$$
$$ds'_\alpha = A'\,d\alpha$$
$$ds'_\beta = B'\,d\beta$$

(9.2.25)

Therefore, from (9.2.24)

$$\varepsilon_{\alpha\alpha} = \frac{A' - A}{A}$$

$$\varepsilon_{\beta\beta} = \frac{B' - B}{B}$$

(9.2.26)

i.e.,

$$A' = A(1+\varepsilon_{aa})$$
$$B' = B(1+\varepsilon_{\beta\beta})$$

(9.2.27)

Equations (9.2.5) can now be written with the help of the above equations (9.2.27) as

$$(1+\varepsilon_{aa})\frac{\vec{M'}_{,a}}{A'} = \frac{\vec{M}_{,a}}{A} + \frac{\vec{U}_{,a}}{A}$$

$$(1+\varepsilon_{\beta\beta})\frac{\vec{M'}_{,\beta}}{B'} = \frac{\vec{M}_{,\beta}}{B} + \frac{\vec{U}_{,\beta}}{B}$$

(9.2.28)

Taking scalar product of the first of the above equations (9.2.28) with $\frac{\vec{M}_{,a}}{A}$, we have

$$(1+\varepsilon_{aa})\frac{\vec{M'}_{,a}}{A'} \cdot \frac{\vec{M}_{,a}}{A} = \frac{\vec{M}_{,a}}{A} \cdot \frac{\vec{M}_{,a}}{A} + \frac{\vec{U}_{,a}}{A} \cdot \frac{\vec{M}_{,a}}{A}$$

$$(1+\varepsilon_{aa})\cos\omega_1 = 1 + \frac{\vec{U}_{,a}}{A} \cdot \frac{\vec{M}_{,a}}{A}$$

(9.2.29)

which for small deformations gives

$$\varepsilon_{aa} = \frac{\vec{U}_{,a}}{A} \cdot \frac{\vec{M}_{,a}}{A}$$

(9.2.30)

In a similar manner

$$\varepsilon_{\beta\beta} = \frac{\vec{U}_{,\beta}}{B} \cdot \frac{\vec{M}_{,\beta}}{B}$$

(9.2.31)

For the shear strain, we have from (9.2.12)

$$\omega = \omega_1 + \omega_2 = -\delta\chi$$

$$= \frac{\vec{U}_{,a}}{A} \cdot \frac{\vec{N}_{(\beta)}}{B} + \frac{\vec{U}_{,\beta}}{B} \cdot \frac{\vec{N}_{(a)}}{A}$$

(9.2.32)

In orthogonal coordinates, the above becomes

$$\omega = -\delta\chi = \frac{\vec{U}_{,a}}{A} \cdot \frac{\vec{M}_{,\beta}}{B} + \frac{\vec{U}_{,\beta}}{B} \cdot \frac{\vec{M}_{,a}}{A}$$

(9.2.32)

Replacing $\vec{N}_{(a)}, \vec{N}_{(\beta)}$ by $\vec{M}_{,a}, \vec{M}_{,\beta}$ in equations (9.2.12), we can write the relations in (9.2.8), (9.2.12), (9.2.29) and (9.2.30) for the auxiliary components of the vectors $\vec{U}_{,a}, \vec{U}_{,\beta}$ as

$$\left|\frac{\vec{U}_{,a}}{A}\right|'_a = \frac{\varepsilon_{aa}}{\sin\chi}; \quad \left|\frac{\vec{U}_{,a}}{A}\right|'_\beta = \frac{\omega_1}{\sin\chi} - \varepsilon_{aa}\cot\chi; \quad \left|\frac{\vec{U}_{,a}}{A}\right|_n = \gamma_1$$

$$\left|\frac{\vec{U}_{,\beta}}{B}\right|'_a = \frac{\omega_2}{\sin\chi} - \varepsilon_{\beta\beta}\cot\chi; \quad \left|\frac{\vec{U}_{,\beta}}{B}\right|'_\beta = \frac{\varepsilon_{\beta\beta}}{\sin\chi}; \quad \left|\frac{\vec{U}_{,\beta}}{B}\right|_n = \gamma_2$$

(9.2.33)

In terms of the basic vectors, the above are

$$\left|\frac{\vec{U}_{,a}}{A}\right|_a = \varepsilon_{aa} - \omega_1\cot\chi; \quad \left|\frac{\vec{U}_{,a}}{A}\right|_\beta = \frac{\omega_1}{\sin\chi}; \quad \left|\frac{\vec{U}_{,a}}{A}\right|_n = \gamma_1$$

$$\left|\frac{\vec{U}_{,\beta}}{B}\right|_a = \frac{\omega_2}{\sin\chi}; \quad \left|\frac{\vec{U}_{,\beta}}{B}\right|_\beta = \varepsilon_{\beta\beta} - \omega_2\cot\chi; \quad \left|\frac{\vec{U}_{,\beta}}{B}\right|_n = \gamma_2$$

(9.2.34)

Hence

$$\frac{\vec{U},_a}{A} = (\varepsilon_{aa} - \omega_1 \cot\chi)\frac{\vec{M},_a}{A} + \frac{\omega_1}{\sin\chi}\frac{\vec{M},_\beta}{B} + \gamma_1 \vec{n}$$

$$\frac{\vec{U},_\beta}{B} = \frac{\omega_2}{\sin\chi}\frac{\vec{M},_a}{A} + (\varepsilon_{\beta\beta} - \omega_2 \cot\chi)\frac{\vec{M},_\beta}{B} + \gamma_2 \vec{n}$$

(9.2.35)

For the middle surface referred to orthogonal coordinates, the above simplifies to

$$\frac{\vec{U},_a}{A} = \varepsilon_{aa}\frac{\vec{M},_a}{A} + \omega_1\frac{\vec{M},_\beta}{B} + \gamma_1 \vec{n}$$

$$\frac{\vec{U},_\beta}{B} = \omega_2\frac{\vec{M},_a}{A} + \varepsilon_{\beta\beta}\frac{\vec{M},_\beta}{B} + \gamma_2 \vec{n}$$

(9.2.36)

Bending Strains

We extend the basic assumptions of classical beam or plate theory and assume that elements which are normal to the middle surface before deformation remain normal to the deformed middle surface during the process of deformation and are not subjected to any stretching. The points of the three dimensional domain is taken by the vector equation

$$\vec{M^*} = \vec{M} + z\vec{n}$$

(9.2.37)

The vector of the deformed three dimensional domain is expressed as

$$\vec{M^{*'}} = \vec{M'} + z\vec{n'}$$

(9.2.37a)

where, as before, $\vec{n'}$ is the unit normal to the deformed middle surface.

Let the middle surface, as before, be assumed to be expressed in an orthogonal system of coordinates. For different values of z, we obtain a family of equidistant similar surfaces. Differentiating equation (9.2.37)

$$\vec{M^*}_{,\alpha} = \vec{M}_{,\alpha} + z\vec{n}_{,\alpha}$$
$$\vec{M^*}_{,\beta} = \vec{M}_{,\beta} + z\vec{n}_{,\beta}$$

(9.2.38)

From the above equations (9.2.38) we can write

$$\vec{M^*}_{,\alpha} \cdot \vec{M^*}_{,\beta} = \vec{M}_{,\alpha} \cdot \vec{M}_{,\beta} + z\left(\vec{M}_{,\alpha} \cdot \vec{n}_{,\beta} + \vec{M}_{,\beta} \cdot \vec{n}_{,\alpha}\right) + z^2 \vec{n}_{,\alpha} \cdot \vec{n}_{,\beta}$$

(9.2.39)

Now, the conditions under which the surfaces of equation (9.2.37) will all be referred to orthogonal coordinates are given by

$$\vec{M}_{,\alpha} \cdot \vec{M}_{,\beta} = AB\cos\chi = 0$$
$$\vec{M}_{,\alpha} \cdot \vec{n}_{,\beta} + \vec{M}_{,\beta} \cdot \vec{n}_{,\alpha} = -2\vec{M} = 0$$
$$\vec{n}_{,\alpha} \cdot \vec{n}_{,\beta} = 0$$

(9.2.40)

Therefore, we find that the middle surface should be referred to the lines of the curvature. Assuming that this condition is satisfied, the differential of arc length on this surface can be written as

$$ds^{*2} = A^{*2}(d\alpha)^2 + B^{*2}(d\beta)^2$$

(9.2.41)

where

$$A^{*2} = \left(\vec{M^*}_{,\alpha}\right)^2 = \left(\vec{M}_{,\alpha}\right)^2 + 2z\vec{M}_{,\alpha} \cdot \vec{n}_{,\alpha} + z^2\left(\vec{n}_{,\alpha}\right)^2$$
$$B^{*2} = \left(\vec{M^*}_{,\beta}\right)^2 = \left(\vec{M}_{,\beta}\right)^2 + 2z\vec{M}_{,\beta} \cdot \vec{n}_{,\beta} + z^2\left(\vec{n}_{,\beta}\right)^2$$

(9.2.42)

444 Dynamics of Plates

Since, for a surface referred to the lines of curvature

$$\left(\vec{M},_a\right)^2 = A^2; \vec{M},_a \cdot \vec{n},_a = -L = \frac{A^2}{R_1}; \left(\vec{n},_a\right)^2 = \left(\frac{A}{R_1}\frac{\vec{M},_a}{A}\right)^2 = \frac{A^2}{R_1^2}$$

$$\left(\vec{M},_\beta\right)^2 = B^2; \vec{M},_\beta \cdot \vec{n},_\beta = -N = \frac{B^2}{R_2}; \left(\vec{n},_\beta\right)^2 = \left(\frac{B}{R_2}\frac{\vec{M},_\beta}{B}\right)^2 = \frac{B^2}{R_2^2}$$

$$(9.2.43)$$

equations (9.2.42) give

$$A^* = A\left(1 + \frac{z}{R_1}\right)$$

$$B^* = B\left(1 + \frac{z}{R_2}\right)$$

$$\chi^* = \frac{\pi}{2}$$

$$(9.2.44)$$

Further using (9.2.43), equations (9.2.38) can be written as

$$\vec{M}^*,_a = \vec{M},_a\left(1 + \frac{z}{R_1}\right)$$

$$\vec{M}^*,_\beta = \vec{M},_\beta\left(1 + \frac{z}{R_2}\right)$$

$$(9.2.45)$$

so that

$$\frac{\vec{M}^*,_a}{A^*} = \frac{\vec{M},_a}{A}$$

$$\frac{\vec{M}^*,_\beta}{B^*} = \frac{\vec{M},_\beta}{B}$$

$$\vec{n}^* = \vec{n}$$

Now, we can write the tangential deformation of the equidistant surfaces of (9.2.37) from (9.2.29), (9.2.30) and (9.2.32) as

$$\varepsilon_{\alpha\alpha}^* = \frac{\vec{U}^*_{,\alpha}}{A^*} \cdot \frac{\vec{M}^*_{,\alpha}}{A^*} = \frac{\vec{U}^*_{,\alpha}}{A\left(1+\frac{z}{R_1}\right)} \cdot \frac{\vec{M}_{,\alpha}}{A}$$

$$\varepsilon_{\beta\beta}^* = \frac{\vec{U}^*_{,\beta}}{B^*} \cdot \frac{\vec{M}^*_{,\beta}}{B^*} = \frac{\vec{U}^*_{,\beta}}{B\left(1+\frac{z}{R_2}\right)} \cdot \frac{\vec{M}_{,\beta}}{B}$$

$$\omega^* = \frac{\vec{U}^*_{,\alpha}}{A^*} \cdot \frac{\vec{M}^*_{,\beta}}{B^*} + \frac{\vec{U}^*_{,\beta}}{B^*} \cdot \frac{\vec{M}^*_{,\alpha}}{A^*}$$

$$= \frac{\vec{U}^*_{,\alpha}}{A\left(1+\frac{z}{R_1}\right)} \cdot \frac{\vec{M}_{,\beta}}{B} + \frac{\vec{U}^*_{,\beta}}{B\left(1+\frac{z}{R_2}\right)} \cdot \frac{\vec{M}_{,\alpha}}{A}$$

(9.2.46)

where $\vec{U}^*$ is the elastic displacement of the equidistant surface given by

$$\vec{U}^* = \vec{M}^{*\prime} - \vec{M}^*$$
$$= \vec{M}' + z\vec{n}' - \vec{M} - z\vec{n}$$
$$= \vec{U} + z\vec{m}$$

(9.2.47)

We can now write equations (9.2.46) as

$$\varepsilon_{\alpha\alpha}^* = \frac{1}{\left(1+\frac{z}{R_1}\right)} \left(\frac{\vec{U}_{,\alpha}}{A} \cdot \frac{\vec{M}_{,\alpha}}{A} + z \frac{\vec{m}_{,\alpha}}{A} \cdot \frac{\vec{M}_{,\alpha}}{A} \right)$$

$$\varepsilon_{\beta\beta}^* = \frac{1}{\left(1+\frac{z}{R_2}\right)} \left(\frac{\vec{U}_{,\beta}}{B} \cdot \frac{\vec{M}_{,\beta}}{B} + z \frac{\vec{m}_{,\beta}}{B} \cdot \frac{\vec{M}_{,\beta}}{B} \right)$$

$$\omega^* = \frac{1}{\left(1+\frac{z}{R_1}\right)} \left(\frac{\vec{U}_{,\alpha}}{A} \cdot \frac{\vec{M}_{,\beta}}{B} + z \frac{\vec{m}_{,\alpha}}{A} \cdot \frac{\vec{M}_{,\beta}}{B} \right)$$

$$+ \frac{1}{\left(1+\frac{z}{R_2}\right)} \left(\frac{\vec{U}_{,\beta}}{B} \cdot \frac{\vec{M}_{,\alpha}}{A} + z \frac{\vec{m}_{,\beta}}{B} \cdot \frac{\vec{M}_{,\alpha}}{A} \right)$$

(9.2.48)

446 Dynamics of Plates

Replacing $\vec{m}$ by $\vec{n} \times \vec{\Omega}$ in the above equation (9.2.48) and after making some transformations, we can arrive at the following.

$$\frac{\vec{m},_{\alpha}}{A} \cdot \frac{\vec{M},_{\alpha}}{A} = -\frac{\vec{M},_{\beta}}{B} \cdot \frac{\vec{\Omega},_{\alpha}}{A}$$

$$\frac{\vec{m},_{\beta}}{B} \cdot \frac{\vec{M},_{\beta}}{B} = \frac{\vec{M},_{\alpha}}{A} \cdot \frac{\vec{\Omega},_{\beta}}{B}$$

$$\frac{\vec{m},_{\alpha}}{A} \cdot \frac{\vec{M},_{\beta}}{B} = -\frac{\delta}{R_1} + \frac{\vec{M},_{\alpha}}{A} \cdot \frac{\vec{\Omega},_{\alpha}}{A}$$

$$\frac{\vec{m},_{\beta}}{B} \cdot \frac{\vec{M},_{\alpha}}{A} = \frac{\delta}{R_2} - \frac{\vec{M},_{\beta}}{B} \cdot \frac{\vec{\Omega},_{\alpha}}{A}$$

(9.2.49)

Now, let

$$\chi_1 = -\frac{\vec{M},_{\beta}}{B} \cdot \frac{\vec{\Omega},_{\alpha}}{A}$$

$$\chi_2 = \frac{\vec{M},_{\alpha}}{A} \cdot \frac{\vec{\Omega},_{\beta}}{B}$$

$$\tau^{(1)} = \frac{\vec{M},_{\alpha}}{A} \cdot \frac{\vec{\Omega},_{\alpha}}{A}$$

$$\tau^{(2)} = \frac{\vec{M},_{\beta}}{B} \cdot \frac{\vec{\Omega},_{\alpha}}{A}$$

(9.2.50)

Making use of the above equations (9.2.49) and (9.2.50), equations (9.2.48) become

$$\varepsilon^*_{\alpha\alpha} = \frac{\varepsilon_{\alpha\alpha} + z\chi_1}{1 + \frac{z}{R_1}}$$

$$\varepsilon^*_{\beta\beta} = \frac{\varepsilon_{\beta\beta} + z\chi_2}{1 + \frac{z}{R_2}}$$

(9.2.51)

$$\omega^* = \frac{\omega + z\left(\tau^{(1)} - \tau^{(2)} + \frac{\omega}{2R_1} + \frac{\omega}{2R_2}\right) + z^2\left(\frac{\tau^{(1)}}{R_2} - \frac{\tau^{(2)}}{R_1}\right)}{\left(1 + \frac{z}{R_1}\right)\left(1 + \frac{z}{R_2}\right)}$$

$$= \frac{\omega + z\left(\tau^{(1)} - \tau^{(2)} + \frac{\omega}{2R_1} + \frac{\omega}{2R_2}\right)}{\left(1 + \frac{z}{R_1}\right)\left(1 + \frac{z}{R_2}\right)} = \frac{\omega + 2z\tau}{\left(1 + \frac{z}{R_1}\right)\left(1 + \frac{z}{R_2}\right)}$$

(9.2.52)

Equations (9.2.51) show that the equidistant surfaces experience stretching over and above that of the middle surface defined by the quantities $\varkappa_1, \varkappa_2$. In equation (9.2.52), z^2 term can be neglected for thin shells, and it shows that the additional shear in the equidistant surface is determined primarily by

$$2\tau = \tau^{(1)} - \tau^{(2)} + \frac{\omega}{2R_1} + \frac{\omega}{2R_2}$$

(9.2.53)

$\varkappa_1, \varkappa_2$ and τ are defined as *bending strains* of the shell. Here, we actually introduced four quantities $\varkappa_1, \varkappa_2$ and $\tau^{(1)}, \tau^{(2)}$ for the bending strains, whereas there are only three quantities needed. The coefficients of the second quadratic form during deformation are

$$\begin{aligned} L' &= -\vec{M}'_{,a} \cdot \vec{n}'_{,a} \\ &= -\left(\vec{M}_{,a} + \vec{U}_{,a}\right) \cdot \left(\vec{n}_{,a} + \vec{m}_{,a}\right) \\ &= -\vec{M}_{,a} \cdot \vec{n}_{,a} - \vec{U}_{,a} \cdot \vec{n}_{,a} - \vec{m}_{,a} \cdot \vec{M}_{,a} \end{aligned}$$

and similarly

$$\begin{aligned} M' &= -\vec{M}_{,a} \cdot \vec{n}_{,\beta} - \vec{U}_{,a} \cdot \vec{n}_{,\beta} - \vec{m}_{,\beta} \cdot \vec{M}_{,a} \\ &= -\vec{M}_{,\beta} \cdot \vec{n}_{,a} - \vec{U}_{,\beta} \cdot \vec{n}_{,a} - \vec{m}_{,a} \cdot \vec{M}_{,\beta} \\ N' &= -\vec{M}_{,\beta} \cdot \vec{n}_{,\beta} - \vec{U}_{,\beta} \cdot \vec{n}_{,\beta} - \vec{m}_{,\beta} \cdot \vec{M}_{,\beta} \end{aligned}$$

(9.2.54)

Let

$$\frac{\vec{V}_{(\alpha)}}{A} = \frac{\vec{U}_{,\alpha}}{A} + \vec{\Omega} \times \frac{\vec{M}_{,\alpha}}{A}$$

$$\frac{\vec{V}_{(\beta)}}{B} = \frac{\vec{U}_{,\beta}}{B} + \vec{\Omega} \times \frac{\vec{M}_{,\beta}}{B}$$

(9.2.55)

Making use of (9.2.3), we have

$$\vec{\Omega} \times \frac{\vec{M}_{,\alpha}}{A} = -\frac{\gamma_2}{\sin\chi}\left(\frac{\vec{M}_{,\alpha}}{A} \times \frac{\vec{M}_{,\alpha}}{A}\right) + \frac{\gamma_1}{\sin\chi}\left(\frac{\vec{M}_{,\beta}}{B} \times \frac{\vec{M}_{,\alpha}}{A}\right)$$

$$+ \delta\left(\vec{n} \times \frac{\vec{M}_{,\alpha}}{A}\right)$$

$$\vec{\Omega} \times \frac{\vec{M}_{,\beta}}{B} = -\frac{\gamma_2}{\sin\chi}\left(\frac{\vec{M}_{,\alpha}}{A} \times \frac{\vec{M}_{,\beta}}{B}\right) + \frac{\gamma_1}{\sin\chi}\left(\frac{\vec{M}_{,\beta}}{B} \times \frac{\vec{M}_{,\beta}}{B}\right)$$

$$+ \delta\left(\vec{n} \times \frac{\vec{M}_{,\beta}}{B}\right)$$

(9.2.56)

Expanding the above, we can show that

$$\vec{\Omega} \times \frac{\vec{M}_{,\alpha}}{A} = -\delta\cot\chi \frac{\vec{M}_{,\alpha}}{A} + \frac{\delta}{\sin\chi}\frac{\vec{M}_{,\beta}}{B} - \gamma_1 \vec{n}$$

$$\vec{\Omega} \times \frac{\vec{M}_{,\beta}}{B} = -\frac{\delta}{\sin\chi}\frac{\vec{M}_{,\alpha}}{A} + \delta\cot\chi\frac{\vec{M}_{,\beta}}{B} - \gamma_2 \vec{n}$$

(9.2.57)

Letting

$$\omega^{(1)} = \frac{1}{2}\omega\sin\chi + \varepsilon_{\alpha\alpha}\cos\chi$$

$$\omega^{(2)} = -\frac{1}{2}\omega\sin\chi - \varepsilon_{\beta\beta}\cos\chi$$

(9.2.58)

and using (9.2.57), we get from equations (9.2.55)

$$\frac{\vec{V}_{(\alpha)}}{A} = \frac{\varepsilon_{\alpha\alpha} - \omega^{(1)} \cos\chi}{\sin^2\chi} \frac{\vec{M}_{,\alpha}}{A} + \frac{\omega^{(1)} - \varepsilon_{\alpha\alpha} \cos\chi}{\sin^2\chi} \frac{\vec{M}_{,\beta}}{B}$$

$$\frac{\vec{V}_{(\beta)}}{B} = -\frac{\omega^{(2)} + \varepsilon_{\beta\beta} \cos\chi}{\sin^2\chi} \frac{\vec{M}_{,\alpha}}{A} + \frac{\varepsilon_{\beta\beta} + \omega^{(2)} \cos\chi}{\sin^2\chi} \frac{\vec{M}_{,\beta}}{B}$$

(9.2.59)

Making use of (9.2.55) and $\vec{m} = \vec{n} \times \vec{\Omega}$, equations (9.2.54) can be shown to be

$$L' = L + \left(\frac{L}{A\sin\chi} \frac{\vec{N}_{(\alpha)}}{A} + \frac{M}{B\sin\chi} \frac{\vec{N}_{(\beta)}}{B} \right) \cdot \vec{V}_{(\alpha)} + A^2 \frac{\vec{\Omega}_{,\alpha}}{A} \cdot \frac{\vec{N}_{(\beta)}}{B}$$

$$M' = M + \left(\frac{M}{A\sin\chi} \frac{\vec{N}_{(\alpha)}}{A} + \frac{N}{B\sin\chi} \frac{\vec{N}_{(\beta)}}{B} \right) \cdot \vec{V}_{(\alpha)} + AB \frac{\vec{\Omega}_{,\beta}}{B} \cdot \frac{\vec{N}_{(\beta)}}{B}$$

$$M' = M + \left(\frac{L}{A\sin\chi} \frac{\vec{N}_{(\alpha)}}{A} + \frac{M}{B\sin\chi} \frac{\vec{N}_{(\beta)}}{B} \right) \cdot \vec{V}_{(\beta)} - AB \frac{\vec{\Omega}_{,\alpha}}{A} \cdot \frac{\vec{N}_{(\alpha)}}{A}$$

$$N' = N + \left(\frac{M}{A\sin\chi} \frac{\vec{N}_{(\alpha)}}{A} + \frac{N}{B\sin\chi} \frac{\vec{N}_{(\beta)}}{B} \right) \cdot \vec{V}_{(\beta)} - B^2 \frac{\vec{\Omega}_{,\beta}}{B} \cdot \frac{\vec{N}_{(\alpha)}}{A}$$

(9.2.60)

Now, with the help of equations (9.2.50) and (9.2.59), the above equations (9.2.60) give the increments of the coefficients of the second quadratic form during the deformation as

$$\delta L = L' - L$$
$$= \frac{L}{\sin\chi}\left(\varepsilon_{\alpha\alpha}\sin\chi - \tfrac{1}{2}\omega\cos\chi\right) + \tfrac{1}{2}\omega\frac{AM}{B\sin\alpha} - \frac{A^2}{\sin\chi}(\varkappa_1 + \tau^{(1)}\cos\chi)$$

$$\delta M = M' - M$$
$$= \frac{M}{\sin\chi}\left(\varepsilon_{\alpha\alpha}\sin\chi - \tfrac{1}{2}\omega\cos\chi\right) + \tfrac{1}{2}\omega\frac{AN}{B\sin\alpha} + \frac{AB}{\sin\chi}(\tau^{(2)} - \varkappa_2\cos\chi)$$

$$= \frac{M}{\sin\chi}\left(\varepsilon_{\beta\beta}\sin\chi - \tfrac{1}{2}\omega\cos\chi\right) + \tfrac{1}{2}\omega\frac{BL}{A\sin\alpha} - \frac{AB}{\sin\chi}(\tau^{(1)} + \varkappa_1\cos\chi)$$

450 Dynamics of Plates

$$\delta N = N' - N$$

$$= \frac{N}{\sin\chi}\left(\varepsilon_{\beta\beta}\sin\chi - \frac{1}{2}\omega\cos\chi\right) + \frac{1}{2}\omega\frac{BM}{A\sin a} - \frac{B^2}{\sin\chi}(x_2 - \tau^{(2)}\cos\chi)$$

(9.2.61)

Note that there are two equations for δM in the above equation (9.2.61), since we introduced four quantities for the bending strains, equating them we have

$$\frac{M}{\sin\chi}\left(\varepsilon_{aa}\sin\chi - \frac{1}{2}\omega\cos\chi\right) + \frac{1}{2}\omega\frac{AN}{B\sin a} + \frac{AB}{\sin\chi}(\tau^{(2)} - x_2\cos\chi)$$

$$= \frac{M}{\sin\chi}\left(\varepsilon_{\beta\beta}\sin\chi - \frac{1}{2}\omega\cos\chi\right) + \frac{1}{2}\omega\frac{BL}{A\sin a} - \frac{AB}{\sin\chi}(\tau^{(1)} + x_1\cos\chi)$$

(9.2.62)

For the middle surface referred to the lines of curvature, $\chi = \frac{\pi}{2}, M = 0$, and making use of formulae for the principal radii of curvature, the above equation gives

$$\tau^{(2)} - \frac{\omega}{2R_2} = -\tau^{(1)} - \frac{\omega}{2R_1}$$

(9.2.63)

If we set

$$\tau^{(2)} = -\tau + \frac{\omega}{2R_2}$$

$$\tau^{(1)} = \tau - \frac{\omega}{2R_1}$$

(9.2.64)

then, equation (9.2.63) becomes an identity and also we satisfy the assumption made earlier in equation (9.2.53). The above equations (9.2.64) can be generalized for the case when the middle surface is refered to an aribitrary curvilinear coordinate system

$$\tau^{(1)} = \tau - x_1\cos\chi + \frac{1}{R_{12}}\left(\varepsilon_{\beta\beta}\sin\chi - \frac{1}{2}\omega\cos\chi\right) - \frac{1}{2}\frac{\omega}{R'_1}$$

$$\tau^{(2)} = -\tau + x_2\cos\chi - \frac{1}{R_{12}}\left(\varepsilon_{aa}\sin\chi - \frac{1}{2}\omega\cos\chi\right) + \frac{1}{2}\frac{\omega}{R'_2}$$

(9.2.65)

In the particular case, when the coordinate system is orthogonal, but not conjugate, then

$$\tau^{(1)} = \tau + \frac{\varepsilon_{\beta\beta}}{R_{12}} - \frac{1}{2}\frac{\omega}{R_1'}$$

$$\tau^{(2)} = -\tau - \frac{\varepsilon_{\alpha\alpha}}{R_{12}} + \frac{1}{2}\frac{\omega}{R_2'}$$

(9.2.66)

From (9.2.65), we can now generalize equation (9.2.53)

$$2\tau = \tau^{(1)} - \tau^{(2)} + (\varkappa_1 + \varkappa_2)\cos\chi + \frac{1}{R_{12}}[\omega\cos\chi - (\varepsilon_{\alpha\alpha} + \varepsilon_{\beta\beta})\sin\chi]$$
$$+ \frac{\omega}{2R_1'} + \frac{\omega}{2R_2'}$$

(9.2.67)

With the help of (9.2.65), the two middle formulae in (9.2.61) can be replaced by

$$\vec{\delta M} = \vec{M'} - \vec{M}$$
$$= -\frac{AB}{\sin\chi}\tau$$

(9.2.68)

Finally, the bending strains for the middle surface referred to the lines of curvature, are given by

$$\varkappa_1 = \frac{L - L'}{A^2} - \frac{\varepsilon_{\alpha\alpha}}{R_1}$$

$$\varkappa_2 = \frac{N - N'}{B^2} - \frac{\varepsilon_{\beta\beta}}{R_2}$$

$$\tau = -\frac{M'}{AB}$$

(9.2.69)

Now, let

$$\zeta_1 = \frac{\vec{\Omega}_{,a}}{A} \cdot \vec{n}$$

452 Dynamics of Plates

$$\zeta_2 = \frac{\vec{\Omega}_{,\beta}}{B} \cdot \vec{n}$$

(9.2.70)

With the help of (9.1.21), (9.2.50) and (9.2.70), we get the following expressions for the auxiliary components of $\vec{\Omega}_{,\alpha}$ and $\vec{\Omega}_{,\beta}$

$$\left|\frac{\vec{\Omega}_{,\alpha}}{A}\right|'_{\alpha} = \frac{\tau^{(1)}}{\sin\chi}; \quad \left|\frac{\vec{\Omega}_{,\alpha}}{A}\right|'_{\beta} = -\frac{\varkappa_1}{\sin\chi}; \quad \left|\frac{\vec{\Omega}_{,\alpha}}{A}\right|'_{n} = \zeta_1$$

$$\left|\frac{\vec{\Omega}_{,\beta}}{B}\right|'_{\alpha} = \frac{\varkappa_2}{\sin\chi}; \quad \left|\frac{\vec{\Omega}_{,\beta}}{B}\right|'_{\beta} = \frac{\tau^{(2)}}{\sin\chi}; \quad \left|\frac{\vec{\Omega}_{,\beta}}{B}\right|'_{n} = \zeta_2$$

(9.2.71)

The basic components of the above vectors are

$$\left|\frac{\vec{\Omega}_{,\alpha}}{A}\right|_{\alpha} = \frac{\tau^{(1)} + \varkappa_1 \cos\chi}{\sin^2\chi}; \quad \left|\frac{\vec{\Omega}_{,\alpha}}{A}\right|_{\beta} = -\frac{\varkappa_1 + \tau^{(1)} \cos\chi}{\sin^2\chi}; \quad \left|\frac{\vec{\Omega}_{,\alpha}}{A}\right|_{n} = \zeta_1$$

$$\left|\frac{\vec{\Omega}_{,\beta}}{B}\right|_{\alpha} = \frac{\varkappa_2 - \tau^{(2)} \cos\chi}{\sin^2\chi}; \quad \left|\frac{\vec{\Omega}_{,\beta}}{B}\right|_{\beta} = \frac{\tau^{(2)} - \varkappa_2 \cos\chi}{\sin^2\chi}; \quad \left|\frac{\vec{\Omega}_{,\beta}}{B}\right|_{n} = \zeta_2$$

(9.2.72)

Hence

$$\frac{\vec{\Omega}_{,\alpha}}{A} = \frac{\tau^{(1)} + \varkappa_1 \cos\chi}{\sin^2\chi} \frac{\vec{M}_{,\alpha}}{A} - \frac{\varkappa_1 + \tau^{(1)} \cos\chi}{\sin^2\chi} \frac{\vec{M}_{,\beta}}{B} + \zeta_1 \vec{n}$$

$$\frac{\vec{\Omega}_{,\beta}}{B} = \frac{\varkappa_2 - \tau^{(2)} \cos\chi}{\sin^2\chi} \frac{\vec{M}_{,\alpha}}{A} + \frac{\tau^{(2)} - \varkappa_2 \cos\chi}{\sin^2\chi} \frac{\vec{M}_{,\beta}}{B} + \zeta_2 \vec{n}$$

(9.2.73)

Strain Displacement Relations

Now we can obtain the strain displacement relations. Consider equation (9.2.30) giving the strain $\varepsilon_{\alpha\alpha}$ in vector form

$$\varepsilon_{\alpha\alpha} = \frac{\vec{U}_{,\alpha}}{A} \cdot \frac{\vec{M}_{,\alpha}}{A}$$

Since the middle surface is assumed to be referred to orthogonal coordinates, the above equation is

$$\varepsilon_{aa} = \left|\frac{\vec{U},a}{A}\right|_a$$

$$= \left|\frac{1}{A}\frac{\partial \vec{U}}{\partial a}\right|_a$$

$$= \frac{1}{A}\frac{\partial}{\partial a}\left|\vec{U}\right|_a + \frac{1}{AB}\frac{\partial A}{\partial \beta}\left|\vec{U}\right|_\beta + \frac{1}{R_1'}\left|\vec{U}\right|_n$$

(9.2.74)

In terms of the components of the elastic displacement, u, v and w resolved along the basic trihedron, the above equation (9.2.74) gives the strain displacement relation for ε_{aa}. Similarly, the scalar form of equations (9.2.31), (9.2.32), (9.2.13), (9.2.8), (9.2.50) and (9.2.70) can be obtained, which are given below.

In-Plane Strains:

$$\varepsilon_{aa} = \frac{1}{A}\frac{\partial u}{\partial a} + \frac{v}{AB}\frac{\partial A}{\partial \beta} - \frac{w}{R_1'}$$

$$\varepsilon_{\beta\beta} = \frac{1}{B}\frac{\partial v}{\partial \beta} + \frac{u}{AB}\frac{\partial B}{\partial a} - \frac{w}{R_2'}$$

$$\varepsilon_{a\beta} = \omega = \frac{A}{B}\frac{\partial}{\partial \beta}\left(\frac{u}{A}\right) + \frac{B}{A}\frac{\partial}{\partial a}\left(\frac{v}{B}\right) + \frac{2w}{R_{12}}$$

(9.2.75)

Angles of Rotation:

$$\omega_1 = \frac{1}{A}\frac{\partial v}{\partial a} - \frac{u}{AB}\frac{\partial A}{\partial \beta} + \frac{w}{R_{12}}$$

$$\omega_2 = \frac{1}{B}\frac{\partial u}{\partial \beta} - \frac{v}{AB}\frac{\partial B}{\partial a} + \frac{w}{R_{12}}$$

$$\delta = \frac{1}{2AB}\left[\frac{\partial}{\partial \beta}(Au) - \frac{\partial}{\partial a}(Bv)\right]$$

(9.2.76)

$$\gamma_1 = -\frac{1}{A}\frac{\partial w}{\partial \alpha} - \frac{u}{R_1'} + \frac{v}{R_{12}}$$

$$\gamma_2 = -\frac{1}{B}\frac{\partial w}{\partial \beta} - \frac{v}{R_2'} + \frac{u}{R_{12}}$$

(9.2.77)

Bending Strains:

$$\varkappa_1 = -\frac{1}{A}\frac{\partial \gamma_1}{\partial \alpha} - \frac{\gamma_2}{AB}\frac{\partial A}{\partial \beta} + \frac{\delta}{R_{12}}$$

$$\varkappa_2 = -\frac{1}{B}\frac{\partial \gamma_2}{\partial \beta} - \frac{\gamma_1}{AB}\frac{\partial B}{\partial \alpha} - \frac{\delta}{R_{12}}$$

$$\tau^{(1)} = -\frac{1}{A}\frac{\partial \gamma_2}{\partial \alpha} + \frac{\gamma_1}{AB}\frac{\partial A}{\partial \beta} + \frac{\delta}{R_1'}$$

$$\tau^{(2)} = \frac{1}{B}\frac{\partial \gamma_1}{\partial \beta} - \frac{\gamma_2}{AB}\frac{\partial B}{\partial \alpha} + \frac{\delta}{R_2'}$$

$$\zeta_1 = \frac{1}{A}\frac{\partial \delta}{\partial \alpha} + \frac{\gamma_2}{R_1'} + \frac{\gamma_1}{R_{12}}$$

$$\zeta_2 = \frac{1}{B}\frac{\partial \delta}{\partial \beta} - \frac{\gamma_1}{R_1'} - \frac{\gamma_2}{R_{12}}$$

(9.2.78)

$$\tau = \tau^{(1)} - \frac{\varepsilon_{\beta\beta}}{R_{12}} + \frac{\omega}{2R_1'}$$

$$= -\frac{1}{A}\frac{\partial \gamma_2}{\partial \alpha} + \frac{\gamma_1}{AB}\frac{\partial A}{\partial \beta} + \frac{\delta}{R_1'} - \frac{\varepsilon_{\beta\beta}}{R_{12}} + \frac{\omega}{2R_1'}$$

(9.2.79)

The bending strains are then represented by

$$\varkappa_1 = \kappa_{\alpha\alpha} = -\frac{1}{A}\frac{\partial \gamma_1}{\partial \alpha} - \frac{\gamma_2}{AB}\frac{\partial A}{\partial \beta} + \frac{\delta}{R_{12}}$$

$$\varkappa_2 = \kappa_{\beta\beta} = -\frac{1}{B}\frac{\partial \gamma_2}{\partial \beta} - \frac{\gamma_1}{AB}\frac{\partial B}{\partial \alpha} - \frac{\delta}{R_{12}}$$

$$\tau = \kappa_{\alpha\beta} = -\frac{1}{A}\frac{\partial \gamma_2}{\partial \alpha} + \frac{\gamma_1}{AB}\frac{\partial A}{\partial \beta} + \frac{\delta}{R_1'} - \frac{\varepsilon_{\beta\beta}}{R_{12}} + \frac{\varepsilon_{\alpha\beta}}{2R_1'}$$

(9.2.80)

9.3 FORCES AND MOMENTS

Consider the middle surface $\vec{M}(\alpha, \beta)$ of a shell, an arbitrary point of the three dimensional space can be represented by the vector equation

$$\vec{M^*} = \vec{M}(\alpha, \beta) + z\vec{n} \qquad (9.3.1)$$

where z varies between $-h$ to $+h$, in the thickness $2h$ of the shell. Consider a normal section cut along the curve Γ as shown in Fig. 9.3.1 and let the stress on this section be represented by the vector $\vec{\sigma}(\alpha, \beta, z)$. The force vector per unit length along this normal section can be written as

$$\vec{R}^{(\Gamma)} ds_\Gamma = \int_{-h}^{h} \vec{\sigma} \, ds_\Gamma^* dz \qquad (9.3.2)$$

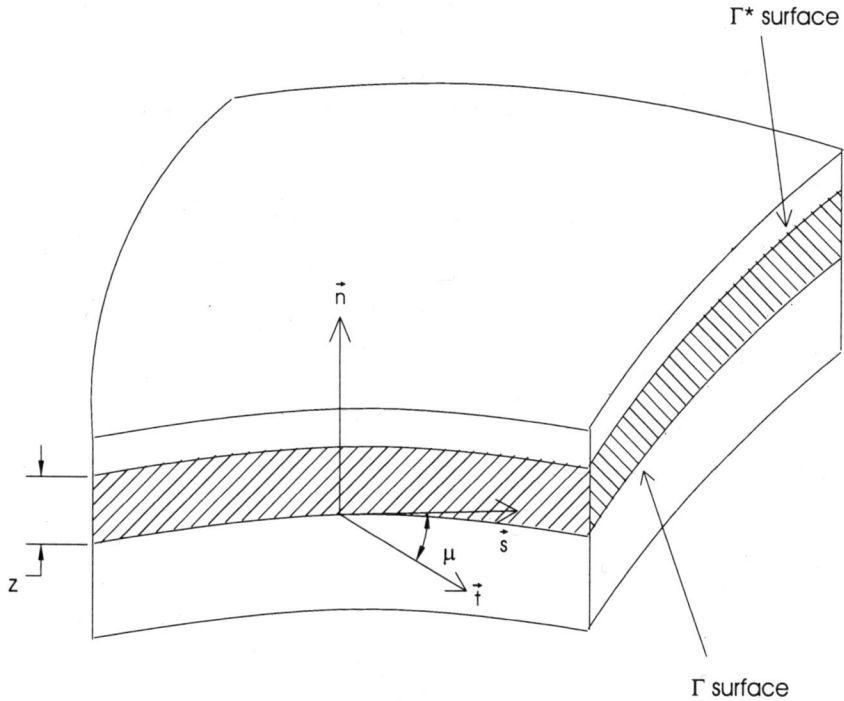

Fig. 9.3.1 Equidistant Surface

where ds_Γ and ds_Γ^* are the differentials of the arc lengths of the curves Γ and Γ^*, respectively. The moment per unit length along this normal section is

$$\vec{Q}^{(\Gamma)} ds_\Gamma = \int_{-h}^{h} (\vec{\sigma} \times z\vec{n}) ds_\Gamma^* dz$$

(9.3.3)

At each point on the curve Γ, a trihedron is set up as shown in Fig. 9.3.1 with $\vec{s}$ tangential to the curve, $\vec{t}$ lying in the tangent plane of the middle surface making an angle μ with $\vec{s}$ and $\vec{n}$ the normal, similar to the basic trihedron considered earlier. The force and moment vectors in (9.3.2) and (9.3.3) can be resolved to give

$$\vec{R}^{(\Gamma)} = -S\vec{s} + T\vec{t} + N\vec{n}$$
$$\vec{Q}^{(\Gamma)} = -G\vec{s} + H\vec{t}$$

(9.3.4)

In the above equations (9.3.4), S is the tangential shear force, T is the normal force, N is the transverse shear force, G is the bending moment and H is the twisting moment all expressed per unit length along the normal section under consideration. When Γ coincides with α- or β-curvilinear coordinates of the middle surface, then the force and moment vectors are denoted as $\vec{R}^{(\alpha)}, \vec{Q}^{(\alpha)}$ and $\vec{R}^{(\beta)}, \vec{Q}^{(\beta)}$ respectively. These force and moment vectors can be resolved along the basic trihedron to give

$$\vec{R}^{(\alpha)} = S_2 \frac{\vec{M}_\alpha}{A} - T_2 \frac{\vec{M}_\beta}{B} + N_2 \vec{n}$$

$$\vec{Q}^{(\alpha)} = G_2 \frac{\vec{M}_\alpha}{A} - H_2 \frac{\vec{M}_\beta}{B}$$

(9.3.5)

$$\vec{R}^{(\beta)} = -T_1 \frac{\vec{M}_\alpha}{A} - S_1 \frac{\vec{M}_\beta}{B} + N_1 \vec{n}$$

$$\vec{Q}^{(\beta)} = -H_1 \frac{\vec{M}_\alpha}{A} - G_1 \frac{\vec{M}_\beta}{B}$$

(9.3.6)

where suffix 2 is used in conjunction with a- curve and suffix 1 with β-curve respectively. These components are shown in Fig. 9.3.2. The ten coefficients $T_1, T_2, S_1, \cdots H_2$ in the above equations define the state of stress of the shell at any point on its middle surface.

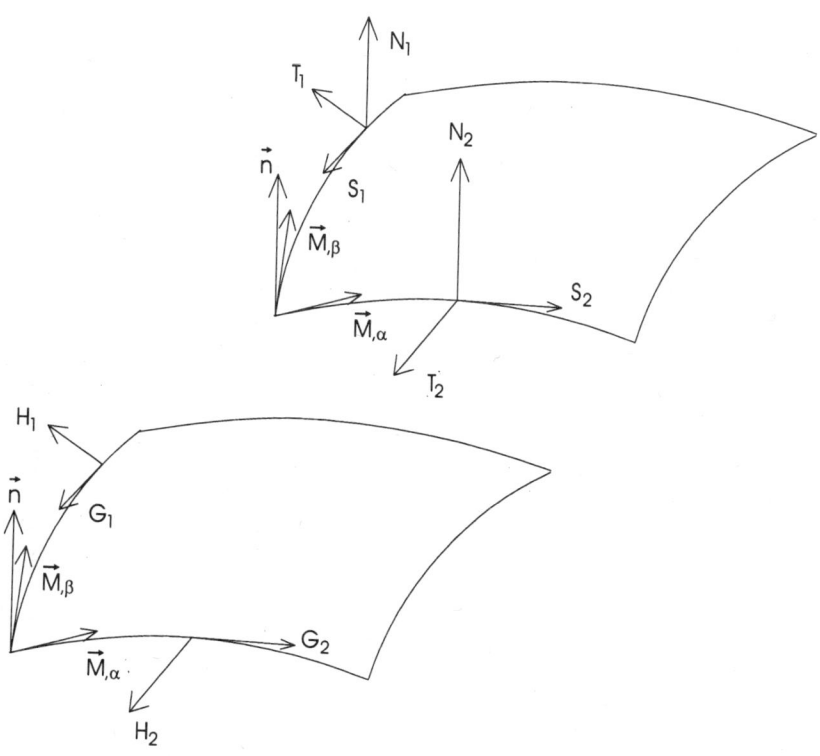

Fig. 9.3.2 Forces and Moments

To determine the forces and moments on any other section, consider an infinitesimally small triangle as shown in Fig. 9.3.3 of lengths ds_a, ds_β and ds_Γ for the sides ab, ac and bc respectively. The sides ab and ac are formed by a- and β-curves respectively and side bc encloses an angle λ with a- curve. The force acting on the side bc is represented by $\vec{R}(\Gamma)$ in the $\vec{s}\ \vec{t}\ \vec{n}$ trihedron as shown in Fig. 9.3.3. For equilibrium of the triangular element, we have

$$\vec{R}^{(a)} ds_a + \vec{R}^{(\beta)} ds_\beta + \vec{R}^{(\Gamma)} ds_\Gamma = 0$$

(9.3.7)

458 Dynamics of Plates

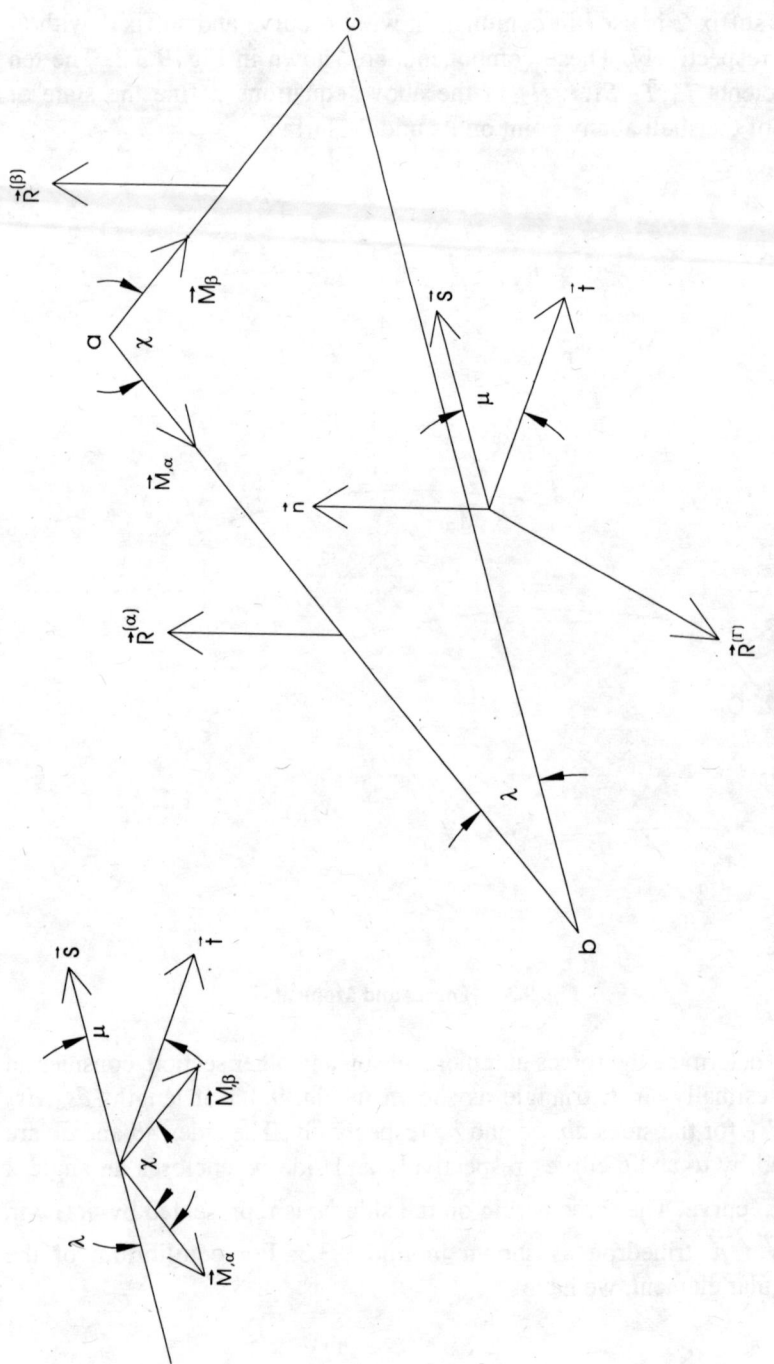

Fig. 9.3.3 Forces and Moments on an Oblique Section

From Fig. (9.3.3), we have

$$\frac{ds_\alpha}{\sin(\chi+\lambda)} = \frac{ds_\beta}{\sin\lambda} = \frac{ds_\Gamma}{\sin\chi}$$

(9.3.8)

Therefore

$$\vec{R}^{(\alpha)}\sin(\chi+\lambda) + \vec{R}^{(\beta)}\sin\lambda + \vec{R}^{(\Gamma)}\sin\chi = 0$$

(9.3.7a)

Making use of (9.3.4) to (9.3.6), the above equation (9.3.7a) can be rearranged as

$$\left(S_2\frac{\vec{M}_\alpha}{A} - T_2\frac{\vec{M}_\beta}{B} + N_2\vec{n}\right)\sin(\chi+\lambda)$$

$$+ \left(-T_1\frac{\vec{M}_\alpha}{A} - S_1\frac{\vec{M}_\beta}{B} + N_1\vec{n}\right)\sin\lambda - \left(S\vec{s} - T\vec{t} - N\vec{n}\right)\sin\chi$$

$$= 0$$

(9.3.7b)

Let us now introduce two auxiliary coordinates $\vec{s}'$, $\vec{t}'$ such that $\vec{s}'$, $\vec{t}'$ and $\vec{s}$, $\vec{t}'$ are orthogonal systems as shown in Fig. 9.3.3. The angles enclosed by vector $\vec{M}_{,\alpha}$ with $\vec{s}'$, $\vec{t}'$ vectors are $(\frac{3\pi}{2} - \lambda - \mu)$ and $(\frac{\pi}{2} - \lambda)$, and by the vector $\vec{M}_{,\beta}$ with $\vec{s}'$, $\vec{t}'$ vectors are $(\frac{3\pi}{2} - \lambda - \mu - \chi)$ and $(\chi - \frac{\pi}{2} + \lambda)$ respectively. Taking the scalar product of equation (9.3.7b) with $\vec{t}'$, $\vec{s}'$ and $\vec{n}$, we can arrive at the following relations for the force components,

$$T = T_1\frac{\sin^2\lambda}{\sin\mu\sin\chi} + T_2\frac{\sin^2(\lambda+\chi)}{\sin\mu\sin\chi} + (S_1 - S_2)\frac{\sin\lambda\sin(\lambda+\chi)}{\sin\mu\sin\chi}$$

$$S = T_1\frac{\sin(\mu+\lambda)\sin\lambda}{\sin\mu\sin\chi} + T_2\frac{\sin(\mu+\chi+\lambda)\sin(\lambda+\chi)}{\sin\mu\sin\chi}$$

$$+ S_1\frac{\sin(\mu+\chi+\lambda)\sin\lambda}{\sin\mu\sin\chi} - S_2\frac{\sin(\chi+\lambda)\sin(\mu+\lambda)}{\sin\mu\sin\chi}$$

$$N = -N_1 \frac{\sin \lambda}{\sin \chi} - N_2 \frac{\sin(\chi+\lambda)}{\sin \chi}$$

(9.3.9)

Similarly, from the equilibrium of moments of the triangular element in Fig. 9.3.3, we get

$$H = H_1 \frac{\sin^2 \lambda}{\sin \mu \sin \chi} + H_2 \frac{\sin^2(\lambda+\chi)}{\sin \mu \sin \chi} + (G_1 - G_2) \frac{\sin \lambda \sin(\lambda+\chi)}{\sin \mu \sin \chi}$$

$$G = H_1 \frac{\sin(\mu+\lambda)\sin \lambda}{\sin \mu \sin \chi} + H_2 \frac{\sin(\mu+\chi+\lambda)\sin(\lambda+\chi)}{\sin \mu \sin \chi}$$

$$+ G_1 \frac{\sin(\mu+\chi+\lambda)\sin \lambda}{\sin \mu \sin \chi} - G_2 \frac{\sin(\chi+\lambda)\sin(\mu+\lambda)}{\sin \mu \sin \chi}$$

(9.3.10)

If the middle surface is referred to an orthogonal system of curvilinear coordinates and also $\vec{t}$ and $\vec{s}$ are orthogonal, the above equations (9.3.9) and (9.3.10) simplify to

$$T = T_1 \sin^2 \lambda + T_2 \cos^2 \lambda + (S_1 - S_2) \sin \lambda \cos \lambda$$
$$S = (T_1 - T_2) \sin \lambda \cos \lambda - S_1 \sin^2 \lambda - S_2 \cos^2 \lambda$$
$$N = -N_1 \sin \lambda - N_2 \cos \lambda$$

(9.3.9a)

$$H = H_1 \sin^2 \lambda + H_2 \cos^2 \lambda + (G_1 - G_2) \sin \lambda \cos \lambda$$
$$G = (H_1 - H_2) \sin \lambda \cos \lambda - G_1 \sin^2 \lambda - G_2 \cos^2 \lambda$$

(9.3.10a)

Stresses

The strains given by equations (9.2.51) and the general relation including z^2 term of (9.2.52) can be written as

$$\varepsilon_{aa}^* = \frac{\varepsilon_{aa} + z\varkappa_1}{1 + \frac{z}{R_1}}$$

$$\varepsilon^*_{\beta\beta} = \frac{\varepsilon_{\beta\beta} + z\chi_2}{1+\frac{z}{R_2}}$$

$$\varepsilon^*_{\alpha\beta} = \frac{\omega(1-z^2 K) + z(2+K'z)\tau}{\left(1+\frac{z}{R_1}\right)\left(1+\frac{z}{R_2}\right)}$$

(9.3.11)

where $K' = \frac{1}{R_1} + \frac{1}{R_2}$ and $K = \frac{1}{R_1 R_2}$. The corresponding stresses are, see Fig. 9.3.3,

$$\sigma^*_{\alpha\alpha} = \frac{E}{1-v^2}\left(\varepsilon^*_{\alpha\alpha} + v\varepsilon^*_{\beta\beta}\right)$$

$$\sigma^*_{\beta\beta} = \frac{E}{1-v^2}\left(\varepsilon^*_{\beta\beta} + v\varepsilon^*_{\alpha\alpha}\right)$$

$$\sigma^*_{\alpha\beta} = \frac{E}{2(1+v)}\varepsilon^*_{\alpha\beta}$$

(9.3.12)

i.e.,

$$\sigma^*_{\alpha\alpha} = \frac{E}{1-v^2}\left[\frac{\varepsilon_{\alpha\alpha} + z\chi_1}{1+\frac{z}{R_1}} + v\frac{\varepsilon_{\beta\beta} + z\chi_2}{1+\frac{z}{R_2}}\right]$$

$$\sigma^*_{\beta\beta} = \frac{E}{1-v^2}\left[\frac{\varepsilon_{\beta\beta} + z\chi_2}{1+\frac{z}{R_2}} + v\frac{\varepsilon_{\alpha\alpha} + z\chi_1}{1+\frac{z}{R_1}}\right]$$

$$\sigma^*_{\alpha\beta} = \frac{E}{2(1+v)}\frac{\omega(1-z^2 K) + z(2+K'z)\tau}{\left(1+\frac{z}{R_1}\right)\left(1+\frac{z}{R_2}\right)}$$

(9.3.13)

Making use of (9.3.2) and (9.3.3), we have

$$\vec{R}^{(\alpha)} ds_\alpha = \int_{-h}^{h} \vec{\sigma}^{(\alpha)} ds^*_\alpha dz \quad \vec{Q}^{(\alpha)} ds_\alpha = \int_{-h}^{h} \left(\vec{\sigma}^{(\alpha)} \times z\vec{n}\right) ds^*_\alpha dz$$

$$\vec{R}^{(\beta)} ds_\beta = \int_{-h}^{h} \vec{\sigma}^{(\beta)} ds^*_\beta dz \quad \vec{Q}^{(\beta)} ds_\beta = \int_{-h}^{h} \left(\vec{\sigma}^{(\beta)} \times z\vec{n}\right) ds^*_\beta dz$$

(9.3.14)

462 Dynamics of Plates

where

$$\vec{\sigma}^{(\alpha)} = -\sigma^*_{\alpha\beta}\frac{\vec{M}^*_\alpha}{A^*} - \sigma^*_{\beta\beta}\frac{\vec{M}^*_\beta}{B^*} - \sigma^*_{z\beta}\vec{n}$$

$$\vec{\sigma}^{(\beta)} = -\sigma^*_{\alpha\alpha}\frac{\vec{M}^*_\alpha}{A^*} - \sigma^*_{\beta\alpha}\frac{\vec{M}^*_\beta}{B^*} - \sigma^*_{z\alpha}\vec{n}$$

(9.3.15)

In view of (9.2.44) and (9.2.45), the above equations in (9.3.15) simplify to

$$\vec{\sigma}^{(\alpha)} = -\sigma^*_{\alpha\beta}\frac{\vec{M}_\alpha}{A} - \sigma^*_{\beta\beta}\frac{\vec{M}_\beta}{B} - \sigma^*_{z\beta}\vec{n}$$

$$\vec{\sigma}^{(\beta)} = -\sigma^*_{\alpha\alpha}\frac{\vec{M}_\alpha}{A} - \sigma^*_{\beta\alpha}\frac{\vec{M}_\beta}{B} - \sigma^*_{z\alpha}\vec{n}$$

(9.3.16)

Now, (9.3.14) can be written with the help of (9.3.5) and (9.3.6) and noting that

$$ds_\alpha = Ad\alpha \quad ds^*_\alpha = A^*d\alpha = \left(1+\frac{z}{R_1}\right)Ad\alpha$$

$$ds_\beta = Bd\beta \quad ds^*_\beta = B^*d\beta = \left(1+\frac{z}{R_2}\right)Bd\beta$$

$$\left(S_2\frac{\vec{M}_\alpha}{A} - T_2\frac{\vec{M}_\beta}{B} + N_2\vec{n}\right)Ad\alpha$$

$$= \int_{-h}^{h}\left(-\sigma^*_{\alpha\beta}\frac{\vec{M}_\alpha}{A} - \sigma^*_{\beta\beta}\frac{\vec{M}_\beta}{B} - \sigma^*_{z\beta}\vec{n}\right)\left(1+\frac{z}{R_1}\right)Ad\alpha dz$$

$$\left(G_2\frac{\vec{M}_\alpha}{A} - H_2\frac{\vec{M}_\beta}{B}\right)Ad\alpha$$

$$= \int_{-h}^{h}\left[\left(-\sigma^*_{\alpha\beta}\frac{\vec{M}_\alpha}{A} - \sigma^*_{\beta\beta}\frac{\vec{M}_\beta}{B} - \sigma^*_{z\beta}\vec{n}\right)\times z\vec{n}\right]\left(1+\frac{z}{R_1}\right)Ad\alpha dz$$

Pre-twisted Plates 463

$$\left(-T_1\frac{\vec{M}_a}{A} - S_1\frac{\vec{M}_\beta}{B} + N_1\vec{n}\right)Bd\beta$$

$$= \int_{-h}^{h}\left(-\sigma_{aa}^*\frac{\vec{M}_a}{A} - \sigma_{\beta a}^*\frac{\vec{M}_\beta}{B} - \sigma_{za}^*\vec{n}\right)\left(1 + \frac{z}{R_2}\right)Bd\beta dz$$

$$\left(-H_1\frac{\vec{M}_a}{A} - G_1\frac{\vec{M}_\beta}{B}\right)Bd\beta$$

$$= \int_{-h}^{h}\left[\left(-\sigma_{aa}^*\frac{\vec{M}_a}{A} - \sigma_{\beta a}^*\frac{\vec{M}_\beta}{B} - \sigma_{za}^*\vec{n}\right)\times z\vec{n}\right]\left(1 + \frac{z}{R_2}\right)Bd\beta dz$$

(9.3.17)

Equating the respective coefficients of $\vec{M}_a, \vec{M}_\beta$ and $\vec{n}$, we get

$$T_1 = \int_{-h}^{h}\sigma_{aa}^*\left(1 + \frac{z}{R_2}\right)dz \quad T_2 = \int_{-h}^{h}\sigma_{\beta\beta}^*\left(1 + \frac{z}{R_1}\right)dz$$

$$S_1 = \int_{-h}^{h}\sigma_{\beta a}^*\left(1 + \frac{z}{R_2}\right)dz \quad S_2 = -\int_{-h}^{h}\sigma_{a\beta}^*\left(1 + \frac{z}{R_1}\right)dz$$

$$N_1 = -\int_{-h}^{h}\sigma_{za}^*\left(1 + \frac{z}{R_2}\right)dz \quad N_2 = -\int_{-h}^{h}\sigma_{z\beta}^*\left(1 + \frac{z}{R_1}\right)dz$$

$$G_1 = -\int_{-h}^{h}\sigma_{aa}^* z\left(1 + \frac{z}{R_2}\right)dz \quad G_2 = -\int_{-h}^{h}\sigma_{\beta\beta}^* z\left(1 + \frac{z}{R_1}\right)dz$$

$$H_1 = \int_{-h}^{h}\sigma_{\beta a}^* z\left(1 + \frac{z}{R_2}\right)dz \quad H_2 = -\int_{-h}^{h}\sigma_{a\beta}^* z\left(1 + \frac{z}{R_1}\right)dz$$

(9.3.18)

Now substituting the relations from equations (9.3.13), and excluding the N terms, the above equations (9.3.18) become

$$T_1 = \frac{E}{1-v^2}\int_{-h}^{h}\frac{\varepsilon_{aa} + z\chi_1}{1+\frac{z}{R_1}}\left(1 + \frac{z}{R_2}\right)dz + \frac{Ev}{1-v^2}\int_{-h}^{h}(\varepsilon_{\beta\beta} + z\chi_2)dz$$

$$T_2 = \frac{E}{1-v^2}\int_{-h}^{h}\frac{\varepsilon_{\beta\beta} + z\chi_2}{1+\frac{z}{R_2}}\left(1 + \frac{z}{R_1}\right)dz + \frac{Ev}{1-v^2}\int_{-h}^{h}(\varepsilon_{aa} + z\chi_1)dz$$

$$S_1 = \frac{E}{2(1+v)} \int_{-h}^{h} \frac{\omega(1-z^2K)+z(2+K'z)\tau}{\left(1+\frac{z}{R_1}\right)} dz$$

$$S_2 = -\frac{E}{2(1+v)} \int_{-h}^{h} \frac{\omega(1-z^2K)+z(2+K'z)\tau}{\left(1+\frac{z}{R_2}\right)} dz$$

$$G_1 = -\frac{E}{1-v^2} \int_{-h}^{h} \frac{\varepsilon_{aa}+z\varkappa_1}{1+\frac{z}{R_1}}\left(1+\frac{z}{R_2}\right)zdz - \frac{Ev}{1-v^2} \int_{-h}^{h} (\varepsilon_{\beta\beta}+z\varkappa_2)zdz$$

$$G_2 = -\frac{E}{1-v^2} \int_{-h}^{h} \frac{\varepsilon_{\beta\beta}+z\varkappa_2}{1+\frac{z}{R_2}}\left(1+\frac{z}{R_1}\right)zdz - \frac{Ev}{1-v^2} \int_{-h}^{h} (\varepsilon_{aa}+z\varkappa_1)zdz$$

$$H_1 = \frac{E}{2(1+v)} \int_{-h}^{h} \frac{\omega(1-z^2K)+z(2+K'z)\tau}{\left(1+\frac{z}{R_1}\right)} zdz$$

$$H_2 = -\frac{E}{2(1+v)} \int_{-h}^{h} \frac{\omega(1-z^2K)+z(2+K'z)\tau}{\left(1+\frac{z}{R_2}\right)} zdz$$

(9.3.19)

Upon evaluating the integrals, one can obtain the following relations.

$$T_1 = \frac{2Eh}{1-v^2}(\varepsilon_{aa}+v\varepsilon_{\beta\beta}) - \frac{2Eh^3}{3(1-v^2)}\left(\frac{1}{R_1}-\frac{1}{R_2}\right)\left(\varkappa_1 - \frac{\varepsilon_{aa}}{R_1}\right)$$

$$T_2 = \frac{2Eh}{1-v^2}(\varepsilon_{\beta\beta}+v\varepsilon_{aa}) - \frac{2Eh^3}{3(1-v^2)}\left(\frac{1}{R_2}-\frac{1}{R_1}\right)\left(\varkappa_2 - \frac{\varepsilon_{\beta\beta}}{R_2}\right)$$

$$S_1 = \frac{Eh\omega}{1+v} - \frac{Eh^3}{3(1+v)}\left(\frac{1}{R_2}-\frac{1}{R_1}\right)\left(\tau - \frac{\omega}{R_2}\right)$$

$$S_2 = -\frac{Eh\omega}{1+v} + \frac{Eh^3}{3(1+v)}\left(\frac{1}{R_1}-\frac{1}{R_2}\right)\left(\tau - \frac{\omega}{R_1}\right)$$

$$G_1 = -\frac{2Eh^3}{3(1-v^2)}(\varkappa_1+v\varkappa_2) + \frac{2Eh^3}{3(1-v^2)}\left(\frac{1}{R_1}-\frac{1}{R_2}\right)\varepsilon_{aa}$$

$$G_2 = -\frac{2Eh^3}{3(1-v^2)}(\varkappa_2+v\varkappa_1) + \frac{2Eh^3}{3(1-v^2)}\left(\frac{1}{R_2}-\frac{1}{R_1}\right)\varepsilon_{\beta\beta}$$

$$H_1 = \frac{2Eh^3}{3(1-v^2)}\left(\tau - \frac{\omega}{R_1}\right) \qquad H_2 = -\frac{2Eh^3}{3(1-v^2)}\left(\tau - \frac{\omega}{R_2}\right)$$

(9.3.20)

In equations (9.3.20) above, the effect of bending deformations of the middle surface of the shell on the forces T_1, T_2, S_1 and S_2 is included and so also the influence of tangential deformations of the middle surface on the moments G_1, G_2, H_1 and H_2 is included.

Approximating $1 + \frac{z}{R_1} \simeq 1, 1 + \frac{z}{R_2} \simeq 1$ and neglecting terms containing z in terms such as $(1 - z^2 K)$, the above relations (9.3.20) simplify to

$$T_1 = \frac{2Eh}{1-v^2}(\varepsilon_{\alpha\alpha} + v\varepsilon_{\beta\beta})$$

$$T_2 = \frac{2Eh}{1-v^2}(\varepsilon_{\beta\beta} + v\varepsilon_{\alpha\alpha})$$

$$S_1 = -S_2 = \frac{Eh\omega}{1+v}$$

$$G_1 = -\frac{2Eh^3}{3(1-v^2)}(\varkappa_1 + v\varkappa_2)$$

$$G_2 = -\frac{2Eh^3}{3(1-v^2)}(\varkappa_2 + v\varkappa_1)$$

$$H_1 = -H_2 = \frac{2Eh^3\tau}{3(1-v^2)}$$

(9.3.21)

With the above simplification, the forces T_1, T_2, S_1 and S_2 are related to the tangential deformations only and the moments G_1, G_2, H_1 and H_2 are related to the bending deformation only. The above formulae are valid only when the middle surface is referred to the lines of curvature.

9.4 STRAIN ENERGY

Consider first the equations of equilibrium of a shell, for this purpose, let us consider a small curvilinear quadrangle $abdc$ shown in Fig. 9.4.1 with

$$ab = ds_\alpha = A d\alpha$$
$$ac = ds_\beta = B d\beta$$

(9.4.1)

The area of the quadrangle is represented by

$$dS = AB \sin\chi \, d\alpha \, d\beta$$

(9.4.2)

The element in Fig. 9.4.1 is subjected to the following forces

466 Dynamics of Plates

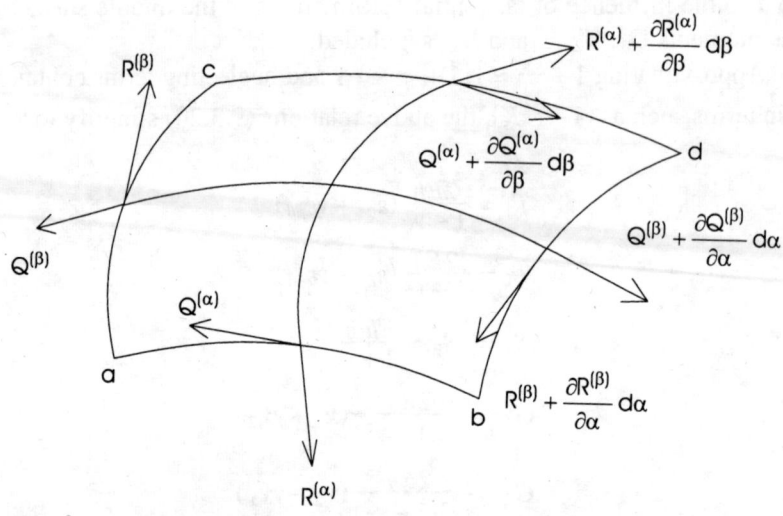

Fig. 9.4.1 Equilibrium of Shell

1. $\vec{R}^{(\alpha)} A d\alpha$ acting on ab

2. $-\left[\vec{R}^{(\alpha)} + \dfrac{\partial \vec{R}^{(\alpha)}}{\partial \beta} d\beta\right]\left(A + \dfrac{\partial A}{\partial \beta} d\beta\right) d\alpha$ acting on cd

3. $\vec{R}^{(\beta)} B d\beta$ acting on ac

4. $-\left[\vec{R}^{(\beta)} + \dfrac{\partial \vec{R}^{(\beta)}}{\partial \alpha} d\alpha\right]\left(B + \dfrac{\partial B}{\partial \alpha} d\alpha\right) d\beta$ acting on bd

5. $\vec{P} AB \sin\chi \, d\alpha \, d\beta$ acting on the area $abcd$

where $\vec{P}$ is the external force represented by

$$\vec{P} = X\frac{\vec{M}_{,\alpha}}{A} + Y\frac{\vec{M}_{,\beta}}{B} - Z\vec{n} \qquad (9.4.3)$$

Further, let

$$\vec{Q} = E\frac{\vec{M}_{,\alpha}}{A} + F\frac{\vec{M}_{,\beta}}{B} \qquad (9.4.4)$$

be the external moment, then we have the following moments acting on the quadrangle element.

1. $\vec{Q}^{(a)} A \, da$ acting on ab

2. $-\left[\vec{Q}^{(a)} A + \frac{\partial}{\partial \beta} \left(\vec{Q}^{(a)} A \right) d\beta \right] da$ acting on cd

3. $\vec{Q}^{(\beta)} B \, d\beta$ acting on ac

4. $-\left[\vec{Q}^{(\beta)} + \frac{\partial}{\partial a} \left(\vec{Q}^{(\beta)} B \right) da \right] d\beta$ acting on bd

5. $\vec{Q} AB \sin \chi \, da \, d\beta$ acting on the area $abcd$

6. $\left(\vec{R}^{(a)} \times \vec{M} \right) A \, da$ from forces acting on ab

7. $-\left[\left(\vec{R}^{(a)} \times \vec{M} \right) A + \frac{\partial}{\partial \beta} A \left(\vec{R}^{(a)} \times \vec{M} \right) d\beta \right] da$ from forces acting on cd

8. $\left(\vec{R}^{(\beta)} \times \vec{M} \right) B \, d\beta$ from forces acting on ac

9. $-\left[\left(\vec{R}^{(\beta)} \times \vec{M} \right) B \, d\beta + \frac{\partial}{\partial a} B \left(\vec{R}^{(\beta)} \times \vec{M} \right) da \right] d\beta$ acting on bd

10. $\left(\vec{P} \times \vec{M} \right) AB \sin \chi \, da \, d\beta$ from external forces acting on the area $abcd$

Summing up the five forces and ten moments listed above, we can obtain the following relations

$$-\frac{\partial}{\partial \beta} \left(A \vec{R}^{(a)} \right) - \frac{\partial}{\partial a} \left(B \vec{R}^{(\beta)} \right) + AB \sin \chi \vec{P} = 0$$

(9.4.5)

$$-\frac{\partial}{\partial \beta} \left(A \vec{R}^{(a)} \times \vec{M} \right) - \frac{\partial}{\partial a} \left(B \vec{R}^{(\beta)} \times \vec{M} \right) - \frac{\partial}{\partial \beta} \left(A \vec{Q}^{(a)} \right)$$
$$- \frac{\partial}{\partial a} \left(B \vec{Q}^{(\beta)} \right) + AB \sin \chi \left(\vec{P} \times \vec{M} \right) + AB \sin \chi \vec{Q}$$
$$= 0$$

(9.4.6)

468 Dynamics of Plates

After some transformations, equation (9.4.6) with the help of (9.4.5) can be written as

$$-\frac{\partial}{\partial \beta}\left(A\vec{Q}^{(\alpha)}\right) - \frac{\partial}{\partial \alpha}\left(B\vec{Q}^{(\beta)}\right) - A\vec{R}^{(\alpha)} \times \vec{M},_\beta$$

$$- B\vec{R}^{(\beta)} \times \vec{M},_\alpha + AB\sin\chi \vec{Q}$$

$$= 0$$

(9.4.7)

Now, let us consider a region G in the parameters (α, β) bounded by a contour g, the work of forces and moments in some state of stress that is denoted by superscript (1) in displacement to another deformed state denoted by superscript (2) can be written as

$$W_{12} = \iint_G \vec{P} \cdot \vec{U}^{(2)} AB\sin\chi \, d\alpha \, d\beta + \iint_G \vec{Q} \cdot \vec{\Omega}^{(2)} AB\sin\chi \, d\alpha \, d\beta$$

$$+ \int_g \vec{U}^{(2)} \left(\vec{R}^{(\alpha)} A \, d\alpha - \vec{R}^{(\beta)} B \, d\beta\right)$$

$$+ \int_g \vec{\Omega}^{(2)} \left(\vec{Q}^{(\alpha)} A \, d\alpha - \vec{Q}^{(\beta)} B \, d\beta\right)$$

(9.4.8)

Taking the scalar product of (9.4.5) with $\vec{U}^{(2)}$ and integrating over the region G we have

$$\iint_G \vec{P} \cdot \vec{U}^{(2)} AB\sin\chi \, d\alpha \, d\beta$$

$$= \iint_G \vec{U}^{(2)} \cdot \left[\frac{\partial}{\partial \beta}\left(A\vec{R}^{(\alpha)}\right) + \frac{\partial}{\partial \alpha}\left(B\vec{R}^{(\beta)}\right)\right] d\alpha \, d\beta$$

(9.4.9)

Integrating by parts

$$\iint_G \vec{P} \cdot \vec{U}^{(2)} AB \sin\chi \, da \, d\beta$$

$$= \int_g \vec{U}^{(2)} \cdot \left[-A\vec{R}^{(a)} \, da + B\vec{R}^{(\beta)} \, d\beta \right]$$

$$- \iint_G \left[A\vec{R}^{(a)} \cdot \vec{U}^{(2)}{}_{,\beta} + B\vec{R}^{(\beta)} \cdot \vec{U}^{(2)}{}_{,a} \right] da \, d\beta$$

(9.4.10)

Similarly, taking the scalar product of (9.4.7) with $\vec{\Omega}^{(2)}$ and integrating over G we have

$$\iint_G \vec{Q} \cdot \vec{\Omega}^{(2)} AB \sin\chi \, da \, d\beta$$

$$= \iint_G \vec{\Omega}^{(2)} \cdot \left[\frac{\partial}{\partial\beta}\left(A\vec{Q}^{(a)}\right) + \frac{\partial}{\partial a}\left(B\vec{Q}^{(\beta)}\right) \right] da \, d\beta$$

$$+ \iint_G \vec{\Omega}^{(2)} \cdot \left(A\vec{R}^{(a)} \times \vec{M}_{,\beta} + B\vec{R}^{(\beta)} \times \vec{M}_{,a} \right) da \, d\beta$$

(9.4.11)

We can integrate the first term on the right hand side and write the above equation (9.4.11) as

$$\iint_G \vec{\Omega}^{(2)} \cdot \left[\frac{\partial}{\partial\beta}\left(A\vec{Q}^{(a)}\right) + \frac{\partial}{\partial a}\left(B\vec{Q}^{(\beta)}\right) \right] da \, d\beta$$

$$= \int_g \vec{\Omega}^{(2)} \cdot \left[-\vec{Q}^{(a)} A \, da + \vec{Q}^{(\beta)} B \, d\beta \right]$$

$$- \iint_G \left[A\vec{Q}^{(a)} \cdot \vec{\Omega}^{(2)}{}_{,\beta} + B\vec{Q}^{(\beta)} \cdot \vec{\Omega}^{(2)}{}_{,a} \right] da \, d\beta$$

(9.4.12)

Now, adding equations (9.4.9) and (9.4.11) and making use of (9.4.10) and (9.4.12), we get

$$\iint_G \vec{P} \cdot \vec{U}^{(2)} AB \sin\chi \, d\alpha \, d\beta$$

$$+ \iint_G \vec{Q} \cdot \vec{\Omega}^{(2)} AB \sin\chi \, d\alpha \, d\beta$$

$$= \int_g \vec{U}^{(2)} \cdot \left[-\vec{R}^{(\alpha)} A \, d\alpha + \vec{R}^{(\beta)} B \, d\beta \right]$$

$$+ \int_g \vec{\Omega}^{(2)} \cdot \left[-\vec{Q}^{(\alpha)} A \, d\alpha + \vec{Q}^{(\beta)} B \, d\beta \right]$$

$$- \iint_G \left[A \vec{R}^{(\alpha)} \cdot \vec{U}^{(2)}{}_{,\beta} + B \vec{R}^{(\beta)} \cdot \vec{U}^{(2)}{}_{,\alpha} \right] d\alpha \, d\beta$$

$$- \iint_G \left[A \vec{Q}^{(\alpha)} \cdot \vec{\Omega}^{(2)}{}_{,\beta} + B \vec{Q}^{(\beta)} \cdot \vec{\Omega}^{(2)}{}_{,\alpha} \right] d\alpha \, d\beta$$

$$+ \iint_G \vec{\Omega}^{(2)} \cdot \left[A \vec{R}^{(\alpha)} \times \vec{M}_{,\beta} + B \vec{R}^{(\beta)} \times \vec{M}_{,\alpha} \right] d\alpha \, d\beta$$

(9.4.13)

The three double integral terms on the right hand side can be expanded by using (9.1.24), (9.3.5) and (9.3.6) to give

$$- \iint_G \left[A \vec{R}^{(\alpha)} \cdot \vec{U}^{(2)}{}_{,\beta} + B \vec{R}^{(\beta)} \cdot \vec{U}^{(2)}{}_{,\alpha} \right] d\alpha \, d\beta$$

$$- \iint_G \left[A \vec{Q}^{(\alpha)} \cdot \vec{\Omega}^{(2)}{}_{,\beta} + B \vec{Q}^{(\beta)} \cdot \vec{\Omega}^{(2)}{}_{,\alpha} \right] d\alpha \, d\beta$$

$$+ \iint_G \vec{\Omega}^{(2)} \cdot \left[A \vec{R}^{(\alpha)} \times \vec{M}_{,\beta} + B \vec{R}^{(\beta)} \times \vec{M}_{,\alpha} \right] d\alpha \, d\beta$$

Pre-twisted Plates 471

$$= \iint_G \begin{bmatrix} T_1 \dfrac{\vec{M}_a}{A} \cdot \dfrac{\vec{U}^{(2)}_{,a}}{A} + T_2 \dfrac{\vec{M}_\beta}{B} \cdot \dfrac{\vec{U}^{(2)}_{,\beta}}{B} \\ +S_1 \left(\dfrac{\vec{M}_\beta}{B} \cdot \dfrac{\vec{U}^{(2)}_{,a}}{A} + \vec{\Omega}^{(2)} \cdot \vec{n} \sin\chi \right) \\ -S_2 \left(\dfrac{\vec{M}_a}{A} \cdot \dfrac{\vec{U}^{(2)}_{,\beta}}{B} - \vec{\Omega}^{(2)} \cdot \vec{n} \sin\chi \right) \\ +G_1 \dfrac{\vec{M}_\beta}{B} \cdot \dfrac{\vec{\Omega}^{(2)}_{,a}}{A} - G_2 \dfrac{\vec{M}_a}{A} \cdot \dfrac{\vec{\Omega}^{(2)}_{,\beta}}{B} \\ +H_1 \dfrac{\vec{M}_a}{A} \cdot \dfrac{\vec{\Omega}^{(2)}_{,a}}{A} + H_2 \dfrac{\vec{M}_\beta}{B} \cdot \dfrac{\vec{\Omega}^{(2)}_{,\beta}}{B} \end{bmatrix} AB\, da\, d\beta$$

$$+ \iint_G \begin{bmatrix} N_1 \left(-\vec{n} \cdot \dfrac{\vec{U}^{(2)}_{,a}}{A} + \dfrac{\vec{N}_{(\beta)}}{B} \cdot \vec{\Omega}^{(2)} \right) \\ + N_2 \left(-\vec{n} \cdot \dfrac{\vec{U}^{(2)}_{,\beta}}{B} - \dfrac{\vec{N}_{(a)}}{A} \cdot \vec{\Omega}^{(2)} \right) \end{bmatrix} AB\, da\, d\beta$$

(9.4.14)

In view of (9.2.9), the second integral on the right hand side of (9.4.14) above becomes zero. Further, letting

$$\varepsilon_{aa} = \frac{\vec{U}_{,a}}{A} \cdot \frac{\vec{M}_{,a}}{A} \qquad \varepsilon_{\beta\beta} = \frac{\vec{U}_{,\beta}}{B} \cdot \frac{\vec{M}_{,\beta}}{B}$$

$$\omega_1 = \frac{\vec{U}_{,a}}{A} \cdot \frac{\vec{M}_{,\beta}}{B} + \vec{\Omega} \cdot \vec{n} \sin\chi$$

$$\omega_2 = -\frac{\vec{U}_{,\beta}}{B} \cdot \frac{\vec{M}_{,a}}{A} + \vec{\Omega} \cdot \vec{n} \sin\chi$$

$$\kappa_{aa} = -\frac{\vec{M}_{,\beta}}{B} \cdot \frac{\vec{\Omega}_{,a}}{A} \qquad \kappa_{\beta\beta} = \frac{\vec{M}_{,a}}{A} \cdot \frac{\vec{\Omega}_{,\beta}}{B}$$

$$\tau^{(1)} = \frac{\vec{M}_{,a}}{A} \cdot \frac{\vec{\Omega}_{,a}}{A} \qquad \tau^{(2)} = \frac{\vec{M}_{,\beta}}{B} \cdot \frac{\vec{\Omega}_{,a}}{A}$$

(9.4.15)

equation (9.4.14) becomes

$$-\iint_G \left[A\vec{R}^{(\alpha)} \cdot \vec{U}^{(2)}_{,\beta} + B\vec{R}^{(\beta)} \cdot \vec{U}^{(2)}_{,\alpha} \right] d\alpha\, d\beta$$

$$-\iint_G \left[A\vec{Q}^{(\alpha)} \cdot \vec{\Omega}^{(2)}_{,\beta} + B\vec{Q}^{(\beta)} \cdot \vec{\Omega}^{(2)}_{,\alpha} \right] d\alpha\, d\beta$$

$$+\iint_G \vec{\Omega}^{(2)} \cdot \left[A\vec{R}^{(\alpha)} \times \vec{M}_{,\beta} + B\vec{R}^{(\beta)} \times \vec{M}_{,\alpha} \right] d\alpha\, d\beta$$

$$=\iint_G \left[\begin{array}{c} T_1\varepsilon_{\alpha\alpha}^{(2)} + T_2\varepsilon_{\beta\beta}^{(2)} + S_1\omega_1^{(2)} + S_2\omega_2^{(2)} - G_1\kappa_{\alpha\alpha}^{(2)} \\ -G_2\kappa_{\beta\beta}^{(2)} + H_1\tau_1^{(2)} + H_2\tau_2^{(2)} \end{array} \right] AB\, d\alpha\, d\beta$$

(9.4.16)

Substituting the above in equation (9.4.13), we get

$$\iint_G \vec{P} \cdot \vec{U}^{(2)} AB \sin\chi\, d\alpha\, d\beta + \iint_G \vec{Q} \cdot \vec{\Omega}^{(2)} AB \sin\chi\, d\alpha\, d\beta$$

$$= \int_g \vec{U}^{(2)} \cdot \left[-\vec{R}^{(\alpha)} A\, d\alpha + \vec{R}^{(\beta)} B\, d\beta \right] + \int_g \vec{\Omega}^{(2)} \cdot \left[-\vec{Q}^{(\alpha)} A\, d\alpha + \vec{Q}^{(\beta)} B\, d\beta \right]$$

$$+\iint_G \left[\begin{array}{c} T_1\varepsilon_{\alpha\alpha}^{(2)} + T_2\varepsilon_{\beta\beta}^{(2)} + S_1\omega_1^{(2)} + S_2\omega_2^{(2)} - G_1\kappa_{\alpha\alpha}^{(2)} \\ -G_2\kappa_{\beta\beta}^{(2)} + H_1\tau_1^{(2)} + H_2\tau_2^{(2)} \end{array} \right] AB\, d\alpha\, d\beta$$

(9.4.17)

Comparing with (9.4.8), the work done by the forces and moments during the deformation is given by

$$\cdot\ W_{12} = \iint_G \left[\begin{array}{c} T_1\varepsilon_{\alpha\alpha}^{(2)} + T_2\varepsilon_{\beta\beta}^{(2)} + S_1\omega_1^{(2)} + S_2\omega_2^{(2)} - G_1\kappa_{\alpha\alpha}^{(2)} \\ -G_2\kappa_{\beta\beta}^{(2)} + H_1\tau_1^{(2)} + H_2\tau_2^{(2)} \end{array} \right] AB\, d\alpha\, d\beta$$

(9.4.18)

Dropping the superscript (2), the above equation (9.4.18) is

$$W_{12} = \iint_G \left[\begin{array}{c} T_1\varepsilon_{\alpha\alpha} + T_2\varepsilon_{\beta\beta} + S_1\omega_1 + S_2\omega_2 - G_1\kappa_{\alpha\alpha} \\ -G_2\kappa_{\beta\beta} + H_1\tau_1 + H_2\tau_2 \end{array} \right] AB\, d\alpha\, d\beta$$

(9.4.19)

Since the work of deformation is equal to half of the work of external forces producing the elastic deformations, the strain energy is given by

$$V = \frac{1}{2} \iint_G \left[\begin{array}{c} T_1\varepsilon_{aa} + T_2\varepsilon_{\beta\beta} + S_1\omega_1 + S_2\omega_2 - G_1\kappa_{aa} \\ -G_2\kappa_{\beta\beta} + H_1\tau_1 + H_2\tau_2 \end{array} \right] AB\, da\, d\beta$$

(9.4.20)

With the help of (9.3.21), the above can be shown to be

$$V = \frac{E}{1-v^2} \iint_G h\left\{(\varepsilon_{aa} + \varepsilon_{\beta\beta})^2 - 2(1-v)\left(\varepsilon_{aa}\varepsilon_{\beta\beta} - \frac{1}{4}\varepsilon_{a\beta}^2\right)\right\} AB\, da\, d\beta$$

$$+ \frac{E}{3(1-v^2)} \iint_G h^3\left\{(\kappa_{aa} + \kappa_{\beta\beta})^2 - 2(1-v)\left(\kappa_{aa}\kappa_{\beta\beta} - \kappa_{a\beta}^2\right)\right\} AB\, da\, d\beta$$

(9.4.21)

The above expression can also be written as

$$V = \frac{E}{1-v^2} \iint_G h \left[\begin{array}{c} \left\{\varepsilon_{aa}^2 + \varepsilon_{\beta\beta}^2 + 2v\varepsilon_{aa}\varepsilon_{\beta\beta} + \frac{1}{2}(1-v)\varepsilon_{a\beta}^2\right\} + \\ \frac{h^2}{3}\left\{\kappa_{aa}^2 + \kappa_{\beta\beta}^2 + 2v\kappa_{aa}\kappa_{\beta\beta} + 2(1-v)\kappa_{a\beta}^2\right\} \end{array} \right] AB\, da\, d\beta$$

(9.4.22)

9.5 PRE-TWISTED RECTANGULAR PLATE

Let us consider a pre-twisted rectangular plate as shown in Fig. 9.5.1. The plate has a width b and is of length l fixed at the edge with the a – curve coinciding with x axis. β – curves are chosen to lie on the surface of the plate in the longitudinal direction and a – curves are across in the width direction. xyz are the rectangular coordinates. The plate is of uniform cross-section twisted uniformly along its length with a pre-twist angle γ. The aspect ratio of the plate is $r = l/b$. The position vector of a point P on an a – curve, at distance $OP = a$ is

$$\vec{M}(a,\beta) = (a\cos\Gamma\beta)\hat{i} + (a\sin\Gamma\beta)\hat{j} + (\beta)\hat{k}$$

(9.5.1)

where

$$\Gamma = \frac{\gamma}{l}$$

(9.5.2)

474 Dynamics of Plates

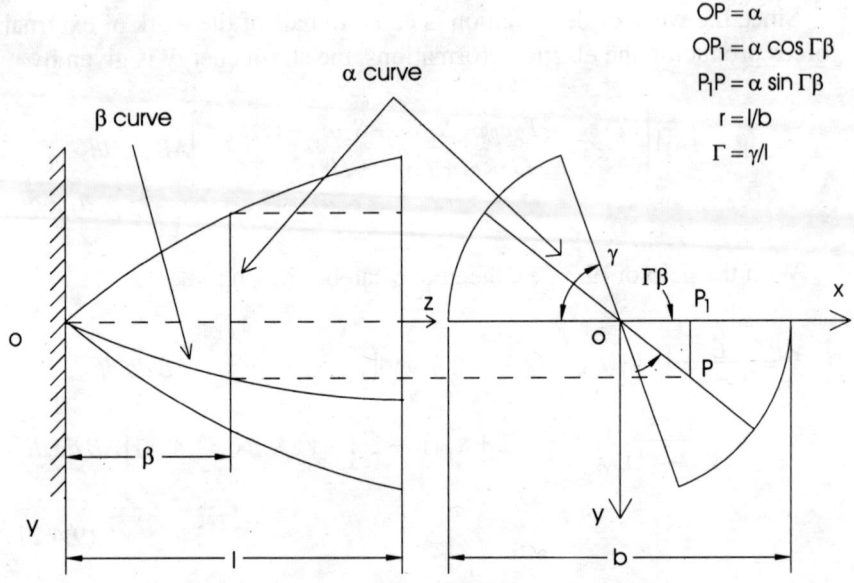

Fig. 9.5.1   Middle Surface of a Pre-twisted Plate

is the unit pre-twist angle and $\hat{i}, \hat{j}, \hat{k}$ are unit vectors in x, y, z directions respectively. The coefficients of the first quadratic form are given by equation (9.1.5) and (9.1.6)

$$A = \left| \vec{M},_\alpha \right| = 1$$

$$B = \left| \vec{M},_\beta \right| = \sqrt{1 + \Gamma^2 \alpha^2}$$

$$\chi = \cos^{-1}\left(\frac{\vec{M},_\alpha}{A} \cdot \frac{\vec{M},_\beta}{B} \right) = \frac{\pi}{2}$$

(9.5.3)

The coefficients of the second quadratic form are given by equations (9.1.26)

$$L = \vec{n} \cdot \vec{M},_{\alpha\alpha} = 0$$

$$M = \vec{n} \cdot \vec{M},_{\alpha\beta} = -\frac{\Gamma}{B}$$

$$N = \vec{n} \cdot \vec{M},_{\beta\beta} = 0$$

(9.5.4)

The radii of curvature of the plate are given by equations (9.1.51) and (9.1.52)

$$\frac{1}{R_1} = -\frac{L}{A^2} = 0$$

$$\frac{1}{R_2} = -\frac{N}{B^2} = 0$$

$$\frac{1}{R_{12}} = \frac{M}{AB} = -\frac{\Gamma}{B^2}$$

(9.5.5)

Since the curvatures along the a- and β-directions are zero, the two curvilinear coordinates coincide with the real directions of asymptotes and the plate is one of negative Gaussian curvature. The Gaussian curvature of the pre-twisted plate is, see (9.1.55)

$$K = \frac{LN - M^2}{A^2 B^2 \sin^2 \chi}$$

$$= -\frac{\Gamma^2}{(1 + \Gamma^2 a^2)^2}$$

(9.5.6)

The three basic vectors from (9.5.1) and (9.5.3) are

$$\frac{\vec{M}_{,a}}{A} = (\cos \Gamma \beta)\hat{i} + (\sin \Gamma \beta)\hat{j}$$

$$\frac{\vec{M}_{,\beta}}{B} = -\left(\frac{\Gamma a}{B} \sin \Gamma \beta\right)\hat{i} + \left(\frac{\Gamma a}{B} \cos \Gamma \beta\right)\hat{j} + \frac{1}{B}\hat{k}$$

$$\vec{n} = \frac{\vec{M}_{,a}}{A} \times \frac{\vec{M}_{,\beta}}{B} = \left(\frac{1}{B} \sin \Gamma \beta\right)\hat{i} - \left(\frac{1}{B} \cos \Gamma \beta\right)\hat{j} + \frac{\Gamma a}{B}\hat{k}$$

(9.5.7)

Let u, v and w be the displacements, then the strains and angles of rotation are given by equations (9.2.75) to (9.2.80)

$$\varepsilon_{aa} = \frac{1}{A}\frac{\partial u}{\partial a} + \frac{v}{AB}\frac{\partial A}{\partial \beta} - \frac{w}{R_1} = \frac{\partial u}{\partial a}$$

$$\varepsilon_{\beta\beta} = \frac{1}{B}\frac{\partial v}{\partial \beta} + \frac{u}{AB}\frac{\partial B}{\partial \alpha} - \frac{w}{R_2} = \frac{1}{B}\frac{\partial v}{\partial \beta} + \frac{\Gamma^2 a u}{B^2}$$

$$\varepsilon_{\alpha\beta} = \frac{A}{B}\frac{\partial}{\partial \beta}\left(\frac{u}{A}\right) + \frac{B}{A}\frac{\partial}{\partial \alpha}\left(\frac{v}{B}\right) + \frac{2w}{R_{12}}$$

$$= \frac{1}{B}\frac{\partial u}{\partial \beta} + \frac{\partial v}{\partial \alpha} - \frac{\Gamma^2 a v}{B^2} - \frac{2\Gamma w}{B^2}$$

(9.5.8)

$$\kappa_{\alpha\alpha} = -\frac{1}{A}\frac{\partial}{\partial \alpha}\left(-\frac{1}{A}\frac{\partial w}{\partial \alpha} - \frac{u}{R_1} + \frac{v}{R_{12}}\right) - \frac{-\frac{1}{B}\frac{\partial w}{\partial \beta} - \frac{v}{R_2} + \frac{u}{R_{12}}}{AB}\frac{\partial A}{\partial \beta}$$

$$+ \frac{\frac{1}{2AB}\left[\frac{\partial}{\partial \beta}(Au) - \frac{\partial}{\partial \alpha}(Bv)\right]}{R_{12}}$$

$$\kappa_{\beta\beta} = -\frac{1}{B}\frac{\partial\left(-\frac{1}{B}\frac{\partial w}{\partial \beta} - \frac{v}{R_2} + \frac{u}{R_{12}}\right)}{\partial \beta} - \frac{-\frac{1}{A}\frac{\partial w}{\partial \alpha} - \frac{u}{R_1} + \frac{v}{R_{12}}}{AB}\frac{\partial B}{\partial \alpha}$$

$$- \frac{\frac{1}{2AB}\left[\frac{\partial}{\partial \beta}(Au) - \frac{\partial}{\partial \alpha}(Bv)\right]}{R_{12}}$$

$$\kappa_{\alpha\beta} = -\frac{1}{A}\frac{\partial\left(-\frac{1}{B}\frac{\partial w}{\partial \beta} - \frac{v}{R_2} + \frac{u}{R_{12}}\right)}{\partial \alpha} + \frac{-\frac{1}{A}\frac{\partial w}{\partial \alpha} - \frac{u}{R_1} + \frac{v}{R_{12}}}{AB}\frac{\partial A}{\partial \beta}$$

$$+ \frac{\frac{1}{2AB}\left[\frac{\partial}{\partial \beta}(Au) - \frac{\partial}{\partial \alpha}(Bv)\right]}{R_1} - \frac{\varepsilon_{\beta\beta}}{R_{12}} + \frac{\varepsilon_{\alpha\beta}}{2R_1}$$

(9.5.9)

Upon substitution of the displacements, curvatures, etc. and simplification, the above equations (9.5.9) simplify to

$$\kappa_{\alpha\alpha} = w,_{\alpha\alpha} - \frac{\Gamma u,_\beta}{2B^3} + \frac{3\Gamma v,_\alpha}{2B^2} - \frac{3\Gamma^3 a v}{2B^4}$$

$$\kappa_{\beta\beta} = \frac{w,_{\beta\beta}}{B^2} + \frac{\Gamma u,_\beta}{2B^3} - \frac{\Gamma v,_\alpha}{2B^2} + \frac{\Gamma^3 a v}{2B^4} + \frac{\Gamma^2 a w,_\alpha}{B^2}$$

$$\kappa_{\alpha\beta} = \frac{w,_{\alpha\beta}}{B} + \frac{\Gamma v,_\beta}{B^3} + \frac{\Gamma u,_\alpha}{B^2} - \frac{\Gamma^3 a u}{B^4} - \frac{\Gamma^2 a w,_\beta}{B^3}$$

(9.5.10)

The angles of rotation given by equations (9.2.77) are

$$\gamma_1 = -\frac{1}{A}\frac{\partial w}{\partial a} - \frac{u}{R_1} + \frac{v}{R_{12}}$$

$$= -\frac{w,_a}{A} - \frac{\Gamma v}{B^2}$$

$$\gamma_2 = -\frac{1}{B}\frac{\partial w}{\partial \beta} - \frac{v}{R_2} + \frac{u}{R_{12}}$$

$$= -\frac{w,_\beta}{B} - \frac{\Gamma u}{B^2}$$

(9.5.11)

Let h be the thickness of the plate, then the strain energy of the plate given by equation (9.4.22) is

$$V = \frac{E}{2(1-v^2)}\iint_G h\left[\begin{array}{l}\{\varepsilon_{aa}^2 + \varepsilon_{\beta\beta}^2 + 2v\varepsilon_{aa}\varepsilon_{\beta\beta} + \frac{1}{2}(1-v)\varepsilon_{a\beta}^2\} + \\ \frac{h^2}{12}\{\kappa_{aa}^2 + \kappa_{\beta\beta}^2 + 2v\kappa_{aa}\kappa_{\beta\beta} + 2(1-v)\kappa_{a\beta}^2\}\end{array}\right]B\,da\,d\beta$$

(9.5.12)

Note that $B = \sqrt{1+\Gamma^2 a^2}$ in the above expressions (9.5.8), (9.5.10), (9.5.11) and (9.5.12). The displacement field of an arbitrary point on an equidistant surface $\overrightarrow{M^*}$ is

$$u^*(a,\beta,z) = u(a,\beta) - \gamma_1 z$$
$$v^*(a,\beta,z) = v(a,\beta) - \gamma_2 z$$
$$w^*(a,\beta,z) = w(a,\beta)$$

(9.5.13)

Therefore, the kinetic energy of the plate can be obtained as

$$T = \iiint_V \frac{1}{2}dm(\dot{u}^{*2} + \dot{v}^{*2} + \dot{w}^{*2})$$

$$= \iint \frac{1}{2}\rho h p^2\left[u^2 + v^2 + w^2 + \frac{h^2}{12}(\gamma_1^2 + \gamma_2^2)\right]AB\,da\,d\beta$$

(9.5.14)

The first three terms give the translatory inertia and the next term gives the rotary inertia of the plate. The Lagrangian function for the plate can now be written as

$$L = \int_{-\frac{b}{2}}^{\frac{b}{2}} \int_0^l \left\{ \frac{EB}{2(1-v^2)} h \left[\begin{array}{c} \{\varepsilon_{\alpha\alpha}^2 + \varepsilon_{\beta\beta}^2 + 2v\varepsilon_{\alpha\alpha}\varepsilon_{\beta\beta} + \frac{1}{2}(1-v)\varepsilon_{\alpha\beta}^2\} \\ + \frac{h^2}{12} \left\{ \begin{array}{c} \kappa_{\alpha\alpha}^2 + \kappa_{\beta\beta}^2 + 2v\kappa_{\alpha\alpha}\kappa_{\beta\beta} \\ +2(1-v)\kappa_{\alpha\beta}^2 \end{array} \right\} \end{array} \right] \right. \left. - \frac{1}{2}\rho Bhp^2 \left[u^2 + v^2 + w^2 + \frac{h^2}{12}(\gamma_1^2 + \gamma_2^2) \right] \right\} d\alpha\, d\beta$$

(9.5.15)

To obtain the natural frequencies of the plate, we can assume the solutions for the displacements as

$$u(\alpha, \beta) = \sum \sum a_{ij} R_i(\alpha) \theta_j(\beta)$$
$$v(\alpha, \beta) = \sum \sum b_{ij} R_i(\alpha) \theta_j(\beta)$$
$$w(\alpha, \beta) = \sum \sum c_{ij} R_i(\alpha) \pi_j(\beta)$$

(9.5.16)

The polynomials in the above equations are assumed as follows.

$$R_1 = 1$$
$$R_2 = \frac{2a}{b}$$
$$R_3 = \frac{1}{2}\left[3\left(\frac{2a}{b}\right)^2 - 1\right]$$
$$R_4 = \frac{1}{2}\left[5\left(\frac{2a}{b}\right)^3 - 3\left(\frac{2a}{b}\right)\right]$$
$$R_5 = \frac{1}{8}\left[35\left(\frac{2a}{b}\right)^4 - 30\left(\frac{2a}{b}\right)^2 + 3\right]$$
$$R_6 = \frac{1}{8}\left[63\left(\frac{2a}{b}\right)^5 - 70\left(\frac{2a}{b}\right)^3 + 15\left(\frac{2a}{b}\right)\right]$$

(9.5.17)

$$\theta_1 = 1.73\frac{\beta}{l}$$

$$\theta_2 = 8.94\left[\left(\frac{\beta}{l}\right)^2 - \frac{3}{4}\left(\frac{\beta}{l}\right)\right]$$

$$\theta_3 = 39.68\left[\left(\frac{\beta}{l}\right)^3 - \frac{3}{4}\left(\frac{\beta}{l}\right)^2 + \frac{2}{5}\left(\frac{\beta}{l}\right)\right]$$

$$\theta_4 = 168\left[\left(\frac{\beta}{l}\right)^4 - 1.815\left(\frac{\beta}{l}\right)^3 + 1.0714\left(\frac{\beta}{l}\right)^2 - 0.1785\left(\frac{\beta}{l}\right)\right]$$

(9.5.18)

$$\pi_1 = 2.236\left(\frac{\beta}{l}\right)^2$$

$$\pi_2 = 15.87\left[\left(\frac{\beta}{l}\right)^3 - \frac{5}{6}\left(\frac{\beta}{l}\right)^2\right]$$

$$\pi_3 = 84\left[\left(\frac{\beta}{l}\right)^4 - \frac{3}{2}\left(\frac{\beta}{l}\right)^3 + 0.535\left(\frac{\beta}{l}\right)^2\right]$$

$$\pi_4 = 398\left[\left(\frac{\beta}{l}\right)^5 - 2.1\left(\frac{\beta}{l}\right)^4 + 1.4\left(\frac{\beta}{l}\right)^3 - 0.291\left(\frac{\beta}{l}\right)^2\right]$$

(9.5.19)

The above equations satisfy the geometric boundary conditions on the fixed edge, viz., $u = v = w = \gamma_1 = 0$. Ritz minimization procedure can be used to obtain the natural frequencies. Fig. 9.5.2 shows the different types of modes, bending modes with $m = 0$, torsional modes with $m = 1$ and plate modes with $m > 1$. n represents the nodal lines parallel to a direction, thus, $0/n$ represents bending modes, $1/n$ represents torsional modes and $2/n$, $3/n$ etc. represent plate modes. Bending modes are symmetrical about the line of axis $a = 0$, torsional modes are anti symmetric, $2/n$, $4/n$ etc. are symmetric while $3/n$, $5/n$ etc. are anti symmetric about $a = 0$.

The symmetry and anti symmetry is governed by the functions $R_i(a)$; $R_1, R_3, R_5 \cdots$ are symmetric, while $R_2, R_4, R_6 \cdots$ are anti symmetric about $a = 0$. A 36 term solution with 4 terms each of θ and π and 3 terms of R gives converged values for the eigen values upto 3/3 (12) modes depending on symmetric or anti symmetric modes. This gives 12 terms each, $i = 1, 3, 5$ and $j = 1$ to 4 for symmetric and $i = 2, 4, 6$ and $j = 1$ to 4 for anti symmetric modes, for u, v and w.

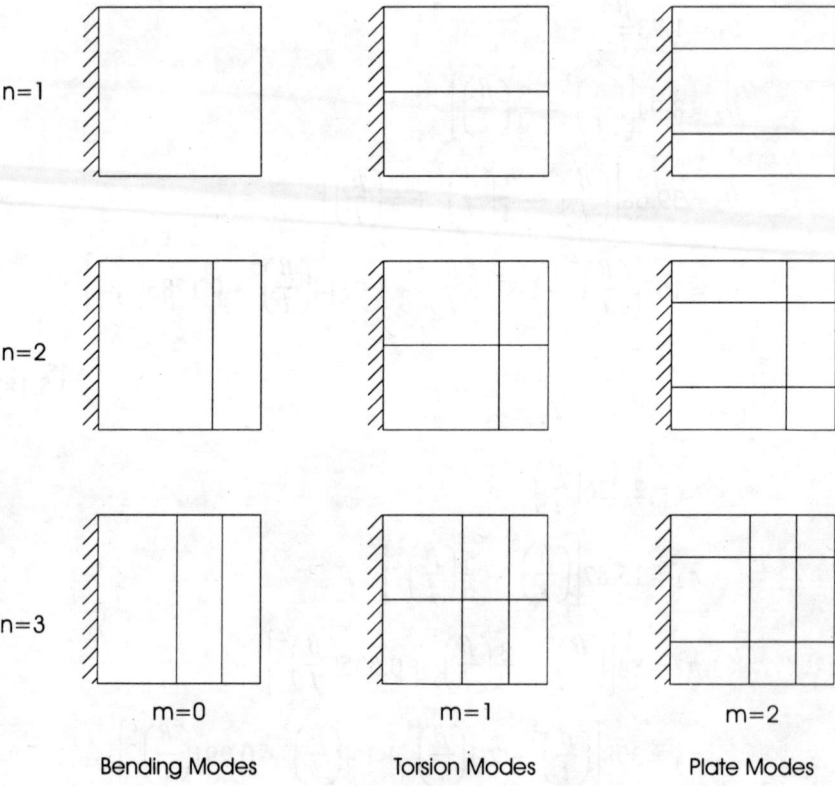

Fig. 9.5.2 Grinsted Nodal Patterns

The natural frequencies of a plate with $r = 0.97$ and pre-twist 41° are:

Table 9.5.1 Bending Modes of a Pre-twisted Plate

$$\bar{p}_r = p_r \sqrt{\frac{12\rho l^4(1-\nu^2)}{Eh^2}}$$

Mode	Frequency
0/1	3.1373
0/2	17.6494
0/3	59.8862

Table 9.5.2 Torsional Modes of a Pre-twisted Plate

$$\bar{p}_r = p_r \sqrt{\frac{12\rho l^4(1-v^2)}{Eh^2}}$$

Mode	Frequency
1/1	15.2652
1/2	34.4975
1/3	69.6379

Table 9.5.3 Plate Modes of a Pre-twisted Plate

$$\bar{p}_r = p_r \sqrt{\frac{12\rho l^4(1-v^2)}{Eh^2}}$$

Mode	Frequency
2/1	25.6428
2/2	55.1055
3/1	60.9640
3/2	91.4839

9.6 ROTATING PRE-TWISTED PLATE

Now, let us consider the effect of centrifugal forces on the cantilever plate considered in the previous section. Let the plate be mounted on a disc of radius R at a stagger angle ϕ as shown in Figs. 9.6.1a and 9.6.1b. The parametric equation of the plate in xyz rotating axes system is already given in the previous section, see (9.5.1). We set up another coordinate axes system $\xi\eta z$ with η axis in the plane of rotation and ξ as the turbine axis. The unit vectors in these two coordinate systems are related by

$$\hat{i} = \hat{I}\sin\phi + \hat{J}\cos\phi$$
$$\hat{j} = \hat{I}\cos\phi - \hat{J}\sin\phi$$
$$\hat{k} = \hat{K}$$

(9.6.1)

Making use of the above equation (9.6.1), the vector equation of the plate middle surface in (9.5.1) and the basic vectors in (9.5.7) can be now written in coordinate axes system $\xi\eta z$ as

$$\vec{M}(\alpha,\beta) = [\alpha\sin(\Gamma\beta+\phi)]\hat{I} + [\alpha\cos(\Gamma\beta+\phi)]\hat{J} + (\beta)\hat{k}$$

(9.6.2)

482 *Dynamics of Plates*

P   undeformed
    position of
    dαdβ element

P_1 deformed
    position of
    dαdβ element

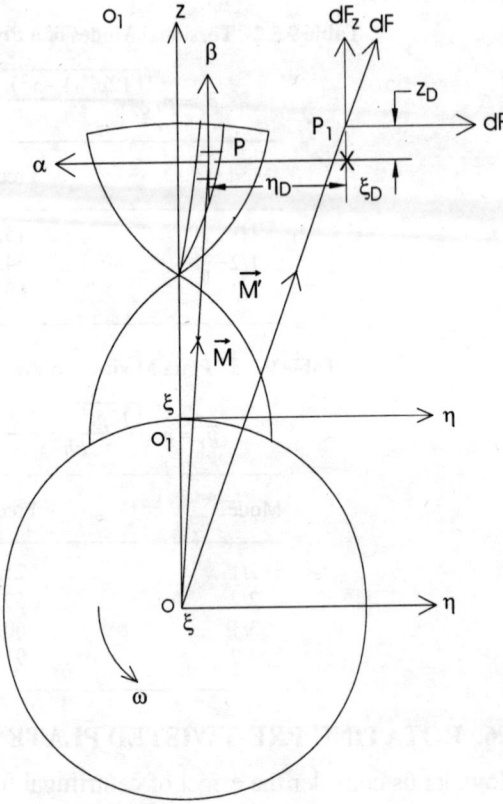

Fig. 9.6.1a Rotating Plate

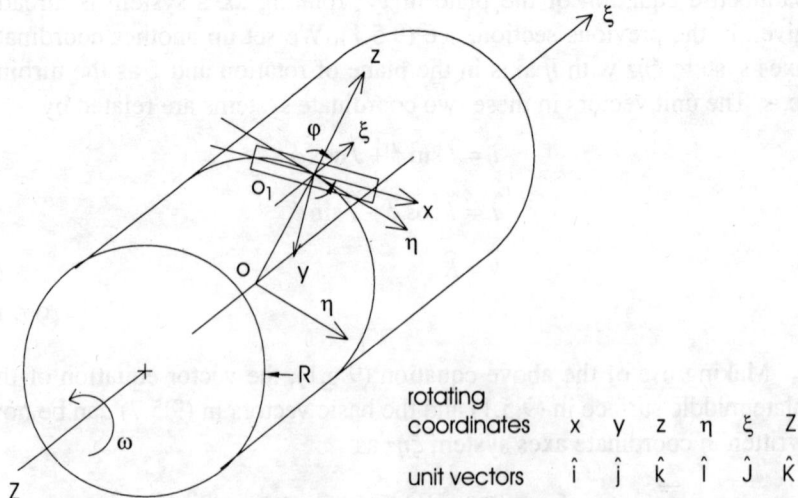

rotating coordinates	x	y	z	η	ξ	Z
unit vectors	$\hat{i}$	$\hat{j}$	$\hat{k}$	$\hat{I}$	$\hat{J}$	$\hat{K}$

Fig. 9.6.1b System of Rotating Coordinates on the Disc

$$\frac{\vec{M},_a}{A} = [\sin(\Gamma\beta+\phi)]\hat{I} + [\cos(\Gamma\beta+\phi)]\hat{J}$$

$$\frac{\vec{M},_\beta}{B} = -\left[\frac{\Gamma a}{B}\cos(\Gamma\beta+\phi)\right]\hat{I} - \left[\frac{\Gamma a}{B}\sin(\Gamma\beta+\phi)\right]\hat{J} + \frac{1}{B}\hat{k}$$

$$\vec{n} = -\left[\frac{1}{B}\cos(\Gamma\beta+\phi)\right]\hat{I} + \left[\frac{1}{B}\sin(\Gamma\beta+\phi)\right]\hat{J} + \frac{\Gamma a}{B}\hat{k}$$

(9.6.3)

Let the displacement vector in the new coordinate system be expressed as

$$\vec{U} = \eta_D \hat{I} + \xi_D \hat{J} + z_D \hat{K}$$

(9.6.4)

where

$$\eta_D = u\sin(\Gamma\beta+\phi) + \frac{w - v\Gamma a}{B}\cos(\Gamma\beta+\phi)$$

$$\xi_D = u\cos(\Gamma\beta+\phi) - \frac{w + v\Gamma a}{B}\sin(\Gamma\beta+\phi)$$

$$z_D = \frac{v - w\Gamma a}{B}$$

(9.6.5)

Equation (9.6.2) can be written with reference to the coordinate system located at the disc center as

$$\vec{M}_o(a,\beta) = [a\sin(\Gamma\beta+\phi)]\hat{I} + [a\cos(\Gamma\beta+\phi)]\hat{J} + (R+\beta)\hat{k}$$

(9.6.2a)

The vector equation of the deformed surface is then given by

$$\vec{M}'(a,\beta) = [a\sin(\Gamma\beta+\phi) + \eta_D]\hat{I}$$
$$+ [a\cos(\Gamma\beta+\phi) + \xi_D]\hat{J} + (R+\beta+z_D)\hat{k}$$

(9.6.6)

The projection of the deformed surface vector in the plane of the disc is, see Fig. 9.6.1a.

$$\overline{OP_1} = \sqrt{[a\sin(\Gamma\beta+\phi)+\eta_D]^2 + (R+\beta+z_D)^2}$$

(9.6.7)

Therefore the centrifugal force of an element $da\,d\beta$ located at this point, see Fig. 9.6.1a is

$$dF = AB\rho h\omega^2 da\,d\beta\,\overline{OP_1}$$

(9.6.8)

The two components of this force in η and z directions are

$$dF_\eta = AB\rho h\omega^2 da\,d\beta[a\sin(\Gamma\beta+\phi)+\eta_D]$$
$$dF_z = AB\rho h\omega^2 da\,d\beta(R+\beta+z_D)$$

(9.6.9)

The gain in strain energy of the element due to the above elemental forces is given by

$$dU_\eta = -\int_0^{\eta_D} dF_\eta\,d\eta_D = -AB\rho h\omega^2 da\,d\beta\left[\eta_D a\sin(\Gamma\beta+\phi) + \frac{\eta_D^2}{2}\right]$$

$$dU_z = -\int_0^{z_D} dF_z\,dz_D = -AB\rho h\omega^2 da\,d\beta\left[(R+\beta)z_D + \frac{z_D^2}{2}\right]$$

(9.6.10)

The net gain in the potential energy of the plate is therefore

$$U_F = \iint(dU_\eta + dU_z)$$
$$= -\omega^2 \int_{-b/2}^{b/2}\int_0^l \left[\eta_D a\sin(\Gamma\beta+\phi) + \frac{\eta_D^2}{2} + (R+\beta)z_D + \frac{z_D^2}{2}\right] AB\rho h\,da\,d\beta$$

(9.6.11)

An inward displacement takes place while the element rotates about a and normal axes as shown in Fig. 9.6.2. This displacement can be written as

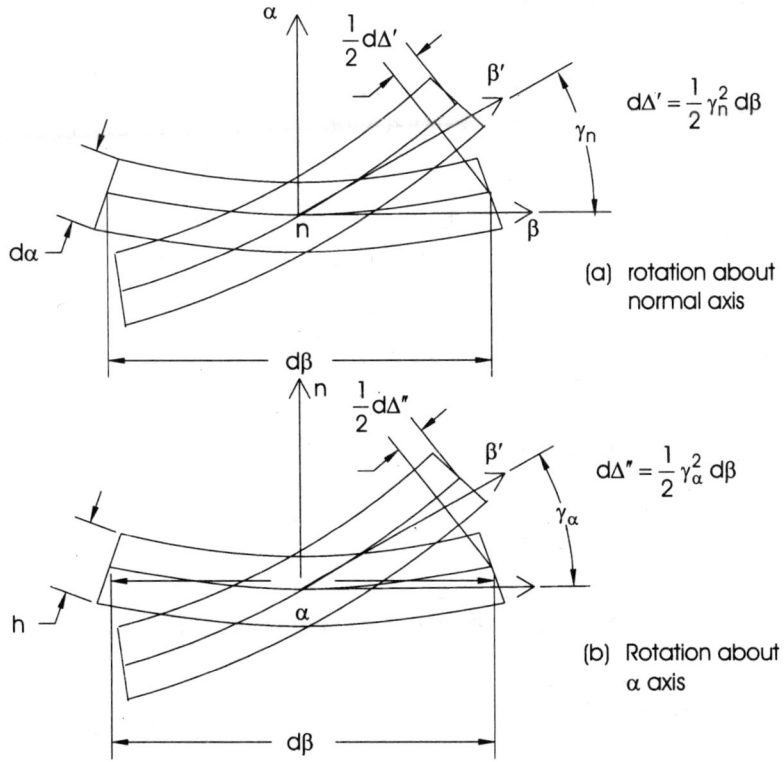

Fig. 9.6.2 Rotations about normal and α axes

$$d\Delta = \tfrac{1}{2}d\beta\left(v_\alpha^2 + v_\beta^2\right)$$
(9.6.12)

The total inward displacement is given by

$$\Delta = \tfrac{1}{2}\int_0^\beta \left(v_\alpha^2 + v_\beta^2\right)d\beta$$
(9.6.13)

In view of the above inward displacement, the displacement v in the axial direction is modified to give (9.6.5) as

$$\eta_D = u\sin(\Gamma\beta+\phi) + \frac{w-(v-\Delta)\Gamma a}{B}\cos(\Gamma\beta+\phi)$$

$$\xi_D = u\cos(\Gamma\beta+\phi) - \frac{w+(v-\Delta)\Gamma a}{B}\sin(\Gamma\beta+\phi)$$

$$z_D = \frac{(v-\Delta)-w\Gamma a}{B}$$

(9.6.5a)

Equation (9.6.11) can now be written as

$$U_F = -\rho h \omega^2 \int_{-b/2}^{b/2}\int_0^l \left[\begin{Bmatrix} u\sin(\Gamma\beta+\phi) \\ +\dfrac{w-(v-\Delta)\Gamma a}{B}\cos(\Gamma\beta+\phi) \end{Bmatrix} a\sin(\Gamma\beta+\phi) \\ +\dfrac{1}{2}\begin{Bmatrix} u\sin(\Gamma\beta+\phi) \\ +\dfrac{w-(v-\Delta)\Gamma a}{B}\cos(\Gamma\beta+\phi) \end{Bmatrix}^2 \\ +(R+\beta)\left\{\dfrac{(v-\Delta)-w\Gamma a}{B}\right\} \\ +\dfrac{1}{2}\left\{\dfrac{(v-\Delta)-w\Gamma a}{B}\right\}^2 \end{array}\right]$$

$\times AB\,da\,d\beta$

(9.6.11)

On further rearranging the above equation (9.6.11), it becomes

$$U_F = -\rho h\omega^2 \int_{-b/2}^{b/2}\int_0^l \left\{ \begin{array}{l} \dfrac{1}{B^2}(v^2+w^2\Gamma^2a^2-2vw\Gamma a)-\dfrac{2\Delta}{B}(R+\beta) \\ +\dfrac{1}{B^2}(w^2+v^2\Gamma^2a^2-2vw\Gamma a)\cos^2(\Gamma\beta+\phi) \\ +u^2\sin^2(\Gamma\beta+\phi) \\ +\dfrac{1}{B}(uw-uv\Gamma a+a^2\Gamma\Delta)\sin 2(\Gamma\beta+\phi) \end{array}\right\} B\,da\,d\beta$$

$$-2\rho h\omega^2 \int_{-b/2}^{b/2}\int_0^l \left\{ \begin{array}{l} \dfrac{1}{B}(v-w\Gamma a)(R+\beta)+au\sin^2(\Gamma\beta+\phi) \\ -\dfrac{1}{2B}(v\Gamma a^2-wa)\sin 2(\Gamma\beta+\phi) \end{array}\right\} B\,da\,d\beta$$

$$-\rho h\omega^2 \int_{-b/2}^{b/2}\int_0^l \dfrac{1}{B^2}\left\{ \begin{array}{l} \dfrac{1}{B^2}[2\Delta(v+w\Gamma a)+\Delta^2]+ \\ \begin{pmatrix}\Delta^2\Gamma^2a^2+2\Delta w\Gamma a \\ -2v\Delta\Gamma^2a^2\end{pmatrix}\cos^2(\Gamma\beta+\phi) \\ +\dfrac{1}{B}(u\Delta\Gamma a)\sin 2(\Gamma\beta+\phi) \end{array}\right\} B\,da\,d\beta$$

(9.6.12)

In the above equation (9.6.12), the first integral terms give the linear terms, the second integral gives the non homogeneous terms for the changes in the mean position (pseudo static deformations) and the last integral gives nonlinear terms. In what follows, we consider only the linear terms given by the first integral.

The Lagrangian expression given in (9.5.15) is now modified to include the effect of centrifugal forces

$$L = \int_{-\frac{h}{2}}^{\frac{h}{2}} \int_0^l \left\{ \begin{array}{l} \dfrac{EB}{2(1-v^2)} h \left[\begin{array}{l} \left\{ \varepsilon_{\alpha\alpha}^2 + \varepsilon_{\beta\beta}^2 + 2v\varepsilon_{\alpha\alpha}\varepsilon_{\beta\beta} + \dfrac{1}{2}(1-v)\varepsilon_{\alpha\beta}^2 \right\} + \\ \dfrac{h^2}{12}\left\{ \kappa_{\alpha\alpha}^2 + \kappa_{\beta\beta}^2 + 2v\kappa_{\alpha\alpha}\kappa_{\beta\beta} + 2(1-v)\kappa_{\alpha\beta}^2 \right\} \end{array} \right] \\ -\dfrac{ph\omega^2}{2B} \left\{ \begin{array}{l} (v^2 + w^2\Gamma^2 a^2 - 2vw\Gamma a) - 2\Delta B(R+\beta) \\ +(w^2 + v^2\Gamma^2 a^2 - 2vw\Gamma a)\cos^2(\Gamma\beta+\phi) \\ +u^2 B^2 \sin^2(\Gamma\beta+\phi) + \\ (uw - uv\Gamma a + a^2\Gamma\Delta)B\sin 2(\Gamma\beta+\phi) \end{array} \right\} \\ -\dfrac{1}{2}\rho Bhp^2 \left[u^2 + v^2 + w^2 + \dfrac{h^2}{12}(\gamma_1^2 + \gamma_2^2) \right] \end{array} \right\} d\alpha d\beta$$

(9.6.13)

We can use the assumed functions given in the previous section 9.5 to determine the eigen values following Ritz method. The fundamental bending frequency of a 30° pre-twisted 4.57 by 2.54 by 0.3175 cm mild steel plate mounted on a disc of 4.445 cm radius at 0° stagger angle is given in Table 9.6.1.

It should be noted that the aspect ratio has a significant influence on the non dimensional natural frequencies, the higher the value of aspect ratio, the plate behaves like a beam. The percentage variation of bending natural frequencies of pre-twisted rotating plates of aspect ratio 1 from 0 to 15000 rpm is given in Table 9.6.2 for different disc radii, pre-twist angles and stagger angles.

Table 9.6.1 I Bending Mode of a Pre-twisted Plate

$$\bar{p}_r = p_r \sqrt{\dfrac{12\rho l^4 (1-v^2)}{Eh^2}}$$

rpm	Frequency
0	3.3178
2000	3.3262
6000	3.3931
10000	3.5225

Table 9.6.2 Percentage Variation in Natural Frequencies of Pre-twisted Rotating Plates from 0 to 15,000 rpm with aspect ratio $r = 1$

Mode	R/b = 1			R/b = 2		R/b = 3			
	0°	21°	41°	0°	21°	0°	21°	41°	← ϕ ↓
0/1	5.200	5.390	5.950	9.620	10.020	13.870	14.460	16.070	0°
	5.890	6.250	7.150	10.290	10.840	14.520	15.250	17.170	30°
	-	-	8.840	-	-	-	-	18.700	60°
	-	-	9.320	-	-	-	-	19.160	90°
0/2	0.980	1.010	1.010	1.580	1.640	2.180	2.260	2.250	0°
	1.000	1.030	1.020	1.600	1.650	2.200	2.270	2.260	30°
	-	-	1.040	-	-	-	-	2.280	60°
	-	-	1.060	-	-	-	-	2.290	90°
0/3	0.360	0.350	0.260	0.580	0.560	0.790	0.770	0.590	0°
	0.360	0.350	0.270	0.580	0.560	0.800	0.770	0.590	30°
	-	-	0.270	-	-	-	-	0.600	60°
	-	-	0.270	-	-	-	-	0.600	90°

In beam theory, torsional modes are not affected by centrifugal forces; in plate theory, however, the centrifugal forces have an influence on torsional modes, though not to the same extent as in bending modes. The effect of centrifugal forces on plate modes, disc radius and stagger angle are given in Tables 9.6.3 to 9.6.5, respectively.

Table 9.6.3 Effect of Centrifugal Forces on Plate Modes

$$\bar{p}_r = p_r \sqrt{\frac{12\rho l^4(1-v^2)}{Eh^2}}$$
$$r = 2.33, \phi = 0°, R/b = 1$$

Mode	0 rpm	5,000 rpm	10,000 rpm	15,000 rpm
2/1	125.2890	125.3890	125.4890	125.7420
2/2	160.8580	161.2110	162.2630	163.9870
2/3	220.0980	220.7010	222.5000	225.4710
3/1	350.3460	350.4260	350.6720	351.1010
3/2	387.6190	387.7790	388.2560	389.0560
3/3	442.0570	442.3590	443.2640	444.7710

Table 9.6.4 Effect of Disc Radius on Plate Modes

$$\bar{p}_r = p_r \sqrt{\frac{12\rho l^4 (1-v^2)}{Eh^2}}$$

$r = 2.33, \gamma = 30°, \phi = 0°, \omega = 10000$ rpm

Mode	R/b = 1	R/b = 2.33
2/1	130.29	130.46
2/2	169.75	170.46
2/3	237.51	238.64
3/1	354.85	355.18
3/2	393.47	393.84
3/3	451.57	452.18

Table 9.6.5 Effect of Stagger Angle on Plate Modes

$$\bar{p}_r = p_r \sqrt{\frac{12\rho l^4 (1-v^2)}{Eh^2}}$$

$r = 2.33, \gamma = 30°, R/b = 2.33, \omega = 10000$ rpm

Mode	$\phi = 0°$	$\phi = 30°$
2/1	130.4590	130.5180
2/2	170.4630	170.5080
2/3	238.6430	238.6730
3/1	355.1830	355.2060
3/2	393.8370	393.8590
3/3	452.1810	452.1980

10

Finite Element Method for Shells

In the previous chapter, we considered a special case of shells, pre-twisted plates which have a negative Gaussian curvature. Other books on shells consider a variety of shells, e.g., cylindrical shells, spherical shells etc. We will concentrate in this book the special case of pre-twisted plates, stationary as well as rotating in developing a finite element method. First, however, we will discuss Ahmad's 8 noded super-parametric element, before developing a finite element based on the shell theory discussed in the previous chapter.

10.1 AHMAD'S SUPER-PARAMETRIC ELEMENT

The basic shape of the element is shown in Fig. 10.1.1. The external faces of the shell are curved, while the sections across the thickness are generated by straight lines. The geometry of the element is described by a pair of points, i_{top}, i_{bottom} each with given Cartesian coordinates. ξ, η are curvilinear coordinates in the middle plane of the shell and ζ is a linear coordinate in the thickness direction. The mid-surface of the parent element in natural coordinates is shown in Fig. 10.1.2.

Shape Functions

Assuming that ξ, η and ζ vary between -1 and $+1$ on the respective faces of the element, the Cartesian coordinates of any point on the shell, geometric shape functions, are expressed as

$$\left\{\begin{array}{c} x \\ y \\ z \end{array}\right\} = \sum_{i=1}^{8} N_i(\xi,\eta) \frac{1+\zeta}{2} \left\{\begin{array}{c} x \\ y \\ z \end{array}\right\}_{top} + \sum_{i=1}^{8} N_i(\xi,\eta) \frac{1-\zeta}{2} \left\{\begin{array}{c} x \\ y \\ z \end{array}\right\}_{bottom}$$

(10.1.1)

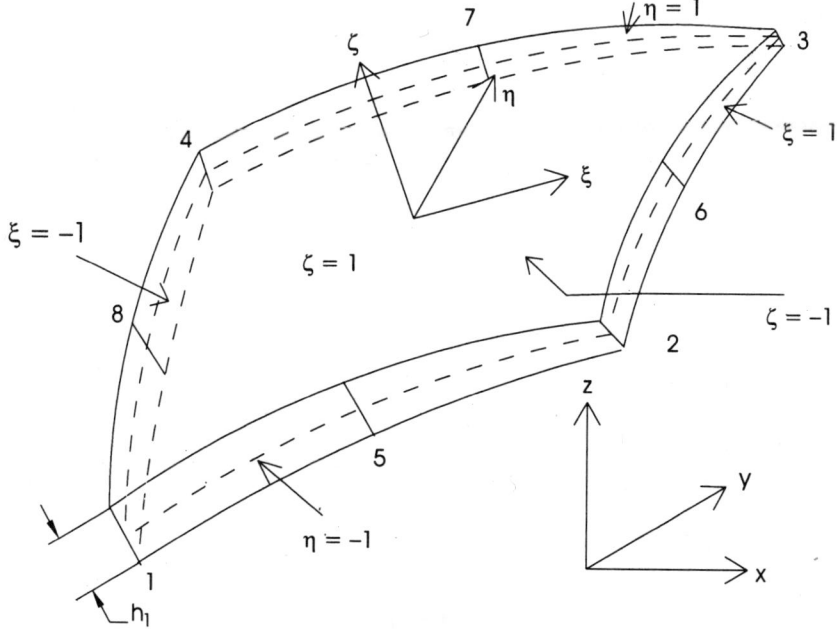

Fig. 10.1.1 Basic Shape of the Shell Element

where

$$N_1(\xi,\eta) = \frac{1}{4}(1-\xi)(1-\eta)(-\xi-\eta-1)$$

$$N_2(\xi,\eta) = \frac{1}{4}(1+\xi)(1-\eta)(\xi-\eta-1)$$

$$N_3(\xi,\eta) = \frac{1}{4}(1+\xi)(1+\eta)(\xi+\eta-1)$$

$$N_4(\xi,\eta) = \frac{1}{4}(1-\xi)(1+\eta)(-\xi+\eta-1)$$

$$N_5(\xi,\eta) = \frac{1}{2}(1-\xi^2)(1-\eta)$$

$$N_6(\xi,\eta) = \frac{1}{2}(1+\xi)(1-\eta^2)$$

$$N_7(\xi,\eta) = \frac{1}{2}(1-\xi^2)(1+\eta)$$

$$N_8(\xi,\eta) = \frac{1}{2}(1-\xi)(1-\eta^2)$$

(10.1.2)

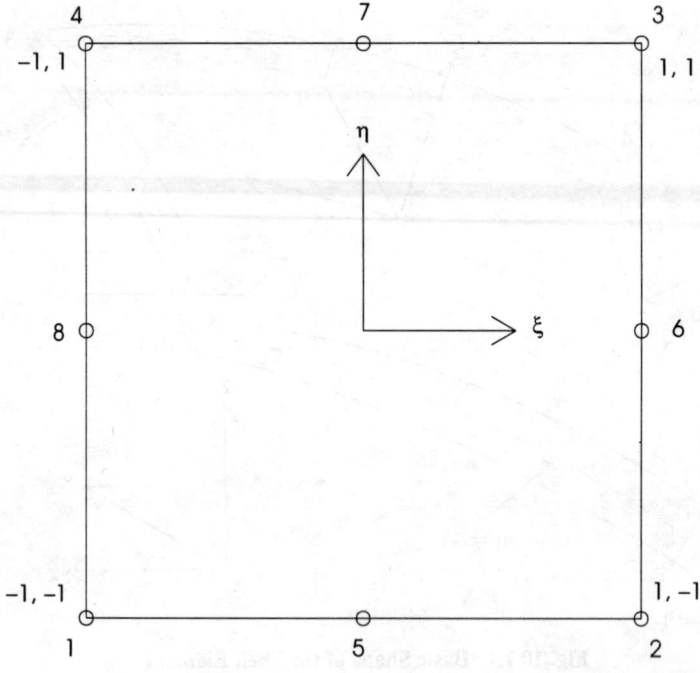

Fig. 10.1.2 Parent Element

Note that $N_i(\xi,\eta)$ takes a value of unity at node i and zero at all other nodes. Equation (10.1.1) can also be expressed in terms of the mid-surface coordinates,

$$\left\{\begin{array}{c} x \\ y \\ z \end{array}\right\} = \sum_{i=1}^{8} N_i(\xi,\eta) \left\{\begin{array}{c} x \\ y \\ z \end{array}\right\}_{mid} + \sum_{i=1}^{8} N_i(\xi,\eta) \frac{\zeta}{2} \tilde{V}_{3i} \tag{10.1.3}$$

where

$$\tilde{V}_{3i} = \left\{\begin{array}{c} x_i \\ y_i \\ z_i \end{array}\right\}_{top} - \left\{\begin{array}{c} x_i \\ y_i \\ z_i \end{array}\right\}_{bottom} \tag{10.1.4}$$

Displacement Field

The displacement field is defined by the three displacements of the mid-surface nodes and two rotations of the nodal vector $\tilde{V}_{3i}$ about its own orthogonal directions.

$$\left\{\begin{array}{c}u\\v\\w\end{array}\right\} = \sum_{i=1}^{8} N_i(\xi,\eta)\left\{\begin{array}{c}u_i\\v_i\\w_i\end{array}\right\} + \sum_{i=1}^{8} N_i(\xi,\eta)\frac{\zeta h_i}{2}\left[\begin{array}{cc}V_{1x}^i & -V_{2x}^i\\V_{1y}^i & -V_{2y}^i\\V_{1z}^i & -V_{2z}^i\end{array}\right]\left\{\begin{array}{c}-\beta_i\\\alpha_i\end{array}\right\}$$

(10.1.5)

where

$$\tilde{V}_{1i} = V_{1x}^i \hat{e}_x + V_{1y}^i \hat{e}_y + V_{1z}^i \hat{e}_z = \frac{\hat{e}_x \times \tilde{V}_{3i}}{|\hat{e}_x \times \tilde{V}_{3i}|}$$

$$\tilde{V}_{2i} = V_{2x}^i \hat{e}_x + V_{2y}^i \hat{e}_y + V_{2z}^i \hat{e}_z = \tilde{V}_{3i} \times \tilde{V}_{1i}$$

(10.1.6)

u_i, v_i, w_i are the displacements at the mid surface nodes and α_i, β_i are the two rotations about the directions $\tilde{V}_{2i}, \tilde{V}_{1i}$ respectively, see Fig. 10.1.3. There are five degrees of freedom per each node given by

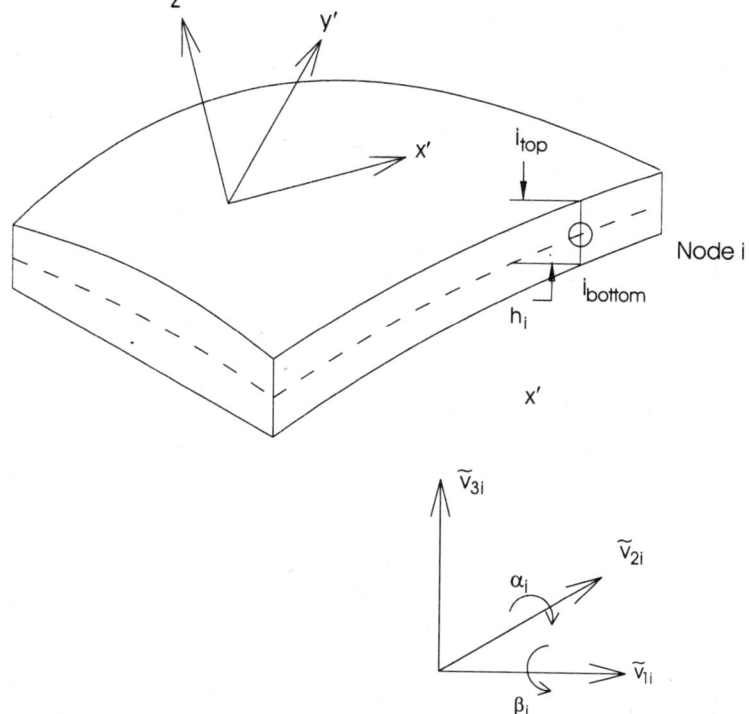

Fig. 10.1.3 Nodal Rotations

494 Dynamics of Plates

$$\{\hat{d}_i\} = \begin{Bmatrix} u_i \\ v_i \\ w_i \\ \alpha_i \\ \beta_i \end{Bmatrix}$$

(10.1.7)

and forty degrees of freedom for the element. Equation (10.1.5) can be written as

$$\begin{Bmatrix} u \\ v \\ w \end{Bmatrix} = [N] \begin{Bmatrix} \hat{d}_1 \\ \hat{d}_2 \\ \cdots \\ \cdots \\ \hat{d}_8 \end{Bmatrix} = [N]\{\hat{d}\}$$

(10.1.8)

where the elements of $[N]$ matrix are obtained by expanding equation (10.1.5).

$$[N] = \begin{bmatrix} [N]_1 & [N]_2 & \cdots & [N]_8 \end{bmatrix}$$

$$[N]_1 = \begin{bmatrix} N_1 & 0 & 0 & -\frac{1}{2}V^1_{2x}N_1h_1\zeta & -\frac{1}{2}V^1_{1x}N_1h_1\zeta \\ 0 & N_1 & 0 & -\frac{1}{2}V^1_{2y}N_1h_1\zeta & -\frac{1}{2}V^1_{1y}N_1h_1\zeta \\ 0 & 0 & N_1 & -\frac{1}{2}V^1_{2z}N_1h_1\zeta & -\frac{1}{2}V^1_{1z}N_1h_1\zeta \end{bmatrix}$$

$$[N]_2 = \begin{bmatrix} N_2 & 0 & 0 & -\frac{1}{2}V^2_{2x}N_2h_2\zeta & -\frac{1}{2}V^2_{1x}N_2h_2\zeta \\ 0 & N_2 & 0 & -\frac{1}{2}V^2_{2y}N_2h_2\zeta & -\frac{1}{2}V^2_{1y}N_2h_2\zeta \\ 0 & 0 & N_3 & -\frac{1}{2}V^2_{2z}N_2h_2\zeta & -\frac{1}{2}V^2_{1z}N_2h_2\zeta \end{bmatrix}$$

...

(10.1.8a)

The total number of degrees of freedom of the element are forty, whereas, the definition in equation (10.1.1) contains only 24 displacements, therefore, the element is of the super-parametric kind.

Strain-Displacement Relations

The components of strains and stresses in directions of orthogonal axes related to the surface $\zeta =$ constant are only considered. Let z' be a normal to this surface at any point and let the other two orthogonal axes be x' and y' which are tangent to z', see Fig. 10.1.3. The strain components are defined by

$$\{\varepsilon'\} = \begin{Bmatrix} \varepsilon'_{xx} \\ \varepsilon'_{yy} \\ \gamma_{x'y'} \\ \gamma_{y'z'} \\ \gamma_{x'z'} \end{Bmatrix} = \begin{Bmatrix} \dfrac{\partial u'}{\partial x'} \\ \dfrac{\partial v'}{\partial y'} \\ \dfrac{\partial u'}{\partial y'} + \dfrac{\partial v'}{\partial x'} \\ \dfrac{\partial v'}{\partial z'} + \dfrac{\partial w'}{\partial y'} \\ \dfrac{\partial u'}{\partial z'} + \dfrac{\partial w'}{\partial x'} \end{Bmatrix}$$

$$= [H] \begin{Bmatrix} \dfrac{\partial u'}{\partial x'} \\ \dfrac{\partial u'}{\partial y'} \\ \dfrac{\partial u'}{\partial z'} \\ \dfrac{\partial v'}{\partial x'} \\ \dfrac{\partial v'}{\partial y'} \\ \dfrac{\partial v'}{\partial z'} \\ \dfrac{\partial w'}{\partial x'} \\ \dfrac{\partial w'}{\partial y'} \\ \dfrac{\partial w'}{\partial z'} \end{Bmatrix}$$

(10.1.9)

where

$$[H] = \begin{bmatrix} 1 & 0 & 0 & 0 & 0 & 0 & 0 & 0 & 0 \\ 0 & 0 & 0 & 0 & 1 & 0 & 0 & 0 & 0 \\ 0 & 1 & 0 & 1 & 0 & 0 & 0 & 0 & 0 \\ 0 & 0 & 0 & 0 & 0 & 1 & 0 & 1 & 0 \\ 0 & 0 & 1 & 0 & 0 & 0 & 1 & 0 & 0 \end{bmatrix}$$

(10.1.10)

Stress-Strain Relation

The stress strain relation is taken as

$$\{\sigma'\} = [D]\{\varepsilon'\} \tag{10.1.11}$$

where

$$[D] = \begin{bmatrix} 1 & v & 0 & 0 & 0 \\ v & 1 & 0 & 0 & 0 \\ 0 & 0 & \frac{1-v}{2} & 0 & 0 \\ 0 & 0 & 0 & \frac{1-v}{2k} & 0 \\ 0 & 0 & 0 & 0 & \frac{1-v}{2k} \end{bmatrix} \tag{10.1.12}$$

Factor k in the above equation (10.1.12) is taken as 1.2 to improve the shear displacement approximation.

Transformation

We need now a transformation from the global coordinates x,y,z to local coordinates x',y',z'. We first define a vector normal to the surface $\zeta =$ constant, by taking the vector product of any two vectors tangent to the surface,

$$\tilde{V}_{z'} = \begin{Bmatrix} \frac{\partial x}{\partial \xi} \\ \frac{\partial y}{\partial \xi} \\ \frac{\partial z}{\partial \xi} \end{Bmatrix} \times \begin{Bmatrix} \frac{\partial x}{\partial \eta} \\ \frac{\partial y}{\partial \eta} \\ \frac{\partial z}{\partial \eta} \end{Bmatrix} \tag{10.1.13}$$

The other two vectors $\tilde{V}_{x'}, \tilde{V}_{y'}$ along x' and y' directions can also be obtained by the process given above and reducing these to unit magnitudes, we can construct a matrix of unit vectors in x',y',z' directions

$$[\theta] = \begin{bmatrix} \hat{V}_{x'} & \hat{V}_{y'} & \hat{V}_{z'} \end{bmatrix}$$

$$= \begin{bmatrix} l_1 & l_2 & l_3 \\ m_1 & m_2 & m_3 \\ n_1 & n_2 & n_3 \end{bmatrix} \tag{10.1.14}$$

We can now write

$$\begin{bmatrix} \frac{\partial u'}{\partial x'} & \frac{\partial v'}{\partial x'} & \frac{\partial w'}{\partial x'} \\ \frac{\partial u'}{\partial y'} & \frac{\partial v'}{\partial y'} & \frac{\partial w'}{\partial y'} \\ \frac{\partial u'}{\partial z'} & \frac{\partial v'}{\partial z'} & \frac{\partial w'}{\partial z'} \end{bmatrix} = [\theta]^T \begin{bmatrix} \frac{\partial u}{\partial x} & \frac{\partial v}{\partial x} & \frac{\partial w}{\partial x} \\ \frac{\partial u}{\partial y} & \frac{\partial v}{\partial y} & \frac{\partial w}{\partial y} \\ \frac{\partial u}{\partial z} & \frac{\partial v}{\partial z} & \frac{\partial w}{\partial z} \end{bmatrix} [\theta]$$

(10.1.15)

i.e.,

$$\begin{Bmatrix} \frac{\partial u'}{\partial x'} \\ \frac{\partial u'}{\partial y'} \\ \frac{\partial u'}{\partial z'} \\ \frac{\partial v'}{\partial x'} \\ \frac{\partial v'}{\partial y'} \\ \frac{\partial v'}{\partial z'} \\ \frac{\partial w'}{\partial x'} \\ \frac{\partial w'}{\partial y'} \\ \frac{\partial w'}{\partial z'} \end{Bmatrix} = [T] \begin{Bmatrix} \frac{\partial u}{\partial x} \\ \frac{\partial u}{\partial y} \\ \frac{\partial u}{\partial z} \\ \frac{\partial v}{\partial x} \\ \frac{\partial v}{\partial y} \\ \frac{\partial v}{\partial z} \\ \frac{\partial w}{\partial x} \\ \frac{\partial w}{\partial y} \\ \frac{\partial w}{\partial z} \end{Bmatrix}$$

(10.1.16)

where

$$[T] = \begin{bmatrix} l_1[\theta]^T & m_1[\theta]^T & n_1[\theta]^T \\ l_2[\theta]^T & m_2[\theta]^T & n_2[\theta]^T \\ l_3[\theta]^T & m_3[\theta]^T & n_3[\theta]^T \end{bmatrix}$$

(10.1.17)

and $l_1, l_2 \cdots$ are directional derivatives. The derivatives of the global displacements u, v, w with respect to the curvilinear coordinates are related to the derivatives with respect to Cartesian coordinates through the Jacobian by

$$\begin{bmatrix} \frac{\partial u}{\partial x} & \frac{\partial v}{\partial x} & \frac{\partial w}{\partial x} \\ \frac{\partial u}{\partial y} & \frac{\partial v}{\partial y} & \frac{\partial w}{\partial y} \\ \frac{\partial u}{\partial z} & \frac{\partial v}{\partial z} & \frac{\partial w}{\partial z} \end{bmatrix} = [J]^{-1} \begin{bmatrix} \frac{\partial u}{\partial \xi} & \frac{\partial v}{\partial \xi} & \frac{\partial w}{\partial \xi} \\ \frac{\partial u}{\partial \eta} & \frac{\partial v}{\partial \eta} & \frac{\partial w}{\partial \eta} \\ \frac{\partial u}{\partial \zeta} & \frac{\partial v}{\partial \zeta} & \frac{\partial w}{\partial \zeta} \end{bmatrix}$$

(10.1.18)

Therefore,

$$\begin{Bmatrix} \frac{\partial u}{\partial x} \\ \frac{\partial u}{\partial y} \\ \frac{\partial u}{\partial z} \\ \frac{\partial v}{\partial x} \\ \frac{\partial v}{\partial y} \\ \frac{\partial v}{\partial z} \\ \frac{\partial w}{\partial x} \\ \frac{\partial w}{\partial y} \\ \frac{\partial w}{\partial z} \end{Bmatrix} = [\Gamma] \begin{Bmatrix} \frac{\partial u}{\partial \xi} \\ \frac{\partial u}{\partial \eta} \\ \frac{\partial u}{\partial \zeta} \\ \frac{\partial v}{\partial \xi} \\ \frac{\partial v}{\partial \eta} \\ \frac{\partial v}{\partial \zeta} \\ \frac{\partial w}{\partial \xi} \\ \frac{\partial w}{\partial \eta} \\ \frac{\partial w}{\partial \zeta} \end{Bmatrix}$$

(10.1.19)

where

$$[\Gamma] = \begin{bmatrix} [J]^{-1} & 0 & 0 \\ 0 & [J]^{-1} & 0 \\ 0 & 0 & [J]^{-1} \end{bmatrix}$$

(10.1.20)

The Jacobian is

$$[J] = \begin{bmatrix} \frac{\partial x}{\partial \xi} & \frac{\partial y}{\partial \xi} & \frac{\partial z}{\partial \xi} \\ \frac{\partial x}{\partial \eta} & \frac{\partial y}{\partial \eta} & \frac{\partial z}{\partial \eta} \\ \frac{\partial x}{\partial \zeta} & \frac{\partial y}{\partial \zeta} & \frac{\partial z}{\partial \zeta} \end{bmatrix}$$

(10.1.21)

where

$$\frac{\partial x}{\partial \xi} = \sum_{i=1}^{8} \frac{\partial N_i}{\partial \xi} x_{\text{mid}}^i + \sum_{i=1}^{8} \frac{1}{2} \frac{\partial N_i}{\partial \xi} h_i \zeta V_{3ix}$$

$$\frac{\partial y}{\partial \xi} = \sum_{i=1}^{8} \frac{\partial N_i}{\partial \xi} y_{\text{mid}}^i + \sum_{i=1}^{8} \frac{1}{2} \frac{\partial N_i}{\partial \xi} h_i \zeta V_{3iy}$$

$$\frac{\partial z}{\partial \xi} = \sum_{i=1}^{8} \frac{\partial N_i}{\partial \xi} z_{\text{mid}}^i + \sum_{i=1}^{8} \frac{1}{2} \frac{\partial N_i}{\partial \xi} h_i \zeta V_{3iz}$$

$$\frac{\partial x}{\partial \eta} = \sum_{i=1}^{8} \frac{\partial N_i}{\partial \eta} x_{\text{mid}}^i + \sum_{i=1}^{8} \frac{1}{2} \frac{\partial N_i}{\partial \eta} h_i \zeta V_{3ix}$$

$$\frac{\partial y}{\partial \eta} = \sum_{i=1}^{8} \frac{\partial N_i}{\partial \eta} y_{\text{mid}}^i + \sum_{i=1}^{8} \frac{1}{2} \frac{\partial N_i}{\partial \eta} h_i \zeta V_{3iy}$$

$$\frac{\partial z}{\partial \eta} = \sum_{i=1}^{8} \frac{\partial N_i}{\partial \eta} z_{\text{mid}}^i + \sum_{i=1}^{8} \frac{1}{2} \frac{\partial N_i}{\partial \eta} h_i \zeta V_{3iz}$$

$$\frac{\partial x}{\partial \zeta} = \sum_{i=1}^{8} \frac{1}{2} N_i h_i \zeta V_{3ix}$$

$$\frac{\partial y}{\partial \zeta} = \sum_{i=1}^{8} \frac{1}{2} N_i h_i \zeta V_{3iy}$$

$$\frac{\partial z}{\partial \zeta} = \sum_{i=1}^{8} \frac{1}{2} N_i h_i \zeta V_{3iz}$$

(10.1.22)

From (10.1.5), we can obtain

$$\begin{Bmatrix} \frac{\partial u}{\partial \xi} \\ \frac{\partial u}{\partial \eta} \\ \frac{\partial u}{\partial \zeta} \\ \frac{\partial v}{\partial \xi} \\ \frac{\partial v}{\partial \eta} \\ \frac{\partial v}{\partial \zeta} \\ \frac{\partial w}{\partial \xi} \\ \frac{\partial w}{\partial \eta} \\ \frac{\partial w}{\partial \zeta} \end{Bmatrix} = [L] \begin{Bmatrix} \hat{d}_1 \\ \hat{d}_2 \\ \cdots \\ \cdots \\ \hat{d}_8 \end{Bmatrix}$$

(10.1.23)

where

$$[L] = \begin{bmatrix} [L_{11}] & [L_{12}] & \cdots & [L_{18}] \\ [L_{21}] & [L_{22}] & \cdots & [L_{28}] \\ [L_{31}] & [L_{32}] & \cdots & [L_{38}] \end{bmatrix}$$

(10.1.24)

$$[L_{11}] = \begin{bmatrix} \dfrac{\partial N_1}{\partial \xi} & 0 & 0 & -\dfrac{1}{2}V^1_{2x}\dfrac{\partial N_1}{\partial \xi}h_1\zeta & -\dfrac{1}{2}V^1_{1x}\dfrac{\partial N_1}{\partial \xi}h_1\zeta \\ \dfrac{\partial N_1}{\partial \eta} & 0 & 0 & -\dfrac{1}{2}V^1_{2x}\dfrac{\partial N_1}{\partial \eta}h_1\zeta & -\dfrac{1}{2}V^1_{1x}\dfrac{\partial N_1}{\partial \eta}h_1\zeta \\ 0 & 0 & 0 & -\dfrac{1}{2}V^1_{2x}N_1h_1 & -\dfrac{1}{2}V^1_{1x}N_1h_1 \end{bmatrix}$$

$$[L_{21}] = \begin{bmatrix} 0 & \dfrac{\partial N_1}{\partial \xi} & 0 & -\dfrac{1}{2}V^1_{2y}\dfrac{\partial N_1}{\partial \xi}h_1\zeta & -\dfrac{1}{2}V^1_{1y}\dfrac{\partial N_1}{\partial \xi}h_1\zeta \\ 0 & \dfrac{\partial N_1}{\partial \eta} & 0 & -\dfrac{1}{2}V^1_{2y}\dfrac{\partial N_1}{\partial \eta}h_1\zeta & -\dfrac{1}{2}V^1_{1y}\dfrac{\partial N_1}{\partial \eta}h_1\zeta \\ 0 & 0 & 0 & -\dfrac{1}{2}V^1_{2y}N_1h_1 & -\dfrac{1}{2}V^1_{1y}N_1h_1 \end{bmatrix}$$

$$[L_{31}] = \begin{bmatrix} 0 & 0 & \dfrac{\partial N_1}{\partial \xi} & -\dfrac{1}{2}V^1_{2z}\dfrac{\partial N_1}{\partial \xi}h_1\zeta & -\dfrac{1}{2}V^1_{1z}\dfrac{\partial N_1}{\partial \xi}h_1\zeta \\ 0 & 0 & \dfrac{\partial N_1}{\partial \eta} & -\dfrac{1}{2}V^1_{2z}\dfrac{\partial N_1}{\partial \eta}h_1\zeta & -\dfrac{1}{2}V^1_{1z}\dfrac{\partial N_1}{\partial \eta}h_1\zeta \\ 0 & 0 & 0 & -\dfrac{1}{2}V^1_{2z}N_1h_1 & -\dfrac{1}{2}V^1_{1z}N_1h_1 \end{bmatrix}$$

$$[L_{12}] = \begin{bmatrix} \dfrac{\partial N_2}{\partial \xi} & 0 & 0 & -\dfrac{1}{2}V^2_{2x}\dfrac{\partial N_2}{\partial \xi}h_2\zeta & -\dfrac{1}{2}V^2_{1x}\dfrac{\partial N_2}{\partial \xi}h_2\zeta \\ \dfrac{\partial N_2}{\partial \eta} & 0 & 0 & -\dfrac{1}{2}V^2_{2x}\dfrac{\partial N_2}{\partial \eta}h_2\zeta & -\dfrac{1}{2}V^2_{1x}\dfrac{\partial N_2}{\partial \eta}h_2\zeta \\ 0 & 0 & 0 & -\dfrac{1}{2}V^2_{2x}N_2h_2 & -\dfrac{1}{2}V^2_{1x}N_2h_2 \end{bmatrix}$$

...

(10.1.25)

From (10.1.9), (10.1.16), (10.1.19) and (10.1.23), we now obtain

$$\{\varepsilon'\} = [B]\begin{Bmatrix} \hat{d}_1 \\ \hat{d}_2 \\ \vdots \\ \vdots \\ \hat{d}_8 \end{Bmatrix}$$

(10.1.26)

where

$$[B] = [H][T][\Gamma][L]$$

(10.1.27)

The elemental stiffness matrix is then given by

$$[\hat{K}] = \int_V [B]^T[D][B]dx\,dy\,dz$$

$$= \int_V [B]^T[D][B]|J|d\xi\,d\eta\,d\zeta$$

(10.1.28)

The elemental mass matrix can be obtained as

$$[\hat{M}] = \int_V \varrho[N]^T[N]dx\,dy\,dz$$

$$= \int_V \varrho[N]^T[N]|J|d\xi\,d\eta\,d\zeta$$

(10.1.29)

Example

Consider a square rectangular cantilever plate 0.05 by 0.05 m with 1 mm thickness with a pre-twist of 20°. Use 16 elements with 65 nodes. Take $E = 710$ GPa, $G = 280$ GPa, Poisson's ratio = 0.25 and density = 2300 kg/m³. As in section 7.6, the plate is divided into 16 elements. The first seven non dimensional natural frequencies $pa^2\sqrt{\dfrac{\varrho h}{D}}$ of the stationary plate are:

Mode No.	Frequency
1. I Bending	3.352
2. II Bending	22.339
3. I Torsion	22.350
4. Plate Mode	32.300
5. II Torsion	45.447
6. Plate Mode	70.712
7. III Bending	73.287

The corresponding mode shapes of the above frequencies are given in Figs. 10.1.4 to 10.1.10.

502 Dynamics of Plates

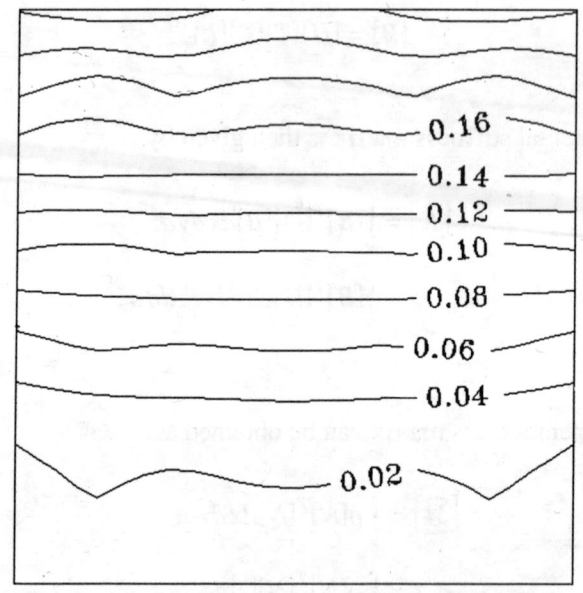

Fig. 10.1.4 The 1st Mode Shape of a Pretwisted Plate

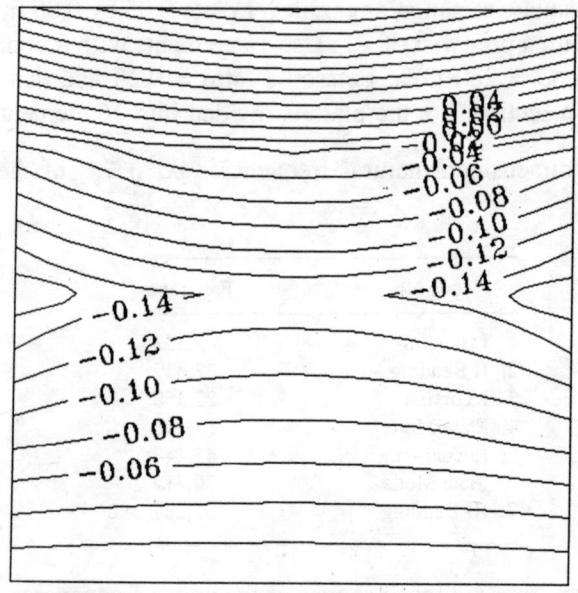

Fig. 10.1.5 The 2nd Mode Shape of a Pretwisted Plate

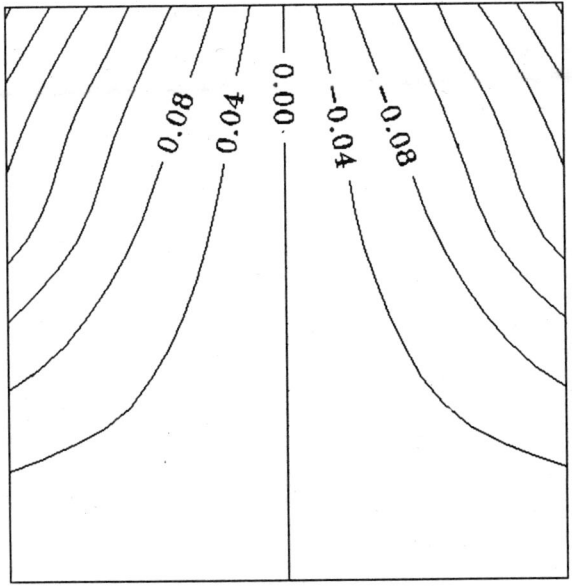

Fig. 10.1.6 The 3rd Mode Shape of a Pretwisted Plate

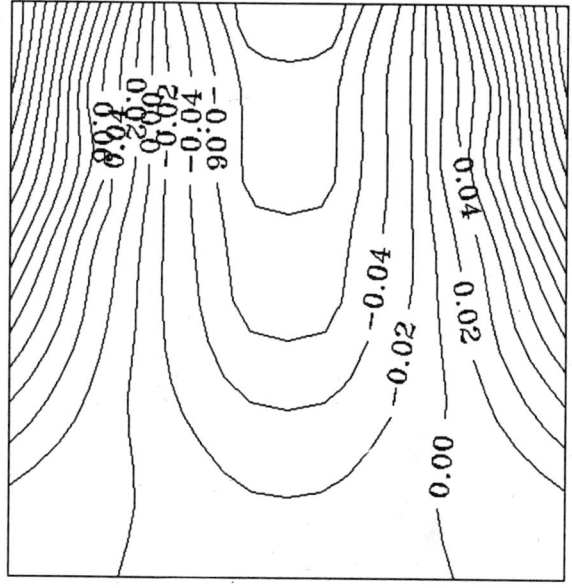

Fig. 10.1.7 The 4th Mode Shape of a Pretwisted Plate

504 *Dynamics of Plates*

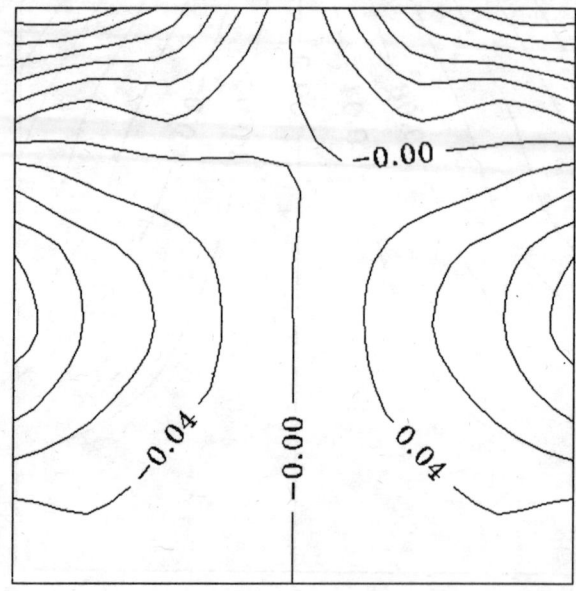

Fig. 10.1.8 The 5th Mode Shape of a Pretwisted Plate

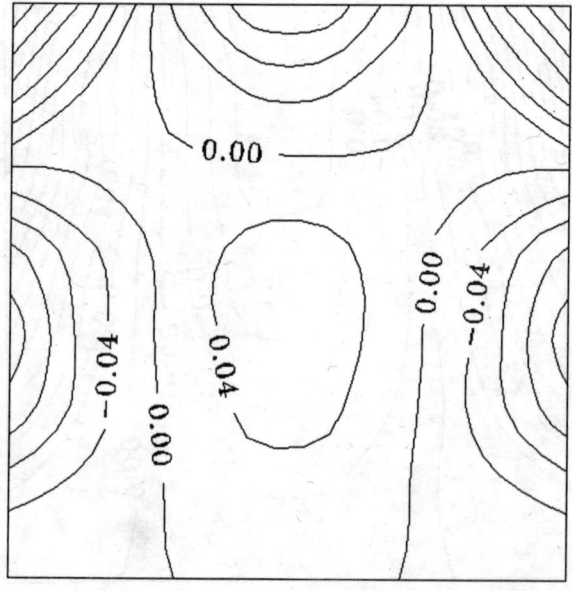

Fig. 10.1.9 The 6th Mode Shape of a Pretwisted Plate

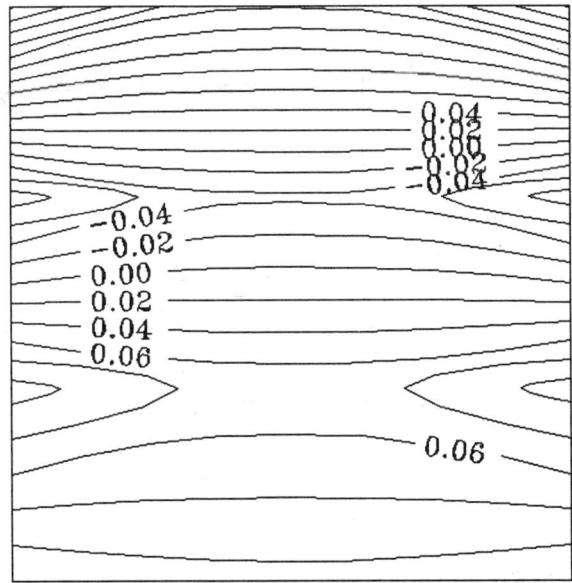

Fig. 10.1.10 The 7th Mode Shape of a Pretwisted Plate

10.2 ROTATING PRE-TWISTED PLATE

Figure 10.2.1 shows a pre-twisted plate mounted on a disc of radius R rotating at an angular velocity ω rad/s. XYZ is the fixed axes system and xyz coordinate system is fixed to the rotating plate as shown. The y axis of the plate is located at a setting angle φ from the tangent to the disc periphery in the XY plane or at a stagger angle $\phi = 90^0 - \varphi$ from the cylinder axis Z. We extend the analysis of previous section by finding the kinetic energy of the rotating plate to find the elemental matrices due to the centrifugal forces. To determine the kinetic energy, we will follow the derivation already given for a straight plate in section 7.7.

Kinetic Energy

The position vector of point P after deformation in the xyz axis system is

$$\overrightarrow{OP} = \begin{Bmatrix} x \\ y \\ z \end{Bmatrix} + \begin{Bmatrix} u \\ v \\ w \end{Bmatrix}$$

$$= \{\bar{x}\} + \{\delta\}$$

(10.2.1)

506 Dynamics of Plates

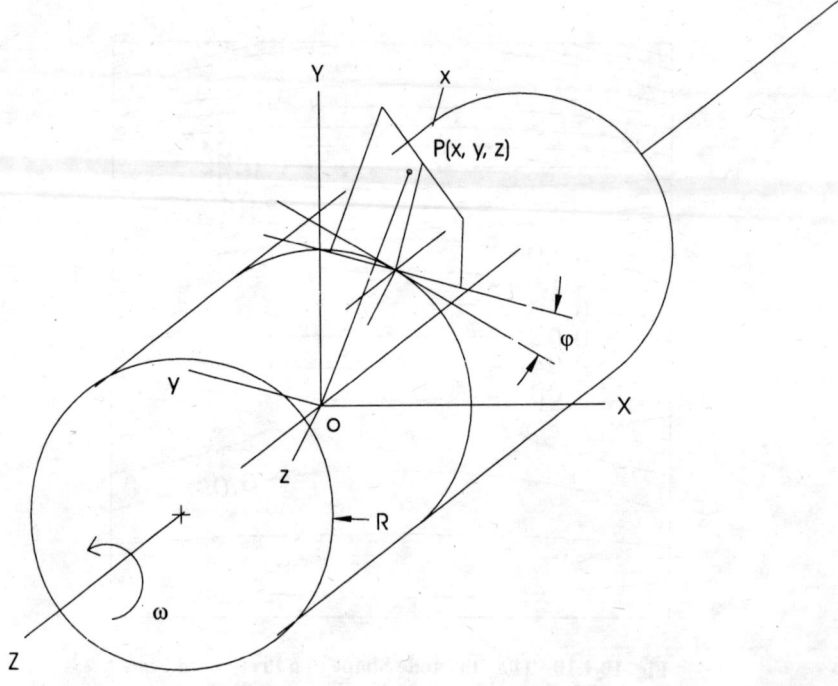

Fig. 10.2.1 A Rotating Pre-twisted Plate

where $\{\bar{x}\}^T = \{\ x\ \ y\ \ z\ \}$ is the coordinates vector of the point P and its displacements are given by the vector $\{\delta\}^T = \{\ u\ \ v\ \ w\ \}$, see equations (10.1.3) and (10.1.8). The angular velocity vector is written as

$$\vec{\omega} = \begin{Bmatrix} \omega_x \\ \omega_y \\ \omega_z \end{Bmatrix} = \begin{Bmatrix} 0 \\ \omega \sin\varphi \\ \omega \cos\varphi \end{Bmatrix} = \begin{Bmatrix} 0 \\ \omega \cos\phi \\ \omega \sin\phi \end{Bmatrix}$$

(10.2.2)

The velocity of point P is

$$\vec{v} = \dot{\overrightarrow{OP}} + \vec{\omega} \times \overrightarrow{OP}$$

$$= \begin{Bmatrix} \dot{u} \\ \dot{v} \\ \dot{w} \end{Bmatrix} + (\omega_y \hat{j} + \omega_z \hat{k}) \times \left[(x+u)\hat{i} + (y+v)\hat{j} + (z+w)\hat{k} \right]$$

$$\left\{\begin{array}{c}\dot{u}\\ \dot{v}\\ \dot{w}\end{array}\right\} + \left[\begin{array}{ccc}0 & -\omega_z & \omega_y\\ \omega_z & 0 & 0\\ -\omega_y & 0 & 0\end{array}\right]\left\{\begin{array}{c}x+u\\ y+v\\ z+w\end{array}\right\}$$

$$= \{\dot{d}\} + [A](\{\bar{x}\} + \{\delta\})$$

(10.2.3)

where

$$[A] = \left[\begin{array}{ccc}0 & -\omega_z & \omega_y\\ \omega_z & 0 & 0\\ -\omega_y & 0 & 0\end{array}\right]$$

(10.2.4)

is the angular velocity matrix. The kinetic energy of the shell can now be written as

$$T = \frac{1}{2} \iiint_V \rho \vec{v} \cdot \vec{v} \, dV$$

$$= \frac{1}{2} \iiint_V \rho [\{\dot{d}\} + [A](\{\bar{x}\} + \{\delta\})] \cdot [\{\dot{d}\} + [A](\{\bar{x}\} + \{\delta\})] dV$$

$$= T_2 + T_1 + T_0$$

(10.2.5)

where

$$T_2 = \frac{1}{2} \iiint_V \rho \{\dot{\delta}\}^T \{\dot{\delta}\} dV$$

$$T_1 = \iiint_V \rho \{\dot{\delta}\}^T [A]\{\delta\} dV + \iiint_V \rho \{\dot{\delta}\}^T [A]\{\bar{x}\} dV$$

$$T_0 = \frac{1}{2} \iiint_V \rho \{\delta\}^T [A]^T [A]\{\delta\} dV + \frac{1}{2} \iiint_V \rho \{\bar{x}\}^T [A]^T [A]\{\bar{x}\} dV$$

$$+ \iiint_V \rho \{\delta\}^T [A]^T [A]\{\bar{x}\} dV$$

(10.2.6)

Elemental Equation

Applying Lagrangian equation for the kinetic energy term in equation (10.2.6) above

$$\frac{d}{dt}\left(\frac{\partial T}{\partial \dot{\delta}}\right) - \frac{\partial T}{\partial \delta}$$
$$= [\widehat{M}]\{\ddot{\hat{\delta}}\} + [\widehat{G}]\{\dot{\hat{\delta}}\} + [\widehat{K}_r]\{\hat{\delta}\} - \{\hat{f}_c\}$$
(10.2.7)

where the various elemental matrices in the above equation are

Mass Matrix:
$$[\widehat{M}] = \iiint_V \rho [N]^T [N] dV$$
(10.2.8)

Coriolis Matrix:
$$[\widehat{G}] = 2 \iiint_V \rho [N]^T [A][N] dV$$
(10.2.9)

Stiffness Matrix:
$$[\widehat{K}_r] = -\iiint_V \rho [N]^T [A]^T [A][N] dV$$
(10.2.10)

Centrifugal Force Vector:
$$\{\hat{f}_c\} = \iiint_V \rho [N]^T [A]^T [A] dV \{\bar{x}\} dV$$
(10.2.11)

To determine the initial stress vector due to centrifugal forces, we first find the initial nodal displacements from the elastic stiffness matrix of (10.1.28)

$$[K_e]\{\delta_c\} = \Sigma\{\hat{f}_c\}$$
$$= \{F_c\}$$
(10.2.12)

From (10.1.11) and (10.1.26), the stress vector is given by
$$\{\tau^0\} = [D][B]\{\delta_c\}$$
(10.2.13)

Finite Element Method for Shells

The strain energy of the system due to initial stress can be written as

$$U_P = \frac{1}{2} \iiint_V [\tau^0]\{\varepsilon\} dV$$

$$= \frac{1}{2} \iiint_V \begin{bmatrix} \tau_{xx} & \tau_{yy} & \tau_{xy} & \tau_{yz} & \tau_{zx} \end{bmatrix} \begin{Bmatrix} \left(\frac{\partial u}{\partial x}\right)^2 + \left(\frac{\partial v}{\partial x}\right)^2 + \left(\frac{\partial w}{\partial x}\right)^2 \\ \left(\frac{\partial u}{\partial y}\right)^2 + \left(\frac{\partial v}{\partial y}\right)^2 + \left(\frac{\partial w}{\partial y}\right)^2 \\ \frac{\partial u}{\partial x}\frac{\partial u}{\partial y} + \frac{\partial v}{\partial x}\frac{\partial v}{\partial y} + \frac{\partial w}{\partial x}\frac{\partial w}{\partial y} \\ \frac{\partial u}{\partial y}\frac{\partial u}{\partial z} + \frac{\partial v}{\partial y}\frac{\partial v}{\partial z} + \frac{\partial w}{\partial y}\frac{\partial w}{\partial z} \\ \frac{\partial u}{\partial x}\frac{\partial u}{\partial z} + \frac{\partial v}{\partial x}\frac{\partial v}{\partial z} + \frac{\partial w}{\partial x}\frac{\partial w}{\partial z} \end{Bmatrix} dV$$

(10.2.14)

The above equation (10.2.14) can be expressed as

$$U_P = \frac{1}{2} \iiint_V \begin{bmatrix} \frac{\partial u}{\partial x} & \frac{\partial u}{\partial y} & \frac{\partial u}{\partial z} & \cdots & \frac{\partial w}{\partial z} \end{bmatrix} \begin{bmatrix} [\tau^1] & 0 & 0 \\ 0 & [\tau^1] & 0 \\ 0 & 0 & [\tau^1] \end{bmatrix} \begin{Bmatrix} \frac{\partial u}{\partial x} \\ \frac{\partial u}{\partial y} \\ \frac{\partial u}{\partial z} \\ \vdots \\ \frac{\partial w}{\partial z} \end{Bmatrix} dV$$

(10.2.15)

where

$$[\tau^1] = \begin{bmatrix} \tau_{xx} & \frac{1}{2}\tau_{xy} & \frac{1}{2}\tau_{xz} \\ \frac{1}{2}\tau_{xy} & \tau_{yy} & \frac{1}{2}\tau_{yz} \\ \frac{1}{2}\tau_{xz} & \frac{1}{2}\tau_{yz} & 0 \end{bmatrix}$$

(10.2.16)

The geometric stiffness matrix is therefore given by

$$[\hat{K}_g] = \frac{1}{2} \iiint_V [B]^T \begin{bmatrix} [\tau^1] & [\tau^1] & [\tau^1] \end{bmatrix} [B] dV$$

(10.2.17)

510 Dynamics of Plates

Assembling the elemental equations, we get the global equations of motion for the rotating plate

$$[M]\{\ddot{d}\} + [G]\{\dot{d}\} + [K]\{d\} = \{F_g\}$$

(10.2.18)

Let the excitation force be harmonic and have three components, then

$$\vec{F} = \begin{Bmatrix} F_x \\ F_y \\ F_z \end{Bmatrix} \cos(\Omega t - \Phi)$$

(10.2.19)

The elemental nodal force is then given by

$$\{\hat{F}_g\} = \left(\iint_A [N]^T \begin{Bmatrix} F_x \\ F_y \\ F_z \end{Bmatrix} dA \right) \cos(\Omega t - \Phi)$$

(10.2.20)

Example
A plate of dimensions 0.05×0.05 m with 1 mm thickness mounted on a disc of radius 0.05 m at a setting angle $0°$ is chosen to calculate the non dimensional natural frequencies. The Young's modulus and shear modulus of the material are taken as $0.71 \times 10^{12}, 0.28 \times 10^{12}$ Pa. The density of the material is taken as 2300 kg/m^3 and Poisson's ratio is assumed 0.25.

As in the previous section, the plate is divided into 16 elements with 65 nodes.

The first five non dimensional natural frequencies $pa^2 \sqrt{\dfrac{\varrho h}{D}}$ of the stationary plate are given in the previous section 10.1.

For a speed of rotation equal to half the first natural frequency of the stationary plate, the first seven non dimensional natural frequencies are:

Mode No.	Frequency
1. I Bending	4.561
2. II Bending	22.260
3. I Torsion	23.108
4. Plate Mode	32.259
5. II Torsion	45.322
6. Plate Mode	70.664
7. III Bending	74.189

10.3 PRE-TWISTED SHELL FINITE ELEMENT

In the previous two sections, we considered a finite element for shells based on Ahmad's super-parametric element. There have been several attempts made to fit in flat plate elements, triangular or quadrilateral ones to generate the mesh for a shell. These methods obviously give erroneous results, the hope however is, the error becomes minimum as the number of elements are increased to fit a given surface. In Ahmad's element essential strains and stresses are defined by erecting a normal z' with two other orthogonal axes x' and y' tangent to it. The elasticity matrix for an isotropic material is assumed to define the stress field. We have seen in section 10.1 that the geometry of the shell plays an important role, particularly the shell curvatures. Pre-twisted plate type shells have a negative Gaussian curvature, and therefore, it is important to consider the accurate thin shell theory of Gol'denveizer. Here a shell element based on the theory given in section 10.1 is developed.

Element Geometry

The pre twisted plate is shown in Fig. 10.3.1a with the global coordinates and located at the root section. The coordinates a and β are measured in the chordwise and spanwise sections respectively of the helicoidal base plane which twists around β such that at any spanwise location the coordinate line a makes an angle $\theta = \theta(\beta)$ with the coordinate line a at the root.

A typical four noded element in local helicoidal coordinates ξ, η, ζ (orthogonal system) is shown in Fig. 10.3.1b. The middle surface is right helicoidal whose spanwise rate is assumed to be linear within the element. The element accounts for variable thickness to permit an accurate physical representation of the shell. Each node has 10 degrees of freedom, three displacements u, v and w and six slopes corresponding to curvilinear ξ, η directions and the term involving the curvature $u, \frac{\partial u}{\partial \xi}, \frac{\partial u}{\partial \eta}, v, \frac{\partial v}{\partial \xi}, \frac{\partial v}{\partial \eta}, w, \frac{\partial w}{\partial \xi}, \frac{\partial w}{\partial \eta}$ and $\frac{\partial^2 w}{\partial \xi \partial \eta}$. The displacement functions assumed are cubic in ξ, η directions and incomplete quartic polynomial in ζ direction. The local coordinate system is located with its origin at the point $a = a^*, \beta = \beta^*$ in the global system. The shape of the shell is defined by specifying the height $h = h(\xi, \eta)$ above the shell together with spanwise rate of twist $\frac{d\theta}{d\eta}$ of the base plane. The shell thickness is approximated globally, rather than locally at each element, within the element itself, it is taken as a linear function of the nodal thickness.

512 Dynamics of Plates

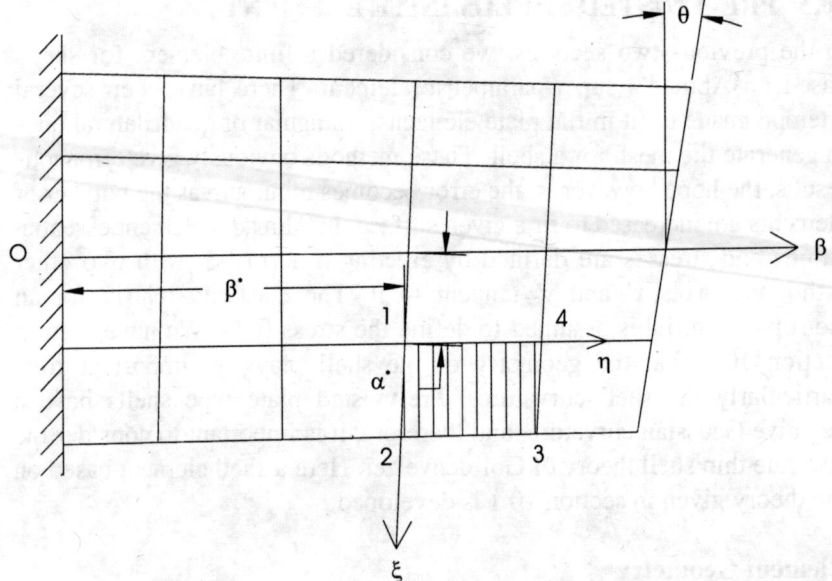

Fig. 10.3.1a Pre-twisted Cantilever Plate

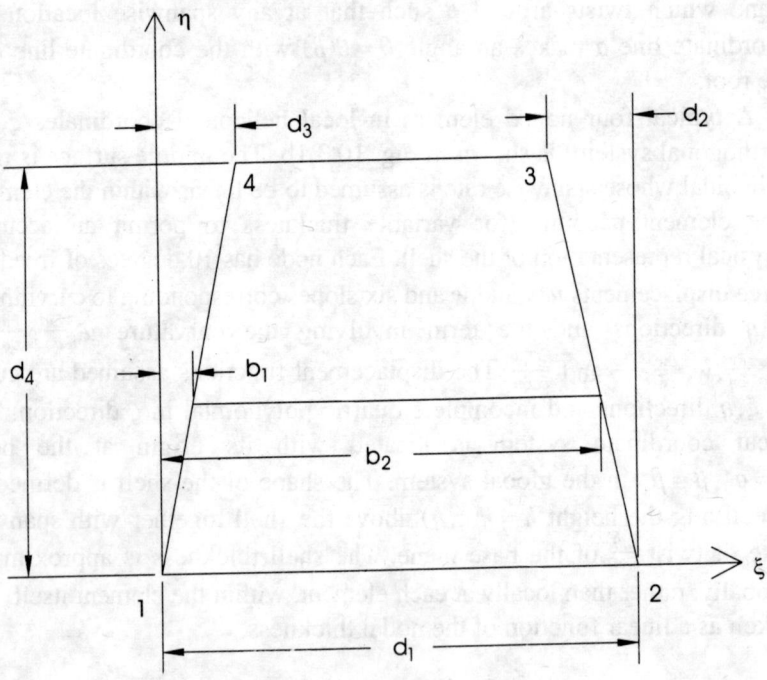

Fig. 10.3.1b Element Geomentry

Displacement Functions

The displacement functions u and v are taken to be cubic in ξ, η directions and incomplete quartic for w in ζ direction.

$$u = a_1 + a_2\xi + a_3\eta + a_4\xi^2 + a_5\xi\eta + a_6\eta^2 + a_7\xi^3$$
$$+ a_8\xi^2\eta + a_9\xi\eta^2 + a_{10}\eta^3 + a_{11}\xi^3\eta + a_{12}\xi\eta^3$$

$$v = a_{13} + a_{14}\xi + a_{15}\eta + a_{16}\xi^2 + a_{17}\xi\eta + a_{18}\eta^2 + a_{19}\xi^3$$
$$+ a_{20}\xi^2\eta + a_{21}\xi\eta^2 + a_{22}\eta^3 + a_{23}\xi^3\eta + a_{24}\xi\eta^3$$

$$w = a_{25} + a_{26}\xi + a_{27}\eta + a_{28}\xi^2 + a_{29}\xi\eta + a_{30}\eta^2 + a_{31}\xi^3$$
$$+ a_{32}\xi^2\eta + a_{33}\xi\eta^2 + a_{34}\eta^3 + a_{35}\xi^3\eta + a_{36}\xi\eta^3 +$$
$$a_{37}\xi^2\eta^2 + a_{38}\xi^2\eta^3 + a_{39}\xi^3\eta^2 + a_{40}\xi^3\eta^3$$

(10.3.1)

We can write the generalized displacements related to the coefficients of the displacement polynomial as

$$\{q\} = [P]\{a\}$$

(10.3.2)

where

$$\{q^T\} = \begin{Bmatrix} u_1 & \frac{\partial u_1}{\partial \xi} & \frac{\partial u_1}{\partial \eta} & v_1 & \frac{\partial v_1}{\partial \xi} & \frac{\partial v_1}{\partial \eta} & w_1 & \frac{\partial w_1}{\partial \xi} & \frac{\partial w_1}{\partial \eta} & \frac{\partial^2 w_1}{\partial \xi \partial \eta} \\ u_2 & \frac{\partial u_2}{\partial \xi} & \frac{\partial u_2}{\partial \eta} & v_2 & \frac{\partial v_2}{\partial \xi} & \frac{\partial v_2}{\partial \eta} & w_2 & \frac{\partial w_2}{\partial \xi} & \frac{\partial w_2}{\partial \eta} & \frac{\partial^2 w_2}{\partial \xi \partial \eta} \\ u_3 & \frac{\partial u_3}{\partial \xi} & \frac{\partial u_3}{\partial \eta} & v_3 & \frac{\partial v_3}{\partial \xi} & \frac{\partial v_3}{\partial \eta} & w_3 & \frac{\partial w_3}{\partial \xi} & \frac{\partial w_3}{\partial \eta} & \frac{\partial^2 w_3}{\partial \xi \partial \eta} \\ u_4 & \frac{\partial u_4}{\partial \xi} & \frac{\partial u_4}{\partial \eta} & v_4 & \frac{\partial v_4}{\partial \xi} & \frac{\partial v_4}{\partial \eta} & w_4 & \frac{\partial w_4}{\partial \xi} & \frac{\partial w_4}{\partial \eta} & \frac{\partial^2 w_4}{\partial \xi \partial \eta} \end{Bmatrix}_{1 \times 40}$$

(10.3.3a)

$$\{a\}^T = \{ a_1 \quad a_2 \quad \cdots \quad a_{40} \}$$

(10.3.3b)

and

$$[P]_{40 \times 40} = \begin{bmatrix} [P_1] & [P_2] & [P_3] & [P_4] \end{bmatrix}$$

(10.3.3c)

514 Dynamics of Plates

with

$$[P_i] = \begin{bmatrix} [R1_i] & 0 & 0 \\ 0 & [R1_i] & 0 \\ 0 & 0 & [R2_i] \end{bmatrix} \qquad (10.3.3d)$$

$$[R1_i] = \begin{bmatrix} 1 & \xi_i & \eta_i & \xi_i^2 & \xi_i\eta_i & \eta_i^2 & \xi_i^3 & \xi_i^2\eta_i & \xi_i\eta_i^2 & \eta_i^3 & \xi_i^3\eta_i & \xi_i\eta_i^3 \\ 0 & 1 & 0 & 2\xi_i & \eta_i & 0 & 3\xi_i^2 & 2\xi_i\eta_i & \eta_i^2 & 0 & 3\xi_i^2\eta_i & \eta_i^3 \\ 0 & 0 & 1 & 0 & \xi_i & 2\eta_i & 0 & \xi_i^2 & 2\xi_i\eta_i & 3\eta_i^2 & \xi_i^3 & 3\xi_i\eta_i^2 \end{bmatrix} \qquad (10.3.3e)$$

$$[R2_i] =$$
$$\begin{bmatrix} 1 & \xi_i & \eta_i & \xi_i^2 & \xi_i\eta_i & \eta_i^2 & \xi_i^3 & \xi_i^2\eta_i & \xi_i\eta_i^2 & \eta_i^3 & \xi_i^3\eta_i & \xi_i^2\eta_i^2 & \xi_i\eta_i^3 \\ 0 & 1 & 0 & 2\xi_i & \eta_i & 0 & 3\xi_i^2 & 2\xi_i\eta_i & \eta_i^2 & 0 & 3\xi_i^2\eta_i & 2\xi_i\eta_i^2 & \eta_i^3 \\ 0 & 0 & 1 & 0 & \xi_i & 2\eta_i & 0 & \xi_i^2 & 2\xi_i\eta_i & 3\eta_i^2 & \xi_i^3 & 2\xi_i^2\eta_i & 3\xi_i\eta_i^2 \\ 0 & 0 & 0 & 0 & 1 & 0 & 0 & 2\xi_i & 2\eta_i & 0 & 3\xi_i^2 & 4\xi_i\eta_i & 3\eta_i^2 \\ & & & & & & & & & & 3\xi_i^3 & 2\xi_i^3\eta_i & 3\xi_i^2\eta_i^2 & 3\xi_i\eta_i^3 \\ & & & & & & & & & & 6\xi_i\eta_i & 6\xi_i^2\eta_i & 9\xi_i^2\eta_i^2 \end{bmatrix}$$

$$(10.3.3f)$$

Note that $\xi_1 = 0$, $\eta_1 = 0$; $\xi_2 = d_1$, $\eta_2 = 0$; $\xi_3 = d_1 - d_2$, $\eta_3 = d_4$; $\xi_4 = d_3$, $\eta_4 = d_4$; in the above equations (10.3.3e) and (10.3.3f).

Differential Geometry

Fig.10.3.2 shows the shell in different axes system adopted here. In the Cartesian coordinate system, X, Y, Z the coordinate lines β and Y are coincident, and the curvilinear coordinate line a coincides with the x axis line at the root of the plate and is twisted through an angle $\theta = \theta(\beta)$ at a distance β from the plate root. The position vector at any point ξ, η within an element, whose origin of the local coordinate system is at a^*, β^* can be written as

$$\vec{R}(a, \beta) = [a \sin(\theta + \phi) + h \cos(\theta + \phi)]\hat{I} +$$
$$[\beta]\hat{J} + [-a \cos(\theta + \phi) + h \sin(\theta + \phi)]\hat{K}$$

(10.3.4)

where $a = a^*$ and $\beta = \beta^* + \eta$.

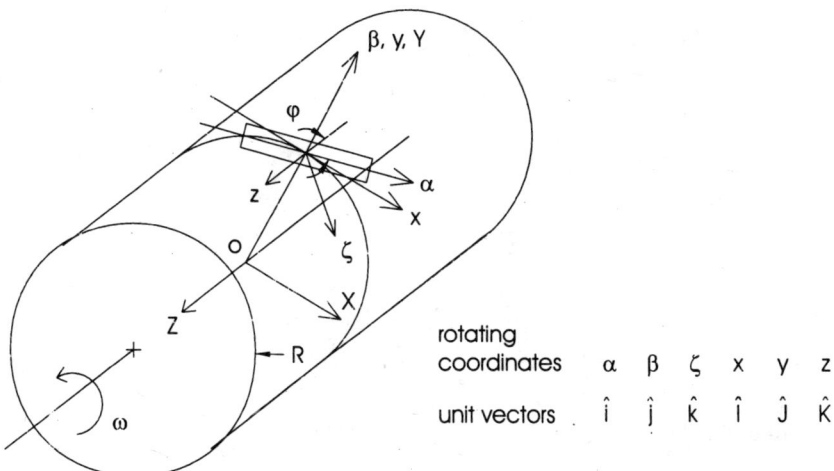

Fig. 10.3.2a Plate Coordinates

The coefficients of first quadratic form and the radii of curvature are given by the following.

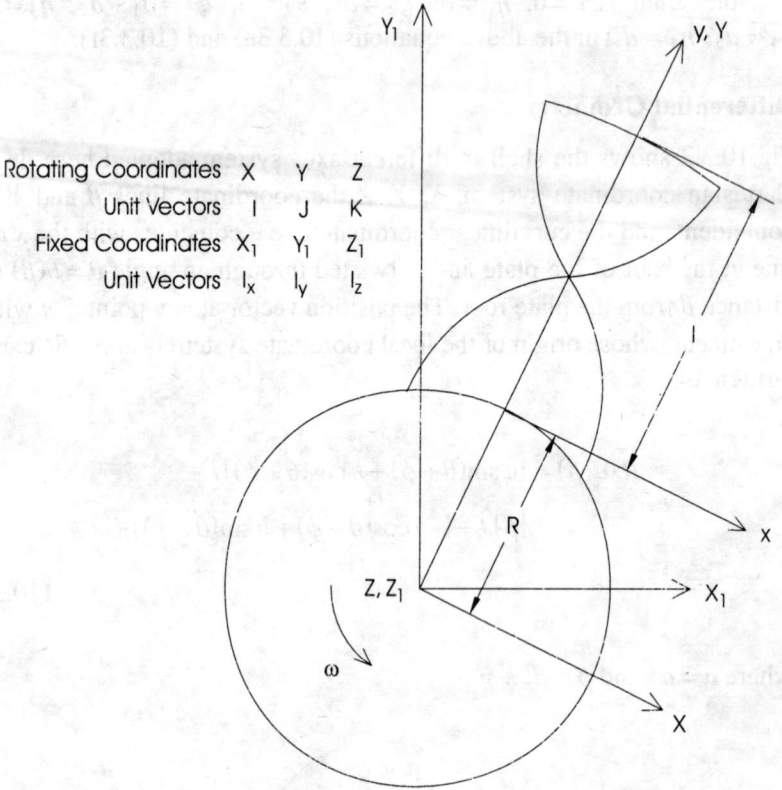

```
Rotating Coordinates    X    Y    Z
Unit Vectors            I    J    K
Fixed Coordinates       X₁   Y₁   Z₁
Unit Vectors            Iₓ   I_y  I_z
```

Fig. 10.3.2b Coordinate Systems of rotating Plate

From equations (9.5.3)

$$A^2 = \vec{R},_\xi \cdot \vec{R},_\xi = 1$$
$$B^2 = \vec{R},_\eta \cdot \vec{R},_\eta = 1 + a^2(\theta,_\eta)^2$$

(10.3.5)

From equation (9.1.9)

$$\hat{n} = \frac{1}{C}\left(\vec{R},_\xi \times \vec{R},_\eta\right)$$
$$C = \left|\vec{R},_\xi \times \vec{R},_\eta\right|$$

(10.3.6)

$$\hat{n} = \frac{1}{C}[\cos(\theta+\phi) - h,_\xi \sin(\theta+\phi)]\hat{I}$$
$$-[a\theta,_\eta + h,_\eta + hh,_\xi \theta,_\eta]\hat{J} + [\sin(\theta+\phi) + h,_\xi \cos(\theta+\phi)]\hat{K}$$
$$C^2 = 1 + a^2(\theta,_\eta)^2$$

(10.3.6a)

From equations (9.1.51) and (9.1.26)

$$\frac{1}{R_1} = \frac{\hat{n}}{A^2} \cdot \vec{R},_{\xi\xi} = \frac{1}{CA^2}h,_{\xi\xi}$$

$$\frac{1}{R_2} = \frac{\hat{n}}{B^2} \cdot \vec{R},_{\eta\eta}$$

$$= \frac{1}{CB^2}\left[ah,_\xi(\theta,_\eta)^2 + hh,_\xi \theta,_{\eta\eta} + a\theta,_{\eta\eta} - h(\theta,_\eta)^2 + h,_{\eta\eta}\right]$$

$$\frac{1}{R_{12}} = \frac{\hat{n}}{AB} \cdot \vec{R},_{\xi\eta} = \frac{1}{CAB}(h,_{\xi\eta} + \theta,_\eta)$$

(10.3.7)

Strain-Displacement Relations
From equations (9.2.75) and (9.2.80), we have

$$\varepsilon_{\xi\xi} = \frac{1}{A}\frac{\partial u}{\partial \xi} + \frac{v}{AB}\frac{\partial A}{\partial \eta} - \frac{w}{R_1}$$

$$\varepsilon_{\eta\eta} = \frac{1}{B}\frac{\partial v}{\partial \eta} + \frac{u}{AB}\frac{\partial B}{\partial \xi} - \frac{w}{R_2}$$

$$\varepsilon_{\xi\eta} = \frac{A}{B}\frac{\partial}{\partial \eta}\left(\frac{u}{A}\right) + \frac{B}{A}\frac{\partial}{\partial \xi}\left(\frac{v}{B}\right) + \frac{2w}{R_{12}}$$

(10.3.8)

$$\kappa_{\xi\xi} = -\frac{1}{A}\frac{\partial \Psi}{\partial \xi} - \frac{\Theta}{AB}\frac{\partial A}{\partial \eta} + \frac{\Phi}{R_{12}}$$

$$\kappa_{\eta\eta} = -\frac{1}{B}\frac{\partial \Theta}{\partial \eta} - \frac{\Psi}{AB}\frac{\partial B}{\partial \xi} - \frac{\Phi}{R_{12}}$$

$$\kappa_{\xi\eta} = -\frac{1}{A}\frac{\partial \Theta}{\partial \xi} + \frac{\Psi}{AB}\frac{\partial A}{\partial \eta} + \frac{\Phi}{R_1} - \frac{\varepsilon_{\eta\eta}}{R_{12}} + \frac{\varepsilon_{\xi\eta}}{2R_1}$$

(10.3.9)

where

$$\Theta = \gamma_2 = -\frac{1}{B}\frac{\partial w}{\partial \eta} - \frac{v}{R_2} + \frac{u}{R_{12}}$$

$$\Psi = \gamma_1 = -\frac{1}{A}\frac{\partial w}{\partial \xi} - \frac{u}{R_1} + \frac{v}{R_{12}}$$

$$\Phi = \delta = \frac{1}{2AB}\left[\frac{\partial}{\partial \eta}(Au) - \frac{\partial}{\partial \xi}(Bv)\right]$$

(10.3.10)

The strain vector can now be expressed in terms of the coefficient vector $\{a\}$

$$\{\varepsilon\} = \left\{\begin{array}{cccccc} \varepsilon_{\xi\xi} & \varepsilon_{\eta\eta} & \varepsilon_{\xi\eta} & \kappa_{\xi\xi} & \kappa_{\eta\eta} & \kappa_{\xi\eta} \end{array}\right\}^T$$

$$= \begin{bmatrix} B_{11} & B_{12} & B_{13} \\ B_{21} & B_{22} & B_{23} \\ B_{31} & B_{32} & B_{33} \\ B_{41} & B_{42} & B_{43} \\ B_{51} & B_{52} & B_{53} \\ B_{61} & B_{62} & B_{63} \end{bmatrix}\{a\}$$

(10.3.11)

where

$$B_{11} = \frac{1}{A}[Q_2]$$

$$B_{12} = \frac{1}{AB}\frac{\partial A}{\partial \eta}[Q_1]$$

$$B_{13} = -\frac{1}{R_1}[Q_4]$$

$$B_{21} = \frac{1}{AB}\frac{\partial B}{\partial \xi}[Q_1]$$

$$B_{22} = \frac{1}{B}[Q_3]$$

$$B_{23} = -\frac{1}{R_2}[Q_4]$$

$$B_{31} = -\frac{1}{AB}\frac{\partial A}{\partial \eta}[Q_1] + \frac{1}{B}[Q_3]$$

$$B_{32} = -\frac{1}{AB}\frac{\partial B}{\partial \xi}[Q_1] + \frac{1}{A}[Q_2]$$

$$B_{33} = \frac{2}{R_{12}}[Q_4]$$

$$B_{41} = \left(\frac{1}{A}\frac{\partial R_1^{-1}}{\partial \xi} - \frac{R_{12}^{-1}}{2AB}\frac{\partial A}{\partial \eta}\right)[Q_1] + \frac{R_1^{-1}}{A}[Q_2] + \frac{R_{12}^{-1}}{2B}[Q_3]$$

$$B_{42} = \left(-\frac{1}{A}\frac{\partial R_{12}^{-1}}{\partial \xi} + \frac{R_2^{-1}}{AB}\frac{\partial A}{\partial \eta} - \frac{R_{12}^{-1}}{2AB}\frac{\partial B}{\partial \xi}\right)[Q_1] - \frac{3R_{12}^{-1}}{2A}[Q_2]$$

$$B_{43} = -\frac{1}{A^3}\frac{\partial A}{\partial \xi}[Q_5] + \frac{1}{AB^2}\frac{\partial A}{\partial \eta}[Q_6] + \frac{1}{A^2}[Q_8]$$

$$B_{51} = \left(-\frac{1}{B}\frac{\partial R_{12}^{-1}}{\partial \eta} + \frac{R_1^{-1}}{AB}\frac{\partial B}{\partial \xi} - \frac{R_{12}^{-1}}{2AB}\frac{\partial A}{\partial \eta}\right)[Q_1] - \frac{3R_{12}^{-1}}{2B}[Q_3]$$

$$B_{52} = \left(\frac{1}{B}\frac{\partial R_2^{-1}}{\partial \eta} - \frac{R_{12}^{-1}}{2AB}\frac{\partial B}{\partial \xi}\right)[Q_1] + \frac{R_{12}^{-1}}{2A}[Q_2] + \frac{R_2^{-1}}{B}[Q_3]$$

$$B_{53} = \frac{1}{A^2B}\frac{\partial B}{\partial \xi}[Q_5] - \frac{1}{B^3}\frac{\partial B}{\partial \eta}[Q_6] + \frac{1}{B^2}[Q_9]$$

$$B_{61} = \left(-\frac{1}{A}\frac{\partial R_{12}^{-1}}{\partial \xi} - \frac{R_1^{-1}}{AB}\frac{\partial A}{\partial \eta} - \frac{R_{12}^{-1}}{AB}\frac{\partial B}{\partial \xi}\right)[Q_1] - \frac{R_{12}^{-1}}{A}[Q_2] + \frac{R_1^{-1}}{B}[Q_3]$$

$$B_{62} = \left(\frac{1}{A}\frac{\partial R_2^{-1}}{\partial \xi} - \frac{R_{12}^{-1}}{AB}\frac{\partial A}{\partial \eta} - \frac{R_1^{-1}}{AB}\frac{\partial B}{\partial \xi}\right)[Q_1] + \frac{R_2^{-1}}{A}[Q_2] - \frac{R_{12}^{-1}}{B}[Q_3]$$

$$B_{63} = (R_1^{-1} + R_2^{-1})R_{12}^{-1}[Q_4] - \frac{1}{A^2B}\frac{\partial A}{\partial \eta}[Q_5] - \frac{1}{AB^2}\frac{\partial B}{\partial \xi}[Q_6] + \frac{1}{AB}[Q_7]$$

(10.3.12)

$$[Q_1] = \begin{bmatrix} 1 & \xi & \eta & \xi^2 & \xi\eta & \eta^2 & \xi^3 & \xi^2\eta & \xi\eta^2 & \eta^3 & \xi^3\eta & \xi\eta^3 \end{bmatrix}$$

$$[Q_2] = \begin{bmatrix} 0 & 1 & 0 & 2\xi & \eta & 0 & 3\xi^2 & 2\xi\eta & \eta^2 & 0 & 3\xi^2\eta & \eta^3 \end{bmatrix}$$

$$[Q_3] = \begin{bmatrix} 0 & 0 & 1 & 0 & \xi & 2\eta & 0 & \xi^2 & 2\xi\eta & 3\eta^2 & \xi^3 & 3\xi\eta^2 \end{bmatrix}$$

$$[Q_4] = \begin{bmatrix} 1 & \xi & \eta & \xi^2 & \xi\eta & \eta^2 & \xi^3 & \xi^2\eta & \xi\eta^2 & \eta^3 & \xi^3\eta & \xi\eta^3 & \xi^2\eta^2 & \xi^2\eta^3 & \xi^3\eta^2 & \xi^3\eta^3 \end{bmatrix}$$

$$[Q_5] = \begin{bmatrix} 0 & 1 & 0 & 2\xi & \eta & 0 & 3\xi^2 & 2\xi\eta & \eta^2 & 0 & 3\xi^2\eta & \eta^3 & 2\xi\eta^2 & 2\xi\eta^3 & 3\xi^2\eta^2 & 3\xi^2\eta^3 \end{bmatrix}$$

$$[Q_6] = \begin{bmatrix} 0 & 0 & 1 & 0 & \xi & 2\eta & 0 & \xi^2 & 2\xi\eta & 3\eta^2 & \xi^3 & 3\xi\eta^2 & 2\xi^2\eta & 3\xi^2\eta^2 & 2\xi^3\eta & 3\xi^3\eta^2 \end{bmatrix}$$

$$[Q_7] = \begin{bmatrix} 0 & 0 & 0 & 0 & 1 & 0 & 0 & 2\xi & 2\eta & 0 & 3\xi^2 & 3\eta^2 & 4\xi\eta & 6\xi\eta^2 & 6\xi^2\eta & 9\xi^2\eta^2 \end{bmatrix}$$

$$[Q_8] = \begin{bmatrix} 0 & 0 & 0 & 2 & 0 & 0 & 6\xi & 2\eta & 0 & 0 & 6\xi\eta & 0 & 2\eta^2 & 2\eta^3 & 6\xi\eta^2 & 6\xi\eta^3 \end{bmatrix}$$

$$[Q_9] = \begin{bmatrix} 0 & 0 & 0 & 0 & 0 & 2 & 0 & 0 & 2\xi & 6\eta & 0 & 6\xi\eta & 2\xi^2 & 6\xi^2\eta & 2\xi^3 & 6\xi^3\eta \end{bmatrix}$$

(10.3.13)

Thickness Approximation

Within the element, the thickness t is assumed to be a linear function of the curvilinear helicoidal coordinates

$$t(\xi,\eta) = e_1 + e_2\xi + e_3\eta + e_4\xi\eta \tag{10.3.14}$$

where the coefficients e_i are determined by specifying the thickness at the four nodes of the element.

Strain Energy

The strain energy is given by equation (9.4.22)

$$V = \frac{E}{2(1-v^2)} \int_0^{d_4} \int_{b_1}^{b_2} t(\xi,\eta)$$

$$\left[\begin{array}{l} \left\{ \varepsilon_{\xi\xi}^2 + \varepsilon_{\eta\eta}^2 + 2v\varepsilon_{\xi\xi}\varepsilon_{\eta\eta} + \frac{1}{2}(1-v)\varepsilon_{\xi\eta}^2 \right\} + \\ \frac{t^2(\xi,\eta)}{12} \left\{ \kappa_{\xi\xi}^2 + \kappa_{\eta\eta}^2 + 2v\kappa_{\xi\xi}\kappa_{\eta\eta} + 2(1-v)\kappa_{\xi\eta}^2 \right\} \end{array} \right] AB\, d\xi\, d\eta$$

(10.3.15)

where

$$b_1 = \frac{d_3\eta}{d_4}$$

$$b_2 = d_1 - \frac{d_2\eta}{d_4}$$

(10.3.16)

Kinetic Energy
The kinetic energy of the element can be written as

$$T = \frac{1}{2}\rho \int_{-t/2}^{t/2} \int_0^{d_4} \int_{b_1}^{b_2} (\dot{u}^2 + \dot{v}^2 + \dot{w}^2 + \zeta^2 \dot{u}^2 + \zeta^2 \dot{v}^2) AB\, d\xi\, d\eta\, d\zeta$$

(10.3.17)

Stiffness Matrix
The strain energy expression in (10.3.15) can be expressed as

$$V = \frac{1}{2} \int_0^{d_4} \int_{b_1}^{b_2} \{\varepsilon\}^T [D] \{\varepsilon\} t AB\, d\xi\, d\eta$$

(10.3.18)

where

$$[D] = \frac{Et}{1-v^2} \begin{bmatrix} 1 & v & 0 & 0 & 0 & 0 \\ v & 1 & 0 & 0 & 0 & 0 \\ 0 & 0 & \frac{1-v}{2} & 0 & 0 & 0 \\ 0 & 0 & 0 & \frac{t^2}{12} & \frac{vt^2}{12} & 0 \\ 0 & 0 & 0 & \frac{vt^2}{12} & \frac{t^2}{12} & 0 \\ 0 & 0 & 0 & 0 & 0 & \frac{(1-v)t^2}{6} \end{bmatrix}$$

(10.3.19)

Substituting equations (10.3.2), (10.3.11) in (10.3.18), we get

$$V = \frac{1}{2}\{q\}^T [\hat{K}] \{q\}$$

(10.3.20)

where

$$[\hat{K}] = [[P]^{-1}]^T \left\{ \int_0^{d_4} \int_{b_1}^{b_2} [B]^T [D] [B] t AB\, d\xi\, d\eta \right\} [P]^{-1}$$

(10.3.21)

The following transformation may be used to rewrite the above equation (10.3.21)

$$\xi = \frac{b_2 - b_1}{2} x_i + \frac{b_2 + b_1}{2}$$

$$\eta = \frac{d_4}{2} x_j + \frac{d_4}{2}$$

$$d\xi = \frac{b_2 - b_1}{2} dx_i$$

$$d\eta = \frac{d_4}{2} dx_j$$

(10.3.22)

so that

$$[\hat{K}] = \frac{1}{4} d_4 [[P]^{-1}]^T \left\{ \int_{-1}^{1} \int_{-1}^{1} [B]^T [D][B](b_2 - b_1) tAB \, dx_i \, dx_j \right\} [P]^{-1}$$

(10.3.23)

We can use n point Gauss-Legendre quadrature procedure and write the above equation (10.3.23) to finally obtain the elemental stiffness matrix as

$$[\hat{K}] = \frac{d_4}{4} [[P]^{-1}]^T \sum_{i=1}^{n} \sum_{j=1}^{n} \left\{ \begin{array}{c} w_i w_j [B(x_i, x_j)]^T [D][B(x_i, x_j)] \\ \times (b_2 - b_1) tA(x_i, x_j) B(x_i, x_j) \end{array} \right\} [P]^{-1}$$

(10.3.24)

Mass Matrix

The kinetic energy given by equation (10.3.17) can be expressed as

$$T = \frac{1}{2} \rho \int_{0}^{d_4} \int_{b_1}^{b_2} \{\dot{U}\}^T [G]\{\dot{U}\} tAB \, d\xi \, d\eta$$

(10.3.25)

where

$$[G] = \begin{bmatrix} 1 & 0 & 0 & 0 & 0 \\ 0 & 1 & 0 & 0 & 0 \\ 0 & 0 & 1 & 0 & 0 \\ 0 & 0 & 0 & \frac{t^2}{12} & 0 \\ 0 & 0 & 0 & 0 & \frac{t^2}{12} \end{bmatrix}$$

(10.3.26)

and the displacement vector is given by using equations (10.3.1) and (10.3.10) as

$$\{U\} = \begin{Bmatrix} u \\ v \\ w \\ \Psi \\ \Theta \end{Bmatrix} = \begin{bmatrix} [Q_1] & 0 & 0 \\ 0 & [Q_1] & 0 \\ 0 & 0 & [Q_4] \\ -\frac{1}{R_1}[Q_1] & \frac{1}{R_{12}}[Q_1] & -\frac{1}{A}[Q_5] \\ \frac{1}{R_{12}}[Q_1] & -\frac{1}{R_2}[Q_1] & -\frac{1}{B}[Q_6] \end{bmatrix} \begin{Bmatrix} a_1 \\ a_2 \\ a_3 \\ \vdots \\ a_{40} \end{Bmatrix}$$

$$= [N_1]\{a\}$$

(10.3.27)

Using equation (10.3.2), the kinetic energy expression in (10.3.25) now becomes

$$T = \frac{1}{2}\{\dot{q}\}^T [\widehat{M}]\{\dot{q}\}$$

(10.3.28)

where the elemental mass matrix is given by

$$[\widehat{M}] = [[P]^{-1}]^T \left\{ \rho \int_0^{d_4} \int_{b_1}^{b_2} [N_1]^T [G][N_1] t AB \, d\xi \, d\eta \right\} [P]^{-1}$$

(10.3.29)

The integration in the above equation can be carried out in a similar manner indicated to that of the stiffness matrix.

Example 10.3.1
Consider a square flat plate with Poisson's ratio 0.3. The non dimensional natural frequencies $\bar{p} = \sqrt{\frac{12\rho p^2 l^4 (1-v^2)}{Et^2}}$ for different mesh sizes is shown in the table below.

Mode	Mesh Size		
	2 X 2	3 X 3	4 X 4
1	3.4832	3.4753	3.4723
2	8.5461	8.5068	8.4970
3	21.3522	21.2513	21.2039
4	27.1951	27.1155	27.0358
5	31.3096	30.9442	30.8310

524 Dynamics of Plates

It can be seen that there is a good convergency and mesh size 4X4 is used in the following examples.

Example 10.3.2
Consider a spanwise tapered plate with $\frac{l}{b} = 1$, $\frac{t_{max}}{t_{min}} = 3.86$, $\frac{t_{max}}{l} = 0.0576$ and $v = 0.3$. The first ten non dimensional frequencies $\bar{p} = \sqrt{\frac{12\rho p^2 l^4 (1-v^2)}{E t_{max}^2}}$ are:

4.0977, 6.9268, 14.8253, 16.4774, 19.5673, 30.1394, 31.3426, 39.0864, 43.1952 and 50.3681.

Example 10.3.3
Consider a chordwise tapered plate with $\frac{l}{b} = 2$, $\frac{t_{max}}{t_{min}} = 3.26$, $\frac{t_{max}}{l} = 0.0352$, $v = 0.3$. The first ten nondimensional frequencies $\bar{p} = \sqrt{\frac{12\rho p^2 l^4 (1-v^2)}{E t_{ave}^2}}$, $t_{ave} = \frac{1}{2}(t_{max} + t_{min})$ are:

3.8717, 16.4699, 23.6640, 48.2926, 60.9771, 63.7356, 89.8464, 95.1069, 105.7207 and 124.7204.

Example 10.3.4
Consider a wedge shaped fan blade 30.5 cm long and 30.5 cm wide in circumferential direction with a radius of curvature 76.2 cm. Cross-section tapers linearly from a thickness 5.2 mm at one free edge to 1.22 mm at other free edge. The first ten non dimensional frequencies are:

79.49, 115.53, 210.85, 265.98, 376.30, 454.72, 503.93, 590.03, 695.70, 790.33.

10.4 ROTATING PRE-TWISTED PLATE FINITE ELEMENT

The finite element given in previous section is now extended to a rotating shell by taking into account the effect of centrifugal forces. Towards this end, the kinetic energy of the shell is written. Referring to Figs. 10.3.1 and 10.3.2, the position vector is given by equation (10.3.4)

$$\vec{R}(\alpha, \beta) = [a\sin(\theta + \phi) + h\cos(\theta + \phi)]\hat{I} + [\beta]\hat{J}$$
$$+ [-a\cos(\theta + \phi) + h\sin(\theta + \phi)]\hat{K}$$

(10.4.1)

where $\alpha = \alpha^* + \xi$ and $\beta = \beta^* + \eta$.

u, v and $-w$ are the displacements along ξ, η and ζ directions, then the displacements of a point on a surface at distance ζ from the middle surface are given by

$$u(\xi,\eta,\zeta) = u(\xi,\eta) + \zeta\Psi$$
$$v(\xi,\eta,\zeta) = v(\xi,\eta) + \zeta\Theta$$
$$w(\xi,\eta,\zeta) = w(\xi,\eta)$$

(10.4.2)

The displacement vector can now be expressed as

$$\vec{U} = u\frac{\vec{R},_\xi}{A} + v\frac{\vec{R},_\eta}{B} - w\vec{n}$$

(10.4.3)

The inward displacement is given by equation (9.6.13) as

$$\Delta = \frac{1}{2}\int_0^\beta (\Theta^2 + \Phi^2)d\beta$$

(10.4.4)

It can be transformed to

$$\Delta = \frac{1}{2}(\Theta^2 + \Phi^2)\int_\beta^l d\beta$$

(10.4.4a)

Equation (10.4.3) is now modified to

$$\vec{U} = u\frac{\vec{R},_\xi}{A} + (v-\Delta)\frac{\vec{R},_\eta}{B} - w\vec{n}$$

(10.4.5)

In Cartesian coordinates, the displacement vector is

$$\vec{U} = x_d\hat{I} + y_d\hat{J} + z_d\hat{K}$$

(10.4.6)

With the help of equation (10.4.1), the derivatives of the position vector $\vec{R}$ can be obtained. Substituting them in equation (10.4.5) and comparing the resulting equation with (10.4.6), we can obtain

526 Dynamics of Plates

$$x_d = \frac{u}{A}[\sin(\theta+\phi) + h,_\xi \cos(\theta+\phi)] +$$

$$+ \frac{(v-\Delta)}{B}[(a\theta,_\eta + h,_\eta)\cos(\theta+\phi) - h\theta,_\eta \sin(\theta+\phi)]$$

$$- \frac{w}{C}[\cos(\theta+\phi) - h,_\xi \sin(\theta+\phi)]$$

$$+ \Psi\frac{\zeta}{A}[\sin(\theta+\phi) + h,_\xi \cos(\theta+\phi)]$$

$$+ \Theta\frac{\zeta}{B}[(a\theta,_\eta + h,_\eta)\cos(\theta+\phi) - h\theta,_\eta \sin(\theta+\phi)]$$

$$y_d = \frac{(v-\Delta)}{B} + \frac{w}{C}(a\theta,_\eta + h,_\eta + hh,_\eta\theta,_\eta) + \Theta\frac{\zeta}{B}$$

$$z_d = \frac{u}{A}[-\cos(\theta+\phi) + h,_\xi \sin(\theta+\phi)] +$$

$$+ \frac{(v-\Delta)}{B}[(a\theta,_\eta + h,_\eta)\sin(\theta+\phi) + h\theta,_\eta \cos(\theta+\phi)]$$

$$- \frac{w}{C}[\sin(\theta+\phi) + h,_\xi \cos(\theta+\phi)]$$

$$+ \Psi\frac{\zeta}{A}[-\cos(\theta+\phi) + h,_\xi \sin(\theta+\phi)]$$

$$+ \Theta\frac{\zeta}{B}[(a\theta,_\eta + h,_\eta)\sin(\theta+\phi) + h\theta,_\eta \cos(\theta+\phi)]$$

(10.4.7)

The vector equation of an equidistant surface from the middle surface is

$$\vec{R}(\xi,\eta,\zeta) = \vec{R}(\xi,\eta) + \zeta\vec{n}$$

(10.4.8)

Substituting equation (10.4.1) and (10.3.6), the above equation (10.4.8) becomes

$$\vec{R}(\xi,\eta,\zeta) = \left[\left(a - \frac{\zeta h,_\xi}{C}\right)\sin(\theta+\phi) + \left(h + \frac{\zeta}{C}\right)\cos(\theta+\phi) + x_d\right]\hat{I}$$

$$+ \left[\beta + R - \frac{\zeta}{C}(a\theta,_\eta + h,_\eta + hh,_\xi\theta,_\eta) + y_d\right]\hat{J} +$$

$$\left[-\left(a - \frac{\zeta h,_\xi}{C}\right)\cos(\theta+\phi) + \left(h + \frac{\zeta}{C}\right)\sin(\theta+\phi) + z_d\right]\hat{K}$$

(10.4.9)

Finite Element Method for Shells 527

See equation (10.3.6a) for C. Let $\hat{I}_x, \hat{I}_y, \hat{I}_z$ be the unit vectors of the fixed coordinate system located at the center of the disc on which the plate is mounted, then

$$\hat{i} = \hat{I}_x \cos \omega t + \hat{I}_y \sin \omega t$$
$$\hat{j} = -\hat{I}_x \sin \omega t + \hat{I}_y \cos \omega t$$
$$\hat{K} = \hat{I}_z$$

(10.4.10)

Equation (10.4.9) can now be written as

$$\vec{R}(\xi,\eta,\zeta)$$
$$= \left[\left\{ \left(a - \frac{\zeta h,_\xi}{C}\right) \sin(\theta+\phi) + \left(h + \frac{\zeta}{C}\right)\cos(\theta+\phi) + x_d \right\} \cos \omega t \right. $$
$$\left. - \left\{ \beta + R - \frac{\zeta}{C}(a\theta,_\eta + h,_\eta + hh,_\xi \theta,_\eta) + y_d \right\} \sin \omega t \right] \hat{I}_x$$

$$+ \left[\left\{ \left(a - \frac{\zeta h,_\xi}{C}\right) \sin(\theta+\phi) + \left(h + \frac{\zeta}{C}\right)\cos(\theta+\phi) + x_d \right\} \sin \omega t \right.$$
$$\left. + \left\{ \beta + R - \frac{\zeta}{C}(a\theta,_\eta + h,_\eta + hh,_\xi \theta,_\eta) + y_d \right\} \cos \omega t \right] \hat{I}_y$$

$$+ \left[-\left(a - \frac{\zeta h,_\xi}{C}\right) \cos(\theta+\phi) + \left(h + \frac{\zeta}{C}\right) \sin(\theta+\phi) + z_d \right] \hat{I}_z$$

(10.4.11)

It may be noted here that h is not the thickness and t represents thickness of the shell as well as time. Differentiating equation (10.4.11) with time, we get

$$\vec{\dot{R}} = [\dot{x}_d \cos \omega t - \dot{y}_d \sin \omega t - \omega\{(x_d + C_1)\sin \omega t + (y_d + C_2)\cos \omega t\}]\hat{I}_x$$
$$+ [\dot{x}_d \sin \omega t + \dot{y}_d \cos \omega t + \omega\{(x_d + C_1)\cos \omega t - (y_d + C_2)\sin \omega t\}]\hat{I}_y$$
$$+ \dot{z}_d \hat{I}_z$$

(10.4.12)

where

$$C_1 = \left(a - \frac{\zeta h, \xi}{C}\right)\sin(\theta + \phi) + \left(h + \frac{\zeta}{C}\right)\cos(\theta + \phi)$$

$$C_2 = \beta + R - \frac{\zeta}{C}(a\theta, _\eta + h, _\eta + hh, _\xi \theta, _\eta)$$

(10.4.13)

The kinetic energy of the plate is now given by

$$T = \frac{1}{2}\rho \int_{-t/2}^{t/2} \int_0^{d_4} \int_{b_1}^{b_2} \vec{R} \cdot \vec{R}\, AB\, d\xi\, d\eta\, d\zeta$$

(10.4.14)

Substituting from (10.4.12) and simplifying, the above equation (10.4.14) becomes

$$T = \frac{1}{2}\rho \int_{-t/2}^{t/2} \int_0^{d_4} \int_{b_1}^{b_2} \left[\begin{array}{c} (\dot{x}_d^2 + \dot{y}_d^2 + \dot{z}_d^2) \\ +2\omega\{(x_d + C_1)\dot{y}_d - (y_d + C_2)\dot{x}_d\} \\ +\omega^2\left(\begin{array}{c} x_d^2 + y_d^2 + 2x_d C_1 \\ +2y_d C_2 + C_1^2 + C_2^2 \end{array}\right) \end{array}\right] AB\, d\xi\, d\eta\, d\zeta$$

(10.4.15)

The kinetic energy due to $(\dot{x}_d^2 + \dot{y}_d^2 + \dot{z}_d^2)$ terms has been considered in deriving the elemental mass matrix for the stationary shell in equation (10.3.29). The second set of terms in the above equation (10.4.15) $2\omega\{(x_d + C_1)\dot{y}_d - (y_d + C_2)\dot{x}_d\}$ give rise to Coriolis terms which are nonlinear, they are not considered here. The third and last set, terms in the above integral $\omega^2(x_d^2 + y_d^2 + 2x_d C_1 + 2y_d C_2 + C_1^2 + C_2^2)$ give the change in the kinetic energy due to rotation. In this third set, the last two terms C_1^2, C_2^2 are independent of any displacement and they are due to rigid body rotation of the plate element; therefore they are not included here. Substituting equations (10.4.7), the remaining terms in equation (10.4.15) lead to

$$T_R = \frac{1}{2}\rho \int_{-t/2}^{t/2} \int_0^{d_4} \int_{b_1}^{b_2} \left[\left\{ \begin{array}{l} u^2 B_1^2 + v^2(B_2^2 + B_4^2) + w^2(B_3^2 + B_5^2) \\ + \zeta^2 \Psi^2 B_1^2 + \zeta^2 \Theta^2 (B_2^2 + B_4^2) + 2uvB_1B_2 \\ + 2uwB_1B_3 + 2u\Psi\zeta B_1^2 + 2u\Theta\zeta B_1B_2 \\ + 2vw(B_2B_3 + B_4B_5) + 2v\Psi\zeta B_1B_2 \\ + 2v\zeta\Theta(B_2^2 + B_4^2) + 2w\Psi\zeta B_1B_3 \\ + 2w\Theta\zeta(B_2B_3 + B_4B_5) + 2\Psi\Theta\zeta^2 B_1B_2 \\ - 2\Delta(B_2C_1 + B_4C_2) \end{array} \right\} \right.$$

$$+ \left\{ \begin{array}{l} 2uB_1C_1 + 2v(B_2C_1 + B_4C_2) \\ + 2w(B_3C_1 + B_5C_2) \\ + 2\Psi\zeta B_1C_1 + 2\Theta\zeta(B_2C_1 + B_4C_2) \end{array} \right\}$$

$$+ \left. \left\{ \begin{array}{l} \Delta^2(B_2^2 + B_4^2) - 2u\Delta B_1B_2 \\ -2v\Delta(B_2^2 + B_4^2) - 2w\Delta(B_2B_3 + B_4B_5) \\ -2\Delta\Psi\zeta B_1B_2 - 2\Delta\Psi\zeta(B_2^2 + B_4^2) \end{array} \right\} \right]$$

$$\times AB\, d\xi\, d\eta\, d\zeta$$

(10.4.16)

where

$$B_1 = \frac{1}{A}[\sin(\theta + \phi) + h,_\xi \cos(\theta + \phi)]$$

$$B_2 = \frac{1}{B}[(a\theta,_\eta + h,_\eta)\cos(\theta + \phi) - h\theta,_\xi \sin(\theta + \phi)]$$

$$B_3 = -\frac{1}{C}[\cos(\theta + \phi) - h,_\xi \sin(\theta + \phi)]$$

$$B_4 = \frac{1}{B}$$

$$B_5 = \frac{1}{C}[a\theta,_\eta + h,_\eta + hh,_\xi \theta,_\eta]$$

(10.4.17)

In equation (10.4.16), the first set of curved bracket terms lead to linear differential equations. The third set of curved bracket terms are of smaller order and nonlinear, they are neglected here. The second set of terms lead to non homogeneous terms in differential equations and are of no significance in the free vibration analysis. They lead to pseudo-static deformations and are also neglected in what follows. The first set of curved bracket terms lead to

$$T_{RI} = \frac{1}{2}\rho\omega^2 \int_0^{d_4} \int_{b_1}^{b_2} \{[U]^T[T][U] - 2\Delta(B_2B_6 + B_4B_7)\}tAB\,d\xi\,d\eta$$

(10.4.18)

where

$$[T] = \begin{bmatrix} B_1^2 & B_1B_2 & B_1B_3 & 0 & 0 \\ B_1B_2 & B_2^2+B_4^2 & B_2B_3+B_4B_5 & 0 & 0 \\ B_1B_3 & B_2B_3+B_4B_5 & B_3^2+B_5^2 & 0 & 0 \\ 0 & 0 & 0 & \frac{t^2}{12}B_1^2 & \frac{t^2}{12}B_1B_2 \\ 0 & 0 & 0 & \frac{t^2}{12}B_1B_2 & \frac{t^2}{12}(B_2^2+B_4^2) \end{bmatrix}$$

(10.4.19)

$$B_6 = a\sin(\theta+\phi) + h\cos(\theta+\phi)$$
$$B_7 = \beta + R$$

(10.4.20)

From (10.3.10), we can write

$$\begin{Bmatrix} \Theta \\ \Phi \end{Bmatrix}$$

$$= \begin{bmatrix} \frac{1}{R_{12}}[Q_1] & -\frac{1}{R_2}[Q_1] & -\frac{1}{B}[Q_6] \\ \frac{1}{2B}[Q_3] & \frac{1}{2AB}B,_a[Q_1] - \frac{1}{2A}[Q_2] & 0 \end{bmatrix} \begin{Bmatrix} a_1 \\ a_2 \\ a_3 \\ \vdots \\ a_{40} \end{Bmatrix}$$

$$= [N_2]\{a\}$$

(10.4.21)

Substituting equations (10.3.27), (10.4.19), (10.4.4a) and (10.4.21) in equation (10.4.18) and using (10.3.2), we get the final expression for the additional kinetic energy due to rotation as

$$T_{RI} = \frac{1}{2}\{q\}^T[[\hat{K}_{RI}] - [\hat{K}_{RII}]]\{q\}$$
$$= \frac{1}{2}\{q\}^T[\hat{K}_R]\{q\}$$

(10.4.22)

where

$$[\hat{K}_{RI}] = [[P]^{-1}]^T \left\{ \rho\omega^2 \int_0^{d_4} \int_{b_1}^{b_2} [N_1]^T [T][N_2] tAB \, d\xi \, d\eta \right\} [P]^{-1}$$

$$[\hat{K}_{RII}] = [[P]^{-1}]^T \left\{ \rho\omega^2 \int_0^{d_4} \int_{b_1}^{b_2} \left[\int_\beta^l 2(B_2 B_6 + B_4 B_7) d\beta \right] \times [N_2]^T [N_2] tAB \, d\xi \, d\eta \right\} [P]^{-1}$$

(10.4.23)

The total kinetic energy is now given by

$$T = \frac{1}{2}\{\dot{q}\}^T [\widehat{M}]\{\dot{q}\} + \frac{1}{2}\{q\}^T [\hat{K}_R]\{q\}$$

(10.4.24)

Including the strain energy term also and applying Lagrange's equation, we get

$$[\widehat{M}]\{\ddot{q}\} + [[\hat{K}] - [\hat{K}_R]]\{q\} = 0$$

(10.4.25)

Example 10.4.1
Consider the square flat plate of example 10.3.1 with $\omega/R = 2.0, R/b = 1.0, \phi = 45°$. The nondimensional natural frequencies $\bar{p} = \sqrt{\frac{12\rho p^2 l^4 (1-v^2)}{Et^2}}$ for different mesh sizes are shown in the table below.

Mode	Mesh Size		
	2×2	3×3	4×4
1	10.8870	10.8187	10.7959
2	13.3487	13.2508	13.2188
3	28.4000	28.2912	28.2061
4	34.7176	34.4371	34.3565
5	39.2956	37.8472	37.1710

As in the previous section 10.3, it can be seen that there is a good convergency and mesh size 4X4 is used in the following examples.

Example 10.4.2
Consider the fan blade of example 10.3.2. This blade is mounted on a disc of radius 30.5 cm at a stagger angle 45° and rotating at 15,000 rpm. The first ten non dimensional frequencies are:

360.25, 365.68, 408.69, 577.70, 937.53, 948.14, 964.55, 1021.63, 1159.38 and 1448.81.

Example 10.4.3

The plate considered in this example has the following data: $l/b = 2$, $t/b = 0.0625$, $r/b = 4$ and $\theta = \phi = 30°$. The non dimensional bending frequencies for this plate at different speeds from 0 to 15,000 rpm are given in the table below.

Mode ω	0	2,000	4,000	7,000	10,000	15,000
$\frac{\omega}{p_1}$	0	0.1515	0.303	0.5302	0.7575	1.1363
0/1	3.4373	3.5794	3.9742	4.8922	6.0306	8.1446
0/2	18.9649	19.0955	19.4815	20.5007	21.9696	25.1447
0/3	61.4638	61.5742	61.9071	62.8329	64.2817	67.8362

References

1. Abromowitz, M. and Cahill, W.F., On the Vibration of a Square Clamped Plate, J. Assoc. Computing Machinery, Vol. 2, July 1955, p. 162
2. Abel, J. F. and Desai, C. S., Comparison of Finite Elements for Plate Bending, J. Struc. Div., ASCE, Vol. 98, No. ST9, September 1972, p. 2143
3. Achenbach, J. D., Keshavan, S. P. and Herrmann, G., Waves in Smoothly Joined Plate and Half-Space, J. Engng. Mech. Div., Proc. ASCE, Vol. 92, No. EM2, April 1966, p. 113
4. Adini, A. and Clough, R. W., *Analysis of Plate Bending by the Finite element Method*, National Science Foundation Report G7337
5. Agarwal, B. D. and Broutman, L. J., *Analysis and Performance of Fiber Composites*, John Wiley, 1980
6. Ahmad, S., Irons, B. M. and Zienkiewicz, O. C., Analysis of Thick and Thin Shell Structures by Curved Finite Elements, Intl. J. Num. Meth. Engng., Vol. 3, 1971, p. 575
7. Aksu, G. and Al-Kaabi, S. A., Free Vibration Analysis of Mindlin Plates with Linearly Varying Thickness, J. Sound Vib., Vol. 119, 1987, p. 189
8. Al-Kaabi, S. A. and Aksu, G., Natural Frequencies of Mindlin Plates of Bilinearly Varying Thickness, J. Sound Vib., Vol. 123, 1988, p. 373
9. Aljanabi, B. S., Hinton, E. and Vuksanovic, Dj., Free Vibration of Mindlin Plates using Finite Element Methods, Part I: Square Plates with Various Edge Conditions, Engng. Computations, Vol. 6, 1989, p. 90
10. Amba Rao, C. L., On the Vibration of A Rectangular Plate Carrying a Concentrated Mass, J. Appld. Mech., Vol. 31, No. 3, 1964, p. 550
11. Ambartsumyan, S. A., *Theory of Anisotropic Plates*, Technomic, 1970
12. Ambartsumyan, S. A., *Theory of Anisotropic Shells*, NASA Report TTF-118, 1964
13. Anderson, R. G., Irons, B. M. and Zienkiewicz, O. C., Vibration and Stability of Plates using Finite Elements, Intl. J. Solids and Struc., Vol. 4, 1968, p. 1031
14. Armenakas, A. E., Gazis, D. C. and Herrmann, G., *Free Vibrations of Circular Cylindrical Shells*, Pergamon, 1969
15. Arnold, R. N. and Bycroft, G. N. and Warburton, G. B., Forced Vibration of a Body on an Infinite Elastic Solid, J. Appld. Mech., Trans. ASME, Vol. 77, 1955, p. 391
16. Ashton, J. E., Natural Modes of Vibration of Tapered Plates, J. Struc. Div., Proc. ASCE, Vol. 95, No. ST4, April 1969, p. 787
17. Ashton, J. E. and Whitney, J. M., *Theory of Laminated Plates*, Technomic, 1970
18. Atnola, L. Ya., Methods of Investigating Elastic Plate Vibrations, Inz. Zhurnal, Vol. 3, No. 2, 1963, p. 312
19. Austin, R. N., Caughfield, D. A., and Plass, H. J. (Jr), Application of Reissner's Variational Principle to the Vibration Analysis of Square Flat Plates with Various Root Support Conditions, Developments in Theoretical and Appld. Mech., Vol. I, Plenum Press, 1963, pp 1-24

20. Averill, R. C. and Reddy, J. N., On the Behavior of Plate Elements Based on the First Order Shear Deformation Theory, Engng. Computations, Vol. 7, No. 1, p. 57, 1990
21. Averill, R. C. and Reddy, J. N., An Assessment of Four Noded Plate Elements Based on A Generalized Third Order Theory, Intl. J. Num. Meth. Engng., Vol. 33, p. 1553, 1992
22. Baharlou, B. and Leissa, A. W., Vibration and Buckling of Generally Laminated Composite Plates with Arbitrary Edge Conditions, Intl. J. Mech. Sci., Vol. 29, 1987, p. 545
23. Barbero, E. J. and Reddy, J. N., Nonlinear Analysis of Composite Laminates Using A Generalized Laminate Plate Theory, AIAA J, Vol. 28, No. 11, p. 1987, 1990
24. Bartlett, C. C., The Vibration and Buckling of A Circular Plate Clamped on Part of Its Boundary and Simply Supported on the Remainder, Quart. J. Mech. Appld. Mech., Vol. 16, Pt. 4, 1963, p. 431
25. Bathe, K. J., *Finite Element Procedures*, Prentice Hall, 1996
26. Bathe, K. J. and Dvorkin, E. N., A Four Node Plate Bending Element Based on Mindlin/Reissner Plate Theory and a Mixed Interpolation, Intl. J. Num. Meth. Engng., Vol. 21, 1985, p. 361
27. Bathe, K. J., Dvorkin, E. N. and Ho, L. W., On Discrete Kirchoff and Isoparametric Shell Elements for Nonlinear Analysis - An Assessment, Computers and Structures, Vol. 16, p. 89, 1983
28. Bathe, K. J. and Wilson, E. L., *Numerical Methods in Finite Element Analysis*, Prentice Hall, 1976
29. Batoz, J. L., Bathe, K. J. and Ho, L. W., A Study of Three-Node Triangular Plate Bending Elements, Intl. J. Num. Meth. Engng., Vol. 15, 1980, p. 1771
30. Bauer, H. F., Nonlinear Response of Elastic Plates to Pulse Excitations, J. Appld. Mech., Vol. 22, No. 4, December 1955, p. 465
31. Bazeley, G. P. et. al., Triangular Elements in Plate Bending - Conforming and Nonconforming Solutions, Proc. Conf. Mat. Meth. Struc. Mech., AFIT, 1965, p. 547
32. Bergan, P. G. and Wang, X., Quadrilateral Plate Bending Elements with Shear Deformation, Comp. Struc., Vol. 19, 1984, p. 25
33. Berger, H. M., A New Approach to the Analysis of Large Deflections of Plates, J. Appld. Mech., Vol. 35, No. 1, March 1968, p. 47
34. Bert, C. W. and Chen, T. L. C., Effect of Shear Deformation on Vibration of Antisymmetric Angle-Ply Laminated Rectangular Plates, Intl. J. Solids Struc., Vol. 14, 1978, p. 465
35. Bert, C. W., Reddy, J. N., Chao, W. C. and Reddy, V. S., Vibration of Thick Rectangular Plates of Bimodulus Composite Materials, J Appld. Mech., Trans. ASME, Vol. 48, 1981, p. 371
36. Bhashyam, G. R. and Gallagher, R. H., An Approach to the Inclusion of Transverse Shear Deformation in Finite Element Plate Bending Analysis, Computers and Structures, Vol. 19, 1984, p. 35
37. Bhat, R. B., Natural Frequencies of Rectangular Plates using Characteristic Polynomials in Rayleigh-Ritz Method, J Sound Vib., Vol. 102, 1985, p. 493
38. Bhat, R. B., Numerical Experiments on the Determination of Natural Frequencies of Transverse Vibrations of Rectangular Plates of Non uniform Thickness, J Sound Vib., Vol. 138, 1990, p. 205
39. Bhat, R. B. and Mundkur, G., Vibration of Plates using Plate Characteristic Functions obtained by Reduction of Partial Differential Equations, J Sound Vib., Vol. 161, No. 1, 1993, p. 157

40. Bhimaraddi, A. and Stevens, L. K., A Higher Order Theory for Free Vibrations of Orthotropic, Homogeneous and Laminated Rectangular Plates, J. Appld. Mech., Vol. 51, p. 195, 1984
41. Bhumbla, R. and Kosmatka, J. B., Stability of Spinning Shear Deformable Laminated Composite Plates, J Sound Vib., Vol. 163, No. 1, 1993, p. 83
42. Bhumbla, R. and Kosmatka, J. B., Behavior of Spinning Pre-Twisted Composite Plates using a Nonlinear Finite Element Approach, AIAA J, Vol. 34, No. 8, p. 1686, 1996
43. Bhumbla, R. Kosmatka, J. B. and Reddy, J. N., Free Vibration Behavior of Spinning Shear Deformable Plates Composed of Composite Materials, AIAA J, Vol. 28, No. 11, 1990, p. 1962
44. Bickley, W. G., Deflections and Vibrations of A Circular Elastic Plate Under Tension, Phil. Mag., Series 7, Vol. 15, No. 100, 1933, p. 776
45. Biggs, J. M., *Structural Dynamics*, McGraw Hill, 1964
46. Bodine, R. Y., The Fundamental Frequencies of A Thin Flat Circular Plate Simply Supported Along A Circle of Arbitrary Radius, ASME J. Appld. Mech., Vol. 26, 1959, p. 666
47. Bogner, F. K., Fox, R. L. and Schmit, A., The Generation of Inter-Element Compatible Stiffness and Mass Matrices by the use of Interpolation Formulas Proc. Conf. Mat. Meth. Struc. Mech., AFIT, 1965, p. 397
48. Boresi, A. P., *Theory of Elasticity*, Prentice Hall, 1965
49. Bossak, M. A. J. and Zienkiewicz, O. C., Free Vibration Analysis of Initially Stressed Solids with particular reference to Centrifugal Force Effects in Rotating Machinery, J. Strain Analysis, Vol. 8, 1973, p. 245
50. Bowlus, J. A., Palazotto, A. N. and Whitney, J. M., Vibration of Symmetrically Laminated Rectangular Plates considering Shear Deformation and Rotary Inertia, Amer. Inst. Aeronautics J., Vol. 25, 1987, p. 1500
51. Brunelle, E. J. and Robertson, S. R., Vibration of an Initially Stressed Moderately Thick Plate, J. Sound Vib., Vol. 45, 1976, p. 405
52. Bycroft, G. N., Frequencies of a Flexible Circular Plate attached to the Surface of a Light Elastic Half-space, J. Appld. Mech., Vol. 26, No. 1, March 1959, p. 13
53. Callahan, W. R., On the Flexural Vibrations of Circular and Elliptical Plates, Quarterly Áppld. Mathematics, Vol. 13, January 1956, p. 371
54. Carmichael, T. E., The Vibration of a Rectangular Plate with Edges Elastically Restrained Against Rotation, Q. J. Mech. and Appld. Math., Vol. 12, Pt. 1, 1959, p. 29
55. Carrington, H., The Frequencies of Vibration of Flat Circular Plates Fixed at the Circumference, Phil. Mag., Vol. 50, No.6, 1925, p. 1261
56. Chamis, C. C., Vibration Characteristics of Composite Fan Blades and Comparisons with Measured Data, J. Aircraft, Vol. 14, No. 7, 1977, p. 644
57. Chatterjee, S. N. and Kulkarni, S. V., Shear Correction Factors for Laminated Plates, AIAA J, Vol. 17, No. 5, p. 498, 1979
58. Chen, A. T. and Yang, T. Y., A 36 DOF Symmetrically Laminated Triangular Element with Shear Deformation and Rotatory Inertia, J Composite Materials, Vol. 22, 1988, p. 341
59. Chen, L. W. and Doong, J. L., Large Amplitude Vibration of an Initially Stressed Moderately Thick Plate, J. Sound Vib., Vol. 89, 1983, p. 499
60. Chen, W. C. and Liu, W. H., Deflection and Free Vibrations of Laminated Plates - Levy Type Solutions, Intl. J. Mech. Sci., Vol. 32, 1990, p. 779
61. Cheung, Y. K. and Zienkiewicz, O. C., Plates and Tanks on Elastic Foundations, Intl. J. Solids and Structures, Vol. 1, No. 4, November 1965, p. 451

References

62. Chladni, E. F. F., *Die Akustik*, Leipzig, 1802
63. Christensen, R. M., *Mechanics of Composite Materials*, John Wiley, 1979
64. Chu, H. and Herrmann, G., Influence of Large Amplitudes on Free Flexural Vibrations of Rectangular Elastic Plates, J. Appld. Mech., Vol. 23, No. 4, December 1956, p. 532
65. Classen, R. W. and Thorne, C. J., Vibrations of Thin Rectangular Isotropic Plates, J. Appld Mech., Trans. ASME, Vol. 28, 1961, p. 304
66. Classen, R. W. and Thorne, C. J., Vibrations of A Rectangular Cantilever Plate, J. Aero. Sci., Vol. 29, No. 11, 1962, p. 1300
67. Clough, R. W. and Felippa, C. A., A Refined Quadrilateral Element for Analysis of Plate Bending, Proc. 2nd Conf. Mat. Meth. Struc. Mech., AFIT, 1968, p. 399
68. Clough, R. W. and Tocher, J. L., Finite Element Stiffness Matrices for Plate Bending, Proc. Conf. Mat. Meth. Struc. Mech., AFIT, 1965, p. 515
69. Cohen, H. and Handelman, G., Vibrations of a Rectangular Plate with Distributed Added Mass, J. Franklin Inst., Vol. 261, No. 3, 1956, p. 319
70. Colwell, R. C., The Vibrations of A Circular Plate, J. Franklin Inst., Vol. 213, No. 1276-77, 1932, p. 373
71. Colwell, R. C. and Hardy, H. C., The Frequencies and Nodal Systems of Circular Plates, Phil. Mag., Series 7, Vol. 24, No. 165, 1937, p. 1041
72. Colwell, R. C., Stewart, J. K. and Arnett, H. D., Symmetrical Sand Figures on Circular Plates, J. Acoust. Soc. Am., Vol. 12, 1940, p. 260
73. Colwell, R. C., Stewart, J. K. and Friend, A. W., Symmetrical Figures on Circular Plates and Membranes, Phil. Mag., Series 7, Vol. 27, 1939, p. 123
74. Conway, H. D., Analysis of Some Circular Plates on Elastic Foundations and Flexural Vibrations of Some Circular Plates, J. Appld. Mech., Vol. 22, No. 2, June 1955, p. 275
75. Conway, H. D., Some Special Solutions for the Flexural Vibration of Discs of Varying Thickness, Ing. Arch., Vol. 26, December 1958, p. 408
76. Cook, R. D., Avoidance of Parasitic Shear in Plane Element, J. Struc. Div., Proc. ASCE, Vol. 101, 1975, p. 1239
77. Cook, R. D., Malkus, D. S. and Plesha, M. E., *Concepts and Applications of Finite Element Analysis*, Wiley, 1989
78. Courant, R. and Hilbert, D., *Methods of Mathematical Physics*, Interscience, 1953
79. Cowper, G. R. et. al., A High Precision Triangular Plate Bending Element, NRC Canada Aeronautical Report LR-514
80. Cowper, G. R., Lindberg, G. M. and Olson, M. D., A Shallow Shell Finite Element of Triangular Shape, Intl. J. Sol. Struc., Vol. 6, No. 8, 1970, p. 1133
81. Cox, H. L., Flexural Vibration of Plates on Uniform Elastic Foundation, J. Royal Aero. Soc., Vol. 58, No. 525, September 1954, p. 651
82. Cox, H. L., *The Buckling of Plates and Shells*, Pergamon, 1961
83. Cox, H. L., Vibration of A Square Plate, Point-Supported at Midpoints of Sides, J. Acoust. Soc. Am., Vol. 27, No. 1, 1955, p. 791
84. Cox, H. L., Vibration of Certain Square Plates Having Similar Adjacent Edges, Q. J. Mech. Appld. Math., Vol. 8, Pt. 4, 1955, p. 454
85. Cox, H. L. and Boxer, J., Vibration of Rectangular Plates Point-Supported at the Corners, Aero Qly., Vol. 11, No. 1, 1960, p. 41
86. Craig, T. J. and Dawe, D. J., Flexural Vibration of Symmetrically Laminated Composite Rectangular Plates including Transverse Shear Effects, Intl. J. Solids and Struc., Vol. 22, 1986, p. 155
87. Crawley, E. F., The Natural Mode of Graphite/Epoxy Cantilever Plates and Shells, J. of Composite Materials, Vol. 13, 1979, p. 195

88. Crawley, E. F. and Dugundji, J., Frequency Determination and Nondimensionalization for Composite Cantilever Plates, J. Sound Vib., Vol. 72, No. 1, 1980, p. 1
89. Das, Y. C., On the Transverse Vibrations of Rectangular Isotropic Plates, J. Aero. Soc. India, Vol. 13, No. 4, 1961, p. 111
90. Das, Y. C. and Navaratna, D. R., Vibrations of A Rectangular Plate with Concentrated Mass, Spring, and Dashpot, J. Appld. Mech., Vol. 30, No. 1, 1963, p. 31
91. Dawe, D. J., A Finite Element Approach to Plate Vibration Problems, J. Mech. Engng. Sci., Vol. 7, 1965, p. 28
92. Dawe, D. J. and Craig, T. J., The Vibration and Stability of Symmetrically Laminated Composite Rectangular Plates subjected to In-plane Stresses, Composites and Struc., Vol. 5, 1985, p. 281
93. Dawe, D. J. and Roufaeil, O. L., Rayleigh-Ritz Vibration Analysis of Mindlin Plates, J. Sound Vib., Vol. 69, 1980, p. 345
94. Deak, A. L. and Pian, T. H. H., Application of Smooth Surface Interpolation to the Finite element Analysis, AIAA J., Vol. 5, 1967, p. 187
95. Deresiewicz, H. and Mindlin, R. D., Axially Symmetric Flexural Vibrations of a Circular Disc, J. Appld. Mech. , Vol. 22, No. 1, March 1955, p. 86
96. Dickinson, S. M., On he Use of Simply Supported Plate Function in Rayleigh-Ritz Method applied to Flexural Vibration of Rectangular Plates, J Sound Vib., Vol. 59, 1978, p. 143
97. Dickinson, S. M. and Blasio, A. Di., On the Use of Orthogonal Polynomials in Rayleigh-Ritz Method for the Study of Flexural Vibration and Buckling of Isotropic and Orthotropic Plates, J Sound Vib., Vol. 108, 1986, p. 51
98. Dickinson, S. M. and Henshell, R. D., Clough-Tocher Triangular Plate Bending Element in Vibration, AIAA J, Vol. 7, 1969, p. 560
99. Dickinson, S. M. and Li, E. K. H., On the Use of Simply Supported Plate Functions in Rayleigh-Ritz Method applied to the Flexural Vibration of Rectangular Plates, J Sound Vib., Vol. 80, 1982, p. 292
100. Dill, E. H. and Pister, K. S., Vibration of Rectangular Plates and Plate Systems, Proc. Third U.S. Natl. Cong. Appld. Mech., 1958, p. 123
101. Di Sciuva, M., A Refined Transverse Shear Deformation Theory for Multi layered Anisotropic Plates, Atti Accad. Sci. Torino, Vol. 118, p. 279, 1984
102. Doong, J. L., Vibration and Stability of an Initially Stressed Thick Plate according to a Higher Order Deformation Theory, J. Sound Vib., Vol. 113, 1987, p. 425
103. Doong, J. L., Lee, C. and Fung, C. P., Vibration and Stability of Laminated Plates based on a Modified Plate Theory, J. Sound and Vib., Vol. 151, 1991, p. 193
104. Dougall, J., An Analytical Theory of the Equilibrium of an Isotropic Elastic Plate, Trans. Royal Society, Edinburgh, Vol. 41, 1904, p. 129
105. Dupuis, G. and Goel, J., A Curved Finite Element for Thin Elastic Shells, Intl. J. Sol. Struc., Vol. 11, 1970, p. 1413
106. Dym, C. L. and Shames, I. H., *Solid Mechanics A Variational Approach*, McGraw Hill, 1973
107. Eisenberg, M. A. and Malvern, L. E., On Finite Element Integration in Natural Coordinates, Intl. J. Num. Meth. Engng., Vol. 7, 1973, p. 574
108. Erdelyi, A. et al, *Tables of Integral Transforms*, Vols. 1 and 2, McGraw Hill Book Co. 1954
109. Ergatoudis, B., Irons, B. M. and Zienkiewicz, O. C., Curved Isoparametric 'Quadrilateral' Elements for Finite Element Analysis, Intl. J. Solids and Struc., Vol. 4, No. 1, 1968, p. 31
110. Filonenko and Boroditch, *Theory of Elasticity*, Peace Publishers
111. Fletcher, H. J., The Frequency of Vibration of Rectangular Isotropic Plates, J. Appld. Mech., Vol. 26, No. 2, 1959, p. 290

112. Fletcher, H. J. and Thorne, C. J., Thin Rectangular Plates on Elastic Foundations, J. Appld. Mech. , Vol. 19, No. 3, September 1952, p. 361
113. Forray, M. J., *Variational Calculus in Science and Engineering*, McGraw Hill, 1968
114. Forsyth, A. R., *Calculus of Variations*, Cambridge University Press, London
115. Forsyth, E. M. and Warburton, G. B., Transient Vibration of Rectangular Plates, J. Mech. Engng. Sci., Vol. 2, No. 4, December 1960, p. 325
116. Fox, C., *Calculus of Variations*, Oxford University Press, 1954
117. Fraeijs De Veubeke B., A Conforming Finite Element for Plate Bending, Intl. J. Solids and Structs., Vol. 4, 1968, p. 95
118. Fung, *Foundations of Solid Mechanics*, Prentice Hall, 1968
119. Gajendar, N., *Bending and Vibration of Elastic Plates*, Ph D Thesis, IIT, Kharagpur, 1965
120. Gajendar, N., Large Amplitude Vibrations of Plates on Elastic Foundations, Intl. J. Nonlinear Mech., Vol. 2, 1967, p. 163
121. Galeetly, G. D., Optimum Design of Thin Circular Plates on an Elastic Foundation, Proc. I Mech. E., Vol. 173, 1959, p. 687
122. Ganapathi, M., Varadan, T. K. and Sarma, B. S., Nonlinear Flexural Vibrations of Laminated Orthotropic Plates, Computers and Struc., Vol. 39, 1991, p. 685
123. Gazis, D. C. and Mindlin, R. D., Extensional Vibrations and Waves in a Circular Disc and a Semi-infinite Plate, J. Appld. Mech., Trans. ASME, Vol. 27, 1960, p. 541
124. Gontkevich, V. S., *Natural Vibrations of Plates and Shells*, A. P. Filippov (Ed), Nauk Dumka (Kiev) 1964; English Translation by Lockheed Missiles and Space Co., Sunnyvale
125. Gontkevich, V. S., The Lower Bounds of Natural Frequencies of A Plate in Flexural Vibrations, Prikl. Mekh., AN UkrSSR, Vol. 6, No. 3, 1960, p. 346
126. Gorman, D. J., *Free Vibration Analysis of Rectangular Plates,* Elsevier, 1982
127. Gorman, D. J., Free Vibration Analysis of Cantilever Plates by Method of Superposition, J. Sound Vib., Vol. 49, 1976, p. 453
128. Gorman, D. J., Free Vibration Analysis of Rectangular Plates with Clamped-Simply Supported Edge Conditions of by the Method of Superposition, J. Appld. Mech., Vol. 44, 1977
129. Gorman, D. J., Free Vibration Analysis of the Completely Free Rectangular Plates by the Method of Superposition, J. Sound Vib., Vol. 57, 1978, p. 437
130. Gorman, D. J., Solutions of Levy Type for the Free Vibration Analysis of Diagonally Supported Rectangular Plates, J. Sound Vib., Vol. 66, 1979, p. 239
131. Gorman, D. J., Accurate Free Vibration Analysis of the Completely Free Orthotropic Rectangular Plate by the Method of Superposition, J Sound Vib., Vol. 165, No 3, 1993, p. 409
132. Gorman, D. J. and Sharma, R. K., A Comprehensive Approach to the Free Vibration Analysis of Rectangular Plates by the Method of Superposition, J. Sound Vib., Vol. 47, 1976, p. 126
133. Gray, A. and Mathews, G. B., *A Treatise on Bessel Functions and Their Applications to Physics,* Macmillan and Co. 1922
134. Grinsted, B., Nodal Pattern Analysis, Proc. Instn. of Mech. Engrs., Series A, Vol. 166, 1952, p. 309
135. Gupta, K. and Rao, J. S., Torsional Vibration of Pre twisted Cantilever Plates, J. Mech. Des., Trans. ASME, Vol. 100, 1978, p. 528
136. Gupta, K. and Rao, J. S., Flexural Vibration of Pre-Twisted Cantilever Plates, J Aero Soc. (India), Vol. 30, 1978, p.131

137. Gupta, K. and Rao, J. S., Inextensional Vibration of Pre-Twisted Cantilever Plates, Proc 7th Canadian Cong Appl Mech., 1979, p.445
138. Gupta, K. and Rao, J. S., Vibrations of Rotating Small Aspect Ratio Blades, 15th Intl. Cong Theo. and Appl Mechanics, Toronto, 1980, p.131
139. Gupta, K. and Rao, J. S., Free Vibrations of Rotating Small Aspect Ratio Pre-Twisted Blades, Natl. Conf. Machines and Mechanisms, Bangalore, 1985, p.125
140. Gupta, K. and Rao, J. S., Free Vibrations of Rotating Small Aspect Ratio Pre-Twisted Blades, Mechanism and Machine Theory, Vol. 22, No. 2, 1987, p.159
141. Gupta, K. K., *STARS-A General Purpose Finite Element Computer Program for Analysis of Engineering Structures*, NASA Reference Publication 1129, 1984
142. Gustafson, P. N., Stokey, W. F. and Zorowski, C. F., An Experimental Study of Natural Vibration of Cantilevered Triangular Plates, J. Aero. Sci., Vol. 20, 1953, p. 331
143. Guyan, R. J., Distributed Mass Matrix for Plate Element Bending, AIAA J, Vol. 3, 1965, p. 567
144. Habata, Y., On the Lateral Vibration of Rectangular Plate Clamped at Four Edges, Trans. JSME, Vol. 13, No. 44, 1947, p. 67
145. Halpin, J. C., Stiffness and Expansion Estimates for Oriented Short Fiber Composites, J. Composite Materials, Vol. 3, p. 732, 1969
146. Hamada, M., A Method of Solving Problems of Vibration, Deflection and Buckling of Rectangular Plates with Clamped or Simply Supported Edges, Bull JSME, Vol. 2, No. 5, 1959, p. 92
147. Hausrath, A. H., Small Deflections of Elastically Supported Plates, Proc. 9th Intl. Cong. Appld. Mech., Vol. 6, 1957, p. 367
148. Hearmon, R. F. S., The Frequency of Vibration of Rectangular Isotropic Plates, J. of Appld. Mech., Vol. 19, 1952, p. 402
149. Hearmon, R. F. S., The Frequency of Flexural Vibration of Rectangular Orthotropic Plates with Clamped or Supported Edges, J. Appld. Mech., Vol. 26, No. 3, 1959, p. 537
150. Hearmon, R. F. S., *An Introduction to Applied Anisotropic Elasticity*, Oxford University Press, 1961
151. Henry, R. and Lalanne, M., Vibration Analysis of Rotating Compressor Blades, J. of Engng. for Industry, ASME, August 1974, p. 1028
152. Herrmann, G., The Influence of Initial Stress on the Dynamic Behavior of Elastic and Viscoelastic Plates, Pub. Intl. Assoc. for Bridge and Struc. Engng., Vol. 16, 1956, p. 275
153. Herrmann, G. and Armenakas, A. E., Vibrations and Stability of Plates under Initial Stress, J. Engng. Mech. Div., Proc. ASCE, Vol. 86, No. EM3, June 1960, p. 65
154. Hidaka, K., Vibration of a Square Plate Clamped At Four Edges, Math. Jap., Vol. 2, 1951, p. 97
155. Hinton, E., *Numerical Methods and Software for Dynamic analysis of Plates and Shells*, Pineridge Press, 1988
156. Hinton, E. and Bicanic, N., A Comparison of Lagrangian and Serendipity Mindlin Plate Elements for Free Vibration Analysis, Computers and Structures, Vol. 10, 1979, p. 483
157. Holl, D. L., Thin Plates on Elastic Foundations, Proc. Fifth Intl. Cong. Appld. Mech., 1938, p. 71
158. Holl, D. L., Dynamic Loads on Thin Plates on Elastic Foundations, Proc. Third Symp. Appld. Mathematics, Vol. 3, 1950, p. 107
159. Hoppmann, W. H. and Greenspon, J., An Experimental Device for Obtaining

Elastic Rotational Constraints on the Boundary of a Plate, Proc. 2nd US Natl. Cong. Appld. Mech., 1954, p. 187

160. Huang, H. C. and Hinton, E., A Nine Node Lagrangian Mindlin Plate Element with Enhanced Shear Interpolation, Engng. Computations, Vol. 1, 1984, p. 369
161. Huffington, N. J., Jr. and Hoppmann II, W. H., On the Transverse Vibrations of Rectangular Orthotropic Plates, J. Appld. Mech., Vol.. 25, No. 3, September 1958, p. 389
162. Hughes, T. J. R. and Cohen, M., The 'Heterosis' Finite Element for Plate Bending, J. Comp. Struc., Vol. 9, 1978, p. 445
163. Hughes, T. J. R., Taylor, R. L. and Kanoknukulchai, W. K., A Simple and Efficient Finite element for Plate Bending, Intl. J. Num. Meth. Engng., Vol. 11, 1977, p. 1529
164. Iguchi, S., Die Eigenwertprobleme fur die Elastiche Rechteckige Platte, Mem. Fac. Eng., Hokkaido Univ., 1938, p. 305
165. Irons, B. M., Quadrature Rules for Brick Based Finite Elements, Intl. J. Num. Meth. Engng., Vol. 3, 1971, p. 293
166. Irons, B. M. and Draper, K. J., Inadequacy of Nodal Connections in A Stiffness Solution for Plate Bending, AIAA J, Vol. 3, 1965, p. 961
167. Iyengar, K. T. S. R. and Pandya, S. K., Vibration of Orthotropic Rectangular Thick Plates, Intl. J. Struc., Vol. 2, 1982, p. 149
168. Jensen, D. W. and Crawley, E. F., Frequency Determination Techniques for Cantilevered Plates with Bending-Torsion Coupling, AIAA J, Vol. 22, No. 3, 1984, p. 415
169. Joga Rao, C. V. and Kantham, C. L., Natural Frequencies of Rectangular Plates with Edges Elastically Restrained Against Rotation, J. Aero Sci., Vol. 24, No. 4, 1957, p. 855
170. Joga Rao, C. V. and Pickett, G., Vibrations of Plates of Irregular Shapes and with Holes, J. Aero. Soc. Ind., Vol. 13, No. 3, 1961, p. 83
171. Joga Rao, C. V. and Vijayakumar, K., On Admissible Functions for Flexural Vibrations and Buckling of Annular Plates, J. Aero. Soc. India, Vol. 15, No. 1, 1963, p. 1
172. Jones, R. M., *Mechanics of Composite Materials*, McGraw Hill, 1975
173. Jones, R. M., Stiffness of Orthotropic Materials and Laminated Fiber-Reinforced Composites, AIAA J, Vol. 12, p. 112, 1974
174. Kaczkowski, Z., Orthotropic Rectangular Plates with Arbitrary Boundary Conditions, Proc. Ninth Intl. Cong. Appld. Mech., Vol. 6, 1957, p. 430
175. Kaczkowski, Z., The Influence of the Shear Forces and the Rotary Inertia on the Vibration of an Anisotropic Plate, Arch. Mech. Stos., Vol. 12, 1960, p. 531
176. Kanazawa, T. and Kawai, T., On the Lateral Vibration on Anisotropic Rectangular Plates, Proc. 2nd Jap. Natl. Cong. Appld. Mech., 1952, p. 333
177. Kantham, C. L., Bending and Vibration of Elastically Restrained Circular Plates, J. Franklin Inst., Vol. 265, No. 6, 1958, p. 483
178. Kantorovich, L. V. and Krylov, V. I., *Approximate Methods of Higher Analysis*, (Translated by Benster, C. D) Interscience, 1958
179. Kapania, R. K. and Raciti, S., Recent Advances in Analysis of Laminated Beams and Plates, Part I: Shear Effects and Buckling, AIAA J, Vol. 27, No. 7, p. 923, 1989
180. Kapania, R. K. and Raciti, S., Recent Advances in Analysis of Laminated Beams and Plates, Part II: Vibrations and Wave Propagation, AIAA J, Vol. 27, No. 7, p. 935, 1989
181. Kato, H., On the Bending and Vibration of Rectangular Plates, J. Soc. Nav. Arch., Vol. 50, 1932, p. 209

182. Kato, T., Fujita, H., Nakata, Y and Newman, M., Estimation of the Frequencies of Thin Elastic Plates with Free Edges, J. Res. Natl. Bur. Std., Vol. 59, No. 3, 1957, p. 169
183. Kaul, R. K. and Tewari, S. G., On the Bounds of Eigenvalues of A Clamped Plate in Tension, J.. Appld. Mech., Vol. 25, No. 1, 1958, p. 52
184. Keilb, R. E., Leissa, A. W. and MacBain, J. C., Vibration of Twisted Cantilever Plates - A Comparison of Theoretical Results, Intl. J. Num. Meth. Engng., Vol. 21, 1985, p. 1365
185. Khdeir, A. A., Free Vibration of Antisymmetric Angle-ply Laminated Plates including Various Boundary Conditions, J Sound Vib., Vol. 122, 1988, p. 377
186. Khdeir, A. A., Free Vibration and Buckling of Symmetric Cross-ply Laminated Plates by an Exact Method, J Sound Vib., Vol. 126, 1988, p. 447
187. Khdeir, A. A., Free Vibration and Buckling of Unsymmetric Cross-ply Laminated Plates using A Refined Theory, J Sound Vib., Vol. 128, 1989, p. 377
188. Khdeir, A. A., Reddy, J. N. and Librescu, L., Levy Type Solutions for Symmetrically Laminated Rectangular Plates Using First Order Shear Deformation Theory, J. Appld. Mech., Vol. 54, p. 640, 1987
189. Kikuchi, F. and Ando, Y., Rectangular Finite element for Plate Bending Analysis Based on Hellinger-Reissner's Variational Principle, J. of Nuclear Sci and Tech, Vol. 9, p. 28, 1972
190. Kirchhoff, G., Uber das Gleichgewicht und die Bewegung Einer Elastichen Schiebe, Math J., (Crelle), Vol. 40, No. 5, 1850, p. 51
191. Kirchhoff, G., Ges. Abhandl., (Leipzig) 1882, p. 259
192. Kirk, C. L., A Note on the Lowest Natural Frequency of A Square Plate Point-Supported at the Corners, J. Roy. Aero. Soc., Vol. 66, 1962, p. 240
193. Kirkhope, J., and Wilson, G. L., Vibration of Circular and Annular Plates Using Finite Elements, Intl. J. Num. Meth. Engng., Vol. 4, No. 2, 1972, p. 181
194. Kishor, B., *Vibration Analysis of Plates on Viscoelastic Foundations*, Ph D Thesis, Indian Institute of Technology, Kharagpur, 1971
195. Kishor, B. and Rao, J. S., Nonlinear Transverse Vibration of An Orthotropic Elastic Plate on Viscoelastic Foundation, Archiwum Budowy Maszyn, Vol. 21, Pt. 1, 1974, p.15
196. Kishor, B. and Rao, J. S., Nonlinear Vibration Analysis of A Plate on Viscoelastic Foundation, Aero Qly, v.25, 1974, p.37
197. Kosmatka, J. B., An Accurate Shear Deformable Six Node Triangular Plate Element for Laminated Composite Structures, Intl. J. Num. Meth. Engng., Vol. 37, No. 3, 1994, p. 431
198. Krauthammer, T., Accuracy of the Finite Element Method near a Curved Boundary, Computers and Structures, Vol. 10, 1979, p. 921
199. Krishna Murty, A. V., Higher Order Theory for Vibration of Thick Plates, AIAA J., Vol. 15, No. 12, p. 1823, 1977
200. Kundu,B. B., The Transverse Vibrations of a Non-Isotropic Plate of Rectangular and Elliptic Boundaries Resting on Elastic Foundation, J. Science and Engng. Res, Vol. 11, Pt. 1, January 1967, p. 107
201. Kurata, M. and Okamura, H., Natural Vibrations of Partially Clamped Plates, J. Engng. Mech. Div., Trans. ASCE, 1963, p. 169
202. Kurlandzki, J., A Method of Solving Problems of Rectangular Plates with Mixed Boundary Conditions, Proc. Vibration Problems, Vol. 2, No. 4, 1961, p. 377
203. Lamb, H. and Southwell, R. V., The Vibrations of A Spinning Disc, Proc. Roy. Soc. (London), series A, Vol. 99, 1921, p. 272
204. Lanczos, C., *The Variational Principles of Mechanics*, University of Toronto Press

205. Langhaar, H. L., *Energy Methods in Applied Mechanics*, John Wiley, 1962
206. Lapid, A. J., Kosmatka, J. B. and Mehmed, O., Behavior of Spinning Laminated Composite Plates with Initial Twist - Experimental Vibrations, Strain and Deflection Results, 34th Structures, Structural Dynamics and Materials Conference, Vol. I, AIAA, Washington D.C., 1993, p. 255
207. Laura, P. A. A. and Duran, R., A Note on Forced Vibration of Clamped Rectangular Plate, J Sound Vib., Vol. 42, 1975, p. 129
208. Leckie, F. A., The Application of Transfer Matrices to Plate Vibrations, Ingr. Arch., Vol. 32, No. 2, 1963, p. 100
209. Lee, S. W., and Pian, T. H. H., Improvement of Plate and Shell Finite Elements by Mixed Formulations, AIAA J, Vol. 16, p. 29, 1978
210. Leissa, A. W., *Vibration of Plates*, NASA SP-160, 1969
211. Leissa, A. W., The Free Vibration of Rectangular Plates, J. Sound Vib., Vol. 31, 1973, p. 257
212. Leissa, A. W., A Method of Analyzing the Vibration of Plates, J. Aerospace Sci., Vol. 29, No. 4, 1962, p. 475
213. Leissa, A. W., Recent Research in Plate Vibrations: Classical Theory, Shock & Vib. Digest, Vol. 9, No. 10, 1977, p. 13
214. Leissa, A. W., Recent Research in Plate Vibrations: Complicating Effects, Shock & Vib. Digest, Vol. 9, No. 11, 1977, p. 21
215. Leissa, A. W., Plate Vibration Research, 1976-80: Classical Theory, Shock & Vib. Digest, Vol. 13, No. 9, 1981, p. 11
216. Leissa, A. W., Plate Vibration Research, 1976-80: Complicating Effects, Shock & Vib. Digest, Vol. 13, No. 9, 1981, p. 19
217. Leissa, A. W., Vibrational Aspects of Rotating Turbomachinery Blades, Appl. Mech. Rev., Vol. 34, 1981, p. 629
218. Leissa, A. W., Recent Research in Plate Vibrations, 1981-85, Part II: Complicating Effects, Shock & Vib. Digest, Vol. 19, No. 3, 1987, p. 10
219. Leissa, A. W. and Ewing, M. S., Comparison of Beam and Shell Theories for the Vibration of Thin Turbomachinery Blades, J. Engng. Power, Trans. ASME, Vol. 105, 1983, p. 383
220. Leissa, A. W., Lee, J. K. and Wang, A. J., Rotating Blade Vibration Analysis using Shell Theory, J. of Engng. for Power, ASME, V. 104, 1982, p. 296
221. Leissa, A. W., Lee, J. K. and Wang, A. J., Vibration of Twisted Rotating Blades, J. of Vib. Acoustics, Stress and Reliability in Des., ASME, V. 106, No. 2, 1984, p. 251
222. Leissa, A. W., Lee, J. K. and Wang, A. J., Vibration of Blades with Variable Thickness and Curvature by Shell Theory, J. of Engng. for Gas Turbines and Power, ASME, V. 106, 1985, p. 11
223. Leissa, A. W., Macbain, J. C. and Kielb, R. E., Vibration of Twisted Cantilever Plates - Summary of Previous and Current Studies, J Sound Vib., 1984, p. 159
224. Leissa, A. W. and Qatu, M. S., Equations of Elastic Deformation of Laminated Composite Shallow Shells, J Appld. Mech., Vol. 58, March 1991, p. 181
225. Lekhnitskii, S. G., *Theory of Elasticity of an Anisotropic Elastic Body*, (Translated by Fern, P) Holden-Day, 1963
226. Lekhnitskii, S. G., *Anisotropic Plates*, (Translated by Tsai, S.W. and Cheron, T.) Gordon and Breach (1968)
227. Levinson, M., An Accurate, Simple Theory of the Statics and Dynamics of Elastic Plates, Mech. Research Communications, Vol. 7, No. 6, p. 343, 1980
228. Levy, S., Large Deflection Theory for Rectangular Plates, Proc. Symp. Appld. Math., Vol. 1, 1949, p. 197

229. Li, N., Forced Vibration Analysis of the Clamped Orthotropic Rectangular Plate by the Superposition Method, J Sound Vib., Vol. 158, No. 2, 1992, p. 307
230. Liew, K. M., Huang, K. C. and Lim, M. K., A Continuum Three Dimensional Vibration Analysis of Thick Rectangular Plates, Intl. J. Solids and Strucs., Vol. 30, 1993, p. 3357
231. Liew, K. M. and Lim, C. W., Vibratory Characteristics of Cantilevered Rectangular Shallow Shells of Variable Thickness, AIAA J., Vol. 32, No. 2, 1994, p. 387
232. Liew, K. M. and Lim, C. W., Vibratory Behavior of Doubly-Curved Shallow Shells of Curvilinear Planform, J Engng. Mech., Trans. ASCE, Vol. 121, No. 12, 1995, p. 1277
233. Liew, K. M. and Lim, C. W., A Ritz Vibration Analysis of Doubly-Curved Rectangular Shallow Shells using A Refined First-Order Theory, Comp Meth in Appld. Mech. and Engng., Vol. 127, 1995, p. 145
234. Liew, K. M. and Lim, C. W., A Higher Order Theory for Vibration analysis of Constrained Curvilinear Shallow Shells, J. Vib. and Control, Vol. 1, No. 1, 1995, p. 15
235. Liew, K. M. and Lim, C. W., Vibration of Doubly-Curved Shallow Shells, Acta Mechanica, Vol. 114, 1996, p. 95
236. Liew, K. M., Lim, C. W. and Kitipornchai, S., Effects of General Laminations and Boundary Constraints on Vibration of Composite Shallow Shells, Composites: Part B, Vol. 27B, 1996, p. 155
237. Liew, K. M., Lim, C. W. and Ong, L. S., Flexural Vibration of Doubly-Tapered Cylindrical Shallow Shells, Intl. J. Mech. Sci., Vol. 36, No. 6, 1994, p. 547
238. Liew, K. M., Lim, M. K., Lim, C. W., Li, D. B. and Zhang, Y. R., Effects of Initial Twist and Thickness Variation on the Vibration Behavior of Shallow Conical Shells, J. Sound Vib., Vol. 180, No. 2, 1995, p. 271
239. Liew, K. M., Lim, C. W. and Ong, L. S., Vibration of Pre twisted Cantilever Shallow Conical Shells, Intl. J. Solids and Struc., Vol. 31, No. 18, 1994, p. 2463
240. Liew, K. M., Xiang, Y. and Kitipornchai, S., Transverse Vibration of Thick Rectangular Plates, I: Comprehensive Sets of Boundary Conditions, Computers and Struc., Vol. 49, 1993, p. 1
241. Liew, K. M., Xiang, Y. and Kitipornchai, S., Research on Thick Plate Vibration: A Literature-Survey, J. Sound Vib., Vol. 180, No. 1, 1995, p. 163
242. Liew, K. M., Xiang, Y. and Kitipornchai, S., Analytical Buckling Solutions for Mindlin Plates involving Free Edges, Intl. J. Mech. Sci., Vol. 38, No. 10, 1996, p. 1127
243. Lim, C. W., Kitipornchai, and Liew, K. M., Vibration of Pre twisted Cantilever Shallow Conical Shells of Non uniform Cross-section, 6th Intl. Symp. Transport Phenomena and Dynamics of Rotating Machinery, February 1996, Honolulu, Vol. I, p. 227
244. Lim, C. W. and Liew, K. M., Effects of Boundary Constraints and Thickness Variations on the Vibratory Response of Rectangular Plates, Thin-Walled Structures, Vol. 17, No. 2, 1993, p. 133
245. Lim, C. W. and Liew, K. M., A *pb*-2 Ritz Formulation for Flexural Vibration of Shallow Cylindrical Shells of Rectangular Planform, J. Sound Vib., Vol. 173, No. 3, 1994, p. 343
246. Lim, C. W. and Liew, K. M., Vibratory Behavior of Shallow Conical Shells by a Global Ritz Formulation, Engng. Struc., Vol. 17, No. 1, 1995, p. 63
247. Lim, C. W. and Liew, K. M., A Higher Order Theory for Vibration of Shear Deformable Cylindrical Shallow Shells, Intl. J. Mech. Sci., Vol. 37, No. 3, 1995, p. 277

248. Lim, C. W. and Liew, K. M., Vibration of Shallow Conical Shells with Shear Flexibility: A First Order Theory, Intl. J. Solids and Struc., Vol. 33, No. 4, 1996, p. 451
249. Lim, C. W., Liew, K. M. and Ong, L. S., Vibration of Shallow Shells by Rayleigh-Ritz Method, Intl. Conf. Computational Methods in Engineering, November 1992, Singapore, Vol. 1, p. 357
250. Lim, C. W., Liew, K. M. and Ong, L. S., Flexural Vibration of Doubly-Connected Doubly-Curved Shallow Shells, Asia-Pacific Vibration Conf. 14-18 November 1993, Kitakyushu, Japan, Vol. 3, p. 1215
251. Lin, C. C. and King, W. W., Free Transverse Vibration of Rectangular Unsymmetrically Laminated Plates, J. Sound Vib., Vol. 36, 1974, p. 91
252. Lindberg, G. M. and Olson, M. D., Convergence Studies of Eigenvalue Solutions using Two Finite Plate Bending Elements, Intl. J. Num. Meth. Engng., Vol. 2, 1970, p. 99
253. Lindberg, G. M., Olson, M. D. and Tulloch, H. A., *Closed Form Finite Element Solutions for Plate Vibration*, NRC Canada Aeronautical Report, LR-518
254. Liu, W. H. and Chang, I. B., Some Studies on Free Vibration of Cantilever Plates with Uniform and Nonuniform Thickness, J. Sound Vib., Vol. 130, 1989, p. 337
255. Livesley, R. K., Some Notes on the Mathematical Theory of a Loaded Elastic Plate Resting on an Elastic Foundation, Q. J. Mech. Appld. Math., Vol. 6, Pt. 1, March 1953, p. 32
256. Lloyd, J. R. and Miklowitz, J., Wave Propagation in an Elastic Beam or Plate on an Elastic Foundation, J. Appld. Mech., Vol. 29, No. 3, September 1962, p. 459
257. Lo, K. H.. Christensen, R. M. and Wu, E. M., A Higher Order Theory of Plate Deformation, Part I: Homogeneous Plates, J. Appld. Mech., Vol. 44, No. 4, p. 663, 1977
258. Lo, K. H.. Christensen, R. M. and Wu, E. M., A Higher Order Theory of Plate Deformation, Part II: Laminated Plates, J. Appld. Mech., Vol. 44, No. 4, p. 669, 1977
259. Love, A. E. H., *The Mathematical Theory of Elasticity*, Dover, 1944
260. Lurie, H., Vibration of Rectangular Plates, J. Aero Sci., Vol. 18, February 1951, p. 139
261. Lurie, H., Lateral Vibrations as Related to Structural Stability, J. Appld. Mech., Vol. 19, No. 2, 1952, p. 195
262. Lynn, P. P. and Dhillon, B. S., Triangular Thick Plate Bending Elements, Proc. 1st Intl. Conf. on Structural Mech. of Reactor Technology, Berlin, Vol. 6, 1971, p. 365
263. MacBain, J. C., Kielb, R. E. and Leissa, A. W., Vibrations of Twisted Cantilever Plates - Experimental Investigation, J. Engng. Gas Turbines Power, Trans. ASME, Vol. 107, 1985, p. 187
264. Macneal, R. H., A Simple Quadrilateral Shell Element, Computers and Structures, Vol. 8, 1978, p. 175
265. Magrab, E. B., Natural Frequencies of Elastically Supported Orthotropic Rectangular Plates, J. Aco. Soc. Amer., Vol. 61, 1977, p. 79
266. Mallick, P. K., *Fiber-Reinforced Composites Materials, Manufacturing and Design* Marcel Dekker, 1993
267. Mansfield, E. H., *The Bending and Stretching of Plates,* Mcmillan, 1964
268. Martin, C. J., Vibrations of A Circular Elastic Plate Under Uniform Tension, Proc. 4th U S Natl. Cong. Appld. Mech., 1962, p. 277
269. Mason. V., Rectangular Finite Elements for Analysis of Plate Vibrations, J. Sound Vib., Vol. 7, 1968, p. 437

270. Mazurkiewicz, Z., The Boundary Conditions and the Equations of Equilibrium and Vibration for an Anisotropic Nonhomegenous Plate, Arch. Mech. Stos., Vol. 11, 1959, p. 729
271. Mazurkiewicz, Z., Free Vibration of an Isotropic Nonhomogenous Rectangular Plate, Bull Acad. Pol. Sci., Vol. 8, 1960, p. 63
272. Mazurkiewicz, Z., Bending Vibration and Buckling of a Rectangular Orthotropic Plate Resting on a Non homogenous Foundation, Bull Acad. Pol. Sci., Vol. 8, 1960, p. 129
273. Mazurkiewicz, Z., The Problem of Bending and Free Vibration of a Simply Supported Isotropic Nonhomogenous Rectangular Plate, Arch. Mech. Stos., Vol. 12, 1960, p. 497
274. Medwadowski, S. J., A Refined Theory of Elastic Orthotropic Plates, J. Appld. Mech., Vol. 25, No. 4, December 1958, p. 437
275. Melosh, R. J., Basis for Derivation of Matrices for the Direct Stiffness Method, AIAA J, Vol. 1, 1963, p. 1631
276. Mikami, T. and Yoshimura, J., Application of he Collocation Method to Vibration Analysis of Rectangular Mindlin Plates, Comp. Struc., Vol. 18, 1984, p. 425
277. Mikhlin, S. G., *Variational Methods in Mathematical Physics,* Mcmillan, 1964
278. Mindlin, R. D., Influence of Rotary Inertia and Shear on Flexural Motions of Isotropic Elastic Plates, J. Appld. Mech., vol. 18, March 1951, p. 31
279. Mindlin, R. D., Schecknow, A. and Deresiewicz, H., Flexural Vibration of Rectangular Plates, J. Appld Mech., Vol. 23, No. 3, September 1956, p. 431
280. Minkarah, I. A. and Hoppmann, W. H., Flexural Vibrations of Cylindrically Aelotropic Circular Plates, J. Acoust. Soc. Am., Vol. 36, No. 3, 1964, p. 470
281. Mizusawa, T. and Leonard, J. W., Vibration and Buckling of Plates with Mixed Boundary Conditions, Engng. Struc., Vol. 12, 1990, p. 285
282. Moiseiwitch, B. L., *Variational Principles,* Interscience
283. Mohan, D. and Kingsbury, H. B., Free Vibration of Generally Orthotropic Plates, J Aco. Soc. America, Vol. 50, 1971, p. 266
284. Munakata, K., On the Vibration and Elastic Stability of a Rectangular Plate Clamped at its Four Edges, J. Math. Phy., Vol. 31, No. 1, April 1952, p. 69
285. Nadai, A., *Die Elastiche Platten,* Julius Springer, 1925
286. Nagaraja, J. V. and Rao, S. S., Vibration of Rectangular Plates, J. Aero Sci., Vol. 20, December 1953, p. 855
287. Naghdi, P. M. and Rowley, J. C., On the Bending of Axially Symmetric Plates on Elastic Foundations, Proc. First Midwestern Cong. Solid Mech., April 1953, p. 119
288. Nakata, Y. and Fujita, H., On Upper and Lower Bounds of the eigen values of A Free Plate, J. Phy. Soc. Japan, Vol. 10, 1955, p. 823
289. Narita, Y. and Leissa, A. W., Frequencies and Mode Shapes of Cantilevered Laminated Composite Plates, J Sound Vib., Vol. 154, No. 1, 1992, p. 161
290. Nash, W. A. and Modeer, J. R., Certain Approximate Analyses of Nonlinear Behavior of Plates and Shallow Shells, Proc Symp. on the Theory of the Elastic Shells, IUTAM, Delft, 1959, p. 331
291. Nelson, R. B. and Lorch, D. R., A Refined Theory for Laminated Orthotropic Plates, J. Appld. Mech., Trans. ASME, Vol. 41, 1974, p. 177
292. Nishimura, T., Studies on Vibration Problems of Flat Plates by Means of Difference Calculus (Vibration of a Square Plate Supported at Four Corners and being Free Along All Edges), Proc Third Japan Natl. Cong. Appld. Mech., May 1954, p. 417
293. Noor, A. K., Free Vibration of Multilayered Composite Plates, Amer. Inst. Aero. Astro. J., Vol. 11, 1973, p. 1038

294. Noor, A. K. and Burton, W. S., Stress and Free Vibration Analyses of Multilayered Composite Plates, Composite Struc., Vol. 11, 1989, p. 183
295. Noor, A. K. and Burton, W. S., Assessment of Shear Deformation Theories for Multi layered Composite Plates, Appld. Mech. Rev., Vol. 42, No. 1, p. 1, 1989
296. Noruoka, M., On Transverse Vibration of Rectangular Flat Plates Clamped at Four Edges, Trans. JSME, Vol. 17, No. 57, 1951, p. 26
297. Novozhilov, V. V., *Theory of Elasticity*, Pergammon, 1961
298. Novozhilov, V. V., *Thin Shell Theory*, P. Nordhoff, 1964
299. Nowacki, W., Free Vibration and Buckling of a Rectangular Plate with Discontinuous Boundary Conditions, Bull Acad. Polonaise Sci., Vol. 3, 1955, p. 159
300. Nowacki, W., Some Problems of Rectangular Plates, Bull Acad. Polonaise Sci., Vol. 9, No. 4, 1961, p. 247
301. Nowacki, W., Application of Difference Equations in the Theory of Plates, Pt. I, Arch. Mech. Stos., Vol. 13, No. 4, 1961, p. 479
302. Nowacki, W., *Dynamics of Elastic Systems*, Chapman and Hall Ltd., London, 1963
303. Nowacki, W. and Kaliski, S., Some Problems of Structural Analysis of Plates with Mixed Boundary Conditions, Proc. Ninth Intl.. Cong. Appld. Mech., Vol. 6, 1957, p. 423
304. Nowacki, W. and Olesiak, Z., Vibration, Buckling and Bending of A Circular Plate Clamped Along Part of Its Periphery and Simply Supported on the Remaining Part, Bull. Acad. Pol. Sci., Vol. 4, No. 4, 1956, p. 247
305. Nowacki, W. and Olesiak, Z., The Problem of A Circular Plate Partially Clamped and Partially Simply Supported Along The Periphery, Arch. Mech. Stos., Vol. 8, 1956, p. 233
306. Nowinski, J. L., Nonlinear Transverse Vibrations of Circular Elastic Plates Built-in at the Boundary, Proc. Fourth U.S. Natl. Cong. Appld. Mech., Vol. 1, 1962, p. 325
307. Nowinski, J. L., Nonlinear Oscillations of Anisotropic Plates under Large Initial Stress, Proc. Tenth Congress on Theo. and Appld. Mech., 1965, p. 17
308. Ochoa, O. O. and Reddy, J. N., *Finite Element Analysis of Composite Laminates*, Kluwer, 1992
309. Oden, J. T., *Mechanics of Elastic Structures*, McGraw Hill, 1967
310. Odman, S. T. A., Studies of Boundary Value Problems. Part II. Characteristic Functions of Rectangular Plates, Proc. NR 24, Swedish Cement and Concrete Research Institute of Technology, 1955, p. 7
311. Olson, M. D. and Lindberg, G. M., Dynamic Analysis of Shallow Shell with a Doubly Curved Triangular Finite element, J Sound Vib., 1971
312. Omprakash, V. and Ramamurti, V., Coupled Free Vibration Characteristics of Rotating Tuned Bladed Disk Systems, J Sound Vib., Vol. 140, No. 3, 1990, p. 413
313. Orris, R. M. and Petyt, M., A Finite Element Study of the Vibration of Trapezoidal Plates, J. Sound Vib., Vol. 27, 1973, p. 325
314. Ota, T. and Hamada, M., Bending and Vibration of A Simply Supported but Partially Clamped Rectangular Plate, Proc. 8th Jap Natl. Cong. Appld. Mech., 1958, p. 103
315. Ota, T. and Hamada, M., Fundamental Frequencies of Simply Supported but Partially Clamped Square Plates, Bull. JSME, Vol. 6, No. 23, 1963, p. 397
316. Pandalai, K. A. V. and Patel, S. A., Natural Frequencies of Orthotropic Circular Plates, AIAA J., Vol. 3, No. 4, 1965, p. 780
317. Park, K. C. and Stanley, G. M., A Curved

318. Pagano, N. J. and Chou, P. C., The Importance of Signs of Shear Stress and Shear Strain in Composites, J. Composite Materials, Vol. 2, p. 332, 1968
319. Petricone, R. and Sisto, F., Vibration Characteristics of Low Aspect Ratio Compressor Blades, J. of Engng. for Power, ASME, Vol. 93, No. 2, 1971, p. 103
320. Phan, N. D. and Reddy, J. N., Analysis of Laminated Composite Plates using a Higher Order Shear Deformation Theory, Intl. J. Num. Meth. Engng., Vol. 12, p. 2201, 1985
321. Pickett, G. and McCormick, F. J., Circular and Rectangular Plates under Lateral Load Supported by an Elastic Solid Foundation, Proc. First U.S. Natl. Cong. Appld. Mech. June 1951, p. 331
322. Pickett, G. and Janes, W. C., Bending under Lateral Load of A Circular Slab on Elastic Solid Foundation, Proc. First Midwestern Conf. Solid Mech., April 1953, p. 112
323. Pickett, G., Badaruddin, S. and Ganguli, S. C., Semi-infinite Pavement Slab Supported by an Elastic Solid Subgrade, Proc. First Cong. Theo. and Appld. Mech., November 1955, p. 51
324. Pipes, L. A., *Appld. Mathematics for Engineers*, McGraw Hill, 1958
325. Pister, K. S. and Westman, R. A., Bending of Plates on Elastic Foundation, J. Appld. Mech., Vol. 29, No. 2, June 1962, p. 369
326. Pister, K. S. and Williams, M. L., Bending of Plates on Viscoelastic Foundation, J. Engng. Mech. Div., Proc. ASCE, Vol. 86, No. EM5, October 1960, p. 31
327. Plass, H. J. (Jr), Gaines, J. H. and Newsom, C. D., Application of Reissner's Variational Principle to Cantilever Plate Deflection and Vibration Problems, J. Appld. Mech., Vol. 29, No. 1, 1962, p. 127
328. Plunkett, R., Natural Frequencies of uniform and non uniform Rectangular Cantilever Plates, J. Mech. Engng. Sci., Vol. 5, 1963, p. 146
329. Poisson, S. D., *L'Equilibre et le Mouvement des Corps Elastiques,* Mem. Acad. Roy. des Sci. de L'Inst. France, Series 2, Vol. 8, 1829, p. 357
330. Popplewell, N. and McDonald, D., Conforming Rectangular and Triangular Plate Bending Elements, J. Sound Vib., Vol. 19, p. 333
331. Prescott, T., *Appld. Elasticity*, Dover, 1961
332. Przemieniecki, J. S., Equivalent Mass Matrices for Rectangular Plates in Bending, AIAA J, Vol. 4, 1966, p. 949
333. Przemieniecki, J. S., *Theory of Matrix Structural Analysis*, McGraw Hill, 1968
334. Putcha, N. S. and Reddy, J. N., A Refined Mixed Shear Flexible Finite Element for the Nonlinear Analysis of Laminated Plates, Computers and Structures, Vol. 22, No. 2, p. 529, 1986
335. Putcha, N. S. and Reddy, J. N., Stability and Natural Vibration Analysis of Laminated Plates by Using a Mixed Element Based on a Refined Plate Theory, J. Sound Vib., Vol. 104, No. 2, p. 285, 1986
336. Pugh, E. D. L., Hinton, E. and Zienkiewicz, O. C., A Study of Quadrilateral Plate Bending Elements with Reduced Integration, Intl. J. Num. Meth. Engng., Vol. 12, 1978, p. 1059
337. Qatu, M. S. and Leissa, A. W., Vibration Studies of Laminated Composite Twisted Cantilever Plates, Intl. J. Mech. Sci., Vol. 33, 1991, p. 927
338. Rajalingam, C. and Bhat, R. B., Axisymmetric Vibration of Circular Plates and Its Analog in Elliptical Plates using Characteristic Orthogonal Polynomials, J Sound Vib., Vol. 161, No. 1, 1993, p. 109
339. Raju, P. N., Vibrations of Annular Plates, J. Aero. Soc. India, Vol. 14, No. 2, 1962, p. 37

340. Ramaiah, G. K., Flexural Vibrations and Elastic Stability of Annular Plates under Uniform In-Plane Tensile Forces along Inner Edge, J. Sound Vib., Vol. 72, 1980, p. 11
341. Ramamurti, V. and Kielb, R. E., Natural Frequencies of Twisted Rotating Plates, J. Sound Vib., V. 97, 1984, p. 429
342. Ramkumar, R. L., Chen, R. L. and Sanders, W. J., Free Vibration Solution for Clamped Orthotropic Plates using Lagrangian Multiplier Technique, Vol. 25, 1987, p. 146
343. Rao, J. S., *Turbomachine Blade Vibration*, John Wiley, 1991
344. Rao, J. S., *Advanced Theory of Vibration*, John Wiley, 1992
345. Rao, J. S., Ramakrishnan, C. V., Gupta, K and Rao, K. K., Vibration Analysis of Rotating Cambered Helicoidal Turbomachine Blades, ASME 97-GT-299, Orlando
346. Rao, J. S., Ramakrishnan, C. V., Gupta, K and Rao, K. K., A Mixed Shell Element for Cambered Helicoidal Blades and Dynamic Stresses Due to Aerodynamic Excitation, 43rd ASME Gas Turbine and Aeroengine Technical Congress, Exposition and Users Symposium, June 2-5, 1998, Stockholm, Sweden.
347. Rao, N. S. V. K., *Variational Approach to Beams and Plates on Elastic Foundation*, Ph D Thesis, IIT, Kanpur, 1969
348. Rao, S. S. and Prasad, A. S., Vibrations of Annular Plates including the Effects of Rotary Inertia and Transverse Shear Deformation, J. Sound and Vib., Vol. 42, 1975, p. 305
349. Rawtani, S. and Dokainish, M. A., Vibration Analysis of Pre-twisted Cantilever Plates, Trans CASI, Vol. 2, 1969, p. 95
350. Rawtani, S. and Dokainish, M. A., Vibration analysis of Rotating Cantilever Plates, Intl. J. for Num. Methods in Engng., Vol. 3, 1971, p. 233
351. Rayleigh, J. W. S., On the Nodal Lines of A Square Plate, Phil. Mag., series 4, No. 304, Vol. 46, 1873, p. 166
352. Rayleigh, J. W. S., *Theory of Sound*, Vols. I and II, Dover, 1894
353. Rayleigh, J. W. S., On The Calculation of Chladni Figures for a Square Plate, Phil. Mag., Vol. 22, 1911, p. 225
354. Reddy, J. N., Free Vibration of Antisymmetric, Angle-Ply Laminated Plates including Transverse Shear Deformation by the Finite element Method, J. Sound Vib., Vol. 66, 1979, p. 565
355. Reddy, J. N., A Penalty-Plate Bending element for the Analysis of Laminated Anisotropic Composite Plates, Intl. J. for Num. Meth in Engng., Vol. 15, p. 1187, 1980
356. Reddy, J. N., On the Solutions to Forced Motions of Rectangular Composite Plates, J. of Appld. Mech., Vol. 49, p. 403, 1982
357. Reddy, J. N., Geometrically Nonlinear Transient Analysis of Laminated Composite Plates, AIAA J, Vol. 21, No. 4, p. 621, 1983
358. Reddy, J. N., *Energy and Variational Methods in Applied Mechanics*, John Wiley, 1984
359. Reddy, J. N., *An Introduction to the Finite Element Method*, McGraw Hill, 1984
360. Reddy, J. N., A Simple Higher-Order Theory for Laminated Composite Plates, J. Appld. Mech., Vol. 51, p. 745, 1984
361. Reddy, J. N., A Refined Nonlinear Theory of Plates with Transverse Shear Deformation, Intl. J. Solids and Struc., Vol. 20, p. 881, 1984
362. Reddy, J. N., A Review on the Literature on Finite Element Modelling of Laminated Composite Plates and Shells, Shock & Vib. Digest, Vol. 17, No. 4, 1985, p. 3

363. Reddy, J. N., On Refined Computational Models of Composite Laminates, Intl. J. for Num. Meth. Engng., Vol. 27, p. 361, 1989
364. Reddy, J. N., A Review of Refined Theories of Laminated Plates, Shock and Vib. Digest, Vol. 22, No. 7, p. 3, 1990
365. Reddy, J. N., A General Non-Linear Third-Order Theory of Plates with Transverse Shear Deformation, J. Nonlinear Mech., Vol. 25, No. 6, p. 677, 1990
366. Reddy, J. N., On Refined Theories of Composite Laminates, Meccanica, Vol. 25, p. 230, 1990
367. Reddy, J. N., Generalization of Two-dimensional Theories of Laminated Composite Plates, Communications in Appld. Num. Meth., Vol. 3, p. 173, 1987
368. Reddy, J. N. and Chao, W. C., A Comparison of Closed-Form and Finite element Solutions of Thick Laminated Anisotropic Rectangular Plates, Nuclear Engng. and Des., Vol. 64, p. 153, 1981
369. Reddy, J. N. and Khdeir, A. A., Buckling and Vibration of Laminated Composite Plates using Various Plate Theories, AIAA J, Vol. 27, No. 12, p. 1808, 1989
370. Reddy, J. N. and Kuppusamy, T., Natural Vibration of Laminated Anisotropic Plates, J. Sound Vib., Vol. 94, 1984, p. 63
371. Reddy, J. N. and Phan, N. D., Stability and Vibration of Isotropic, Orthotropic and Laminated Plates according to a Higher Order Shear Deformation Theory, J. Sound Vib., Vol. 98, 1985, p. 157
372. Reddy, J. N. and Tsay, C. S., Mixed Rectangular Finite Elements for Plate Bending, Proc. Oklahoma Academy of Science, Vol. 57, p. 144, 1977
373. Reid, W. P., Free Vibrations of A Circular Plate, J. Soc. Ind. Appld. Math., Vol. 10, No. 4, 1962, p. 668
374. Reismann, H., Bending of Circular and Ring Shaped Plates on Elastic Foundations, J. Appld. Mech., Vol. 21, No. 2, June 1954, p. 129
375. Reismann, H., Forced Vibrations of a Circular Plate, J. Appld. Mech., Vol. 26, No. 4, December 1959, p. 526
376. Reismann, H., *Elastic Plates: Theory and Application,* John Wiley, 1988
377. Reissner, E., On a Variational Theorem in Elasticity, J. Math Phy., Vol. 29, 1940, p. 90
378. Reissner, E., On the Theory of Bending of Elastic Plates, J. Math. Phy., Vol. 23, No. 4, November 1944, p. 184
379. Reissner, E., The Effect of Transverse Shear Deformation on the Bending of Elastic Plates, J. Appld. Mech., Trans. ASME, Vol. 12, No. 2, June 1945, p. A.69
380. Reissner, E., On Bending of Elastic Plates, Q. App. Math., Vol. 5, April 1947, p. 55
381. Reissner, E., On Axisymmetrical Vibrations of Circular Plates of Uniform Thickness, including the Effects of Transverse Shear Deformation and Rotatory Inertia, J. Acoust. Soc. Amer., Vol. 26, March 1954, p. 252
382. Reissner, E., Stresses in Elastic Plates over Flexible Subgrades, Proc. ASCE, Vol. 122, 1957, p. 627
383. Reissner, E., A Note on Deflections of Plates on a Viscoelastic Foundation, J. Appld. Mech., Vol. 25, No. 1, March 1958, p. 144
384. Reissner, E., Note on the Formulation of the Problem of Plate on an Elastic Foundation, Acta Mechanica, Vol. 4, No. 1, 1967, p. 88
385. Reissner, E. and Stavsky, Y., Bending and Stretching of Certain Types of Heterogeneous Aelotropic Elastic Plates, J. Appld. Mech., Vol. 28, 1961, p. 402
386. Reissner, E. and Stein, M., Torsion and Transverse Bending of Cantilever Plates, NACA TN 2369, 1951
387. Ren, J. G., A New Theory of Laminated Plate, Composites Sci. and Tech., Vol. 26, No. 3, p. 225, 1986

388. Ren, J. G., Bending of Simply Supported, Antisymmetrically Laminated Rectangular Plate Under Transverse Loading, Composites Sci. and Tech., Vol. 28, No. 3, p. 231, 1987
389. Ren, J. G. and Hinton, E., The Finite Element Analysis of Homogeneous and Laminated Composite Plates Using a Simple Higher Order Theory, Communications in Appld. Num. Meth., Vol. 2, No. 2, p. 217, 1986
390. Ritz, W., Theorie der Transversalschwingungen einer Quadratischen Platte mit Freien Randern, Ann Phy., Vol. 28, 1909, p. 737
391. Roberson, R. E., Transverse Vibrations of a Free Circular Plate Carrying a Concentrated Mass, J. Appld. Mech., Vol. 18, No. 3, September 1951, p. 280
392. Roberson, R. E., Vibrations of a Clamped Circular Plate Carrying a Concentrated Mass, J. Appld. Mech., Vol. 18, No. 4, December 1951, p. 349
393. Rock, T. A. and Hinton, E., Free Vibration and Transient Response of Thick and Thin Plates using the Finite Element Method, Intl. J. Earthquake Engng. Struc. Dyn. Vol. 3, 1974, p. 51
394. Rock, T. A. and Hinton, E., A Finite Element Method for the Free Vibration of Plates allowing for Transverse Shear Deformation, Computers and Structures, Vol. 6, 1976, p. 37
395. Roufaeil, O. L. and Dawe, D. J., Rayleigh-Ritz Vibration Analysis of Rectangular Mindlin Plates subjected to Membrane Stresses, J. Sound Vib., Vol. 85, 1982, p. 263
396. Sakata, T. and Hosokawa, K., Vibrations of Clamped Orthotropic Rectangular Plate, J. Sound Vib., Vol. 125, 1988, p. 429
397. Sawko, F. and Merriman, P. A., An Annular Segment Finite Element for Plate Bending, Intl. J. Num. Meth. Engng., Vol. 3, No. 1, 1971, p. 119
398. Schechter, R. S., *The Variational Method in Engineering*, McGraw Hill, 1967
399. Schleicher, F., *Circular Plates on Elastic Foundation* (in German), Springer, 1926
400. Schmidt. R., A Refined Nonlinear Theory of Plates with Transverse Shear Deformation, J. Industrial Math Soc., Vol. 27, No. 1, p. 23, 1977
401. Seide, P., An Improved Approximate Theory for the Bending of Laminated Plates, Mechanics Today, Vol. 5, p. 451, 1980
402. Serebryanyi, R. V., The Deflection of a Thin Semi-Infinite Plate Resting on an Elastic Layer of Finite Thickness, Soviet Phy.-Doklady, Vol. 4, No. 2, October 1959, p. 460
403. Sezawa, K., Free Vibration of A Clamped Square Plate, J. Aero. Res. Inst., No. 6, 1924, p. 29
404. Sezawa, K., On the Lateral Vibration of A Rectangular Plate Clamped at Four Edges, Rept. No. 70, Aero. Res. Inst., Tokyo Univ., 1931, p. 61
405. Shames, I. H., *Mechanics of Deformable Bodies*, Prentice Hall, 1964
406. Sharma, P. C., Dynamic Response of Rectangular Plates on Elastic Foundations, Bull Ind. Soc. Earthquake Tech., Vol. 2, No. 2, July 1965, p. 51
407. Sharma, R. L., Dependence of the Frequency Spectrum of Circular Discs on Poisson's Ratio, J. Appld Mech., Vol. 24, No. 1, March 1957, p. 53
408. Shiau, L. C. and Chang, J. T., Transverse Shear Effect on Vibration of Laminated Plate using Higher-Order Plate Element, Computers and Struc., Vol. 39, 1991, p. 735
409. Shiau, L. C. and Wu, T. Y., A High Precision Higher Order Triangular Element for Free Vibration of General Laminated Plates, J Sound Vib., Vol. 161, No. 2, 1993, p. 265
410. Shiau, T. N. and Chang, S. J., Optimization of Rotating Blades with Dynamic-Behaviour Constraints, J. of Aerospace Engng., Vol. 4, 1991, p. 127

411. Shiau, T. N., Yu, Y. D. and Kuo, C. P., Vibration and Optimum Design of Rotating Laminated Blades, to appear in An Intl. J of Composites Engng Part B
412. Shuleshko, P., Buckling of Rectangular Plates uniformly compressed in Two Perpendicular Directions with one Free Edge and Opposite Edge Elastically Restrained, J Appld. Mech., Trans. ASME, Vol. 78, 1956, p. 359
413. Shuleshko, P., Buckling of Rectangular Plates with Two Unsupported Edges, J Appld. Mech., Trans. ASME, Vol. 79, 1957, p. 537
414. Singh, G. and Rao, Y. V. K. S., Fibre Orientation Effects on the Vibration of Thick Composite Plates, Composite Struc., Vol. 6, 1986, p. 319
415. Sinha, S. N., Large Deflections of Plates on Elastic Foundations, J. Engng. Mech. Div., Proc. ASCE, Vol. 89, No. EM1, (Pt. 1) February 1963, p. 1
416. Sirkar, R., Bending of a Rectangular Plate Resting on Elastic Foundation under Simultaneous Action of Lateral Load and Force in the Middle of the Plate, J. Sci Engng. Res., Vol. 13, Pt. 1, January 1969, p. 85
417. Sivakumaran, K. S., Natural Frequencies of Symmetrically Laminated Rectangular Plates with Free Edges, Composite Struc., Vol. 7, 1987, p. 191
418. Smirnov, V. I., *Integral Equations and Partial Differential Equations, Pergamon, 1964*
419. Sneddon, I. N., *Fourier Transforms*, McGraw Hill Book Co. 1951
420. Sokolnikoff, I. S., *Mathematical Theory of Elasticity*, McGraw Hill, 1956
421. Solecki, R., Vibration of Plates with Concentrated Masses, Bull. Acad. Pol. Sci., Series Sci. & Tech., Vol. 9, No. 4, 1961, p. 209
422. Soni, S. R. and Amba Rao, C. L., On Radially Symmetric Vibrations of Orthotropic Non-uniform Disks including Shear Deformation, J. Sound Vib., Vol. 42, 1975, p. 57
423. Sonneman, G., On Correction of Buckling and Vibration of Plates, Proc 1st Midwest. Conf. Solid Mech., 1953, p. 124
424. Southwell, R. V., On The Free Transverse Vibrations of a Uniform Circular Disc Clamped at Its Center and on the Effect of Rotation, Proc. Roy. Soc., (London), Series A, Vol. 101, 1922, p. 133
425. Spiker, R. L. and Munir, N. I., The Hybrid Stress Model for Thin Plates, Intl. J. Num. Meth. Engng., Vol. 15, p. 1239, 1980
426. Sreenivasamurthy, S. and Ramamurti, V., Effect of Tip Mass on the Natural Frequencies of a Rotating Pre-twisted Cantilever Plate, J. Sound Vib., Vol. 70, No. 4, 1980, p. 598
427. Sreenivasamurthy, S. and Ramamurti, V., A Parametric Study of Vibration of Rotating Pre-twisted and Tapered Low Aspect Ratio Cantilever Plates, J. Sound Vib., Vol. 76, No. 3, 1981, p. 311
428. Sreenivasamurthy, S. and Ramamurti, V., Coriolis Effect on the Vibration of Flat Rotating Low Aspect Ratio Cantilever Plates, J. of Strain Analysis, Vol. 16, No. 2, 1981, p. 97
429. Sreenivasamurthy, S. and Ramamurti, V., A Parametric Study of Vibration of Rotating Pre-twisted and Tapered Low Aspect Ratio Cantilever Plates, J. Sound Vib., Vol. 76, 1981, p. 311
430. Srinivas, S., Joga Rao, C. V. and Rao, A. K., An Exact Analysis for Vibration of Simply Supported Homogeneous and Laminated Thick Rectangular Plates, J. Sound Vib., Vol. 12, 1970, p. 187
431. Srinivas, S. and Rao, A. K., Bending, Vibration and Buckling of Simply Supported Thick Orthotropic Rectangular Plates and Laminates, Intl. J. Solids and Struc., Vol. 6, 1970, p. 1463
432. Stewart, J. K. and Colwell, R. C., The Calculation of Chladni Patterns, J. Acoust. Soc. Am., Vol. 11, 1939, p. 147

433. Subramanian, N. R. and Kumaraswamy, M. P., Antisymmetric Vibrations of a rectangular Plate with Distributed Added Mass, J. Aero. Soc. India, Vol. 12, No. 3, 1960, p. 63
434. Sun, C. T. and Chin, H., Analysis of Asymmetric Composite Laminates, AIAA J, Vol. 26, No. 6, 1988, p. 714
435. Suzuki, S., On the Transverse Vibrations of Rectangular Flat Plates Clamped at Four Edges, Trans. JSME, Vol. 13, No. 44, 1947, p. I-58
436. Stanisic, M. M., Free Vibration of a Rectangular Plate with Damping Considered, Q. App Math., Vol. 12, January 1956, p. 361
437. Stanisic, M. M., An Approximate Method Applied to the Solution of the Problem of Vibrating Rectangular Plates, J. Aero. Sci., Vol. 24, No. 2, 1957, p. 159
438. Stoker, J. J., *Nonlinear Vibrations*, Interscience, 1950
439. Stokey, W. F. and Zorowski, C. F., Normal Vibrations of A Uniform Plate Carrying any Number of Finite Masses, J. Appld. Mech., Vol. 26, No. 2, 1959, p. 210
440. Szabo, B. A. and Lee, G. C., Derivation of Stiffness Matrices for Problems in Plane Elasticity by Galerkin s Method, Intl. J. Num. Meth Engng., Vol. 1, p. 301, 1969
441. Szabo, I., The Axisymmetrically Loaded Circular Plate Elastically Supported, Ing. Arch., Vol. 19, No. 2, 1951, p. 128
442. Szabo, I., Contributions to the Theory of the Axisymmetrically Loaded Circular Plates, Specially in the case of Elastic Support, Ing. Arch., Vol. 19, No. 6, 1951, p. 342
443. Szilard, R., *Theory and Analysis of Plates*, Prentice Hall, 1974
444. Tadjbaksh, I. and Saibel, E., On the Large Elastic Deflections of Plates, Z Angew Math Mech., Vol. 40, No. 5/6, May/June 1960, p. 259
445. Tang, S. C., Dynamic Response of Elastic Plates on Viscoelastic Foundations to Moving Loads, Ing. Arch., Vol. 36, No. 3, 1967, p. 155
446. Tessler, A. and Hughes, T. J. R., A Three-Node Mindlin Plate Element with Improved Transverse Shear, Computer Methods in Applied Mechanics and Engineering, Vol. 50, No. 1, 1985, p. 71
447. Timoshenko, S. P., *Strength of Materials*, Part II, McGraw Hill, 1956
448. Timoshenko, S. P. and Gere, J. M., *Theory of Elastic Stability*, McGraw Hill, 1961
449. Timoshenko, S. P. and Goodier, J. N., *Theory of Elasticity*, McGraw Hill, 1951
450. Timoshenko, S. P. and Woinowsky-Kreiger, S., *Theory of Plates and Shells*, McGraw Hill, 1959.
451. Timoshenko, S. P. and Young, D. H., *Vibration Problems in Engineering*, Van Nostrand, 1955
452. Tobias, S. A., Nonlinear Forced Vibrations of Circular Discs, Engineering, Vol. 186, 1958, p. 51
453. Tocher, J. L., *Analysis of Plate Bending using Triangular Elements*, Ph. D. Thesis, University of California, Berkeley, 1962
454. Tocher, J. L. and Japur, K. K., Comment on 'Basis for Derivation of Matrices for the Direct Stiffness', AIAA J, Vol. 3, 1965, p. 1215
455. Tomotika, S., The Transverse Vibration of A Square Plate Clamped at Four Edges, Phil. Mag., Series 7, Vol. 21, No. 142, 1936, p. 745
456. Tsai, S. W. and Pagano, N. J., Invariant Properties of Composite Materials, Composite Materials Workshop, Technomic, p. 233, 1968
457. Tsay, C. S. and Reddy, J. N., Bending, Stability and Free Vibration of Thin Orthotropic Plates by Simplified Mixed Finite Elements, J. Sound Vib., Vol. 59, p. 307, 1978
458. Turner, M. J., et. al., Stiffness and Deflection Analysis of Complex Structures, J. Aero. Sci., Vol. 23, No. 9, 1956, p. 805

459. Turvey, G., *Buckling and Post Buckling of Composite Plates*, Chapman and Hall, 1995
460. Uflyand, Ya. S., The Propagation of Waves in the Transverse Vibrations of Bars and Plates, Akad. Nauk SSSR Prikl. Math. Mech., Vol. 12, 1948, p. 287
461. Ungar, E. E., Maximum Stresses in Beams and Plates Vibrating at Resonance, Trans. ASME, J. Engng. Ind., Vol. 84B, 1962, p. 149
462. Veletsos, A. S. and Newmark, N. M., Determination of Natural Frequencies of Continuous Plates Hinged Along Two Opposite Edges, J. Appld. Mech., Vol. 23, No. 1, 1956, p. 97
463. Venkatesan, S. and Kunukkasseril, V. X., Free Vibrations of Layered Circular Plates, J. Sound Vib., Vol. 60, 1978, p. 511
464. Vet, M., Natural Frequencies of Thin Rectangular Plates, Mach. Des., Vol. 37, No. 13, 1965, p. 183
465. Vijayakumar, K. and Ramiah, G. K., Analysis of Vibration of Square Plates by the Rayleigh-Ritz Method with Asymptotic Solutions from a Modified Bolotin Method, J. Sound Vib., Vol. 56, 1978, p. 127
466. Vlasov, V. Z., *General Theory of Shells and Its Applications in Engineering*, NASA TTF-99, April 1964
467. Vlasov, V. Z. and Leontiev, U. N., *Beams, Plates and Shells on Elastic Foundations*, Translated from Russian, Israel Programme for Scientific Translations, Jerusalem, 1966
468. Vogel, S. M. and Skinner, D. W., Natural Frequencies of Transversely Vibrating Uniform Annular Plates, J. Appld. Mech., Vol. 32, 1965, p. 926
469. Volterra, E., On the General Problem of a Plate Supported by Elastic Soil, Accad. Lincei, Vol. 2, May 1947, p. 596
470. Volterra, E., Method of Internal Constraints and Its Applications, J. Engng. Mech. Div., Proc. ASCE, Vol. 87, No. EM4, August 1961, p. 103
471. Volterra, E. and Zachamanoglou, E. C., *Dynamics of Vibrations*, Merril, 1965
472. Wah, T., Natural Frequencies of Plate-Mass Systems, Proc. Ind. Soc. Theo. and Appld. Mech., 1961, p. 157
473. Wah, T., Vibration of Circular Plates, J. Acoust. Soc. Amer., Vol. 34, No. 3, March 1962, p. 275
474. Wah, T., Large Amplitude Flexural Vibrations of Rectangular Plates, Intl. J. Mech. Sci., Vol. 5, 1963, p. 425
475. Wah, T., Vibration of Circular Plates at Large Amplitudes, J. Engng. Mech. Div., Proc. ASCE, Vol. 89, No. EM5, October 1963, p. 1
476. Walker, K. P., Vibrations of Cambered Helicoidal Fan Blades, J. Sound Vib., Vol. 59, 1978, p. 35
477. Waller, M. D., Vibrations of Free Circular Plates, Proc. Phy. Soc. (London), Vol. 50, 1938, p. 70
478. Waller, M. D., Fundamental Vibration of A Rectangular Plate, Nature, Vol. 143, No. 3610, 1939, p. 27
479. Waller, M. D., Vibration of Free Rectangular Plates, Proc. Phy. Soc. (London), series B, Vol. 62, No. 353, 1949, p. 277
480. Waller, M. D., Concerning Combined and Degenerate Vibrations of Plates, Acoustica, Vol. 3, 1953, p. 370
481. Wang, J. T. S., Shaw, D. and Mahrenholtz, O., Vibration of Rotating Rectangular Plates, J. Sound Vib., Vol. 112, No. 3, 1987, p. 455
482. Warburton, G. B., The Vibration of Rectangular Plates, Chartered Mech. Engr., Vol. 1, January 1954, p. 22
483. Warburton, G. B., The Vibration of Rectangular Plates, Proc. I Mech. E., Vol. 168, 1954, p. 371

484. Warburton, G. B., *The Dynamical Behaviour of Structures*, Pergamon, 1976
485. Washizu, K., *Variational Methods in Elasticity and Plasticity*, Pergamon, 1968
486. Way, S., Bending of Circular Plates with Large Deflections, Trans. ASME, Vol. 56, 1934, p. 627
487. Weinstein, A. and Chien, W. Z., On the Vibrations of A Clamped Plate Under Tension, Q. Appld. Math., Vol. 1, 1943, p. 61
488. Weinstock, R., *Calculus of Variations*, McGraw Hill, 1952
489. Whiteman, J. R., (Ed) *The Mathematics of Finite Elements and Applications*, Academic Press, 1976
490. Whitney, J. M., Basic Mechanics of Fiber Reinforced Composite Materials, Textile Research Journal, vol. 37, 1967, p. 1056
491. Whitney, J. M., The Effect of Transverse Shear Deformation in the Bending of Laminated Plates, J. Composite Materials, Vol. 3, p. 534, 1969
492. Whitney, J. M., The Effect of Boundary Conditions on the Response of Laminated Composites, J. Comp. Materials, Vol. 4, 1970, p. 192
493. Whitney, J. M., Shear Correction Factors for Orthotropic Laminates under Static Load, J. Appld. Mech., Vol. 40, No. 1, p. 302, 1973
494. Whitney, J. M., *Structural Analysis of Laminated Anisotropic Plates*, Technomic, 1987
495. Whitney, J. M. and Pagano, N. J., Shear Deformation in Heterogeneous Anisotropic Plates, J. Appld. Mech., Vol. 37, No. 4, p. 1031, 1970
496. Wilde, P., Rectangular Anisotropic Plates with Clamped Edges, Arch. Mech. Stos., Vol. 12, 1960, p. 241
497. Wilson, R. E. and Plunkett, R., Vibration of Cantilever Plates with Rectangular and Wedge-Shaped Cross-Sections, DF53GL17, General electric Co., 1953
498. Woinowski-Krieger, S., Bending of an Infinite Orthotropic Plate on an Elastic Foundation, Ing. Arch., Vol. 29, No. 1, 1960, p. 22
499. Wood, A. B., An Experimental Determination of the Frequencies of Free Circular Plates, Proc. Phy. Soc., (London) Vol. 47, No. 5, 1935, p. 794
500. Wu, C. and Vinson, J. R., Influence of Large Amplitudes, Transverse Shear Deformation and Rotatory Inertia on Lateral Vibrations of Transversely Isotropic Plates, J. Appld. Mech., Vol. 36, No. 2, June 1969, p. 254
501. Xiang, Y., Wang, C. M. and Kitipornchai, S., Exact Vibration Solution for Initially Stressed Mindlin Plates on Pasternak Foundations, Intl. J. Mech. Sci., Vol. 36, 1994, p. 311
502. Xiang, Y., Wang, C. M., Liew, K. M. and Kitipornchai, S., Mindlin Plate Buckling with pre buckling In-Plane Deformation, J. Engng. Mech., ASCE, Vol. 119, 1993, p. 1
503. Yamaki, N., Influence of Large Amplitudes on Flexural Vibrations of Elastic Plates, Z Angew. Math. Mech., Vol. 41, December 1961, p. 501
504. Yeh, G. C., Bending of A Rectangular Plate on an Elastic Foundation with Two Adjacent Edges Fixed and the Others Free, Proc. Second U.S. Natl. Cong. Appld. Mech., 1954, p. 375
505. Young, D., Vibration of Rectangular Plates by the Ritz Method, J. Appld. Mech., ol. 17, December 1950, p. 448
506. Yu, S. D. and Cleghorn, W. L., Generic Free Vibration of Orthotropic Rectangular Plates with Clamped and Simply Supported Edges, J. Sound Vib., Vol. 163, No. 3, 1993, p. 439
507. Yu, Y. Y., On the Generalized Ber, Bei, Ker and Kei Functions with Applications to Plate Problems, Q. J. Mech. Appld. Math., Vol. 10, No. 2, May 1957, p. 254

508. Yuan, F. G. and Miller, R. E., A Cubic Triangular Finite Element for Flat Plates with Shear, Intl. J. Num. Meth. Engng., Vol. 28, No. 1, 1989, p. 109
509. Zeissig, C., Ein Einfacher Fall der Transversalen Schwingungeneiner Rechteckigen Elastischen Platte, Ann. Physik, Vol. 64, 1898, p. 361
510. Ziegler, H., The Influence of Inplane Deformation on the Buckling Loads of Isotropic Plates, Ing-Arch., Vol. 53, 1983, p. 61
511. Zienkiewicz, O. C., *The Finite Element Method*, McGraw Hill, 1977
512. Zienkiewicz, O. C., Taylor, R. L. and Too, J. M., Reduced Integration Technique in General Analysis of Plates and Shells, Intl. J. Num. Meth. Engng., Vol. 3, p. 275, 1971

Index

ACM element 167
Adini, Clough and Melosh element 167
admissible path 6
Ahmad's super-parametric element 490
Airy stress function 191
angle ply lamina 349, 359
anisotrophic plates 312
antisymmetric laminate 359
asymptotic directions 431
asymptotic lines 431
auxiliary trihedra 419
auxiliary vectors 421
average curvature 50
axial stress resultants 183, 361
axisymmetric vibration 120

balanced laminate 359
basic trihedra 418
basic vectors 418
bending moment 45
bending moments for laminated plates 362, 370
bending strains (Gol'denveizer) 442
Berger's approximation 211, 221, 390
Bessel equations 119
boundary conditions 12
buckling 203, 210

Christoffel symbols 423
circular plates 110, 192, 221, 234
clamped plates (circular) 115
clamped plates (rectangular) 84, 342
Codazzi equations 426
coefficients of first quadratic form 418
coefficients of mutual influence 352
coefficients of second quadratic form 423
completely isotropic material 319
compliance matrix 313
compliance matrix (orthotropic lamina) 354
composite plate 319, 346
conformal element 267

conjugate directions 431
conjugate net of curves 431
consistent mass matrix 163, 251, 272
constant strain triangular element 239
Coriolis matrix 309
corner reactions 65
cross-ply laminates 359
curvature 46, 47

deflected surface 54
delta operator 15
directional cosines 162
discontinuous fiber lamina 350
displacement function 154, 267, 290
divergence theorem 8
Duffing's equation 220, 226, 233, 237, 398, 405
Dupin's Indicatrix 430

eight noded isoparametric element 289, 406
elastic displacements of a shell 432
elastic rotations of a sheel 433, 434
elastic stiffness matrix 301
elemental equation 158
elemental mass matrix 166, 176
elemental stiffness matrix 159, 174, 248, 270, 286, 295, 300
engineering constants 320
error in differential equation 32, 217
Euler-Lagrange equation 11
extensional rigidity 183
extremum property 5

fiber reinforced laminae 346
finite element method 152
finite elements
    ACM element 167
    Ahmad's super-parametric element 490

558 *Index*

eight noded isoparametric element 289, 406
isoparametric element (quadrilateral) 280
isoparametric element (eight noded) 289
isoparametric element (Mindlin) 295
membrane element 239, 265
Mindlin plate element 295, 406
orthotropic plate finite element 406
pre-twisted shell finite element 511
quadrilateral element 280
rectangular element 167
rectangular membrane element 265
rotating orthotropic laminated plate element 411
rotating plate element 305
rotating pre-twisted plate finite element 505, 524
Tocher's element 153
triangular element 153
triangular element with in-plane stresses 275
triangular membrane element 239
first quadratic form 417
first variation 5
flexural rigidity 45
forced vibration 77
forcing vector (elemental) 159
forcing vector (global) 162
free boundary circular plates 131
free vibration 74, 119
Frenet's formula 429
functional 5

Gakerkin method 31, 217, 224, 228, 235, 395, 404
Gauss equation 426
Gaussian curvature 430, 432
Gaussian quadrature formula 288
Gauss' theorem 8
general eigen value problem 90
generalized Hooke's law 312
geometric stiffness matrix 301, 509
global equation 161, 167
Green strain tensor 180
Green's theorem 19
Gol'denveizer theory of shells 415
Gol'denveizer strain displacement relations 432, 415

Hamilton's principle 64
harmonic balancing method 220
helicoidal coordinate system 511

initially stressed plate 199
in-plane force 184-369
in-plane strains (Gol'denveizer) 438
integral with two space coordinates 5
integral with two space coordinates and time 24
inter laminar shear stresses 371
invariants in stiffness matrix 366
isoparametric element (quadrilateral) 280
isoparametric element (eight noded) 289
isoparametric element (Mindlin) 295

Jacobian matrix 285, 293, 498

Kantorovich method 39
Kelvin and Tait derivation 64
Kelvin type viscoelastic foundation 103, 215, 221, 392
kinematric boundary conditions 12
kinetic energy 64
Kirchoff's note on boundary conditions 63
Kirchoff strain displacement relations 179

Lagrange method 25
lamina stresses 371
laminate constitutive relations 363
laminate equations 367
laminate geometry 359
laminated plate 319, 346, 358
laminated plate equations of motion 374
laminated plate strain energy 375
layered plate 319
Levy's method 34, 332
lines of curvature 431
long rectangular plate 42

mass matrix 166, 167
material symmetry 317
membrane element 239, 265
mid plane symmetric laminate 358
Mindlin plate element 295, 406
Mindlin plate theory 91, 136
modal method 79, 125
mode shapes 70
monoclinic symmetry 317

moving auxiliary trihedron 421

Nadai solution 80
natural boundary conditions 12
natural coordinates 280
natrual frequencies 70, 124, 131, 135, 193, 197, 203, 336, 339, 344
Navier's solution 66, 330, 379
nodal circles 123
nodal degrees of freedom 154
nodal diameters 123
nodal force vector 158, 175
nonconformal element 155, 167
nonlinear strain displacement relations 180
nonlinear differential equation 217

orthotropic material 318
orthotropic plate 330, 390
orthotropicplate finite element 406
orthotropic plate on viscoelastic foundation 390

plate 42
Poisson free edge conditions 63
polar coordinates 110
potential functional 246
pre-twist angle 415
pre-twisted plates 415
pre-twisted rectangular plate 473
pre-twisted shell finite element 511
principal curvatures 51
principal directions 431
principal planes of curvatures 51
principal radii of curvature 431
principal vectors 418

quadrilateral element 280

radius of curvature 428
radius of torsion 432
radius vector of a surface 415
randomly oriented fiber lamina 351
rectangular element 167
rectangular membrane 192
rectangular membrane element 265
rectangular plates 42, 199, 227, 330, 338
reduced load 187
reduced stiffness matrix 321
Ritz method 28, 247, 339, 342, 378

rotating orthotropic laminated plate element 411
rotating plate element 305
rotating pre-twisted plate 480
rotating pre-twisted plate finite element 505, 524

second quadratic form 423, 429
setting angle 305
shape function 162, 170
shear coupling 323
shear coupling coefficients 352
shear force 59
shear resultants 370
shell finite element 491, 511
shells 415
simply supported plates (circular) 127
simply supported angle ply laminates 384
simply supported cross ply laminates 378
simply supported plates (rectangular) 66, 99
simply supported and clamped plates 338
slopes of middle surface 47
Sophie Germain theory 47
specially orthotropic lamina 353
specially orthotropic plates 326, 369
stagger angle 305
standard linear viscoelastic foundation 142
state of stress in a sheel 437
stiffness matrix (elemental) 159, 174, 248, 270, 286, 295, 300, 522
stiffness matrix (generalized Hooke's law) 313
stiffness matrix (global) 162
stiffness matrix invariants 366
stiffness matrix (orthotropic lamina) 355
stiffness transformations 321
strain displacement relations 157, 180, 244, 268, 293, 298, 360, 432, 453, 495, 517
strain energy in polar coordinates 114
strain energy in pure bending 56
strain energy of a sheel 465
strain invariants 213, 221, 392
stress function equations 372
stress strain relations 157
stress strain relations for thin lamina 352, 361
subparametric element 280
super-parametric element 280, 494
surface theory 415

symmetric laminates 368

theory of a surface 415
Tocher's element 153
total variation 16
transformation laws 313
transformation matrix 162
transformation matrix (strains) 314
transformation matrix (stresses) 315
transversely isotropic material 319
triangular element 153
triangular element with in-plane stresses 275
triangular membrane element 239
twisting moment 54
twisting of the surface 50

two fold symmetry 317

unidirectional continuous lamina 347, 358

variational principles 5
varied path 6
vector of elastic rotations 434
viscoelastic foundation 103, 142
Voigt notation 312
Von Karman orthotropic plate 398
Von Karman plate on viscoelastic foundation 227, 234
Von Karman's theory 179

weight factors 288